Advances in Cardiovascular Engineering

NATO ASI Series

Advanced Science Institutes Series

A series presenting the results of activities sponsored by the NATO Science Committee, which aims at the dissemination of advanced scientific and technological knowledge, with a view to strengthening links between scientific communities.

The series is published by an international board of publishers in conjunction with the NATO Scientific Affairs Division

A	Life Sciences	Plenum Publishing Corporation
B	Physics	New York and London
C	Mathematical and Physical Sciences	Kluwer Academic Publishers
D	Behavioral and Social Sciences	Dordrecht, Boston, and London
E	Applied Sciences	
F	Computer and Systems Sciences	Springer-Verlag
G	Ecological Sciences	Berlin, Heidelberg, New York, London,
H	Cell Biology	Paris, Tokyo, Hong Kong, and Barcelona
I	Global Environmental Change	

Recent Volumes in this Series

Volume 230—Biological Control of Plant Diseases: Progress and Challenges for the Future
edited by E. C. Tjamos, G. C. Papavizas, and R. J. Cook

Volume 231—Formation and Differentiation of Early Embryonic Mesoderm
edited by Ruth Bellairs, Esmond J. Sanders, and James W. Lash

Volume 232—Oncogene and Transgenics Correlates of Cancer Risk Assessments
edited by Constantine Zervos

Volume 233—T Lymphocytes: Structure, Functions, Choices
edited by Franco Celada and Benvenuto Pernis

Volume 234—Development of the Central Nervous System in Vertebrates
edited by S. C. Sharma and A. M. Goffinet

Volume 235—Advances in Cardiovascular Engineering
edited by Ned H. C. Hwang, Vincent T. Turitto, and Michael R. T. Yen

Volume 236—Rhythms in Fishes
edited by M. A. Ali

Series A: Life Sciences

Advances in Cardiovascular Engineering

Edited by

Ned H. C. Hwang
University of Miami
Coral Gables, Florida

Vincent T. Turitto
Memphis State University
Memphis, Tennessee

and

Michael R. T. Yen
Memphis State University
Memphis, Tennessee

Springer Science+Business Media, LLC

Proceedings of a NATO Advanced Study Institute
on New Frontiers in Cardiovascular Engineering,
held December 4–14, 1991,
in Málaga, Spain

NATO-PCO-DATA BASE

The electronic index to the NATO ASI Series provides full bibliographical references (with keywords and/or abstracts) to more than 30,000 contributions from international scientists published in all sections of the NATO ASI Series. Access to the NATO-PCO-DATA BASE is possible in two ways:

—via online FILE 128 (NATO-PCO-DATA BASE) hosted by ESRIN, Via Galileo Galilei, I-00044 Frascati, Italy

Additional material to this book can be downloaded from http://extras.springer.com

Library of Congress Cataloging-in-Publication Data

Advances in cardiovascular engineering / edited by Ned H.C. Hwang
Vincent T. Turitto , and Michael R.T. Yen.
    p.   cm.
  "Proceedings of a NATO Advanced Study Institute on New Frontiers
in Cardiovascular Engineering, held December 4-14, 1991, in Málaga,
Spain."--T.p. verso.
  "Published in cooperation with NATO Scientific Affairs Division."
  Includes bibliographical references and index.
  ISBN 978-1-4419-3228-0   ISBN 978-1-4757-4421-7 (eBook)
  DOI 10.1007/978-1-4757-4421-7
  1. Cardiovascular--Diseases--Treatment--Congresses. 2. Biomedical
engineering--Congresses. 3. Microcirculation--Congresses.
4. Rheology (Biology)--Congresses. 5. Prosthetic heart valves-
-Congresses.   I. NATO Advanced Study Institute on New Frontiers in
Cardiovascular Engineering (1991 : Málaga, Spain) II. North
Atlantic Treaty Organization. Scientific Affairs Division.
III. Hwang, Ned H. C. IV. Turitto, Vincent T. V. Yen, Michael R.
T.
  [DNLM: 1. Biomedical Engineering--congresses. 2. Heart Valve
Prosthesis--congresses. 3. Microcirculation--physiology-
-congresses. 4. Rheology--congresses. 5. Thrombosis--congresses.
WG 169 A2445 1991]
RC669.A37 1993
617.4'1059--dc20
DNLM/DLC
for Library of Congress   92-48928
                                                              CIP

ISBN 978-1-4419-3228-0

© 1992 Springer Science+Business Media New York
Originally published by Plenum Press, New York in 1992

All rights reserved

No part of this book may be reproduced, stored in a retrieval system, or transmitted in any form or by any means, electronic, mechanical, photocopying, microfilming, recording, or otherwise, without written persmission from the Publisher

PREFACE

Advances of cardiovascular engineering prompt one to consider innovative device technology - that is, the development of new replacement heart valves or engineering of a totally implantable energy source for an artificial heart. However, these kinds of advances have often proved unable to achieve a long-lasting benefit as the cardiovascular field has matured so fast. Cardiovascular engineering has matured to the point where a major innovation must not only function, but must continuously function better than existing devices. This is difficult to accomplish in the complex cardiovasculature system, in which energy source, biocompatibility, compliance, and functionality all must be considered. The maturation of the field is evident from the fact that many engineered prosthetic systems perform well - for example, heart valves function for long periods of time, large-vessel vascular grafts are quite adequate, extracorporeal membrane oxygenation has significantly prolonged the feasible length of heart bypass and other surgical operations, and total artificial hearts can be used as a bridge to transplant without serious complications, yet none of these systems is as good as the natural ones it replaces. The reasons for this are many and incompletely understood. The next stage of progress must be better understandings of the various components of vasculature and their response to alterations by our devices, be they at the micro- or macro-circulatory levels, in the blood, or associated with the vascular wall.

This conference attempted to tie together various aspects of the cardiovascular system in the hope that a better understanding of each subsystem will lead to an improved appreciation of the overall system. We focused on microcirculation, blood rheology, biomaterials and their relationship to thrombosis in conjunction with the various devices used currently by the surgical community. All these fields were represented by some of their leading experts who were assigned the task of relating not only the state-of-the-art science as it exists today, but also areas where there are significant gaps in our knowledge to be bridged.

Defining the "frontiers of our knowledge" will hopefully serve as a guide to push those back. The participants, both contributors and attendees, were encouraged to "cross-communicate," to reach into the area that was not their own in order to forge links that will stabilize our knowledge at new levels. Such communication is an essential part of these workshops which are sponsored by the NATO Scientific Affairs Division. The meeting certainly abided by the goals of communication; at some particularly long sessions we may have overindulged. Unfortunately, such post- or intra-presentation discussion is virtually impossible to incorporate into the book, which is often a state-of-the-art assessment devoid of the excitement that comes with discussion. For this shortcoming, the editors are apologetic; we can only hope that those of you who read the chapters will attend or perhaps lead similar such meetings in the future. We have enjoyed our task and look forward to the next one.

The Editors

CONTENTS

INTRODUCTION: OLD CONCEPTS-NEW CHALLENGES 1
   Benjamin W. Zweifach

MICROCIRCULATION AS A DYNAMIC ENTITY 7
   Benjamin W. Zweifach

CONCENTRATION AND VELOCITY PROFILES OF BLOOD CELLS
     IN MICROCIRCULATION 25
   Robert S. Reneman, Bea Woldhuis, Mirjam G.A. oude Egbrink,
   Dick W. Slaaf, and Geert Jan Tangelder

INTERCELLULAR COLLISIONS AND THEIR EFFECT ON
     MICROCIRCULATORY TRANSPORT 41
   Harry L. Goldsmith

MODEL STUDIES OF THE RHEOLOGY OF BLOOD IN MICROVESSELS 65
   Michael R.T. Yen

FLUID DYNAMICS AND THROMBOSIS 91
   Steven M. Slack, Winnie Cui, and Vincent T. Turitto

BLOOD RHEOLOGY, BLOOD FLOW AND DISEASE 103
   Gordon D.O. Lowe

BIOMECHANICAL ASPECTS OF TISSUE OF GROWTH
     AND ENGINEERING 109
   Y.C. Fung

FLOW AND VASCULAR GEOMETRY 127
   Harry L. Goldsmith and Takeshi Karino

EX VIVO MODELS FOR STUDYING THROMBOSIS: SPECIAL EMPHASIS
     ON SHEAR RATE DEPENDENT BLOOD-COLLAGEN
     INTERACTIONS 151
   Kjell S. Sakariassen, Helge E. Roald, and J. Aznar Salatti

CHANGES IN VASCULAR GEOMETRY IN ATHEROSCLEROTIC
    PLAQUE RUPTURE AND ITS RELATIONSHIP TO
    THROMBOSIS IN ACUTE VASCULAR EVENTS 175
  Lina Badimon and Juan Jose Badimon

CURRENT STATUS OF BIOMATERIALS: USE AND DEVELOPMENT 189
  Paul Didisheim

CARDIAC VALVE REPLACEMENT WITH MECHANICAL PROSTHESES:
    CURRENT STATUS AND TRENDS 197
  F. Javier Teijeira and Adel A. Mikhail

FLOW THROUGH MECHANICAL HEART VALVES AND THROMBOSIS:
    VISUALIZATION BY WASHING TEST 229
  Bernard Brami

DESIGN, DEVELOPMENT, AND TESTING OF BLOOD PUMPS 237
  H. Reul

LASER MEASUREMENTS IN CARDIOVASCULAR FLOW
    DYNAMICS RESEARCH 259
  Shi-Kang Wang and Ned H.C. Hwang

FLOW MODELS STUDIES OF HEART VALVES 299
  Klaus Affeld, Klaus Schichl, and Andreas Ziemann

HEART TRANSPLANTATION: THE PRESENT AND FUTURE -
    THE "REINA SOFIA" HOSPITAL
    (CORDOBA, SPAIN) EXPERIENCE 325
  *Manuel Concha, Manuel Anguita, Anastasio Montero,
  José M. Arizón, Federico Vallés, José M. Latre,
  Amelia Jiménez, and Fernando López-Rubio*

CURRENT ADVANCES IN CARDIAC ASSIST DEVICES 345
  Jean-Raoul Monties, Patrick Havlik, and Thierry Mesana

MECHANICAL SUPPORT FOR THE FAILING HEART 363
  Peter P. McKeown and Stephen G. Kovacs

VASCULAR GRAFTS: CLINICAL AND HEMODYNAMIC APPLICATIONS ... 385
  Travis J. Phifer and Ned H.C. Hwang

NEW CONCEPTS AND DEVELOPMENT OF VASCULAR
    GRAFT PROTHESES 417
  Yauharu Noishiki

CONTRIBUTORS 435

INDEX 439

INTRODUCTION: OLD CONCEPTS - NEW CHALLENGES

Benjamin W. Zweifach

Ames-Bioengineering
University of California, San Diego
La Jolla, California 92093-0412
U.S.A.

Scientific research is in an era of expansion in which much of the dogma that we have been using as a frame of reference is being replaced by information at an extraordinarily complex level of molecular organization. The difficulties encountered in any given discipline are further compounded when as bioengineers an attempt is made to interpret data from the life sciences on the basis of the tools of a physical science. Our focus at this Conference is on Circulatory Dynamics and if one has to crystalize the ultimate organizational purpose of the Cardiovascular System, it is to nourish the population of cells that make up the various organs and tissues of the body. In that context, the heart and large arterial blood vessels serve only an ancillary role. The actual dynamics of this life-sustaining activity are accomplished by functional adjustments in the peripheral vessels lying within the tissue that is being supplied with blood. It is therefore fitting that this Conference should begin by focussing on the so-called "microcirculation."

The famous French physiologist, Claude Bernard, coined the term - homeostasis - to focus attention on the basic need in bodily physiology to maintain at a stable uniform level the "milieu interiéur," the internal environment in which the tissue cells are immersed so that they can carry out physico-chemical activities in a predictable way[1]. This concept was reenforced by W.B.Cannon in his monograph on The Wisdom of The Body dealing with the autonomic nervous system[2]. He concluded that irrespective of other bodily needs, it was the primary purpose of this major coordinating system to ensure that the interstitial environment was buffered despite the continuously changing bodily activities. It is this fundamental process that is the domain of the microcirculation.

What does the term "microcirculation" encompass? It has long been appreciated that the arterial blood vessels are conduits which branch repeatedly to form smaller hierarchical subdivisions. These minute blood vessels were first described by Malpighi with the aid of a crude microscope and the term "capilla," hair like, was used to refer to the capillaries or to the capillary bed[3].

During the past century it has become evident that all of the microscopic sized vessels were not the same and that the network of vessels in different tissues were arranged in distinctive patterns. Nonetheless, in view of the common nutritive objectives of all of these microcirculatory networks, there was a trend to generalize and to assume that the microvascular beds in all tissues shared common design features. Modules depicting the terminal vascular bed were set up on the basis of particular tissues that were available for direct microscopy under in vivo conditions.

Today, much of our information in this area is derived from extracts of living cells or tissues grown outside of the body. Although challenging and highly provocative, the information deals with cells that have been removed from the intrinsic controls utilized in maintaining cardiovascular homeostasis. The complex nature of these redundant molecular pathways and the overlapping array of controls in the body make a theoretical analysis the only practical approach. Such endeavors can only be successful if data dealing with the fundamental components of the vascular system are made available for the vascular tree along a continuous pathway and not as fragmentary, isolated measurements.

The unique contribution of the field of microcirculation to our understanding of pathophysiology derives from the fact that it represents the only physical access that the cells of the body have to the external environment. As molecular information has become available, a firmer grasp of the behavior of the microcirculatory system has opened up new dimensions in our understanding of basic aspects of systemic disorders such as diabetes, hypertension, stroke and shock.

The capillary network was viewed initially as analogous to a set of irrigation ditches which could be flooded with blood by opening and closing of arteriolar floodgates. Perhaps the single most stimulating concept in the past 100 years was the recognition that the microcirculation is a discrete organ system with its own characteristic structural and functional organization. Because of its strategic location within the parenchymal tissue, the microcirculation is responsive not only the biochemical activity of the cellular building blocks of the tissue but to the control mechanisms concerned with bodily homeostasis.

Early thinking concerning the organization of the microcirculatory network was formulated on the basis of a simple hierarchical branching in which the successive segments became smaller until vessels of capillary dimensions are reached. Such an overly simplified arrangement is rarely encountered. The regular in-series deployment of the larger vessels is replaced within the tissue proper by numerous in parallel offshoots together with artery to artery interconnections so that even identification of a particular segmental branch order becomes arbitrary[4].

August Krogh, whose thinking dominated the field in the 1920's and 30's, based his concepts of the local control of the terminal vascular bed on the availability of oxygen to skeletal muscle[5]. He developed the so-called Krogh cylinder model in which blood-tissue exchange depended upon the spatial deployment of the blood capillaries. He envisaged a reduction in the number of openly perfused capillaries under conditions of rest and the "recruitment" of additional capillaries upon demand during periods of exercise. Two mechanisms were postulated to account for this phenomenon, a slow narrowing of the capillaries by an active contraction of the endothelial cells or by the action of specialized pericytes with contractile properties. The presence of such primitive pericapillary cells had been suggested by the French embryologist, Charles Rouget[6].

As investigators began to use a variety of mammalian tissue preparations for vital microscopy, considerable doubt was cast on the active participation of the true capillaries in local adjustments of tissue blood flow. It became clear that early investigators had identified capillaries under in vivo conditions on the basis of diameter alone and thus included mistakenly data on the terminal arterioles and precapillaries into a generalized interpretation of capillary vasomotor activity[7]. Pericytes are difficult to identify in living tissues and there was no evidence that the capillaries responded to any of the conventional musculotropic agents, or reacted to a massive discharge of the sympathetic nervous system.

A number of recent developments have rekindled interest in some form of endothelial cell contractility and in a possible contractile function for pericytes in some tissues, such as the retina. Molecular probes have shown the presence of the contractile proteins actin and myosin in endothelial cells. By itself such a finding is hardly decisive since these proteins are present in a whole array of cells such as fibroblasts, macrophages and leukocytes. Mediators, which increase vascular permeability, such as histamine, kinins, prostaglandins, appear to act by widening the spacing between the surfaces of contiguous endothelial cells[8]. Whether this type of limited movement is the result of a swelling of the the cells or represents some form of active shape change remains uncertain. Observations on the liver show what appears to be an active protrusion of the endothelial cell nucleus into the lumen that is sufficient to obstruct flow along the sinusoidal vessels[9]. It is possible that changes in endothelial cell shape of some kind may be a contributory factor to the redistribution of blood through the delicate capillaries.

The concept of recruitment carries with it a number of implications with respect to the overall design of microvascular networks, particularly since there is a need for the activation of strategically located controls to provide for a systematic increase or decrease in the number of exchange vessels with an active flow. A majority of the precapillary branches that lead directly into the capillaries originate as abrupt, side-branches and form an in-parallel circuit to the parent stem.

A number of early investigators, some 100 years ago, noted side-arm branches with muscle cells in the junctional region. Tannenberg[10] assigned the term "Pfortnerzellen" to these structures. Such a deployment of vessels allows for an alternating ebb and flow of the bloodstream with the possibility for spatial as well as temporal recruitment of capillaries. It is unfortunate that overemphasis of negative findings with respect to the universality of sphincters led to a failure to recognize the need for some type of definitive structural alignment of in-series and in-parallel circuit controls as a basic feature of nutritive type microvascular networks[11].

Sir Thomas Lewis[7] added another dimension to local control of the capillary circulation by his extraordinarily perceptive studies of the vessels in the skin of man. He preferred the term "minute blood vessels" in dealing with the capillary sized components in the skin. Lewis's detailed analyses of the triple reaction or the flare response to simple stroking of the skin brought into focus the interplay of nervous and humoral factors in local responses of the small blood vessels.

About the same time, the importance of time-dependent aspects of vascular perfusion was recognized through the vivid description by Beatrice Carrier[12] of spontaneous vasomotor activity in the vessels of the wing of the bat. Vasomotion has been shown to have a direct impact on a whole array of exchange processes. For example, the continuous shifting of pressure and flow within a network may represent a basic mechanism for modulating transcapillary fluid exchange in line with the Starling Constitutive Equation[13].

The endothelial cell for many years was considered more of an oddity than a typical cell with distinctive metabolic functions. It's role in exchange was attributed to the thinness of the endothelial membrane rather than to any intrinsic vital properties. Unlike other cellular membranes in which the cells have an active role in the selective permeability of the barrier, the capillary wall appears to behave as a passive filter sieving off water soluble molecules on the basis of their molecular dimensions. Since thinness by itself cannot account for the unusual properties of the endothelial barrier, it was necessary to postulate either that the endothelial cell is different from other cells, or that other structural attributes characterize endothelial membranes. Both of these possibilities have been explored by different workers. Ludwig[14] in 1861 had already recognized that the blood pressure served to drive fluid out of the capillaries. The return of fluid and materials back into the circulation was not fully understood until E.H. Starling[15] demonstrated the interplay of osmotic forces as well as hydrostatic factors in the exchange between the two major bodily compartments.

Except for the demonstration by electron microscopy of uptake of large molecular aggregates by the endothelial cells, little was known concerning the pathways used by which the various materials permeated the capillary barrier. In view of the demonstration that perfusion with calcium-free solutions led to an increased filtration rate, the concept of an intercellular pathway was advanced by Chambers and Zweifach[16] as the equivalent of the small pore channels for transcapillary exchange. A model based upon changes in the physico-chemical makeup of the intercellular matrix material could account for the perviousness of the barrier to small solutes. With injury, a further loosening of the intercellular matrix could lead to the leakage of macromolecules.

The use of electron microscopy brought into our thinking the role of the numerous endothelial organellae as potential transport vehicles for all kinds of molecules[17]. Because of the difficulty of translating static structural information into a dynamic phenomenon, it has not been possible to obtain a consensus as to the contribution of endothelial cell vesicles to the transcapillary flux of solutes. Current thinking includes an inner lining layer of glycoproteins as an important factor in maintaining the properties of the endothelial barrier. With the increased recognition of dynamic factors such as blood shear rate on endothelial cell properties, not only with respect to exchange but to the adhesion and outward migration of leukocytes as well as the aggregation of platelets[18], the possibility exists for highly localized modifications of microvascular function.

It is apparent that the microcirculation is a labile organic entity with a significant capacity to be remodeled in a positive direction. Adaptive changes in arteriolar length and dimensions serve to meet flow requirements; increased capillarization will provide an increased surface area for exchange; venular widening or tortuosity appears to be secondary to changes in transmural pressure, etc. Some of these anatomical changes are permanent in nature. Others are only temporary and reversible. This type of morphometric adaptation can readily be detected by recording segmental changes in microcirculatory topology. Distinctive changes of this kind have been found to develop well in advance of conventional angiopathies. The fact that striking changes in both aging and in disease frequently appear and occur in the venular channels is probably a reflection of a disturbance in pressure-flow relationships that leads to a maldistribution of blood which is then compounded on the low shear side of the microvascular bed because of blood rheological factors.

For the most part, abnormal manifestations have been described in general terms; e.g., ischemic trends vs. a hyperemic trend, hyperplastic vs atrophic networks, on the assumption that a comparable set of changes exists throughout the network. The data clearly indicate that network remodeling is not uniform, but develops initially in focal portions a network and then spreads to other portions until more and more of the network is involved. The reason why certain vessels show a predilection to change may be purely random.

We still have much to learn concerning the properties of the cellular building blocks of the microcirculation. This delicately poised functional unit provides a sensitive index not only for the efficacy of local homeostasis but for bodily homeostatic adjustments. Alterations in the microcirculation during disease reflect not only the metabolic status of a particular organ but provide a unique insight into systemic control mechanisms. Hopefully a blending of in-vivo and in-vitro experimental data will allow us to utilize this laboratory tool to supplement and strengthen our clinical base of information.

It is obvious that we have come a long way. As new pieces of information become available, things become more complex as they must before simple answers are available.

REFERENCES

1. Bernard CL: Leçons de Pathologie experimentale, J. B. Bailbiere et Fils, Paris, 1880.
2. Cannon WB: The Wisdom of the Body, W. W. Norton, New York, 1932.
3. Malpighi M: De Pulmonibus: Observationes Anatomical, Bologna, 1661.
4. Zweifach BW: General principles governing behavior of the microcirculation, Amer. J. Med., 23:684, 1957.
5. Krogh A: The supply of oxygen to the tissues and the regulation of the capillary circulation, J. Physiol. (London), 52:457, 1919.
6. Rouget C: Mémoire sur les développment, la structure et les proprietes physiologiques des capillaires sanguins et lymphatiiques, Arch. Physiol. Norm. Pathol., 5:603, 1873.
7. Lewis T: The Blood Vessesls of the Human Skin and Their Responses, Shaw, Ltd., London, 1927.
8. Crone C: Modulation of solute permeability in microvascular endothelium, Fed. Proc., 45:77, 1986.
9. Oda M, Azuma T, Watanabe N, Nishizahi Y, Nishida J, Ishii K, Suzuku H, Kaneko H, Somatsu H, Tsukada N, TsuchiyaM: Regulatory mechanisms of hepatic microcirculation-Involvement of contraction and dilation of sinudoids and sinusocidal endothelial fenestrae, In: Progress in Applied Microcirculation (Eds., Messmer K, Hammersen F.), Karger, Basel, 17 [Gastrointentinal Microcirculation], 103, 1990.
10. Tannenberg J: Ueber die Capillartätigkeit, Zeit. f. allg Path u Path. Anat., 36:374, 1925.
11. Schmid-Schönbein GW, Skalak R, Chien S: Hemodynamics in arterial networks, In: The Resistance Vasculature (Eds., Bevan JA, Halpern W, Mulvany MJ), Humana Press, New Jersey, 1991.

12. Carrier EB: Observations of living cells in the bat's wing, In: Physiological Papers Dedicated to A. Krogh, 1-9, Levin and Munkgaard, Copenhagen, 1926.
13. Landis EM: Micro-injection studies of capillary permeability. II. Relationship between capillary pressure and the rate at which fluid passes through the capillary walls of single capillaries, Amer. J. Physiol., 82:217, 1927.
14. Ludwig C: Die physiologischen Leistungen des Blutdruckes, Airzel, Leipzig, 1865.
15. Starling EH: On the absorption of fluids from the connective tissue spaces, J. Physiol. (London), 312, 1896.
16. Chambers R, Zweifach BW: Capillary endothelial cement in relation to permeability, J. Cell. Comp. Physiol., 15:255, 1940.
17. Simionescu M, Ghitescu L, Fixman A, Simionescu N: How plasma macromolecules croos the endothelium, N.I.P.S., 2:97, 1987.
18. Mehta LL, Lawton DL, Nicolini FA, Ross HM, Player DW: Effects of activated polymorphonuclear leukocytes on vascular smooth muscle tone, Amer. J. Physiol., 261 [Ht. Circ. Physiol. 30]:H-327, 1991.

MICROCIRCULATION AS A DYNAMIC ENTITY

Benjamin W. Zweifach

AMES-Bioengineering
University of California, San Diego
La Jolla, California 92093-0412
U.S.A.

ABSTRACT

A substantial array of data on the molecular level, together with the application of new immunological probes, have reinforced the concept that the microcirculatory segment of the vascular system contributes in a significant way to the pathophysiologic manifestations of a number of major systemic disorders such as hypertension, diabetes, sepsis and shock. The design features and functional attributes which enable this segment of the vascular system to operate as an independent organic entity have intrigued investigators for many years.

The peripheral circulation is concerned with two levels of circulatory homeostasis - to maintain the systemic blood pressure in a prescribed range and to ensure the stability of the internal environment [the tissue interstitium] in the face of the continually shifting range of bodily activities. The capacity of the microvascular subdivisions of the tree to act as an independent organic entity is a consequence of structural as well as functional features.

The point where the supply arteries for the various organs of the body enter the tissue proper can be taken as the dividing line between the macro and micro-circulation. The paramount objective becomes the local distribution of blood in accord with changing needs of the parenchymal cells. In the ensuing hierarchical branching, the arterial vessels begin to interconnect with one another to form an arcade type of network from which the majority of the terminal arterioles are distributed as side-arm offshoots, a situation that in essence provides for a variable in-parallel circuit.

There is considerable non-uniformity in the design pattern and control features among the various tissues depending upon whether the tissue has a relatively stable level of metabolism as opposed to tissues which exhibit extremes in metabolic requirements when activated. Most terminal vascular beds exhibit intrinsic features that enable them to operate as an independent functional unit.

Key factors in the maintenance of tissue homeostasis are the selective distribution of the available volumetric flow in conjunction with the modulation of pressure levels relative to the blood-tissue exchange process. The latter involves an interplay of factors affecting volumetric flow delivery and the number of capillary vessels that are perfused so as to ensure a surface area compatible with an effective blood-tissue exchange. The feedback controls in this regard exert their effects through a local resetting of smooth muscle cell tone, an interaction of shear rate with the lining endothelial cells, as well as blood-borne and parenchymal cell cytokines.

INTRODUCTION

The primary purpose of the circulatory system, the two way exchange of materials between the bloodstream and the parenchymal tissue compartment, is accomplished in the microcirculatory portion of the cardio-vascular apparatus. The vascular tree is essentially a delivery system which distributes blood on a regional basis in line with the metabolic activities of a particular structure. Once the blood is apportioned to the parenchymal tissue, the objective is to distribute the volumetric flow among the myriads of terminal ramifications so as to optimize the effective exchange of materials between the blood and tissue compartments by matching the volume of blood in transit with an exchange surface area that is appropriate for the nutritive function of the tissue[1]. The integration of these two parameters is essential to meet the considerable range of metabolic activities encountered under physiological demands.

Just as good health is synonymous with an efficient microcirculation, so is microcirculatory insufficiency a mirror of disease. Over the years, much has been learned about cardiovascular disease at different levels of organization of the circulatory system - at one extreme, the involvement of the heart and the large blood vessels, and at the other, biochemical derangements at the cellular level [Table 1]. Much uncertainty still exists concerning the precise feedback that links the systemic and local levels of homeostasis at these two extremes of bodily organization.

Table 1. Functional Attributes of Microcirculation

- nutritional needs of tissue
- stable interstitial milieu
- selective distribution within network
- endothelial cell modulation of pressure vs flow
- exchange surface area in proportion to flow level
- removal of parenchymal cell byproducts

There is a considerable body of evidence indicating that at the peripheral microcirculatory level the system operates independently as a discrete functional unit to maintain local tissue homeostasis. A subdivision into a macro vs micro circulation implies that a separate set of regulatory influences originating in the parenchymal tissue itself comes into play to modulate tissue perfusion in accord with the needs of particular cell populations. The boundary limits for this organic entity are usually taken to be the point where the regional feed arteries penetrate the tissue that is being supplied with blood. Because of their intermediate location, these feed or supply arterial vessels [75-125 μm] represent a bridging structure in which central as well as local controls overlap and serve as a major site for the modulation of peripheral vascular resistance.

A prominent feature of the successively smaller branchings of the feed artery is the progressive attenuation of the smooth muscle cell component of the vessel wall from an investment of 3-4 muscle cells down to a single layer of muscle cells in the smallest arterioles. Coincident with this change in muscularity, there is a progressive thinning of connective tissue content of the arterial wall until the well-defined inner elastin lamella becomes discontinuous so that in the 20-25 μm arterioles only occasional elastin fibers are found.

Table 2. Functional Alignment of Terminal Vascular Bed

MACROCIRCULATION
- central reservoir
- pipelines
- regional apportionment circulating volume

FEED ARTERY TO TISSUE
- volume distribution to given organ [A1]

MICROCIRCULATION
- regional reservoir (arcade arterioles) [A2]
- volume apportionment (transverse arterioles) [A3]
- pressure adjustment (terminal arterioles) [A4]
- number of capillaries with active flow
  (local modulation of smooth muscle tone)
- blood-tissue exchange
  (permeability + surface area)

These morphometric modifications lead not only to a change in the thickness of the wall relative to lumen dimensions, but result in a substantial redistribution of the fraction of the wall strain borne by the various structural components when the vessel is distended by the intravascular pressure. The pressure and flow properties of the blood circulating within the microvasculature are adjusted in a distinctive way by the hierarchical segments that make up the terminal vascular bed. If one follows the subdivisions of the feed arteries that supply a particular tissue, the arrangement of the major branches in the form of an interanastomosing set of arterial arcades in essence sets up a reservoir from which blood can be withdrawn by way of a series of in-parallel side-arm offshoots that operationally can be thought of as spigots. Such arterioles are highly responsive to fluctuations in pressure so that their excursions in caliber serve to feed a variable volume of blood into the capillary network [Table 2]. The final step determining the actual adjustment would appear to be the starting position along the corresponding Tension/Length curve at which the smooth muscle cell contractile elements are set. This intrinsic set-point, in the case of the arteriolar smooth muscle cells, is continuously modified by a family of vasoactive modalities of local origin, providing for a degree of independence from the systemic circulation[2].

GENERAL DESIGN FEATURES OF MICROVASCULAR NETWORK

The functional activities of the vascular system are concerned with two major categories of control - at the systemic level of organization a set of controls maintains the central blood pressure in a prescribed range and ensures regional distribution of blood in accord with the output of the heart. This type of adjustment is accomplished by a baroreceptor meshwork[3] in the walls of strategic branchings of the LBV which acts via the nervous system on both the heart rate and peripheral vascular dimensions to buffer undue variations in systemic blood pressure. The hierarchical vascular segments are affected to a different degree by such SNS stimuli by virtue of the reduction in muscularity of the successively smaller branchings. Individual arteriolar segments fine-tune further pressure and flow conditions in a stepwise manner along the arterial to venous pathway[4].

The larger arterial pipelines into which the blood is ejected by the action of the heart serve as elastic conduits for the regional distribution of a comparatively fixed volume of blood. The walls of the large arterial conduits are designed to withstand the stress imposed by the systemic blood pressure, being thickest where the pressure is highest. Under conditions where the systemic pressure has been shifted to a higher level, the physical makeup of the vessel wall becomes reinforced accordingly[5]. The deformation of the wall of the larger arteries imposed by the blood pressure is borne largely by the elastic lamella and the collagen fibers. In contrast, at the level of the 10-20 μm terminal arterioles, the vascular smooth muscle cells are the principal stress-bearing component. This latter contingency contributes to the amplified myogenic response in the arteriolar segments of the microcirculation.

As the peripheral ramifications of the arterial vessels enter the tissue proper, they begin to interconnect with comparably sized vessels at a number of hierarchical levels to set up a series of arteriole to arteriole anastomoses that in effect ensure a more uniform spatial apportionment of blood within the tissue. Arteriole to arteriole interconnections among the successively smaller arterioles in turn distribute large numbers of side-arm offshoots which ramify into discrete masses of capillaries[6]. An arrangement of this kind, in which the majority of branchings are in the form of in-parallel circuits arising from a series of arterial meshes, provides a structural framework for the separate modulation of two broad physiologic functions - the selective distribution of pressure on a segmental level and the local adjustment of capillary perfusion.

The arterioles that make up the proximal region of the microcirculation [the parent trunk together with its immediate branchings] are endowed with a single layer of smooth muscle down to the point where the true capillaries are given off. The capillary vessels are devoid of typical muscle cells but have numerous, multiple branching pericytes on the outer surface. These pericytes are believed to represent either immature muscle cells or phagocytic cells[7]. Acutely induced perturbations in capillary dimensions have been reported in only rare instances. The effluent portion of the network is structurally undistinguished, except for the presence of an attenuated endothelium and a correspondingly heavier basement membrane in the successively wider postcapillary vessels.

The geometric constraints of the segmental ramifications determine the hydraulic resistance to blood flow. Homeostatic adjustments of systemic origin require only a modest change in vessel dimensions [5-10%] in order to modulate the regional distribution of blood. The distribution of blood within the microvasculature proper is however influenced to an increasing extent by factors that reflect the local biochemical milieu. Pressure and flow levels within the tissue are adjusted in individual terminal arterioles through an active resetting of vessel caliber and by virtue of the in-series and in-parallel deployment of the branch arterioles. Coincident with the changes in the structural makeup of the successive arterial segments, there is a shift in the functional activities of these hierarchical subdivisions. Any of a number of phenomenological characteristics [tone, vessel diameter, active vs. passive responses] can be modulated separately in the successive segments.

Within the immediate precapillary domain a new set of functional requirements must be met - the matching of capillary perfusion levels with fluctuations in parenchymal tissue metabolic demands and the maintenance of an effective exchange of materials between the blood and interstitial tissue compartments. In both instances, homeostasis is dependent upon structural as well as functional attributes.

REGIONAL VARIATIONS IN M.C. DESIGN

During embryogenesis, the initial randomly aligned capillary mesh is remodeled into distinctive segmental hierarchies as the heart develops and the level of blood flow is increased. The vascular segments in the network undergo a structural adaptation in line with the changing pressure-flow conditions to which they are exposed as the tissue matures and grows in size. This capacity for structural adaptation persists throughout life. The design characteristics of particular terminal vascular networks are dependent upon the physical constraints imposed by the tissue parenchymal architecture, as well as by the range of metabolic activity characteristic for that tissue [Figure 1].

Conventional schema depicting the arterial branching at the tissue level have been set up on the basis of a repeated dichotomous subdivision into comparably sized branches. Such a deployment is rarely seen except in membranous structures such as some mesenteries, the bulbar conjuctiva, and the cheek pouch of the hamster[8,9]. In most tissues arterial to arterial interanastomoses are present from which large numbers of small twigs are delivered as in parallel circuits. This latter configuration provides a structural framework that allows for a selective change in pressure and flow across a given branch segment by way of a local feedback. The more peripheral transverse arterioles, because of their thin wall and small caliber, frequently show a complete closure - in essence a mechanism for recruitment or derecruitment of groups of capillaries. By the time that the blood has been routed through the sequence of arteriolar ramifications, the systemic blood pressure has been reduced by 75-80%. The precise nature of the overlapping feedback mechanisms in this regard remain speculative.

A number of schematics depicting a modular design for the microcirculatory network in general have been advanced on the basis of studies in thin flat tissues where the entire array of vessels from artery to venule can be reconstructed. The various patterns of microvascular distribution that are encountered depend, however, upon the extremes in metabolic activity exhibited by particular tissues [Figure 1]. A theoretical model of the microcirculation can be justified more readily in a functional context because of the considerable degree of independence from larger blood vessel adjustments.

A geometric feature with special functional implications is the presence of a comparatively long arterial stem from which varying numbers off side-arm offshoots are deployed. The arteriolar stem arrangement has two variants - either the parent

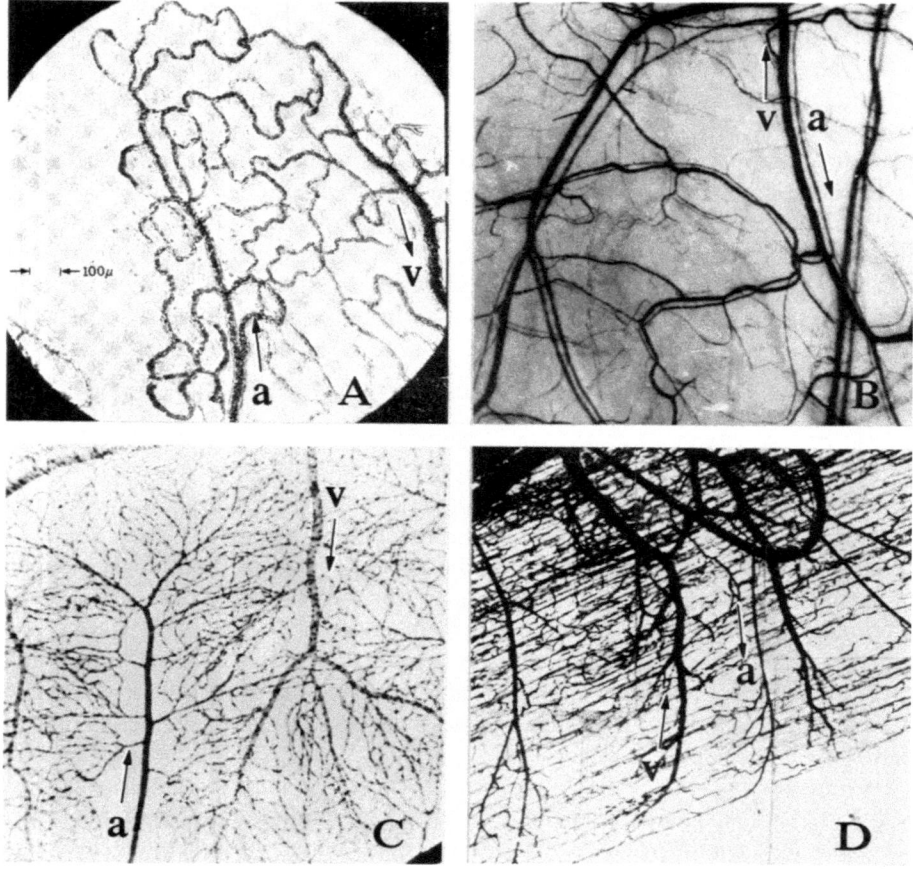

Figure 1. Examples of the design characteristics of the microvascular network in four representative tissues. The differences appear to be related to the extent to which the metabolic needs of particular tissues fluctuate in line with their functional activity. A. Omentum, a randomly dichotomous branching array. B. Conjunctiva of man showing a more regularly arranged branching with paired arterioles and venules. C. Retina, alternating arterial and venous stems with characteristic side-arm branchings. D. Skeletal muscle with arteriole to arteriole arcade interconnections and the majority of offshoots as in parallel circuits.

vessel becomes tapered as successive branches are given off until the stem is essentially of capillary dimensions[10,11] or two such arterioles interconnect head-on to form an arcade type mesh, as in the case of skeletal muscle or in the mesentery of many animals. In all cases, the successively smaller branchings of the arterial tree are associated with a progressive loss of their smooth muscle component and a diminished SNS innervation.

The meshwork of exchange capillaries in adult tissues is a labile entity which can be expanded in a functional as well as a structural framework. In addition, under conditions where individual capillaries remain without an active perfusion for periods of time, the vessels frequently regress and are removed. In regenerating tissue, the capillary sized vessels can be remodelled into larger, heavier walled conduits when flow is maintained at above normal levels by a higher driving pressure.

Tissues which display a stable level of metabolism are usually supplied by a dichotomously branching meshwork[12] that seems to be deployed in a randomly arranged pattern [conjunctiva, omentum, urinary bladder]. Such tissues are characterized by the presence of an active capillary circulation at all times. One or more of the randomly arranged capillary pathways usually show a rapid, continuous flow [mesentery of rat and dog, wing of the bat].

In the heart, the dichotomous branchings of the coronary arterioles proceed from the epicardium to form a three dimensional array of randomly distributed subdivisions in a tissue where blood flow needs to remain at a consistently high level[13].

In other tissues, long stem-like feeders and outflow channels are aligned in an alternating sequence and interconnected by side-arm branchings. Such a pattern is seen in the retina where each long parent arteriole gives rise to as many as 10-12 side-arm branches but has no arteriole to arteriole interconnections.

In skeletal muscle, the terminal vascular network is characterized by a series of arcade interconnections between arterioles[14]. These arcades deliver substantial numbers of twig-like sidearm offshoots, a configuration that represents a unique departure in several respects. The arrangement allows for circuits which can be included or excluded from the active circulation without disrupting the blood supply to a given area. Inasmuch as the parent trunk at the level of the arteriole to arteriole arcades is large with respect to the daughter branch, the configuration sets up a restricted entry condition across which a sharp drop in pressure and flow can be induced.

Early investigators assigned the term "precapillary sphincter" to such configurations, giving rise to the generalization that the perfusion of individual capillaries could be controlled by vasomotor adjustments at such entry points. Since the parent and the offshoot are both muscular vessels, active changes can be introduced in a selective manner in both components. In the case of skeletal muscle, the A3 segmental branches, the transverse arterioles originate from large arcade meshes, whereas the A4 segments, the terminal arterioles arise from a much finer arcade mesh. Each of these arteriolar offshoots in turn gives rise on average to from 4 to 10 capillaries. In the wall of the intestine during digestion, flow can be diverted to the inner mucosal layer as opposed to the medial muscularis by the presence of several plexiform interconnections on the arterial and venous sides[15]. A somewhat analogous situation is found in the skin where as many as three distinctive plexuses can be present and serve to divert blood either towards the surface or away from the surface to maximize or to minimize heat transfer[16].

MICROVASCULAR TONE AS A DYNAMIC ENTITY

A key factor in the active control of the successive arterial segments is the modulation of the physico-chemical state of the contractile proteins that make up the smooth muscle cell effector unit. A characteristic of smooth muscle cells in general is their ability to maintain the intrinsic contraction machinery in a sustained state of partial shortening - a feature referred to as basal tone[17]. This configuration is advantageous since it permits an adjustment in a forward or in a reverse direction in so far as vessel caliber is concerned.

Tone as a physiological entity represents an active force. Its implementation and

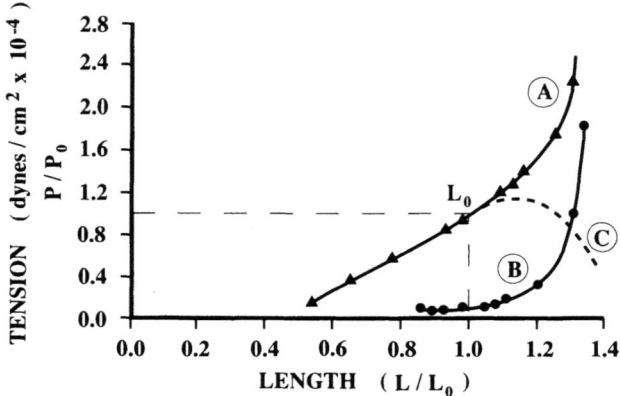

Figure 2. Schematic depicting tension/length relationship for a large artery [from plot for hog carotid[22]]. The total tension (solid triangles) as calculated from the La Place Law in relation to a pressure perturbation [P] above and below normal levels [P0] is shown by the curve A. Passive changes in L above and below L0 [circles] are indicated on curve B. The broken line curve C shows the active tension fraction as L0 is increased passively. The actual starting point [L0] along this tension/length curve in situ is shifted by the prevailing pressure and the modulating influence of flow shear rate on endothelial cell mediator release.

modulation are, however, highly dependent upon an array of passive physical constraints. The smooth muscle cell is embedded in a compact layer of connective tissue that is an integral part of the vessel wall so that the resting length of the cell [L0] is coupled to and in many instances may be subservient to the visco-elastic properties of the vessel. Different factors, that operate to adjust the level of vessel tone by manipulating the L0 length setting, can also be activated to produce a finite shortening or relaxation of the smooth muscle cell [Figure 2]. Inasmuch as the actual mechanical shortening of the smooth muscle cell in final analysis is brought about by a length adjustment of the cytoplasmic contractile filaments, the induced response to vasoactive stimuli in general is determined by the basal tone set point of the cell. The identification and relative contribution of factors that can determine and modify this basal state are difficult to incorporate into in situ adjustments, since these include locally derived mediators from the parenchymal cell environment, as well as mediators of endothelial cell origin.

In the case of vascular smooth muscle, the working tone has been found to be referable to both extrinsic and intrinsic factors. Thus, even when all extrinsic stimuli are removed or blocked with pharmacological agents, the smooth muscle cells continue to be maintained in a partially shortened configuration that is operationally referred to as basal tone[18]. Superimposed onto the basal level of tone are extrinsic mechanisms that range from neurogenic stimuli to blood-borne humoral agents to cytokines. Measurements of tone in situ thus reflect a sustained state of partial contraction under the influence of a variable set of physiologic mechanisms.

When the vessel wall is subjected to passive stretch by an increase in intravascular pressure, the vessel responds in an active mode by what is termed a "myogenic" adjustment. In terms of in situ controls, the arterioles represent the vascular segment where such myogenic regulation has its greatest influence. The term "autoregulation" has been introduced for the active response pattern that serves to keep blood flow at the tissue level within a narrow range in the face of perturbations in the driving pressure and tissue oxygenation.

The L0 starting length at which the smooth muscle cells are maintained determines the actual response to a given perturbation[19]. This set-point has been found to be under the local control of at least two sets of factors - an endothelial cell derived sequence, as well as a chemical feedback from the local environment. In the case of vascular smooth muscle, the L0 starting configuration is not fixed but can be reset in response to locally derived vasotropic materials. The coupling of the stimulus to the mechanical displacement involves the regulation of the uptake and release of $Na+$, $K+$ and $Ca++$ ions, especially where receptor mediated reactions are involved[20]. Regional differences in this regard, as well as segmental differences, are found. The relative contribution of separate modalities will determine the actual degree to which a particular vessel is narrowed or dilated.

The operational activity of the arteriolar subdivisions is strongly influenced by shifts in basal tone. As indicated, the ultimate consideration at the cellular level is the length setting of the contractile protein filaments at the time that the stimulus is introduced. A braking action imposed by endothelial cell derived mediators, such as EDRF[21] becomes less important in the terminal ramifications of the arterial tree where the shear rate has fallen off considerably. These thin-walled arterioles are especially sensitive to changes in intravascular pressure, showing the greatest proportional response to a given shift in pressure.

As indicated, adjustments of the vascular system are routinely analyzed in terms of tone and autoregulation, which are not necessarily interchangeable entities. The change in caliber in response to a shift in vessel wall tension encompasses a myogenically induced adjustment which will vary depending upon the duration of the period of displacement [Table 3]. Autoregulatory phenomena are routinely measured with respect to adjustments of blood flow, although it is not clear whether flow rate per se adequately reflects the imbalance in parenchymal cell metabolic activity[23].

Table 3. Determinants of Vascular Tone

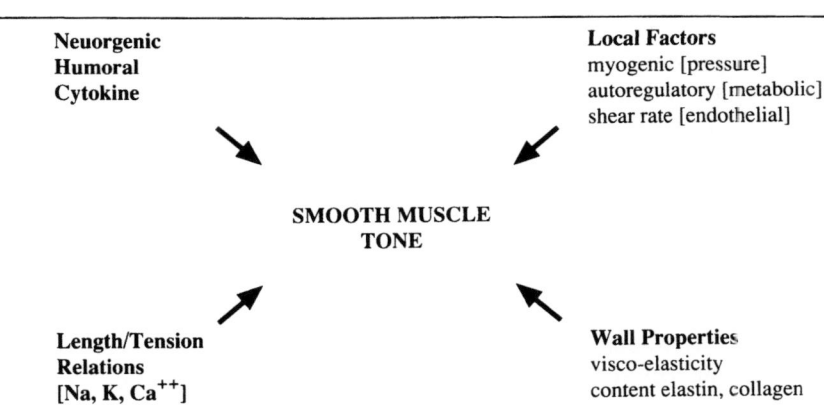

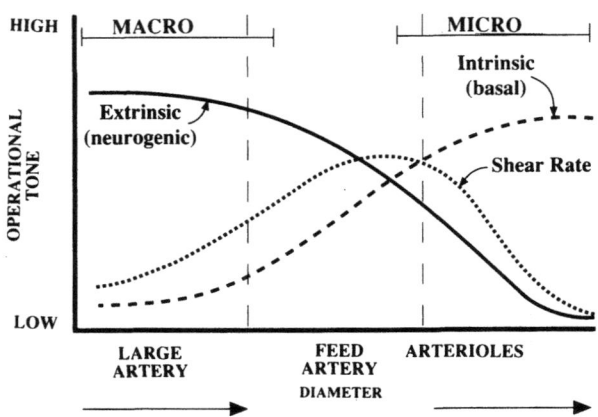

Figure 3. Schema indicating the factors influencing the state of tone in the arterial vessels. A distinction is made between the intrinsic tone remaining after all extracellular regulators have been blocked or removed, as opposed to the tone imposed by specific mechanisms. Note that the modulating role of the blood shear rate by way of vendothelial cell-mediated factors is most pronounced in the midsized arteries and their immediate branchings. Factors originating within the parenchymal tissue operate largely by way of the terminal arterioles.

The involvement of an autoregulatory reaction implies a situation where changes in the level of volumetric flow are not matched by a commensurate adjustment of blood-tissue exchange. There is evidence that the feedback for this phenomenon may occur through the intermediary of other contiguous cells [pericytes, mast cells, endothelial cells and circulating leukocytes]. In the context of the homeostatic imbalance relative to parenchymal cell needs, the inadequate delivery of nutritive materials related to oxidative mechanisms would appear to be the underlying perturbation. Since vascular adjustments entail a modulation of vessel diameter, the effect on blood flow will be non-linear requiring some involvement of tissue metabolic factors.

Even isolated arteries display a state of tone and respond to mechanical deformation by an active myogenic response[24]. In contrast, vessels which display little measurable basal tone respond to mechanical deformation by a passive widening and essentially exhibit no active shortening or narrowing in the face of a reduced intravascular pressure.

It is apparent that a family of myogenic response curves will be generated for an arterial vessel following a given stimulus depending upon the prevailing intravascular pressure. The working set point for the contractile elements per se can be modified in an upward or downward direction by intrinsic as well as extrinsic mediators [Table 3].

The head of pressure under which individual arteriolar segments are maintained sets up a unique wall stress which in turn leads to an appropriate myogenic adjustment to restore the length of the smooth muscle cells to their starting L0 configuration. Concurrently, the velocity of the blood in the vessel sets up shear forces with respect to the endothelial cell surface that affect the chemical transduction by the endothelium of mediators which have an action on smooth muscle and thereby affect the tone setting for these muscle cells[25]. Thus, a downregulation of the endothelium-derived braking action will allow for further shortening of the smooth muscle cell contractile filaments. Blood-borne materials with a direct action on the smooth muscle cell have been identified and include numerous peptides and cytokines[26]. Many of these vasoactive materials appear also to exert an intermediate effect on endothelial cells as well.

Intrinsic mechanisms operate by way of two possible pathways - endothelial cell derived mediators can modify the contractile response and specific chemical modalities can either enhance or blunt the degree of cell shortening. Steady-state tone varies regionally -the extent to which vessel diameter is modified depending upon variations in pressure per se [myogenic stimulus]. In the case of the larger blood vessels, the myogenic component is a comparatively minor determinant of vessel caliber.

The endothelial cell derived mediators at the feed artery level represent an important extrinsic factor to the extent that shear rate dependent effects become the principal regulatory feature. In the transverse arterioles, the shear effect is less important, neurogenic influences are marginal and the basal tone of the smooth muscle cell is now coupled directly to changes in the interstitial milieu. Proximal to this segmental branching level, there is little or no input from the tissue milieu [Figure 3]. Changes in the tissue environment become more decisive in the precapillary region through their effect on basal tone so that the L0 setting is a critical factor.

FUNCTIONAL ATTRIBUTES OF SUCCESSIVE SEGMENTS

Basically the three distinctive facets of microcirculatory function are the regulation of pressure, flow and exchange. Such modulation is tied to the changing metabolic demands of the parenchymal cells. Although pressure and flow are dependent physical variables, they are handled separately at the level of the microcirculation[18]. Structural as well as functional mechanisms are involved. Each of these three entities has been treated in the past in an overly simplified manner without due regard to the segmental differences in the branchings of the terminal vascular bed. Microvascular dynamics cannot be examined solely in the light of lumped group values or in an analogue mode.

Most vascular networks have the capacity to autoregulate in the face of perturbations in pressure and blood flow presumably in line with the need to restore perfusion to some reference steady-state value. Inasmuch as such perturbations can be of either systemic origin or locally invoked, the related adjustments may vary accordingly.

The question remains as to whether the larger primary arcades as opposed to the small secondary arcade arterioles can be considered to represent scaled-down functional versions of one another, or whether they serve different functions in view of the fact that the proximal and distal portions of the microvascular branching array do not respond in a uniform way to receptor dependent agonists.

The distinctive branching pattern at the microvascular level makes available a mechanism for distributing blood flow through a succession of arterioles in which pressure can then be further modulated in a number of prescribed steps commensurate with local needs. This latter contingency is achieved through locally

Table 4. Characteristics of Micro-Circulatory Functional Unit

- a side-branch configuration [in parallel circuit]
- offshoots are much smaller than parent stem
- restricted entry conditions into offshoot
- sphincter type control is possible [high branch shear rate]
- volume distribution can be varied over a wide range by on-off action of individual arteriolar spigots

elaborated mediators that readjust the setting of the smooth muscle cell contractile filaments [Table 4].

In developing models of microvascular homeostasis, the possibility has been advanced that such networks have common features which enable them to serve as a discrete organic unit. Operational design characteristics for such an entity are introduced by way of differences in number, diameter, length, relative surface area, etc. It is obvious that a much higher flow rate is potentially possible at the more proximal transverse arteriole level than in the smaller terminal side-arm offshoots. The vascular alignment is such that when the entrance into a side-branch offshoot is widened, the resulting flow is distributed through all of the ramifications of that branch. When flow is increased following dilation, the distribution among the capillaries becomes somewhat non-uniform depending on the caliber and length of the available pathways.

The end type arterioles and their branchings can be looked upon as a basic Functional Unit for local homeostasis [Table 4]. By virtue of their deployment as side-arm offshoots, such functional micro-units can bypass selected vessels completely and thereby utilize time-dependent modulation for meeting homeostatic needs.

The maintenance of local homeostasis requires the delivery via the transverse arterioles of a volume of blood that is sufficient to perfuse the entire array of precapillaries distributed by that arteriolar Functional Unit. In essence, the parent arcade network serves as a reservoir from which aliquots of blood are withdrawn as needed by adjusting the dimensions of individual small side-arm offshoots.

In this light, the Functional Units formed by the transverse and terminal arterioles at consecutive levels fulfill somewhat different functions, dimensional adjustments of the larger transverse offshoots are flow-related, whereas at the level of the small arteriole configuration the vessels are more responsive to pressure perturbations. The final local adjustment is a composite of three variables - pressure distribution, shear rate [flow related factors] and the nutritional effectiveness [oxygen] of the perfusion [Figure 3].

Models dealing with the properties of the arteriolar wall usually depict an active smooth muscle cell component in parallel with the passive elastic component. The relative contribution of each of these elements varies so that in cases where the strain carried by the elastic component is greater than that handled by the smooth muscle cell, a change in muscle cell tone would have a comparatively small effect on the overall compliance of the vessel[27]. In the case of the large arterial conduits, diameter is determined in the main by the degree of neurogenic tone imposed by the SNS. A potential exists for such neurogenically mediated influences to be blunted or amplified locally by endothelial cell-derived mediators.

PRESSURE DISTRIBUTION

Peripheral vascular pressure distribution plots show a linear relation with respect to vessel diameter among the vessels that are relegated for analysis into a given hierarchical segment. On the other hand, when a cubic spline curve fitting is made for the pressure-diameter relations for the entire sequence of arteriolar subdivisions, the non linear curve exhibits two distinct reflection points which correspond to the transverse arterioles [A2] and the terminal arterioles [A3] respectively, regions where an abrupt reduction in pressure is introduced by virtue of the side-arm disposition of these offshoots[28]. At both of these sites, the bloodstream is characteristically diverted from a wide parent vessel into narrow offshoots. At both of these inflection points, the parent trunk, as well as the daughter branch, are muscular vessels capable of independent modulation of caliber. Such an in-parallel alignment is especially striking among the arteriolar branchings in skeletal muscle [Figure 4].

The neurogenic feedback associated with the buffering of central blood pressure acts primarily on the large arteries to maintain a uniform regional apportionment of the volume of blood delivered by the heart to the major arterial conduits. The primary function of these supply vessels is to ensure the availability of a volume of blood in line with the overall needs of that particular structure. The proximal feed arteries are the principal microvascular site where such mediated adjustments act. These vessels are on average between 100-150 μm wide in a skeletal muscle preparation such as the spinotrapezius muscle of the rat. In this context they represent the site where the autoregulation of local resistance is prominent.

There is only a modest [20-25%] reduction in blood pressure along the more proximal larger arteries down to the level of the Feed Artery branching. More distally the pressure is then substantially reduced by virtue of a series of active myogenic adjustments. The extent to which a particular vessel is narrowed, however, is strongly influenced by the initial length setting of the contractile proteins under steady-state conditions [basal tone].

As indicated, the transverse arterioles and their branches are highly pressure sensitive so that they can serve to bring pressure levels into a range that is compatible with blood-tissue exchange. Such in-parallel circuits can be narrowed and excluded selectively from the active circulation. Although there is evidence for a response to sympathetic nerve stimulation in these delicate terminal muscular vessels[29], the dominant regulatory feature at the immediate precapillary level is provided by tissue factors that allow for a graded capillary perfusion. The terminal precapillary branchings can be thus be conceived as strategic structures for increasing or decreasing the number of capillaries that are actively perfused.

BLOOD - TISSUE EXCHANGE

Let us examine briefly what the physiological maintenance of interstitial homeostasis entails. Exchange between the blood and tissue compartments occurs primarily across a physical barrier, the walls of the 4-10 μm wide capillaries which are endothelial tubes supported by a thin connective tissue membrane or basal lamina. Depending upon the architectural constraint of the parenchyma in individual organ systems, the capillaries are spaced at sufficiently close intervals to permit a uniform extravascular distribution of blood-borne materials.

The actual pathway taken by the bloodstream within a meshwork of exchange capillaries is determined to a large extent by the structural alignment of the vascular

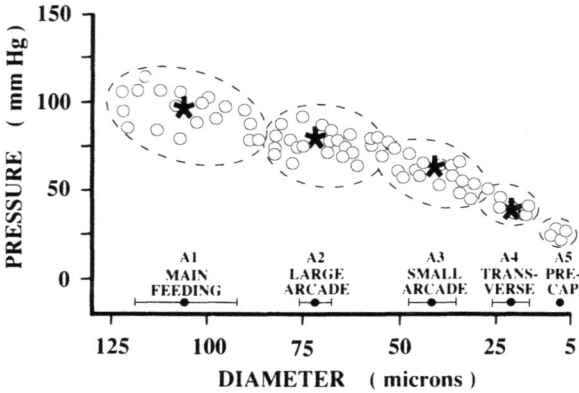

Figure 4. Distribution of arterial pressures in successive hierarchical branchings of the Feed artery in a skeletal muscle preparation of the rat [spinotrapezius]. The open circles are individual measurements, the asterisk represents the average for the group. There is a 40-50% reduction in pressure down to the level of the precapillaries. The actual pressure profile is not linear with respect to diameter but shows two inflection points in the arteriolar branching region corresponding to the transverse arteriolar configuration and the branching of the terminal

conduits in which the blood is being circulated. A shift in the level of volumetric flow rate is not by itself the key factor for ensuring an effective blood-tissue exchange. The surface area available for exchange, together with the nature of the barrier between the blood and tissue compartments, are equally important[30]. In terms of acute adjustments, the surface area available for exchange is the entity that has to be modulated.

Particular facets of the exchange process have been selected as end-points for the activation of such homeostatic readjustments. These include oxygen per se, lactic acid, adenosine, pH and fluid translocation. There is evidence that under conditions where an imbalance persists, the permeability of the exchange barrier itself can be modified. Inasmuch as pressure per se represents the driving force for the convective movement of water out of and into the capillary bloodstream, some tie-in must exist between the hydraulic forces in the capillaries and the actual surface area made available for exchange. Although a shear rate dependent interaction with the endothelial cells has recently been proposed[31] as a basic factor influencing the vasomotor activity of the arterial vessels, no comparable tie-in with the permeability characteristics of the capillary wall has been demonstrated.

In order to analyze more directly the efficacy of in situ blood-tissue exchange, the term exchange must be expanded. Blood-tissue exchange involves a wide range of molecular species from oxygen to complex macromolecules, fluid filtration and absorption as well as the removal of metabolic by-products from the interstitium. Considerable emphasis has been placed on oxygen delivery, the term hypoxia being used almost synonymously with a subnormal perfusion. There is no clear consensus as to just which end-point serves as the feedback for tissue perfusion adjustments. Renkin[32] used the term P.S. coefficient to indicate a relationship between the permeability of the wall per se and the surface area for exchange in an overall generalized formulation, but such a relationship is difficult to measure in a direct way.

How can a set of muscular vessels of this type selectively adjust flow into side-offshoots in response to changing needs for blood flow and match this changing volumetric rate with a surface area appropriate for exchange? An important determinant in this regard is a shift in the factors that serve to modulate smooth muscle cell tone in successive segments. There is evidence that such a feed-back may operate through the intermediary of an effect on other cells in the immediate vicinity of the smooth muscle cells [pericytes, endothelial cells, mast cells and even WBC in the circulation].

ABNORMAL STATES

Microcirculatory homeostasis can be undermined by intrinsic derangements within the network proper, as well as by way of extrinsic perturbations, such as hemorrhage, hypertension, sepsis. Although systemic disorders have been well documented on a physiological or organ level, intrinsic derangements at the tissue level are only sketchily defined. Analyses dealing with functional behavior of the whole organ have dealt with criteria that reflect the effectiveness of tissue perfusion in maintaining oxidative metabolism. Quantitative estimates of perturbations in individual vessels are, however, difficult to integrate into the operation of the network as an organic entity.

Microcirculatory damage in a clinical context has been documented for the most part, either as a change in vascular permeability due to alterations in the capillary barrier, or by visual manifestations detected by intravital microscopy[33]. Vascular abnormalities have usually been attributed to a material failure in the vessel wall [e.g., aneurysms], excessive dilation of venules, areas of ischemia [leukocyte plugging], maldistribution of microvessel blood flow, a shift in the autoregulatory threshold or in the response to reactive hyperemia, coagulation, thrombus formation, RBC aggregation, etc.

A number of major systemic disorders have been shown to be consistently associated with characteristic microvascular derangements. These have been documented in vivo for man in a number of tissues that are accessible for in vivo microscopy - the bulbar conjunctiva of the eye, the nail-fold of the digits and the skin[34,35]. The pathological changes that are recorded may reflect either a genetic predisposition on a restricted regional scale or an intrinsic defect in the terminal blood vessels in general. At issue is whether these abnormalities are to be considered primary or secondary manifestations - a reflection of changes in parenchymal cell metabolism or of vascular origin per se.

With regard to the latter, it has been demonstrated, for example, that changes in the status of the circulating leukocytes [activation] can serve as a point of departure for an analysis of the progression of the disease by focusing on the factors that lead to leukocyte activation and the pathophysiological consequences for vascular endothelium associated with the resulting activation of leukocytes[36].

SUMMARY

The dimensions of the large arterial conduits are under the control of a neural feedback which exerts its effect by way of several types of adrenergic receptor sites and is modified by humoral agents circulating in the bloodstream.

The prominent internal elastic lamella in the 50 - 25 μm arterioles bears the brunt of the strain imposed by distending action of the blood pressure. The steady-state setting of the L0 starting length for the muscle cells is directly affected by the prevailing intravascular pressure. Inasmuch as the tension/length characteristics of the smooth muscle cells in the various arterial branching hierarchies differ considerably, the position of the starting length [L0] on this curve will have a strong influence on the actual changes in the caliber of the vessel induced by a given perturbation. The initial hierarchical ramifications of the microcirculation, the arcade arterioles, which basically ensure a more uniform spatial distribution of the blood, maintain a relatively fixed diameter under steady-state conditions.

In the <u>feed artery</u>, which represents an intermediate segment that bridges the macro and the micro portions of the circulation, endothelial cell modulation becomes increasingly important while the neurogenic influence wanes. The vessels are under the influence on the one hand of regulatory controls concerned with the buffering of the systemic blood pressure via the SNS and on the other with adjustment of local tissue perfusion in line with tissue metabolic requirements.

At the level of the <u>mid-sized arterioles</u>, the prevailing vessel tone is set largely by the intervention of endothelial cell derived mediators in concert with humoral blood-borne factors. The final physical adjustment, the narrowing or dilation of the vessel, is triggered in turn by the distending action of the prevailing blood pressure. By means of a resetting of L0 relations, differences in vessel diameter are possible even in vessels which are under the same head of pressure. More distally, as the structural makeup of the smaller branchings is modified, the dominant variable is no longer flow per se [in contrast to the implications of the metabolic autoregulatory concept], but pressure.

In the <u>terminal arterioles</u>, where the dominant component is a single layer of smooth muscle cells, an elastic lamella is no longer present, only occasional longitudinal elastin fibrils are seen. At this level, it is the degree of smooth muscle cell basal tone per se that is the principal determinant of vessel dimensions. Locally derived factors modulate the cell L0 setting and thereby the actual physical readjustment. Under conditions of subnormal flow, some of the intracellular energy dependent pathways are diverted into glycolytic reactions [anerobic]- e.g.,they involve xanthine oxidase and xanthine derivatives which appear to serve as a feedback for modulation of smooth muscle cell tone in the terminal arterioles. The degree to which an oxygen deficit has been incurred can in this way influence the extent to which the arterial pressure is reduced across a particular segment.

There is an abrupt increase in the shear rate of the blood at the junctional origin of the arteriolar offshoot. It has been suggested that the accompanying local high shear rate blunts the production of the endothelial cell derived EDRF so that the smooth muscle component is more easily narrowed. The junctional portion of the terminal arterioles as they branch from the arcade network displays a prominent myogenic response. Because of the sharp fall in pressure across this region of high resistance, a complete closure of the 20-25 μm arterioles occurs frequently. The feedback mechanism for local control in this region operates not only by adjusting vessel caliber, but by changing the number of open vessels and the time-dependent vasomotion activity.

REFERENCES

1. Zweifach BW, Lipowsky HH: Pressure-flow relations, In: Blood and Lymph

Microcirculation, Ch. 7, In: Handbook of Physiology. The Cardiovascular System (Eds., Renkin EM, Michel CC), Vol. 11, Sec. 5, Amer. Physiol. Soc., Washington, D.C., 1985.
2. Lash JM, Bohlen HG, Waite L: Mechanical characteristics and active tension generation in rat intestinal arterioles, Amer. J. Physiol., 260 [Ht. Circ. Physiol. 29]:H-1561, 1991.
3. Chapleau MK, Hajduczok G, Abboud FM: Paracrine modulation of baroreceptor activity by vascular endothelium, N.I.P.S. 6:210, 1991.
4. Fleming BP, Barrow KW, Howes TW, Smith JK: Response of the microcirculation in rat cremaster muscle to peripheral and central sympathetic stimulation, Circ. Res., 6 [suppl. II]:II-26, 1987.
5. Plunkett WC, Overbeck HW: Arteriolar wall thickening in hypertensive rats unrelated to pressure or sympathoadrenergic influences, Circ. Res., 63:937, 1988.
6. Engelson ET, Skalak TC, Schmid-Schönbein GW: The microvasculature in skeletal muscle. I. Arteriole network topology, Microvasc. Res., 30:28, 1985.
7. Tilton RG, Kilo C, Williamson JR, Murch DW: Differences in pericyte contractile function in rat cardiac and skeletal muscle microvasculature, Microvasc. Res., 18:336, 1979.
8. Zweifach BW: Biomechanics of the microcirculation, In: Frontier in Biomechanics (Eds., Schmid-Schönbein GW, Woo S, Zweifach BW), Springer-Verlag, New York, 233, 1986.
9. Davis MJ, Ferrer PN, Gore RW: Vascular anatomy and hydrostatic pressure profile in the hamster cheek pouch, Amer. J. Physiol., 250 [Ht. Circ. Physiol. 19]:H-291, 1986.
10. Kuwabara T, Cogan DG: Studies of retinal vascular patterns. I. Normal architecture, Arch. Ophth., 64:904, 1960.
11. Nicoll PA, Webb RL: Blood circulation in the subcutaneous tissue of the living bat's wing, Ann. N.Y. Acad. Sci., 46:697, 1946.
12. Grafflin AL, Bagley EH: Studies of peripheral vascular beds, Bull. J. Hopkins Hosp., 92:47, 1953.
13. Kassab GH, Rider CA, Fung YC: Morphometry of the coronary vasculature of the pig, Circ. Res., 1992, In Press.
14. Zweifach BW, Kovalcheck S, DeLano FA, Chen PC: Micropressure-flow relationships in a skeletal muscle of spontaneously hypertensive rats, Hypertension, 3:601, 1981.
15. Bohlen HG: Intestinal microvascular adaptation during maturation of spontaneously hypertensive rats, Hypertension, 5:739, 1983.
16. Fagrell B, Tooke J, Ostergren J: Vital microscopy for evaluating skin microcirculation in man, Progr. Appl. Microcirc., 6:129, 1984.
17. Bevan JA, Laher I: Pressure and flow-dependent vascular tone, FASEB J., 5:2267, 1991.
18. Folkow B: The resistance vasculature: Functional importance in the circulation, Ch. 2, In: The Resistance Vasculature (Eds., Bevan JA, Halpern W, Mulvany MJ), Humana Press, New Jersey, 1991.
19. Meininger GA, Mack CA, Fehr JL, Bohlen HG: Myogenic vasoregulation overrides local metabolic control in resting skeletal muscle, Circ. Res., 60:861, 1987.
20. Meininger GA, Zaweija DC, Falcone JC, Hill MA, Davey JP: Calcium measurement in isolated arterioles during myogenic and agonist stimulation, Amer. J. Physiol., 261 [Ht. Circ. Physiol. 30]:H950, 1991.
21. Furchgott RF, Vanhoutte PM: Endothelium derived relaxing and contracting factors, FASEB J., 3:2007, 1989.

22. Jackson PA, Duling BR: Myogenic response and wall mechanics of arterioles, Amer. J. Physiol., 257 [Ht. Circ. Physiol. 26]:H-1147, 1989.
23. Johnson PC: Autoregulation of blood, Circ. Res., 59:483, 1986.
24. Grande PO, Borgstrom P, Mellander S: On the nature of basal vascular tone in cat skeletal muscle and its dependence on transmural pressure stimuli, Acta Physiol. Scand., 102:365, 1979.
25. Koller A, Kaley G: Endothelial regulation of wall shear stress and blood flow in skeletal muscle microcirculation, Amer. J. Physiol., 260 [Ht. Circ. Physiol. 29]:H862, 1991.
26. Rubanyi GM, Botelho LH: Endothelins, FASEB J., 5:2713, 1991.
27. Schmid-Schönbein GW, Skalak R, Chien S: Hemodynamics in arteriolar networks, Ch. 19, In: The Resistance Vasculature (Ed., Bevan JA, Halpern W, Mulvany MJ), The Humana Press, New Jersey, 319, 1991.
28. Zweifach BW: Functional Behavior of the Microcirculation, C.C. Thomas, Springfield, Ill, 1961.
29. Marshall JM: The influence of the sympathetic nervous system on individual vessels of the microcirculation of skeletal muscle in the rat, J. Physiol.[London], 332:169, 1982.
30. Renkin EM, Curry FE: Transport of water and solutes across capillary endothelium, Ch. 1, In: Membrane Transport in Biology. IV.B. Transport Organs (Eds., Giebisch G, Testeson DC, Ussing HH), Amer. Physiol. Soc, Washington, D.C., 1978.
31. Tolins JP, Shultz PJ, Raiz L: Role of endothelium derived relaxing factor in regulation of vascular tone and remodeling. Update on humoral regulation of vascular tone, Hypertension, 17:909, 1991.
32. Ditzel J, Sagild U: Morphologic and hemodynamic changes in the smaller blood vessels in diabetes mellitus. II. The degeneration and hemodynamic changes in the bulbar conjunctiva of normotensive diabetic patients, New Engl. J. Med., 250:587, 1954.
33. Bollinger A, Frey J, Jager K, Furier J, Seglias J, Siegenthaler W: Patterns of diffusion through the skin capillaries in patients with long term diabetes, New Engl. J. Med., 307:1305, 1982.
34. Chen PCY, Kovalcheck SW, Zweifach BW: Analysis of microvascular network in bulbar conjunctiva by image processing, Int. J. Microcirc. Clin. Exp., 6:245, 1987.
35. Kunitomo N: Microcirculation of Human Conjunctiva, Igaku, Tosho, Shupen, Tokyo, 1974.
36. Schmid-Schönbein GW, Seifige D, DeLano FA, Shen K, Zweifach BW: Leukocyte counts and activation in spontaneously hypertensive and normotensive rats, Hypertension, 17:323, 1991.

CONCENTRATION AND VELOCITY PROFILES OF BLOOD CELLS IN THE MICROCIRCULATION

Robert S. Reneman, Bea Woldhuis, Mirjam G.A. oude Egbrink, Dick W. Slaaf and Geert Jan Tangelder

Departments of Physiology and Biophysics
Cardiovascular Research Institute Maastricht
University of Limburg
Maastricht, The Netherlands

INTRODUCTION

Although this chapter basically deals with the concentration and velocity profiles of blood cells in general, most of the data presented concern blood platelets. These small blood cells, with a density close to that of blood plasma, can be nicely used as natural markers of flow by fluorescently labeling them in vivo allowing their localization and the assessment of blood flow velocities at various sites in microvessels. Because of their role in hemostasis, thrombosis, and maintaining endothelial cell integrity, blood platelets can be expected to come in contact with the vessel wall. Recent in vivo studies have shown that blood platelets indeed do come close to the wall in arterioles[1,2], but less so in venules. In the latter microvessels a relatively large zone near the vessel wall from which blood platelets seem to be expelled, has to be appreciated[2].

To be able to interpret blood platelet-vessel wall interactions in microvessels adequately, quantitative information about wall shear rate and wall shear stress is required. Wall shear rate is generally calculated based on the assumption of a parabolic velocity profile[3]. In vivo studies, however, have shown that in arterioles the velocity profiles differ significantly from a parabola; they are blunt in both systole and diastole[4] leading to higher wall shear rate values, and, hence, wall shear stress values, than expected on the basis of a parabolic velocity profile[5]. Since in arterioles platelets do come as close to the vessel wall as 0.5 μm, wall shear rate can be estimated rather accurately in these microvessels. The estimation will be less accurate in venules because larger extrapolations to zero flow are required due to the relative exclusion of platelets near the venular wall. From the ratio of the wall shear rate value derived from the actual velocity profile and that expected in the case of a parabolic velocity profile, an estimate of the relative apparent blood viscosity can be made. The wall shear stress and relative apparent blood viscosity values derived from the velocity profiles can be compared with those based upon direct measurements in vivo with more invasive techniques[6,7].

In this chapter a survey will be given of the concentration distribution of platelets in arterioles and venules, the velocity profiles as recorded in arterioles and venules (the latter data being preliminary) and the wall shear rate values in arterioles as estimated from the actual velocity profiles in these microvessels. The wall shear stress and relative apparent blood viscosity values, as derived from the measured and calculated parameters, as well as the orientation of the platelets when flowing in arterioles, will be discussed briefly. Where pertinent attention is paid to red blood cell and leukocyte rheology, and to the possible difference in endothelial cell function between arterioles and venules.

CONCENTRATION DISTRIBUTION OF BLOOD PLATELETS IN ARTERIOLES AND VENULES

Labeling and Localization of Blood Platelets In Vivo

The concentration distribution is determined with the use of fluorescently labeled blood platelets. The in vivo labeling of platelets[8] and their localization in a thin optical section[9] have been described in detail before. In short blood platelets are preferentially labeled by intravenous injection (lasting 20-30 s) of 2.5 ml of a solution (18.2 mM) of the fluorochrome acridine red (absorption and emission peak at 525 and 625 nm, respectively). With this procedure nearly all blood platelets are labeled without activation of the platelets by the dye or induction of gross ultrastructural changes. Moreover, their ability to adhere and aggregate is maintained[8]. Leukocytes and the vessel wall are labeled as well, but not red blood cells. The blood platelets flowing in mesenteric arterioles and venules are visualized by intravital fluorescence video microscopy with a Leitz x100 or x50 salt-water immersion objective (numerical apertures 1.20 and 1.0, respectively). Final optical magnification is 200 x. Incident flashed illumination with flashes of short duration (<0.1 ms) is used to prevent smearing of the image due to moving of the platelets.

The flashes are given in the blanking period of the TV camera every 6th field, corresponding to a flash interval of 120 ms. This interval is sufficiently long to ensure that each instantaneous picture contains no information from the previous flash. On the other hand, the flash frequency is high enough to obtain an adequate number of pictures within a relatively short period of time and is sufficiently far away from the heart rate (range 110-200/min) and the respiratory rate (range 30-60/min) of the anesthetized animal to obtain pictures that are randomly dispersed throughout both cycles. In all cases a high flash power range is selected and adjusted prior to the experiments with the vernier intensity control to avoid blooming of the TV camera. A picture obtained in this way is presented in Figure 1.

The position of the platelets is determined with the region of sharp focus about the median plane of the vessel, i.e. the focal position yielding the widest vessel diameter. Only blood platelets flowing within a thin optical section (5-7 μm) are used for analysis[9]. To localize a platelet within this optical section, the TV fields of the

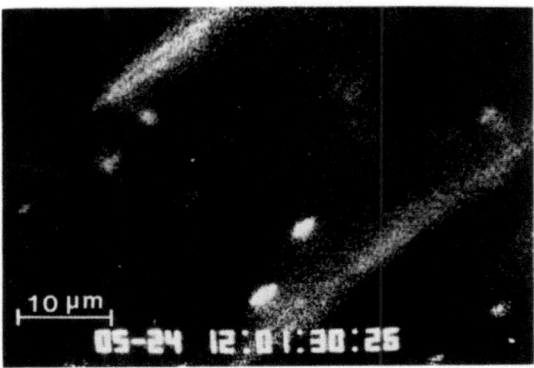

Figure 1. Fluorescent blood platelets flowing in an arteriole. Level of sharp focus in median plane of the vessel. Vessel wall is also labeled by dye that permits orientation within the microvasculature. Picture taken from video monitor.

video recordings are digitized (512 by 512 pixels) by image processing on a personal computer (IBM). The intensity of each pixel may range from 0 (black) to 255 (white). In a band of 3 pixels wide, drawn through the center of the image of a platelet and the lumen of the vessel, the intensity curve is plotted. A platelet is considered to be within the optical section when the intensity peak of the platelet exceeds the background noise. In this analysis, the thickness of the optical section remains constant over the whole vessel width[1]. The centroids of the images of the blood platelets present within the optical section during the period of analysis are marked on a transparency positioned over the monitor screen. According to the direction of flow, the left and right vessel walls are drawn and a vessel segment selected with a length of approximately 20-30 μm and without upstream branch points within at least 10 vessel diameters. For each blood platelet marked, the distance (d) between the centroid of the platelet and the left vessel wall, as well as the vessel diameter at that specific point (2R), are determined and the relative position of the platelet (d/2R) is calculated. In the different vessels, the number of platelets counted range from 63 to 144 (112±26;x±sd) in arterioles and from 60 to 259 (148±62;x±sd) in venules.

Determination of Concentration Distribution in Arterioles and Venules

For presentation of the blood platelet concentration distributions, relative frequency distributions are used. As an example, the distribution for the experiment shown in Figure 2A is presented in Figure 2B. To obtain such a histogram, the blood platelets are assigned to 20 segments, according to their relative position along the width of the vessel. The number of twenty segments is chosen to obtain an estimate of the general distribution (see below) with sufficient detail given the number of blood platelets counted. To obtain a general frequency distribution for both groups of microvessels, individual histograms are pooled. Before pooling, each individual distribution of counts (left Y-axis in Figure 2B) is normalized with respect to the mean

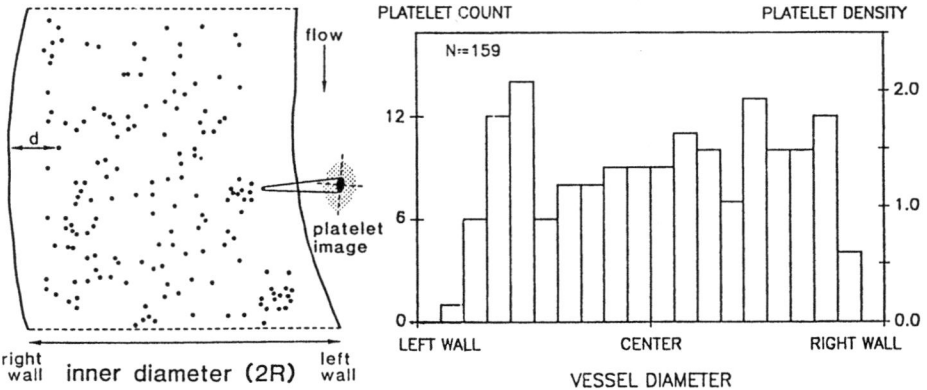

Figure 2. (A) Schematic presentation of platelet centroid posititions (dots) present within the optical section in 160 TV-fiels in a control venule, i.e. no dextran sulfate used; (B) concomitant platelet count histogram (left Y-axis) and normalized distribution (right Y-axis). An image of a platelet (shaded) and representation of its centroid is shown schematically in A; d = distance between the centroid of a platelet and the left wall. Vessel diameter = 31 μm. After Woldhuis et al, Am. J. Physiol. 1992. With permission of the American Physiological Society.

count per unit volume, i.e mean density (right Y-axis in Figure 2B), in that vessel. This is especially necessary for proper comparison between arterioles and venules, since their distributions differ and different segments contribute to the mean density to a different extent. This analysis has been described in detail elsewhere[2].

The Shape of the Concentration Distribution in Arterioles and Venules

General blood platelet density distributions for arterioles and venules (15-33 μm in diameter), normalized with respect to the mean density in the vessel, are presented in Figure 3. The mean value in the center of arterioles is 0.55 ± 0.33 and in the center of venules 1.04 ± 0.45, indicating that approximately two times more platelets are located in the center of venules as compared to arterioles. The relatively low concentration of blood platelets in the center of arterioles can probably be explained by the tendencey of red blood cells to stream at higher velocities, leading to platelet expulsion in this part of the vessel (see section Shape of Velocity Profiles). Figure 3 also shows that the distributions differ near the vessel wall. Unlike in arterioles, in venules there is a relative exclusion of blood platelets from the segments close to the vessel wall. This difference is more clear when the left and right segments near the vessel wall are pooled. The average blood platelet densities close to the vessel wall, as obtained from these pooled data, are shown in Figure 4. The peak densities shown have a similar height in venules and arterioles, but their relative radial positions differ. In arterioles, blood platelet density rises steeply in the two segments close to the wall, reaching a peak relatively close to the wall, while in venules this increase is less steep. The blood platelet densitites are significantly different between arterioles and venules in the two segments between 0.5 and 0.7 and in the segment between 0.8 and 0.9 of the vessel radius[2].

Since, in venules, leukocytes are often moving along the endothelium at a

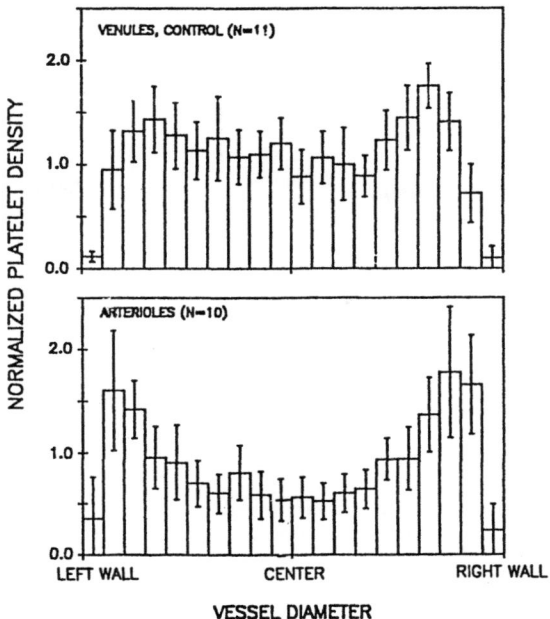

Figure 3. Normalized platelet density distribution with 95% confidence limits in venules and arterioles. After Woldhuis et al, Am. J. Physiol. 1992. With permission of the American Physiological Society.

velocity significantly lower than that of other blood cells (see Figure 9), we have investigated whether this leukocyte rolling can be held responsible for the relative exclusion of blood platelets near the wall in venules. Inhibition of leukocyte rolling with sulfated dextrans[10], however, does not affect the size of this exclusion zone[2]. It is tempting to speculate that the relative platelet exclusion in venules is caused by differences in wall structure and/or function between arterioles and venules, because the differences in blood platelet distribution are also not caused by differences in geometry between these microvessels[2], and the velocity ranges in these two vessel types greatly overlap. Whether the higher blood platelet concentration in the center of venules is caused by a reduced tendency of red blood cells to stream in this part of the vessel in venules, is subject to further investigation (see section Shape of Velocity Profiles).

VELOCITY PROFILES IN ARTERIOLES AND VENULES

The Assessment of Velocity Profiles

Until recently, most of the information available about velocity profiles in microvessels has been obtained in vitro, using ghost cell suspensions[11,12] or blood[13-15]. These studies yield conflicting results as far as the shape of the velocity profile is concerned; flat, blunted and parabolic profiles are found. Besides, these studies have

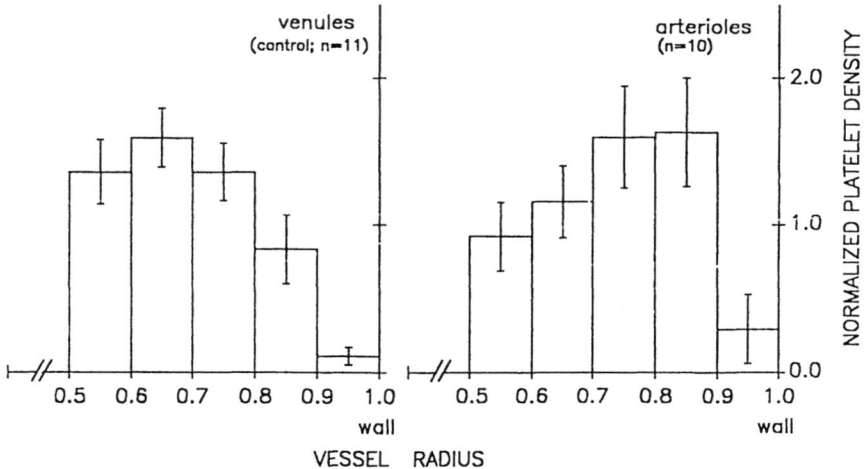

Figure 4. Normalized platelet densitites with 95% confidence limits in the 5 segments closest to the vessel wall. Distributions obtained by pooling left and right segments. After Woldhuis et al, Am. J. Physiol. 1992. With permission of the American Physiological Society.

been performed in glass tubes with a diameter larger than 30 μm.

Over the years assessment of velocity profiles in microvessels in vivo has been hampered by the absence of techniques to measure velocities in these vessels with adequate spatial resolution. Recently, however, a method has been developed to accurately determine velocity profiles in mesenteric arterioles and venules[4]. In this technique fluorescently labeled blood platelets, which are small disc shaped cells, are used as natural markers of flow. The techniques used to label and localize the blood platelets have been described previously. In the assessment of velocity profiles, pairs of flashes, rather than one single flash, are given. The flash pairs are given with a short, preset time interval between the two flashes, yielding in one video field two images of the same blood platelet displaced over a certain distance for the given time interval (Figure 5). The first flash is given in the blanking period of the TV camera.

In each experiment, the time interval between the first and the second flash (range 1-5 ms) is selected in such a way that the two concomitant images of a blood platelet show no or only little overlap. Within a measuring period (30-60 s), all flash pairs are recorded at a selected moment in the cardiac cyle (~ 20 msec) by triggering the light flashes by the R wave of the electrocardiogram (ECG). A preset delay is used to determine velocity profiles in both systole and diastole[4]. Instantaneous red blood cell velocity, as assessed photometrically with a prism grating system[16], is used to select the value of the preset delay.

Video recordings are analyzed frame by frame. To determine the velocity profile, the centroids of the images of a blood platelet are identified by eye and the following parameters measured with the use of vernier calipers: 1) the displacement of the blood platelet in the preset time interval, yielding its velocity, and 2) its relative radial position in the vessel, defined as the mean of the radial positions of the two images relative to the vessel radius. Only reasonable sharp images are used. This means that, with the high numerical aperture of the objective lens and the total optical

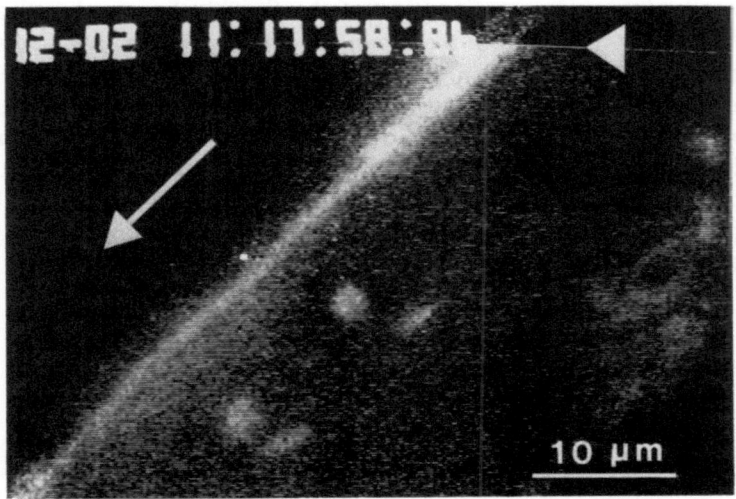

Figure 5. Fluorescently labeled blood platelets flowing in an arteriole, as observed with dual-flash illumination, yielding in one TV picture two images of the same blood platelet displaced over a certain distance during the time interval between the two flashes. The first flash is given in the blanking period of the TV camera. Arrowhead indicates the moment of the second flash (4 msec after the first one). Arrow indicates the direction of flow. After Tangelder et al, 1986. With permission of the American Heart Association.

magnification used, the data are obtained from an optical section about the median plane of the vessel with a depth of ~5 μm or less[9]. Because the centroid of a blood platelet image is used for the measurements, no data points can be obtained closer to the wall than 0.5 μm, because of the physical size[17] and orientation[18] of the blood platelets. A platelet-velocity profile obtained in this way in a rabbit mesenteric arteriole is presented in Figure 6. All velocity profiles are recorded at vessel sites without upstream branch points within at least 6 vessel diameters. At this distance from an upstream side branch, disturbance of the velocity profile due to the bifurcation, if any, will be small. In the microcirculation the Reynolds numbers are far less than unity. Therefore, inertial forces are negligible and the entrance length will be about one vessel diameter[19]. Curves immediately up or downstream of the site of measurement cannot always be avoided, since in the unstretched mesentery most arterioles meander with curves in opposite directions shortly following each other.

The Shape of Velocity Profiles

In arterioles the experimental platelet-velocity profiles can be adequately described with the following equation modified after Roevros[20]:

$$V(r) = V_{max} \left(1 - \left| a \left(\frac{r}{R} \right) + b \right|^K \right), a > 0, \qquad (1)$$

where V(r) is the velocity at radial position r, the vertical stripes denote absolute values, V_{max} is the maximal velocity in the vessel, R is the radius of the vessel, a is a scale factor allowing a non-zero intercept of the fit with the vessel wall and b is a parameter correcting for a shift of the top of the profile away from the vessel center. If the scale factor a is smaller than 1, the intercept of the fit with the vessel wall will be positive and will increase towards the value of V_{max} when a approaches zero. If K = 2, a parabolic velocity distribution is obtained for a = 1. An increase in K yields a progressively flatter profile. The ratio of V_{max} and the mean velocity of the profile (V_{mean}) can also be used as an index of the degree of blunting of a profile, being 2 in the case of a parabolic profile and 1 in the case of complete plug flow, i.e. all layers of fluid are travelling at the same speed.

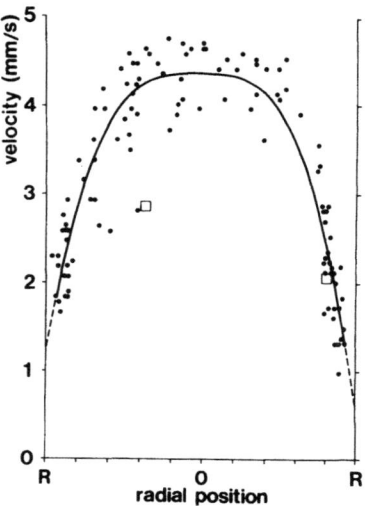

Figure 6. Velocity profile of blood platelets flowing in an arteriole with a diameter of 23 µm. Each dot represents the velocity of one platelet. The open squares indicate the velocities of the two leukocytes which could be observed during determination of the platelet-velocity profile. In addition, the best fit is shown as obtained with equation (1). Ordinates at R indicate position of vessel walls. R = vessel radius, O = center of the vessel. The dashed part of the fit is the extrapolated part. After Tangelder et al, 1988. With permission of the American Physiological Society.

As previously described[4], the ratio of V_{max} and V_{mean} can be approximated, with the use of the following equation:

$$\frac{V_{max}}{V_{mean}} = \frac{K + 2}{K + 2 - 2(a)^K} \qquad (2)$$

The best fits are obtained when the no-slip condition is ignored (a≠1), leading to a positive intercept with the vessel wall (Figure 6). The latter is irrealistic because flow velocity at the vessel wall is likely to be zero due to the no-slip condition[19]. This indicates that the velocity profiles in arterioles have to be described with a two phase model and a steep velocity gradient near the wall may be assumed (Figure 7). By means of this method of analysis, K values varying between 2.3 and 4.0, and V_{max} and V_{mean} ratios ranging from 1.39 to 1.54 are found in arterioles with a diameter ranging from 17 to 32 μm. These findings show that the velocity profiles in arterioles can be considered to be flattened parabolas. The finding that values of b are small (<4% deviation from the vessel center) indicates that the velocity profiles are rather symmetric.

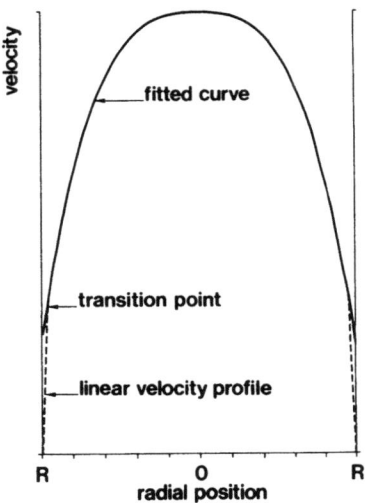

Figure 7. Schematic representation of the curves used to describe the velocity distribution in the arterioles and to calculate the least estimates of wall shear rate in these vessels. At the wall velocity is zero (no-slip condition). In the thin layer of fluid near the wall, which has a lower apparent viscosity than fluid in the remainder of the vessel, the mean velocity gradient is determined, resulting in a linear velocity profile (dashed lines). In the remainder of the vessel the velocity profile is described by the fitted platelet velocity profile (solid line). After Tangelder et al, 1988. With permission of the American Physiological Society.

In arterioles, the shape of the velocity profile of FITC-labeled red blood cells[4] is similar to that of blood platelets, but most of the red blood cells are travelling at streamlines with higher velocities, i.e. more towards the center of the vessel (Figure 8). This tendency of red blood cells to occupy higher velocity streamlines probably explains the relatively low blood platelet density in the center of arterioles[21-24]. At the present state of the art it is not known whether the relatively higher density of platelets in the center of venules can be explained by a relative absence of red blood cells in this part of these microvessels, because, to our best knowledge, hematocrit distributions over the cross-sectional area of microvessels have not been measured. The tendency of red blood cells to form aggregates might also play a role[21], provided that differences in red blood cell aggregate formation exist between arterioles and venules.

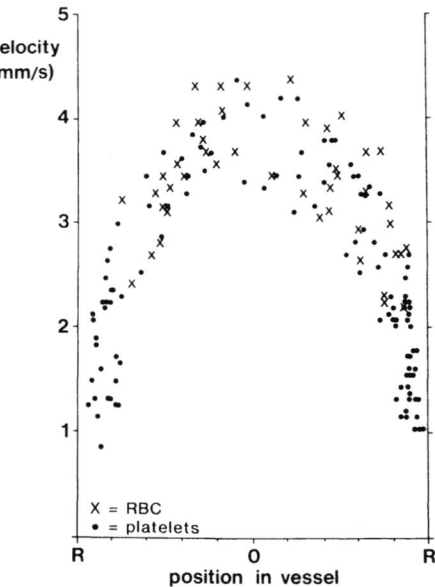

Figure 8. Velocity profile of blood platelets and red blood cells (RBC) flowing in an arteriole with a diameter of 24 μm during the systolic phase. Number of blood platelets and red blood cells is 117 and 47, respectively. The flash interval is 4 msec. After Tangelder et al, 1986. With permission of the American Heart Association.

Near the arteriolar wall blood platelets tend to align with flow showing their largest surface area towards the vessel wall (Figure 5). At the wall they can be seen tumbling along it, but seldomly showing adherence, at least under normal circumstances. In the center of these vessels the orientation of the blood platelets is more random[18].

Preliminary experiments in our laboratories indicate that the blood platelet velocity profiles in rabbit mesenteric venules are also flattened parabolas.

In arterioles leukocytes are generally travelling in the bloodstream (squares in Figure 6) at the same velocity as red blood cells. In venules on the contrary, leukocytes tend to roll along the vascular wall at a velocity significantly lower than the velocity at which other blood cells are moving in these microvessels (Figure 9). This leukocyte rolling in venules is a well-known phenomenon in exposed mesentery[10,25,26]. It is considered to be a first step in leukocyte-endothelial cell interaction upon endothelial cell stimulation, for example, by inflammatory mediators.

WALL SHEAR RATE IN VIVO

By describing the velocity profiles, as recorded in arterioles, by the best fit through the measuring points with the use of equation (1) and a linear interpolation from the point closest to the wall at which the velocity can be measured (transition point) to zero flow at the wall (Figure 7), a least estimate of wall shear rate (WSR)

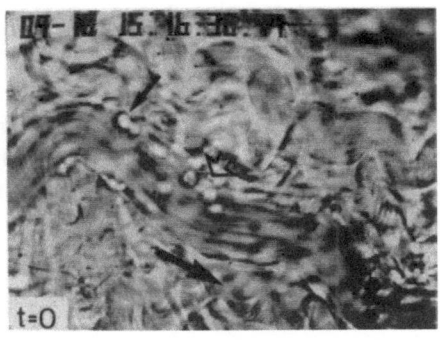
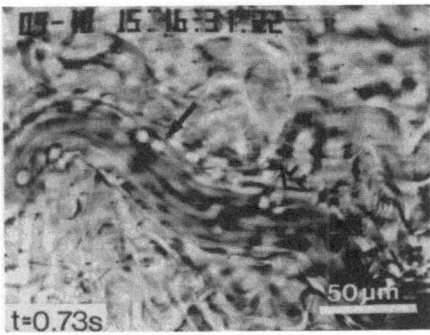

Figure 9. Leukocytes rolling along the venular wall in rabbit mesentery. Note the displacement of the leukocytes marked with the small and open arrows when comparing the top and bottom illustrations. The large arrow indicates the flow direction.

can be determined by means of the following equation[5]:

$$WSR = \frac{2V(x)}{D(1 - |x|)} \qquad (3)$$

where V(x) is the velocity at x, which is the relative radial position of the transition point, and D is the vessel diameter.

The least estimates of average wall shear rate, i.e. the mean of the values obtained at both vessel walls, range from 472 - 4712 s^{-1} (median: 1700 s^{-1}) for centerline red blood cell velocities ranging from 1.3 to 14.4 min.s^{-1} (median: 3.4 mm.s^{-1}). These data are derived from 12 velocity profiles in 9 arterioles. The values of the least estimates of wall shear rate at opposite vessel walls may differ from the mean value for that vessel by 2 to 44%. These differences are caused by 1) unequal thickness of the thin layer of fluid near the wall on opposite sides and 2) the slight asymmetry of the velocity profiles with a shift of their peaks away from the vessel center line. Although the latter effect is small (<4%), its influence on wall shear rate cannot be neglected.

The wall shear rate values expected in the case of a parabolic velocity distribution (WSR-P), but with the same volume flow in the vessel, can be derived with the following equation:

$$WSR\text{-}P = \frac{4\, V_{max}\ parabola}{D} = \frac{8\, V_{mean}}{D} \qquad (4)$$

The extent to which the wall shear rate values, as derived from the actual velocity profiles in vivo, are higher than those calculated assuming a parabolic velocity distribution for the same volume flow is given by the relation:

$$\frac{WSR}{WSR\text{-}P} = 0.25 * \frac{V(x)}{1-|x|} * V_{mean} \qquad (5)$$

At peak red blood cell velocities ranging from 1.3 to 14.4 mm.s-1 (median 3.4 mm.s-1) the estimated wall shear rate values are at least 1.46 to 3.94 (median: 2.12) times higher than those calculated on the basis of a parabolic velocity profile.

ESTIMATION OF RELATIVE APPARENT VISCOSITY

The ratio WSR/WSR-P, i.e., the ratio of the least estimate of wall shear rate derived from the actual velocity profile, and the wall shear rate expected in the case of a parabolic velocity distribution but with the same volume flow (WSR-P), might also be used as a least estimate of the apparent viscosity of blood in an arteriole relative to the viscosity of plasma. In the experimental situation, the local viscosity at the wall reflects the plasma viscosity, because of the low local haematocrit at the wall[19]. The estimate of wall shear stress is equal to wall shear rate (WSR) times plasma viscosity. Let us now consider a microvessel through which the blood is flowing as a Newtonian fluid with the same volume flow. In that case the velocity profile is parabolic. Consequently, the wall shear stress in this situation is equal to wall shear rate of parabolic flow (WSR-P) times apparent blood viscosity. For both situations the pressure drop in the vessel is proportional to both wall shear stress and apparent viscosity[6,19]. Therefore, the ratio of the apparent viscosity of blood and the viscosity of plasma, being the relative apparent viscosity of blood, is equal to the ratio of the wall shear rates in the blood and plasma in the flow situations mentioned above, and, hence, to WSR/WSR-P.

Lipowsky et al[6] have estimated apparent blood viscosity in microvessels of the cat mesentery with the use of measurements of pressure gradients. For arterioles between 18 and 32 μm in diameter, these investigators found a median apparent viscosity of ~3 cP. When a mean plasma viscosity for cats of 1.28 cP is used, a value found by Lipowsky and colleagues[7] in a later study, the average relative apparent viscosity in these arterioles can be calculated to be ~2.34, which is close to the median value of 2.12 for the ratio WSR/WSR-P as described in the previous section and well within the limits of its 95% confidence interval (1.80 - 2.44). As indicated by the latter study of Lipowsky et al[7], values of relative apparent viscosity obtained in vitro are lower, presumably because of the hemodilution used.

LEAST ESTIMATION OF WALL SHEAR STRESS

From plasma viscosity (η_p) and the least estimate of wall shear rate (WSR), a least estimate of wall shear stress (WSS) can be obtaineded with the use of the following equation:

$$WSS = WSR \times \eta_p \qquad (6)$$

The value of η_p (0.0107 dynes.s.cm^{-2}) used in this calculation is an average for the rabbits used in our studies. The median least estimate of wall shear stress in arterioles as estimated in this way, using the wall shear rate values described previously, is 18.2 dynes.cm^{-2} (range: 5.1 - 50.4 dynes.cm^{-2}). This value is substantially lower than the one directly determined by Lipowsky and colleagues[6] in cat mesenteric arterioles by means of micropressure measurements upstream and downstream. These investigators found a wall shear stress of 47.1 ± 23.4 dynes.cm^{-2} (x ± sd). In their study, however, the diameters of the arterioles were larger and the velocities higher than in our study. Consequently, reduced velocity, defined as mean velocity divided by vessel diameter, was significantly higher in the study of Lipowsky and co-investigators than in ours (on the average 87 vs 208 s^{-1}). Assuming that the shape of the velocity profile is independent of the velocity level, as was found in the arterioles investigated[4], this difference in reduced velocity explains the difference in wall shear stress values as found by Lipowsky and co-investigators and our group. At a 2.4 times higher reduced velocity the average wall shear stress value in our study would have amounted to 43.7 dynes.cm^{-2}, a value close to the one measured by Lipowsky and colleagues.

CONCLUSIVE REMARKS

At the present state of the art one may conclude that under normal circumstances blood platelets come indeed in close contact with the wall in arterioles. They can be seen tumbling along the wall, but near the wall platelets generally tend to align with flow. Towards the center of the microvessel, however, their orientation is more random. In venules there is a zone near the wall of relative exclusion for blood platelets. Recent findings indicate that this exclusion is not caused by the presence of rolling leukocytes. These differences between arterioles and venules can most likely be ascribed to differences in wall structure and/or function between these microvessels. Differences in function between the walls of arterioles and venules have also been observed for the thromboembolic reaction following wall puncture, being significantly more pronounced in arterioles[27,28].

To estimate in vivo and in vitro findings concerning blood cell-endothelial cell interaction at their true value, one has to be informed of the hemodynamic conditions in microvessels in vivo. Information about wall shear rate and wall shear stress is especially important, because of the increasing evidence that these parameters are important determinants of endothelial cell function[29] and may influence blood platelet and red blood cell function. To be able to reliably assess wall shear rate, detailed description of the shape of the velocity profile is required. It has been shown that in arterioles the velocity profile is a flattened parabola. The profile does not develop to a full parabola due to the viscous forces in these microvessels. Therefore, wall shear rate is significantly higher (2-4 times) than expected on the basis of a parabolic velocity profile. The relative apparent viscosity, as estimated from the ratio of the wall shear rate based upon the actual velocity profile and that expected on the basis of a parabola is rather similar to the relative apparent viscosity as measured directly. This also holds for the wall shear stress values estimated from wall shear rate and plasma viscosity, which are close to the values directly assessed from upstream and downstream pressure measurements.

The importance of assessing wall shear rate in vivo is illustrated by the following observation. In a preliminary study on the thromboembolic reactions in arterioles following wall puncture, it was interesting to see that the embolic reaction is practically independent of the level of wall shear rate. Even wall shear rate values of 16,000 s^{-1}, as found within a stenosis, did not by itself induce blood platelet aggregation. This might be explained by enhanced production of antiplatelet substances, like PGI_2, by the endothelial cells at these relatively high wall shear rate values. These thromboembolic studies also revealed that activated platelets inhibit leukocyte rolling, either directly or through an effect on endothelial cells[30].

In arterioles the shape of the red blood cell velocity profile is similar to that of blood platelets, albeit that red blood cells tend to travel at the higher velocities near the center of the vessel. This streaming of red blood cells probably explains the low blood platelet density in the center of arterioles.

Preliminary studies indicate that in venules the blood platelet velocity profiles are also flattened parabolas.

In arterioles leukocytes are generally moving in the bloodstream at the same velocity as red blood cells. In venules, however, leukocytes tend to roll along the vessel wall at significantly lower velocities than the other blood cells are travelling in these microvessels. Because this rolling is only observed in venules, and many arterioles and venules can be found with the same diameter and velocity[27,30], this difference between both types of microvessels also points to a possible functional difference between arteriolar and venular endothelium. This rolling phenomenon is considered to be the first step in leukocyte-endothelial cell interaction upon endothelial cell stimulation, for example, by inflammatory mediators.

ACKNOWLEDGEMENT

The authors are indebted to Jos Heemskerk and Karin Van Brussel for their help in preparing the manuscript.

REFERENCES

1. Tangelder GJ, Teirlinck HC, Slaaf DW, Reneman RS: Distribution of blood platelets flowing in arterioles, Am. J. Physiol., 248:H318, 1985.
2. Woldhuis B, Tangelder GJ, Slaaf DW, Reneman RS: Concentration profile of blood platelets differs in arterioles and venules, Am. J. Physiol., 262:H1217, 1992.
3. Turitto VT: Blood viscosity, mass transport, and thrombogenesis, In: Progress in Hemostasis and Thrombosis 6 (Ed., Speat TH), Grune and Stratton, New York, 1982.
4. Tangelder GJ, Slaaf DW, Muijtjens AMM, Arts T, oude Egbrink MGA, Reneman RS: Velocity profiles of blood platelets and red blood cells flowing in arterioles of the rabbit mesentery, Circ. Res., 59:505, 1986.

5. Tangelder GJ, Slaaf DW, Arts T, Reneman RS: Wall shear rate in arterioles in vivo: least estimates from platelet velocity profiles, Am. J. Physiol., 254:H1059, 1988.
6. Lipowsky HH, Kovalcheck S, Zweifach BW: The distribution of blood rheological parameters in the microvasculature of cat mesentery, Circ. Res., 43:738, 1978.
7. Lipowsky HH, Usami S, Chien S: In vivo measurements of "apparent viscosity" and microvessel hematocrit in the mesentery of the cat, Microvasc. Res., 19:297, 1980.
8. Tangelder GJ, Slaaf DW, Reneman RS: Fluorescent labeling of blood platelets in vivo, Thromb. Res., 28:803, 1982.
9. Tangelder GJ, Slaaf DW, Teirlinck HC, Alewijnse R, Reneman RS: Localization within a thin optical section of fluorescent blood platelets flowing in a microvessel, Microvasc. Res., 23:214, 1982.
10. Tangelder GJ, Arfors K-E: Inhibition of leukocyte rolling in venules by protamine and sulfated polysaccharides, Blood, 77:1565, 1991.
11. Goldsmith HL: The flow of model particles and blood cells and its relation to thrombogenesis, In: Progress in Hemostasis and Thrombosis 1 (Ed., Speat TH), Grune and Stratton, New York, 1972.
12. Goldsmith HL, Marlow JC: Flow behavior of erythrocytes, II, Particle motions in concentrated suspensions of ghost cells, J. Colloid. Int. Sci., 71:383, 1979.
13. Baker M, Wayland H: On-line volume flow rate and velocity profile measurement for blood in microvessels, Microvasc. Res., 7:131, 1974.
14. Bugliarello G, Hayden JW: Detailed characteristics of the flow of blood in vitro, Trans. Soc. Rheology, 7:209, 1963.
15. Bugliarello G, Sevilla J: Velocity distribution and other characteristics of steady and pulsatile blood flow in fine glass tubes, Biorheology, 7:85, 1970.
16. Slaaf DW, Rood JPSM, Tangelder GJ, Jeurens TJM, Alewijnse R, Reneman RS, Arts T: A bidirectional optical (BDO) three-stage prism grating system for on-line measurement of red blood cell velocity in microvessels, Microvasc. Res., 22:110, 1981.
17. Frojmovic MM, Panjwani R: Geometry of normal mammalian platelets by quantitative microscopic studies, Biophys. J., 16:1071, 1976.
18. Teirlinck HC, Tangelder GJ, Slaaf DW, Muijtjens AMM, Arts T, Reneman RS: Orientation and diameter distribution of rabbit blood platelets flowing in small arterioles, Biorheology, 21:317, 1984.
19. Caro CG, Pedley TJ, Schroter RC, Seed WA: The Mechanisms of the Circulation, Oxford University Press, Oxford, 1978.
20. Roevros JMJG: Analogue processing of CW Doppler flowmeter signals to determine average frequency shift momentaneously without the use of a wave analyser, In: Cardiovascular Applications of Ultrasound (Ed., Reneman RS), North-Holland Publishing Company, Amsterdam, 1974.
21. Palmer AA: Platelet and leukocyte skimming, Bibl. Anat., 9:300, 1967.
22. Turitto VT, Baumgartner HR: Platelet interaction with subendothelium in a perfusion system: physical role of red blood cells, Microvasc. Res., 9:335, 1975.
23. Eckstein EC, Tilles AW, Millero FJ: Conditions for the occurrence of large near-wall excesses of small particles during blood flow, Microvasc. Res., 36:31, 1988.
24. Tilles AW, Eckstein EC: The near-wall excess of platelet-sized particles in

blood flow: Its dependence on hematocrit and wall shear rate, Microvasc. Res., 33:211, 1987.
25. Atherton A, Born GVR: Quantitative investigations of the adhesiveness of circulating polymorphonuclear leukocytes to blood vessel walls, J. Physiol., 222:447, 1972.
26. House SD, Lipowsky HH: Leukocyte-endothelium adhesion: microhemodynamics in the mesentery of the cat, Microvasc. Res., 34:363, 1987.
27. Oude Egbrink MGA, Tangelder GJ, Slaaf DW, Reneman RS: Thromboembolic reaction following wall puncture in arterioles and venules of the rabbit mesentery, Thromb. Haemost., 59:23, 1988.
28. Oude Egbrink MGA, Tangelder GJ, Slaaf DW, Reneman RS: Effect of blood gases and pH on thromboembolic reactions in rabbit mesenteric microvessels, Eur. J. Physiol., 414:324, 1989.
29. Pohl U, Holtz J, Busse R, Bassenge E: Crucial role of endothelium in the vasodilator response to increased flow in vivo, Hypertension, 8:37, 1986.
30. Oude Egbrink MGA, Tangelder GJ, Slaaf DW, Reneman RS: Influence of platelet-vessel wall interactions on leukocyte rolling in vivo, Circ. Res., 70:355, 1992.

INTERCELLULAR COLLISIONS AND THEIR EFFECT ON MICROCIRCULATORY TRANSPORT

Harry L. Goldsmith

McGill University Medical Clinic
Montreal General Hospital
Montreal, Quebec
Canada

INTRODUCTION

Shear-induced interactions between cells in flowing blood are involved in several important phenomena, such as the formation of aggregates, the dispersion of plasma and cells within vessels, the enhanced collision of cells with each other and the vessel wall, and changes in the distribution of cells within the vessels. This paper is concerned with an analysis of such interactions, particularly as they occur in small vessels. In order to better understand the phenomena, we begin by describing the simplest interaction, that of a two-body collision between spherical particles, and how fluid mechanical and colloid chemical theories applied to such collisions can be used to better understand interactions between red cells and between platelets. The effects of two-body collisions on particle trajectories are described, before proceeding to consider multi-body interactions in model suspensions and in blood, where the red cells exercise a considerable influence on platelet and white cell flow behavior.

TWO-BODY COLLISIONS: PARTICLE TRAJECTORIES AND FORCES

Interactions Between Charged Rigid Spheres

We consider two equal-sized neutrally buoyant, non-conducting, charged rigid spheres of radius b suspended in an incompressible Newtonian fluid of viscosity η containing electrolyte. The suspension is subjected to laminar viscous flow in a circular tube of radius R_o in which the fluid velocity in the axial, X_3-direction at a radial distance R, is given by:

$$u_3(R) = \frac{2Q}{\pi R_o^4} (R_o^2 - R^2) \qquad (1)$$

where Q is the volume flow rate. The fluid mechanical problem of predicting the trajectories of the spheres in a simple shear flow was originally solved for the case where there are no interaction forces between the particles.[1-3] It was then extended to include the case where interaction forces, $F_{int}(h)$, other than hydrodynamic operate as sphere surfaces approach each other to within a distance h < 100 nm.[4] Assuming the interaction forces act along the line joining the centers of the spheres, theory shows that the relative velocity of the centers separated by a distance r is given by:[4]

$$\frac{dr}{dt} = A(r^*)\, Gb\, \sin^2\theta_1\, \sin 2\phi_1 + \frac{C(r^*)\, F_{int}(h)}{3\pi\eta b} \qquad (2)$$

and the angular motion of the doublet axis by:

$$\frac{d\theta_1}{dt} = \frac{1}{4} B(r^*)\, G\, \sin 2\theta_1\, \sin 2\phi_1 \qquad (3)$$

$$\frac{d\phi_1}{dt} = \frac{1}{2} G\, [1 + B(r^*)\cos 2\theta_1] \qquad (4)$$

Here, G is the shear rate and θ_1 and ϕ_1 the respective polar and azimuthal angles relative to X_1, the vorticity axis, as shown in Fig. 1 for a two-body collision in Poiseuille flow. $A(r^*)$ and $C(r^*)$ are known dimensionless functions of $r^* = r/b$,[3] and $B(r^*)$ is the angular velocity coefficient[5] related to the equivalent ellipsoidal axis ratio $r_e(r^*)$ by:

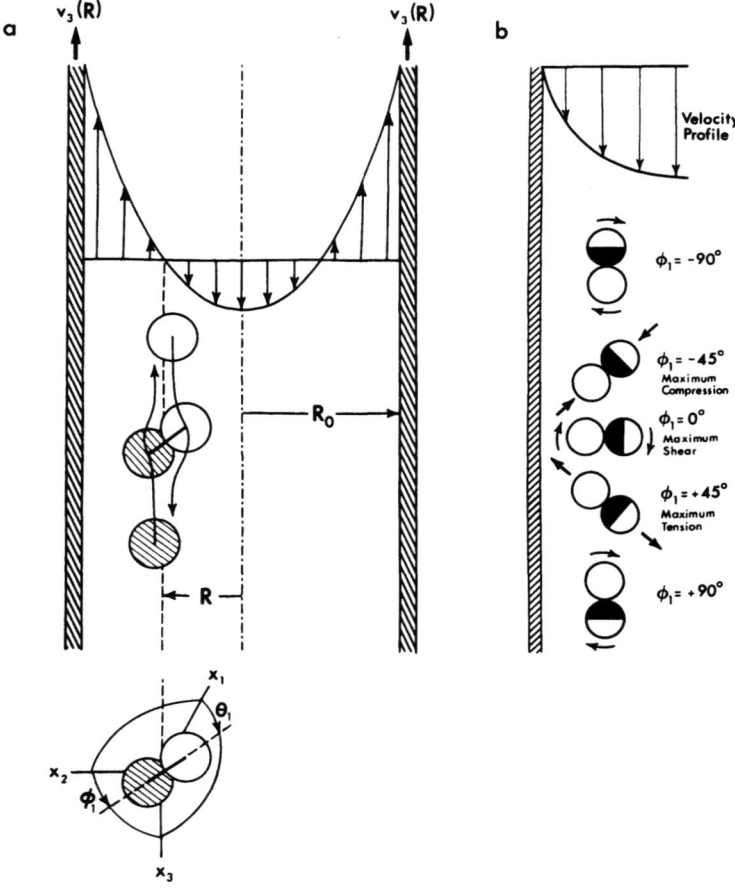

Figure 1. (a) Two-body collision between rigid spheres in Poiseuille flow showing the formation of a transient doublet in the median plane of a tube of radius R_o being moved upward with a velocity $v_3(R)$, equal to that of the downward flowing doublet whose center is at a radial distance R from the tube axis. Cartesian (x_i) and polar coordinates θ_1, ϕ_1 are constructed at the midpoint of the doublet axis. (b) Rotation over half orbit of a permanent doublet with $\theta_1 = \pi/2$ showing the ϕ_1-orientations corresponding to maximal normal and shear stress. (From Tha et al)[6]

$$B(r^*) = \frac{r_e^2(r^*) - 1}{r_e^2(r^*) + 1} \tag{5}$$

Rearrangement of Eq. (2) gives the force equation:

$$\frac{3\pi\eta b}{C(r^*)} \frac{dr}{dt} = \frac{A(r^*)}{C(r^*)} 3\pi\eta G b^2 \sin^2\theta_1 \sin 2\phi_1 + F_{int}(h) \tag{6}$$

The term on the left represents the hydrodynamic drag force between spheres resisting approach of the particles. The first term on the right is the normal hydrodynamic force between the spheres acting along the line of their centers = F_3, and is maximal for $\theta_1 = \pi/2$, i.e. when rotation is limited to the X_2X_3-plane (Fig. 1b), at $\phi_1 = -\pi/4$ where the force is compressive, and at $\phi_1 = \pi/4$ where the force is tensile. Equations (2) - (6) have been shown to apply in Poiseuille flow providing the ratio of particle to tube radius, $b/R_o << 1$.[7,8]

In the absence of interaction forces, $F_{int}(h) = 0$, the trajectories of approach and recession are symmetrical about $\phi_1 = 0$, as observed experimentally[9,10] and shown in Fig. 1a. In the case of charged spheres surrounded by a diffuse ionic electrical double layer, the paths of approach and recession are no longer symmetrical, $F_{int}(h) \neq 0$, when particle

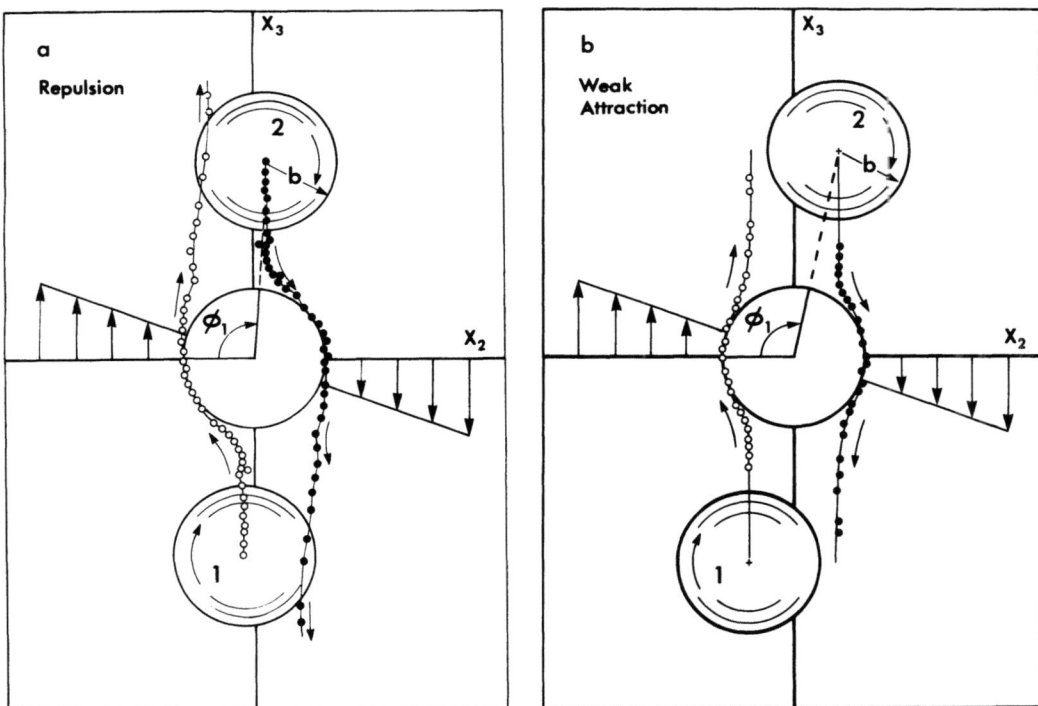

Figure 2. Trajectories of shear-induced collisions between 2.6 μm polystyrene latex spheres in 50% aqueous glycerol undergoing Poiseuille flow showing the projection on the X_2X_3-plane of the paths of the sphere centers from the mid point between them. At the center is the exclusion sphere which cannot be penetrated for collisions occurring in the X_2X_3-plane. (a) 1 mM KCl - a significant increase in the separation of centers after the collision due to electrostatic repulsion. (b) 10 mM KCl - shrinkage of electrical double layer allowing van der Waals attraction to occur with a slight decrease in separation of sphere centers. (From Takamura et al)[8]

surfaces are within 100 nm of each other and subject to double layer repulsion and van der Waals attraction as illustrated in Fig. 2 for two-body collisions of 2 µm latex spheres in aqueous glycerol solution containing KCl.

The analysis of the asymmetry of collisions such as those shown in Fig. 2, showed that net repulsive or attractive forces as small as 10^{-13} N (0.01 µdyne) could be detected. Applying colloid stability theory,[11,12] assuming that $F_{int}(h)$ is the sum of double layer repulsive and van der Waals attractive forces, enabled values of the Hamaker constant for polystyrene to be computed.[8] In fact, the assumption of additivity of hydrodynamic and electrostatic forces is incorrect, since the presence of the double layer changes the velocity field around the sphere.[13] Equations (2) - (4) must then be modified to include coupling (or electroviscous) forces and torques. It turns out that, if experiments are conducted under conditions at which the double layer thickness is very small compared to the particle radius, as in the experiments shown in Fig. 2, these terms are negligible.

Some collisions even resulted in the formation of permanent doublets, the spheres of which were in a secondary potential energy minimum at distances h between 15 and 17 nm, and were capable of independent rotation as shown by measurements of the period of rotation, T, of the doublets.[7] The period T is given by integration of Eq. (4) giving the angular velocity of a rigid prolate ellipsoid[14] of equivalent axis ratio $r_e(r^*)$:

$$\tan \phi_1 = r_e(r^*) \tan \left(\frac{2\pi t}{T}\right) \qquad (7)$$

where T is defined by

$$T = \frac{2\pi}{G}\left[r_e(r^*) + r_e^{-1}(r^*)\right] \qquad (8)$$

It has been shown that the dimensionless period of rotation, TG, for doublets of non-touching spheres depends not only on the distance of separation of sphere surfaces but also on whether they are capable of rotation relative to each other;[3] $r_e(r^*)$ at a given r^* is markedly greater for independently rotating than for rigidly coupled spheres.[3,15]

In the presence of polyelectrolyte, permanent doublets formed which rotated as dumbbells of rigidly linked spheres, presumably because polymer bridge formation had occurred.[7,16] For such a doublet in shear flow, $dr/dt = 0$, and if it can be induced to break up, then:

$$F_{int}(h) = -F_3 = \frac{A(r^*)}{C(r^*)} 3\pi\eta Gb^2 \sin^2\theta_1 \sin2\phi_1 \qquad (9)$$

Forces Acting Between Spheres

Bonds linking rigid spheres cross-linked by polymer are subject not only to hydrodynamic forces acting along the doublet axis, but also to forces acting normal to it, i.e. shear forces. We have used a general method for calculating the forces and torques, and the translational and rotational velocities of rigid neutrally buoyant spheres suspended in viscous fluids in the creeping flow regime, given by Arp and Mason.[3] The method is based on the matrix formulation of hydrodynamic resistances in creeping flow by Brenner and O'Neill.[17] The solution of the Brenner O'Neill force-torque vector equation in terms of the particle and external field coordinates led to the derivation of the following expressions for the normal force acting along, and the shear force acting perpendicular to, the axis of the doublet of spheres:

Normal Force: $\qquad F_3 = \alpha_3(h)\eta Gb^2 \sin^2\theta_1 \sin2\phi_1 \qquad (10)$

Shear Force: $\sqrt{F_1^2 + F_2^2} = \alpha_{12}(h)\eta Gb^2 \left[(\cos2\theta_2 \cos\phi_2)^2 + (\cos\theta_2 \sin\phi_2)^2\right]^{\frac{1}{2}} \qquad (11)$

where θ_2 and ϕ_2 are the polar angles with respect to the X_2-axis of the external flow field, F_1 and F_2 are the forces acting in directions normal to the doublet axis, and $\alpha_{12}(h)$ and $\alpha_3(h)$ are coefficients, functions of h.[3,6]

Application to Biological Systems

The above expressions for the respective normal and shear forces acting along, and perpendicular to, the axis of a doublet of rigid spheres were used to determine the hydrodynamic forces required to separate two red cell spheres of antigenic type B cross-linked by the corresponding polyclonal,[18] or monoclonal[19] IgM antibody. Cells were sphered and swollen in isotonic buffered glycerol containing sodium dodecyl sulfate, fixed in 0.08% glutaraldehyde, and suspended in either aqueous glycerol or sucrose containing 0.15 M NaCl and anti-B antibody at concentrations from 44 - 214 mM (polyclonal) and 0.16 to 0.62 nM (monoclonal). The cells were observed through a microscope flowing in 150 - 175 μm tubes by gravity feed between two reservoirs. Using a traveling microtube apparatus,[7,18] doublets of red cell spheres, formed through shear-induced collisions, were tracked in a constantly accelerating flow, and translational and rotational motions recorded on videotape until breakup occurred. From a frame by frame replay of the tape, the radial position, velocity, θ_1- and ϕ_1-orientations of the doublets were obtained and values of the normal and shear forces of separation computed from Eqs. (10) and (11) using values of $\alpha_{12} = 7.02$ and $\alpha_3 = 19.33$ based on a minimum separation distance h = 20 nm.[7] Measurements of the mean dimensionless period of rotation of the doublets, TG, in steady flow yielded a narrow distribution about the value 15.64 predicted for rigidly-linked doublets.[3,7]

There was a large scatter in the forces of separation at each antibody concentration; however, as shown in the Table below for the monoclonal antibody experiments, differences in the mean values increased significantly with antibody concentration. There was also a significant effect of suspending fluid viscosity on the distribution of separation forces, these being lower at the lower sucrose viscosities since doublets with many cross-bridges were not seen to break up before the limiting velocity of the microtube was reached; at high sucrose viscosities, doublets with few cross-bridges broke up before they could be tracked. It is impossible to determine from these experiments whether normal or shear forces are responsible for particle breakup.

It was first assumed that the distribution of forces reflected a distribution in the number of antibody-antigen cross-bridges, the lowest values (~ 0.01 nN) being of the same order as that (0.04 nN) predicted for the rupture of a single antigen-antiserum bond.[20] Subsequent work[21] showed that doublets may not break up the instant a critical force to break all bonds is reached; they may break up under a constant applied force, if given time. This agrees with observations of the stochastic nature of intercellular bonds from micropipette aspiration experiments.[22] Also, it is possible that the hydrodynamic forces are acting to uproot the antigenic molecules from the cell membrane rather than breaking antigen-antibody bonds.[22]

TABLE I. NORMAL HYDRODYNAMIC FORCE OF BREAK-UP

[Antibody] nM	No. of Doublets	F_3 ± S.D. nN
0.15	48	0.048 ± 0.033
		$p < 0.001$
0.30	59	0.074 ± 0.040
		$p < 0.001$
0.60	83	0.098 ± 0.038

TWO-BODY COLLISIONS: KINETICS OF AGGREGATION

Orthokinetic Aggregation

The rate at which equal-sized spheres collide with each other in shear flow, the two body collision frequency, j_s, can be calculated from the governing trajectory equation, Eq. (2). A reference sphere is placed at the origin of the flow field and the collision cross-section through which all particle centers collide with the reference sphere is computed. It turns out that j_s computed from the trajectory equation is very close to that predicted by Smoluchowski,[24] who assumed that all spheres move along linear trajectories until they collide with the reference sphere, and the collision cross-section is then simply a sphere of radius 2b. The two-body collision frequency for monodisperse suspensions of rigid spheres in simple shear flow is then given by:

$$j_s = \frac{32}{3} Gb^3 N, \quad (12)$$

where N is the number concentration. Provided $b/R_o \ll 1$, Eq. (12) has been shown to apply in Poiseuille flow.[10] Using the mean tube shear rate, the total number of two-body collisions per sec per unit volume of suspension is then:

$$J_s = \frac{1}{2} N j_s = \frac{4\Phi \overline{G} N}{\pi}, \quad (13)$$

where Φ is the volume fraction of suspended particles.

Smoluchowski equated the two-body collision frequency with the two-body capture frequency. The actual rate of doublet formation can be compared to the collision frequency by defining an orthokinetic collision capture efficiency, $\alpha_o = j_c/j_s$ in Eq. (12) where j_c is the collision capture frequency, which accounts for the influence of both interaction and hydrodynamic forces on particle capture.[23] If $\alpha_o = 1$, then $j_c = j_s$ and every collision results in capture; however, in the absence of attractive forces permanent capture is impossible.

In the absence of aggregate break-up or the formation of higher order aggregates, the kinetics of aggregation are first order with respect to the total particle concentration, N_∞,[25]

$$\frac{dN_\infty}{dt} = -\frac{4\Phi \alpha_o \overline{G} N_\infty}{\pi} \quad (14)$$

Integration of Eq. (14) yields:

$$\ln \frac{N_\infty(t)}{N_\infty(0)} = -\frac{4\Phi \alpha_o \overline{G} t}{\pi} \quad (15)$$

where $N_\infty(0)$ and $N_\infty(t)$ are the total particle concentrations at time 0 and t, respectively. Thus, the total particle concentration decays exponentially and a plot of $\ln N_\infty(t)$ vs t should give a straight line, the slope of which yields α_o.

Application of colloid stability theory to compute capture efficiencies in the case where electrostatic interactions are negligible shows that α_o increases linearly with increasing ratio of van der Waals to hydrodynamic force:[13]

$$\alpha_o \propto \left(\frac{A}{36\pi\eta Gb^3}\right)^{0.18} \quad (16)$$

where A is the Hamaker constant. This is not true of unequal-sized particles because these do not readily approach to within distances at which the colloidal forces are appreciable. Thus, in shear flow the coagulation of equal-sized particles is favored over that of unequal-sized particles.[13]

It is evident from the trajectory equations that the capture efficiency is expected to decrease with increasing shear rate, as the interaction term, which remains constant, becomes negligible in comparison to the hydrodynamic term. There is experimental evidence for the decrease of α_o with shear rate when repulsive forces are negligible.[23,26,27]

Application to Biological Systems: Collisions Between Sphered Red Cells

Two-body collisions between sphered and swollen red cells, prepared as described in the previous section, were observed and analyzed in Poiseuille flow. The trajectories were markedly asymmetric with considerable repulsion between the cells, so that the relative velocities, $|dr/dt|$, during recession were much greater than during approach, as was the case for electrostatic repulsion, in Fig. 2. Analysis of the trajectories showed that the minimum distance of approach $\approx$ 50 nm, a distance large compared to that at which double layer interactions become important.[28]

Application of colloid stability theory to red cells has been difficult. Thus, calculations using a model of a lipid layer coated with charged mucoprotein have shown that there is a secondary energy minimum with the interaction potential $\sim 10^2 kT/\mu m^2$, more than enough to hold cells together.[29] Even in the absence of fibrinogen, therefore, red cells would be expected to be aggregated at low or zero shear rates, contrary to observation. When the Smoluchowski equation for the electrophoretic mobility of a charged colloidal particle is applied to the red cell, the surface charge is underestimated by a factor of 2 to 3.[30] To account for this low mobility, Levine *et al.*[31] have proposed a model which takes into account the extracellular mass coating the membrane bilayer, the glycocalyx. It appears that it is the glycocalyx, providing a layer of 8 - 10 nm thickness, which prevents the close approach of two colliding cells in shear flow.

Application to Biological Systems: Aggregation of Platelets in Tube Flow

Basic to all considerations of thrombus formation is the fact that a velocity gradient is necessary for aggregation of platelets, the primary constituents of a thrombus, to occur. To study this aspect of the problem we have undertaken studies of the effect of shear rate and red blood cells on the aggregation of ADP-stimulated platelets in steady flow through circular cylindrical tubes.[32-35] Shear rate is the most important physical parameter governing platelet aggregation in flowing suspensions. It determines the platelet collision frequency, the shear and normal stresses which activate single cells and break up aggregates, and the interaction time of cell-cell and cell-surface collisions. The predilection of white platelet thrombi to form in regions of high wall shear rate in the arterial circulation emphasizes the need to focus on the effect of shear rate on platelet aggregation in well-defined flow. Time-averaged systemic arterial wall shear rate in humans is estimated to range from 50 - 2000 s^{-1} based on a parabolic velocity profile for whole blood.[36]

Measurement of aggregation: Venous blood was drawn from healthy volunteers into plastic syringes containing sodium citrate at a predetermined concentration. Platelet-rich plasma (PRP) containing approximately 3×10^5 cells was prepared by centrifuging the blood at 100g and adjusting the platelet concentration by adding plasma. In the case of whole blood, the citrate concentration was adjusted according to the donor hematocrit to give a final concentration of 0.62% in plasma.

Platelet-rich plasma (or whole blood) and ADP were simultaneously infused into a small cylindrical mixing chamber using independent syringe pumps at a fixed volume ratio PRP (or whole blood) : ADP = 9:1. After rapid mixing, the suspension exited the chamber through lengths of 0.595 or 0.380 mm radius polythene tubing corresponding to meant transit times $\bar{t} = \Delta X_3/\bar{U}$ between 1 and 86 s, where ΔX_3 (from 2 cm to 15.3 m) is the distance down the flow tube and $\bar{U}$ the mean linear velocity. Total volumetric flow rates, Q, were preset from

13 to 155 μl s^{-1} to generate volume flow averaged mean tube shear rates, $\overline{G}$, in Poiseuille flow from 41.9 to 1920 s^{-1}. Tube Reynolds numbers ranged from 8 to 148. Control runs in which Tyrodes instead of ADP was infused, were also carried out.

The aggregation reaction was instantaneously and permanently arrested by collecting known volumes of the effluent into 0.5% isotonic glutaraldehyde. In the case of whole blood, the effluent, fixed in glutaraldehyde, was layered onto isotonic Percoll solution and centrifuged to separate red cells from single platelets and aggregates.

The number concentration and volume of single platelets and aggregates were measured using an electronic particle counter in conjunction with a logarithmic amplifier and a 100 channel pulse height analyzer to generate continuous 250 class log-volume histograms over the volume range 1 - 10^5 μm^3.$^{(32)}$ Computer integration of the log-volume histograms yielded the number concentration and volume fraction of particles between set lower and upper volume limits. Individual histograms from multiple donors were averaged yielding a histogram of the mean class volume fraction normalized to the maximum class content at $\overline{t} = 0$.$^{(33)}$ The mean, modal and median single platelet volume were calculated from the mean and standard deviation of the log-volume distribution, assuming a normal distribution of the latter.$^{(37)}$

Aggregation in Platelet-Rich Plasma. Figure 3 shows the single platelet number concentration after $\overline{t} = 43$ s exposure to 0.2 μM ADP normalized to a control at $\overline{t} = 0$ s as a function of the mean tube shear rate. The extent of aggregation of platelets from female donors was significantly greater than that of platelets from male donors ($p < 0.001$) over the range of shear rate $41.9 \leq \overline{G} \leq 1920$ s^{-1}.$^{(33,34)}$ The sex difference was greatest at $\overline{G} = 335$ s^{-1} where 76% of single platelets from the female donors but only 49% of those from male donors had aggregated. Significant changes ($p < 0.001$) in the single platelet concentration as a function of mean tube shear rate produced a similar pattern of aggregation for both groups of donors. Aggregation increased as the shear rate increased up to a maximum at $\overline{G} = 168$ and 335 s^{-1} for male and female donors, respectively. Thereafter, it decreased linearly with increasing shear rate down to a minimum at $\overline{G} = 1000$ and 1335 s^{-1} for male and female donors, respectively. Further increases in shear rate produced an increase in aggregation for both sexes. The two aggregation curves intersect at $\overline{G} = 1335$ s^{-1} where aggregation had begun to increase for the male donors while it was still decreasing for the female donors.

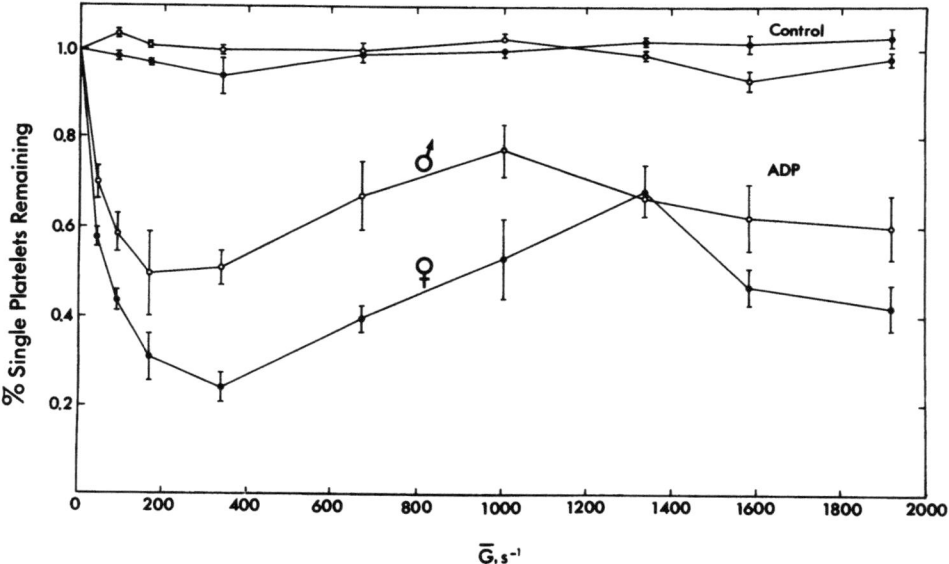

Figure 3. Mean values of the % single platelets remaining after t = 43 s exposure to 0.2 μM ADP, as a function of mean tube shear rate for male and female donors. In the control runs, physiological salt solution was infused, instead of ADP. (From Bell et al.)$^{(33)}$

The evolving pattern of aggregate growth for the female donors is shown in Fig. 4 where the normalized average class volume fraction is plotted against particle volume at successive mean transit times. The decrease in single platelet concentration (1 - 30 μm³) was accompanied by a sequential rise and fall of aggregates of successively increasing volume. The steadily decreasing rate of aggregation with time shown in Fig. 4 at $\overline{G}$ = 41.9 s⁻¹ was associated with the formation of aggregates having a broad spectrum of size at $\overline{t}$ = 86 s. At $\overline{G}$ = 335 s⁻¹, no distinct aggregate peaks were present prior to $\overline{t}$ = 8.6 s but by $\overline{t}$ = 21 s⁻¹ aggregates of relatively discrete size had appeared. As aggregation continued, the upper limit of aggregate size increased. By $\overline{t}$ = 86 s, most aggregates were present in one large group, a significant proportion of which exceeded 10⁵ μm³, the largest volume measured. Although

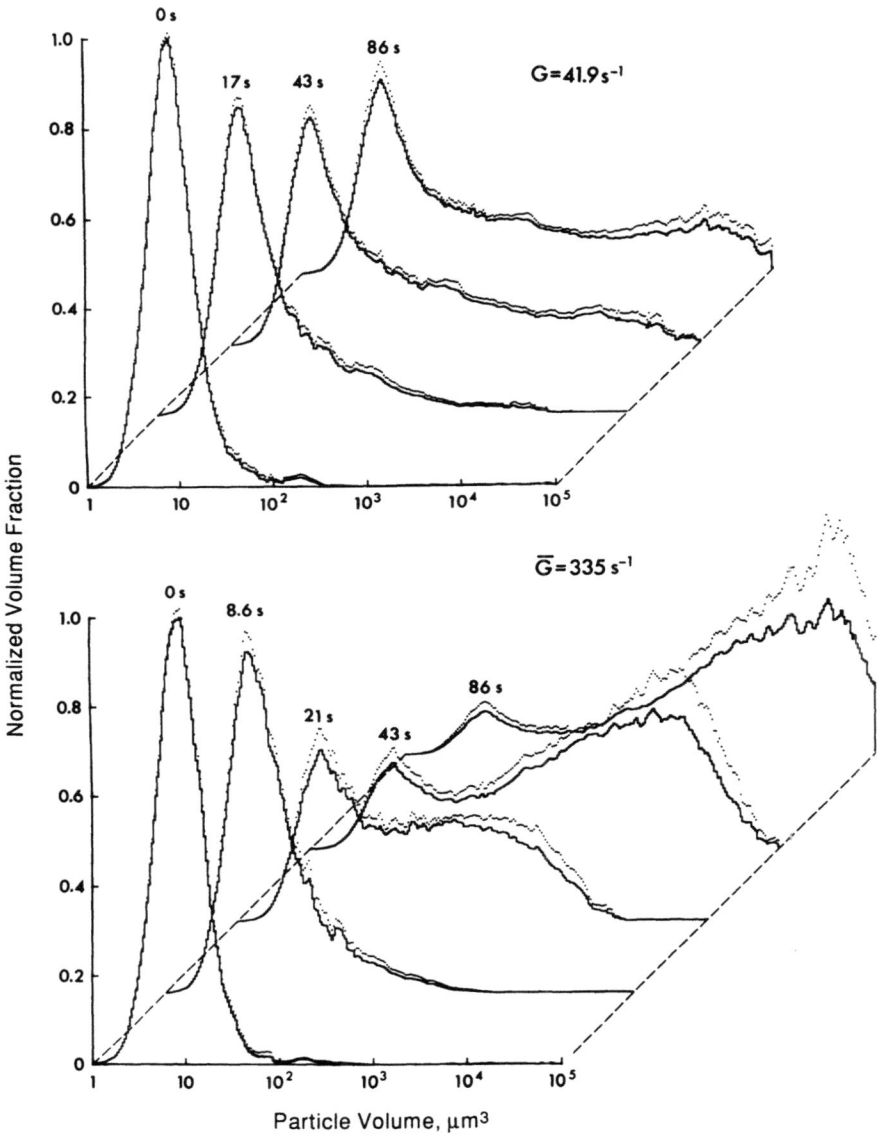

Figure 4. Aggregate growth in citrated-PRP at $\overline{G}$ = 41.9 s⁻¹ (upper) and 335 s⁻¹ (lower). Three-dimensional plot of the mean normalized volume fraction (± SEM, dotted lines) versus particle volume at mean transit times from 0 to 86 s. The time axis is not drawn to scale. (From Bell *et al.*)[33]

there was considerable aggregation at $\overline{G}$ = 1920 s^{-1} (not shown), the aggregates were smaller and occupied a size range considerably narrower than at the same mean transit time at $\overline{G}$ = 335 s^{-1}. A similar pattern of aggregate growth was exhibited by the male donors whose aggregate size was much reduced compared to that of female donors at the same $\bar{t}$ and mean tube shear rate.

Application of Two-Body Collision Theory: Collision Capture Efficiency. Theory applicable to dilute suspensions of rigid spheres, given above, was used to calculate the two-body collision efficiency in PRP in the initial stages of the reaction when the number of multiplets was small. In the present work the influence of Brownian motion on aggregation can be neglected due to the large value of the Péclet number, Pe = $\overline{G}b^2/D_t$ > 1200 where D_t is the translational diffusion coefficient for a single sphere calculated from the Stokes-Einstein equation. Thus, the measurement of the total particle concentration over the early stages of aggregation provides a value for α_0.

A plot of Eq. (15) using the data for female donors is shown in Fig. 5. Although the extent of aggregation was greatest at $\overline{G}$ = 335 s^{-1}, the highest rate of aggregation occurred within the first $\bar{t}$ = 2 s at $\overline{G}$ = 41.9 s^{-1}, where the single platelet concentration decreased at a mean rate of 4.2% s^{-1}. At this shear rate, the rate of aggregation steadily decreased with increasing mean transit time. At higher $\overline{G}$, there was an initial lag phase followed by progressively increasing then decreasing rates of aggregation giving rise to the sigmoid curves in Fig. 5. The length of the lag phase increased with increasing mean tube shear rate reaching $\bar{t}$ ~ 11 s at $\overline{G}$ = 1335 s^{-1}. In addition, as the mean tube shear rate was raised, not only did the maximum rate of aggregation decrease, but it occurred at progressively increasing mean transit times. Male donors exhibited a pattern of aggregation similar to that of the female donors at the same shear rate but always with much longer lag phases and reduced rates of aggregation.

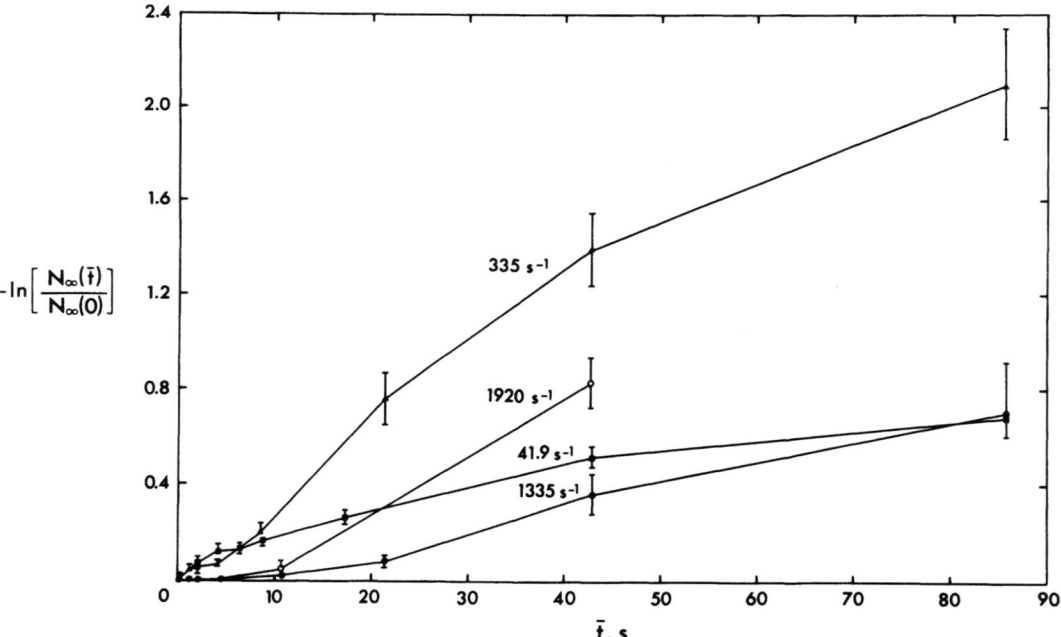

Figure 5. Decrease in total particle concentration with mean transit time at 0.2 µM ADP. Plot of Eq. (15) for the female donors at the mean shear rates shown on the graph. (From Bell et al.)[33]

Figure 6 shows the collision capture efficiencies, averaged over three time intervals, plotted against log $\bar{t}\bar{G}$. In the time interval from $\bar{t} = 0$ to 4.3 s, both the collision efficiency and its rate of decrease fell rapidly with increasing shear rate from a maximum of 0.26 at $\bar{G} = 41.9$ s^{-1}. A similar decline in α_o in the interval from $\bar{t} = 4.3$ to 8.6 s was interrupted between $\bar{G} = 168$ and 335 s^{-1} before resuming at a higher rate of decrease beyond $\bar{G} = 335$ s^{-1}. In contrast, during the time interval from $\bar{t} = 8.6$ to 21 s, α_o decreased by only 12% up to $\bar{G} = 168$ s^{-1}, but decreased sharply thereafter. Throughout all 3 time intervals, α_o either decreased or remained constant as mean tube shear rate increased up to $\bar{G} = 1335$ s^{-1} where $\alpha_o \leq 0.002$. At $\bar{G} = 1920$ s^{-1}, there was a small but significant increase in α_o which, together with the high collision frequency, was sufficient to support the high rate of aggregation shown in Fig. 5.

As described above, theory predicts that when net attractive forces operate between colliding colloidal particles, the two-body collision efficiency decreases with increasing shear stress. This has been shown experimentally at shear stresses < 0.1 Nm^{-2} for human platelets exposed to 1 µM ADP.[39] In the present experiments, the mean tube shear stress ranged from 0.08 to 3.5 Nm^{-2}. As the mean tube shear rate increases, the extent of aggregation is the result of a balance between an increased frequency of collision and an increased fluid shear stress. High collision rates support a high rate of aggregation in the absence of shear stresses sufficient to inhibit doublet formation. Beyond an optimum shear rate for aggregation, higher fluid shear stresses can prevent stable platelet-platelet bond formation and aggregation decreases. Short interaction times may limit stable doublet formation at high shear rates but, since this time is inversely proportional to the shear rate, it is impossible to

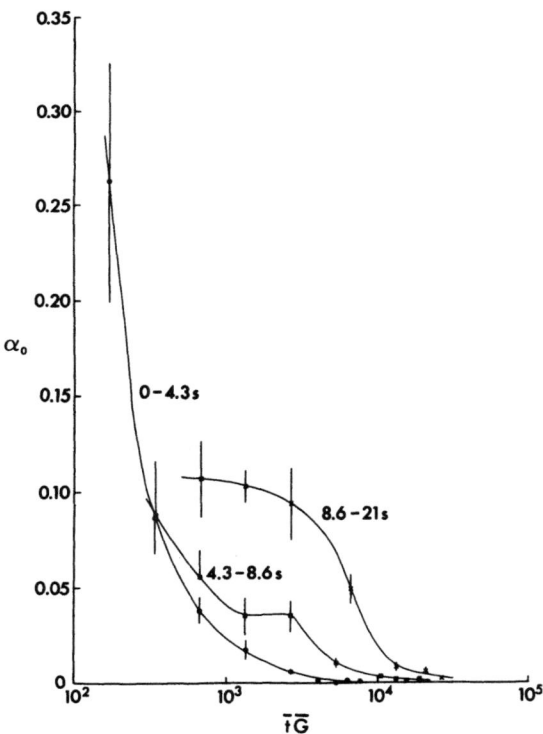

Figure 6. The collision capture efficiency, α_o (± SEM) plotted against log tG for the data shown in Fig. 5. The interruption in the decrease in α_o with increasing mean tube shear rate for mean transit times between 4.3 and 8.6 s is interpreted as indicating a transition from a weak to a strong platelet-platelet bond. (From Bell et al.)[33]

separate the effects of short interaction time and high shear stress in limiting stable bond formation. The two-body collision capture efficiency, however, takes both effects into account by measuring the fraction of total platelet-platelet collisions that result in stable doublet formation.

The collision capture efficiencies were computed from changes in the total particle concentration using Eq. (15) which assumes that the decrease in total particle concentration with time is due only to the formation of doublets. This assumption was tested by calculating the fraction of successful two-body collisions from the initial rate of single platelet decrease from Eq. (13) using the initial number concentration and volume fraction of single platelets. At 0.2 μm ADP and $\bar{G} = 41.9$ s^{-1}, initial rates of single platelet decrease of 2.7 and 4.2% s^{-1} yielded collision efficiencies of 0.14 and 0.21 for female and male donors, respectively. These values are remarkably similar to the collision efficiencies, 0.12 and 0.26, calculated according to Eq. (15) between $\bar{t} = 0$ and 4.3 s, well into the aggregation reaction. Actually, one would expect the collision efficiency calculated at long times to yield lower values due to the decrease in particle concentration with increasing aggregation. Thus, it is unlikely that the observed increase in collision efficiency with time in the present experiment is artifactual. It should be noted, however, that the presence of pseudopods on the activated platelets has been estimated to double the effective collision diameter of the cells.[39] This will result in an 8-fold increase in collision frequency and a corresponding decrease in α_o, and implies that the aggregation reaction is capable of propagation at very low collision capture efficiencies.

Heterogeneity of Platelet-Platelet Bonds. The results show that the highest rate of aggregation and two-body collision efficiency (26%) occurred at the lowest shear rate but only for the first few seconds of the reaction. Although the fluid shear stress is low, the collision rate is not sufficient to sustain a high rate of aggregation. At higher shear rates there was generally a decrease in the fraction of efficient collisions but a high collision frequency accounts for the higher rates of aggregation. At the maximum shear rate in the present experiments, collision capture efficiencies between 0.6×10^{-3} and 2.3×10^{-3} were measured. The large decrease in collision efficiency at relatively low shear rates (Fig. 6) and the persistence of a nonzero collision efficiency at high shear rates suggest that more than one type of platelet-platelet bond mediates ADP-induced aggregation.

The increasing rates of aggregation with time as depicted by the sigmoid aggregation curves (Fig. 5) indicate that collision efficiency increases with time, even after a delay of up to 11 s in the onset of aggregation. Indeed, not only do calculations of collision efficiency confirm this but they also indicate that the heterogeneity among platelet bonds is time-dependent (Fig. 6). At early transit times the high rate of aggregation at low shear rates is sustained by a weak bond that is easily disrupted at higher shear rates, resulting in a corresponding shear-dependent lag phase. Even at low shear rates the strength of this bond gradually diminishes with increasing transit time and, in conjunction with a low collision rate, produces a steadily decreasing rate of aggregation. At high shear rates, the increasing rate of aggregation at times beyond the lag phase reveals the emergence of a second stronger bond. Longer times are required before each bond is sufficiently strong, or is present in sufficient numbers, to support aggregation. The two types of bonds coexist at intermediate transit times where the weak bond is disrupted at low shear rates but higher shear rates are required to disrupt the stronger bonds. The interruption in the decrease in collision efficiency with increasing mean tube shear rate between 168 and 335 s^{-1} at mean transit times between 4.3 and 8.6 s (Fig. 6) points to a transition from weak to strong bond. At very long exposure times the strong bonds are maximally expressed through either strength or numbers, and only high shear rates are sufficient to disrupt them.

The cross-linking of bivalent fibrinogen molecules between activated glycoprotein IIb-IIIa (GPIIb-IIIa) complexes in the platelet membrane is the mechanism believed to underlie the ADP-induced aggregation of platelets.[40] The steadily increasing lag phase preceding aggregation with increasing shear rate in the present work points to a latency in the strength of the platelet-platelet cross-bridge. For any colloid whose aggregation is mediated by polymer cross-linking, unoccupied binding sites must be available on both surfaces for cross-linking to occur. In the case of platelets, the high concentration of fibrinogen in plasma ($\sim 2 \times 10^8$ molecules per platelet) would be expected to saturate all binding sites on platelets (~ 5×10^4 receptor sites per cell[41]) before cross-linking could occur. In fact, cross-linking would require the simultaneous binding of opposite ends of the bivalent fibrinogen molecule to two platelets immediately after activation of GPIIb-IIIa complexes, and prior to saturation of these receptors with free fibrinogen. This scenario seems unlikely. It is more likely that all GPIIb-IIIa complexes would be saturated with free fibrinogen long before two platelets could simultaneously bind a single fibrinogen molecule. Instead, a model of aggregation requires either a low affinity for fibrinogen binding by activated platelets, and the subsequent continuous breaking and forming of new platelet-fibrinogen bonds, or the time-dependent exposure of new bonds that permits cross-linking during the interaction time of collision. There is evidence that both high and low affinity binding sites for fibrinogen exist on ADP stimulated platelets,[42] and that binding increases with time.[42,43] In addition, fibrinogen itself appears to have a binding sequence in the carboxy terminus of the γ-chain that recognizes the GPIIb-IIIa complex, but is distinct from the arginine-glycine-aspartic acid- (RGD) containing sequence in the α chain.[44,45] Thus, the relatively slow kinetics of fibrinogen binding and the existence of heterogeneity in the affinity of fibrinogen provide a mechanism for platelet aggregation in the presence of high concentrations of the cross-linking ligand. Furthermore, the increase in the two-body collision efficiency with time and the formation of a high shear rate-resistant bond can be explained in terms of a time-dependent increase in fibrinogen binding.

There is at present, however, no conclusive proof that fibrinogen mediates aggregation by directly cross-linking activated platelets. An alternate mechanism proposes that GPIIb-IIIa receptor clustering is a prerequisite for fibrinogen binding and platelet aggregation.[46,47] It is possible that the role of fibrinogen is to stabilize such clusters, permitting them to interact in some complementary manner between activated platelets. Thus, fibrinogen cross-linking between platelets *per se* would not be necessary for aggregation. If the platelets were initially aggregated by a mechanism independent of fibrinogen cross-linking but which maintained close contact, cross-linking could follow as fibrinogen binding sites were expressed. Coller[48] has proposed that both platelet binding of fibrinogen and the small radius of curvature of the pseudopods would be sufficient to lower the electrostatic repulsion between the similarly charged platelets and permit aggregation through van der Waals attraction. Since such aggregation is not mediated by specific fibrinogen cross-linking, it may not be resistant to high shear rates. It does, however, provide a mechanism for the relatively weak aggregation observed at short transit times in the present work.

Aggregation in Whole Blood. At all mean tube shear rates, and at both 0.2 and 1 μM ADP, the rate and extent of aggregation in whole blood were always much greater than those in PRP.[35] At 0.2 μM ADP, $\bar{G} = 41.9$ s^{-1}, the rate of aggregation was highest over the first $\bar{t} = 1.7$ s where 26% of single platelets aggregated, more than 7× the rate in PRP. By $\bar{t} = 43$ s, only 13% of platelets remained unaggregated in whole blood as compared to 64% in PRP. At 1.0 μM ADP, the initial rate of aggregation in whole blood of 37% s^{-1} was 9× the mean value for the donors in PRP.[33] As shown in Fig. 7, the rate of aggregation at $\bar{G} = 335$ s^{-1} was also much greater than that in PRP at both 0.2 and 1.0 μM ADP. The pattern of aggregation at $\bar{G} = 1920$ s^{-1} in whole blood was almost identical to that at $\bar{G} = 335$ s^{-1}.

In whole blood, the rapid formation and size of the aggregates was most remarkable. At 1 μM ADP and $\bar{G} = 41.9$ s^{-1}, the visible flake-like aggregates were red in color due to red cells trapped within the interior. At $\bar{G} > 335$ s^{-1}, aggregate growth was initially slower, with successive formation of distinct populations of increasing size, an effect even more pronounced at $\bar{G} = 1920$ s^{-1}.

Mechanism of Red Cell-Enhanced Platelet Aggregation: Chemical Effects. There was significant aggregation at $\bar{G} = 41.9$ s^{-1} (Fig. 7), and to a lesser extent at 335 s^{-1} in control runs with whole blood in which Tyrodes solution, instead of ADP, was infused. The effect was not seen in PRP. Although tests showed that there was no release of ADP from platelets and no hemolysis of red cells,(35) small amounts of ADP released from intact red cells could lead to the formation of platelet aggregates, only stable at low shear rates. Indeed, as shown in Fig. 8, in the presence of the enzyme CP-CPK, which converts ADP to ATP, aggregation in the absence of added ADP was abolished at all shear rates.

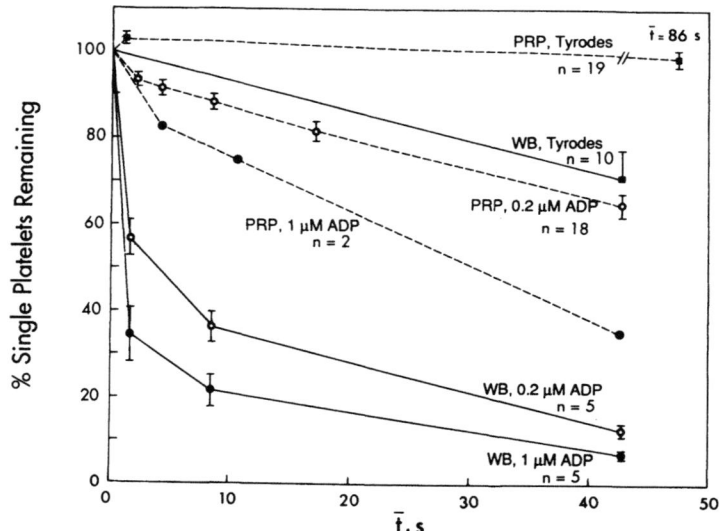

Figure 7. Aggregation in whole blood (WB). Mean values of the % single platelets remaining (± SEM, n = number of donors) at $\overline{G} = 41.9$ s^{-1} versus mean transit time $\overline{t}$ at 0.2 and 1.0 µM ADP. There was significant aggregation in the WB control runs in which Tyrodes instead of ADP was infused. (From Bell et al.)(35)

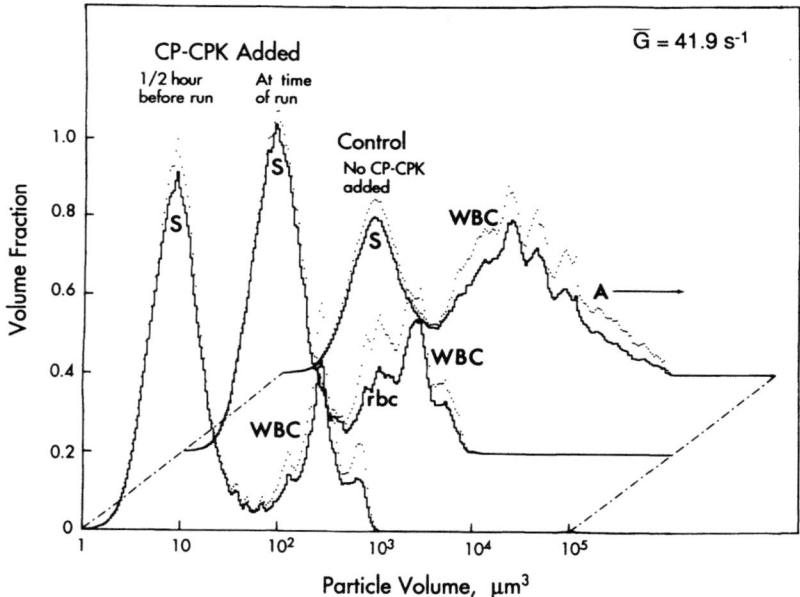

Figure 8. Addition of CP-CPK abolishes aggregation in the absence of added ADP. Three-dimensional plot of the mean normalized volume fraction (± SEM, dotted lines) versus particle volume at a mean transit times of 43 s. Shown are the singlet platelet peaks (S), contaminating white cell peaks (WBC) and, in the absence of the enzyme, aggregates (A).

Mechanism of Red Cell-Enhanced Platelet Aggregation: Physical Effects.
The presence of red cells at volume concentrations from 40 to 50% considerably disturbs the motions of the formed elements and the plasma. One might assume that in such concentrated suspensions, the red cells would impede the diffusion of other cells and solutes. This is true when the blood is stationary, but the opposite is observed in flow. The continued collisions between, and deformation of the red cells in flowing blood lead to a continual radial displacement of their paths and an alternate method of solute mixing, similar on a macroscopic scale to the intermolecular collisions which result in Brownian motion. Red cells markedly increase the diffusivity of platelets in the plasma of blood in tube flow The effective translational diffusion coefficient, D_t, increases by two orders of magnitude from 10^{-9} to $> 10^{-7}$ cm^2 s^{-1}.[49] That this occurs because of an increase in the lateral dispersion of platelets caused by the erratic motions of the continuously colliding and deforming red cells, has been visually demonstrated in optically transparent suspensions of red cell ghosts flowing through small tubes.[50,51] Figure 9 shows the observed radial displacements of tracer 2.0 µm diameter latex spheres and platelets in a 42% ghost cell suspension flowing through 75 and 104 µm diameter tubes, respectively. It is evident that the amplitudes of the displacements of the tracer particles are large, and can lead to collisions, with the vessel wall, of particles initially far removed from the boundary.

One can estimate the increased frequency of two-body collisions due to lateral dispersion of platelets by the red cells by treating the dispersion as a Brownian motion. The two-body collision frequency due to translational Brownian motion of rigid spheres, $j_d = 16\pi bND_t$,[24] and the ratio of shear-induced [Eq. (12)] to Brownian motion-induced collision frequency is:

$$\frac{j_s}{j_d} = \frac{2\overline{G}b^2}{3\pi D_t}. \qquad (17)$$

Assuming $D_t = 2 \times 10^{-7}$ cm^2 s^{-1} in whole blood, and an equivalent sphere radius $b = 1.2$ µm based on a mean platelet volume = 7.5 µm^3, $j_s/j_d = 0.64$ and 5.2 at $\overline{G} = 41.9$ and 335 s^{-1}, respectively, corresponding to increases of 156 and 19% in the collision frequency. However, results from the present work show that collision frequencies in whole blood should be 7 and 15 times greater than in PRP at $\overline{G} = 335$ and 41.9 s^{-1}, respectively, if the collision efficiency remained unchanged. An explanation for the higher rates of aggregation in whole blood may thus be sought in terms of a much greater collision efficiency, perhaps resulting from an increased velocity of approach and/or time of interaction during collision due to the presence of the red cells. We are testing this hypothesis using 40% suspensions of red cell ghosts containing 3×10^5 /µl plasma of 2 µm latex spheres serving as models of platelets. The suspensions flow through a 100 µm diameter tube and cine films of two-body collisions between spheres are analyzed to obtain the doublet lifetime, τ_{meas}, during which the particles are in apparent contact. Measured τ are compared with τ_{calc} predicted by theory assuming the spheres rotate together as a spheroid of axis ratio = 1.98:[7,52]

$$\tau_{calc} = \frac{5}{G(R)} \tan^{-1}(\tfrac{1}{2}\tan\phi_1^o) \qquad (18)$$

Here, $-\phi_1^o$ is the azimuthal angle of orientation of the doublet axis, referred to X_1 as the polar, and vorticity axis, when the spheres first make contact (apparent angle of collision). In the absence of three-body interactions and electrostatic or van der Waals forces between sphere surfaces, the particles rotate as a rigid dumbbell until at an angle $+\phi_1^o$ (the reflection of the apparent angle of collision) they separate. Analysis of 170 collisions at G(R) from 5 - 40 s^{-1} has shown that the mean measured τ is $2.0 \times$ greater than predicted from theory. A histogram of the distribution in τ_{meas}/τ_{calc} is shown in Fig. 9.

It should also be noted that the collision frequency itself would be higher in tubular vessels if, at the periphery, where the shear rate is highest, the platelet concentration is two-fold greater than at the centre of the vessel, as observed in small arteries, and described below.[53] It should be emphasized, however, that an increase in collision efficiency due to locally released ADP or other metabolites from red cells cannot be entirely excluded.[54]

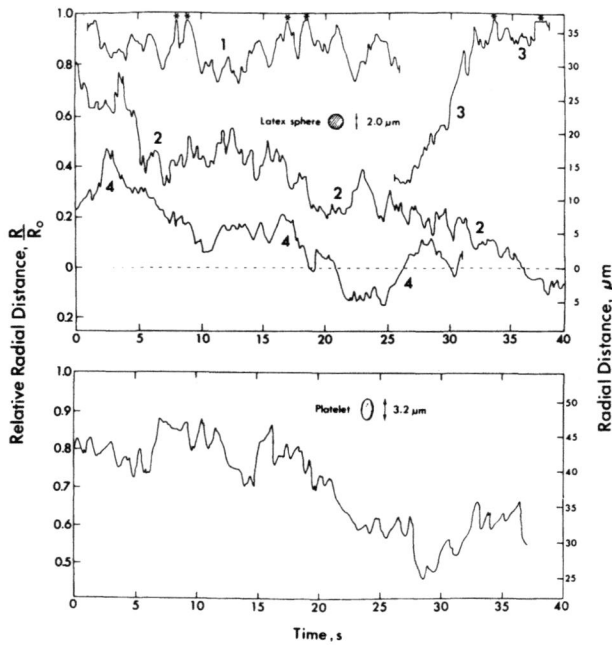

Figure 9. Radial displacements of a tracer 2 μm diameter latex sphere in a 75 μm diameter tube (upper), and a human platelet in a 104 μm diameter tube, suspended in a 42% ghost cell suspension. Plot of the relative radial position (left) and the radial distance from the tube axis (right) against time. Cells marked along paths 1 and 3 collided with the wall at times indicated by the asterisks.

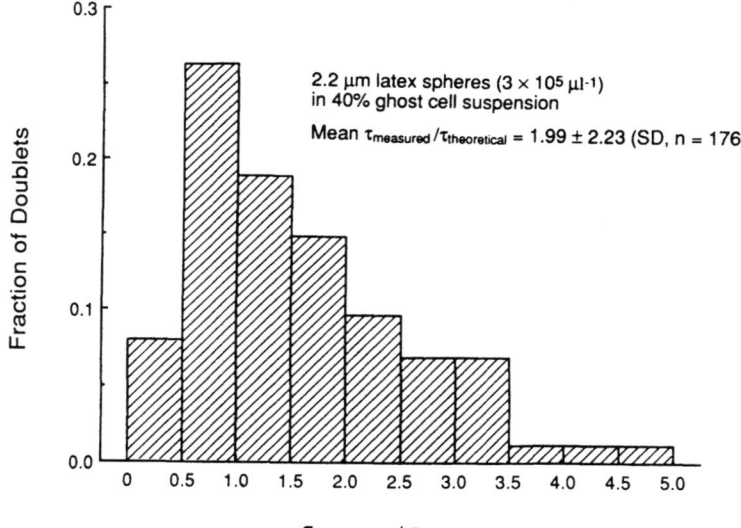

Figure 10. Histogram showing the distribution in the ratio of measured to calculated lifetimes of doublets formed by two-body collisions between 2 μm latex spheres in a 40% suspension of red cell ghosts flowing through a 100 μm tube. The values of τ_{calc} were obtained from Eq. (17) using the measured apparent collision angle ϕ_1^o and the shear rate $G(R)$ at the doublet center of rotation.

EFFECT OF RED CELLS ON PLATELET AND LEUKOCYTE MOTIONS

Wall Adhesion of Platelets. As a consequence of the radial dispersion of the other formed elements by the motions of the red cells, such as those shown in Fig. 9, the frequency of wall encounters of platelets increases with increasing hematocrit. The effect has been correlated with the observed increase in adhesion to subendothelium[55,56] and other surfaces[57,58] as the hematocrit is increased. The measured increase in diffusivity of platelets in blood as compared with that in plasma,[49] and the resulting increased cell-wall interactions are of considerable importance in thrombotic processes in regions of disturbed flow in large vessels.[59] An even more striking increase in platelet adhesion to subendothelium has been noted at at the highest wall shear rate (2600 s^{-1}).[60] The extent to which red cell deformation or size may affect the increase in the effective diffusion constant of cells and solute has also been studied. The transport of molecular solute was found to increase slightly with both cell size and rigidity,[61,62] and the wall adhesion of platelets was observed to increase substantially with red cell size.[62] Augmented solute transport in the shear flow of concentrated suspensions of model particles and red blood cells has been thoroughly reviewed by Zydney and Colton.[63] They propose a model of augmented solute transport based on shear-induced particle migrations and the concomitant dispersive fluid motion induced by these particle migrations. Augmented solute transport is defined as $(D_{eff}/D_{SF}) - 1$ where D_{eff} is the effective solute diffusion constant measured in the sheared suspension, and D_{SF} is the solute diffusion constant in stationary flow. If particle rotations are assumed to be unimportant, $D_{eff} = D_{SF} + D_p$, D_p being the particle diffusion constant. Augmented diffusion is predicted to vary linearly with the Peclet number, $Pe = b^2 G/D_{SF}$.

Redistribution of Platelets in Tube Flow. In addition to disturbing the motion of the other formed elements in blood, red cells appear to affect the distribution of leukocytes and platelets because of their radial migration away from the wall and aggregation in the core of the vessel where shear rates are very low. In the case of platelets in arterioles,[53,64,65] and latex microspheres 1 to 2.5 µm diameter in tubes of ~ 200 µm diameter,[66-68] particle number concentrations near the vessel wall have been shown to be higher than in the core of the flowing suspension ("near wall excess"). The effect, which requires the presence of red cells, has recently been modeled by adding a lateral drift term to the convective diffusion equation for platelet transport in flowing blood.[69] The reason for the net outward drift is believed to arise from the inward migration of the more rapidly migrating red cells,[50] which has been shown, both *in vivo*[70] and *in vitro*[71] to result in a marginal "plasma layer" of lower red cell concentration, whose width (~ 2 red cell diameters) continuously fluctuates and should be viewed in a statistical sense. The outward lateral drift of platelets or microspheres occurs because of a net flux of particles from a region of higher red cell concentration and hence higher red cell collision rate to a region of lower red cell concentration and lower red cell collision rate. Such a drift has been shown to occur in alloys.[72] The fact that the location of the near wall excess of microspheres occurs a few microns away from the wall is due to the fact that the particles are physically and fluid dynamically repelled from the wall.[52] It should be noted that the above hypothesis does not attribute the motion of platelets toward the wall to their exclusion by red cells in the interior of the tube. As described below for leukocytes, when there is red cell aggregation, such exclusion does occur.

Redistribution of Leukocytes in Tube Flow. It is the aggregation of red cells that accounts for the major redistribution of leukocytes towards the periphery of the vessel. This so-called margination is a well-documented phenomenon in the microcirculation, associated with low blood flow states such as occur in shock and in inflammation. Leukocyte distribution in blood flowing through glass tubes has been extensively studied. At high flow rates, the cells are found to be axially distributed,[73,74] although not to the same extent as the red cells.[75] However, when red cells are caused to aggregate, either through the addition of fibrinogen[73] or high molecular weight Dextran, or by reducing the flow rate,[74,75] leukocytes accumulate at the vessel periphery. Figure 11 illustrates the effect observed in vertically positioned 100 µm diameter tubes through measurements of leukocyte concentration in samples of whole blood withdrawn from the tubes. Providing there is no screening effect of the cells entering from the infusing reservoir, at the steady state, mass balance demands that the reservoir and discharge concentrations are equal, and that the ratio of tube, n_t, to reservoir number concentration of cells, n_o, is equal to the the ratio of mean blood velocity, $\overline{U}$, to mean leukocyte, $\overline{u}_L$, or red cell velocity, $\overline{u}_{rbc}$:[76]

$$\frac{n_t}{n_o} = \frac{\overline{U}}{\overline{u}_{L,\,rbc}} \qquad (19)$$

Figure 11 shows that, at low mean tube shear rates, defined as $\overline{G} = 2\overline{U}/R_o$, the concentration of leukocytes in the tube was always greater than in the infusing reservoir, implying that $\overline{u}_L < \overline{U}$ and that the leukocytes were concentrated in the tube periphery. The opposite was true for the red cells, which were more axially transported in the tube. At high $\overline{G}$, both leukocyte and red cell concentrations were lower in the tube than in the reservoir.

Movie films taken of blood flow in 100 - 340 μm tubes clearly show that the effect is due to the outward displacement of white cells by the inwardly migrating network of packed red cell rouleaux, which exclude plasma containing both white cells and platelets towards the periphery. The effect, as seen at 20% hematocrit in a 100 μm diameter tube, is illustrated in Fig. 12. Margination of leukocytes is no longer found when the experiments are carried out in washed red cell suspensions in which rouleau formation is totally suppressed.[75]

Changes in Hydrodynamic Resistance due to Red Cell Aggregation. As pointed out above, aggregation of red cells in blood flowing through small tubes at low shear rates leads to lateral migration of rouleaux into an axial core surrounded by a cell-depleted peripheral layer. Such two-phase flow was suspected,[78,79] or known to be accompanied by a decrease in hydrodynamic resistance to flow.[80,81] That two-phase flow in a tube may lead to a decrease in hydrodynamic resistance was also suggested by earlier work on the oscillatory flow of concentrated suspensions of rigid spheres,[82] and subsequently confirmed experimentally.[83]

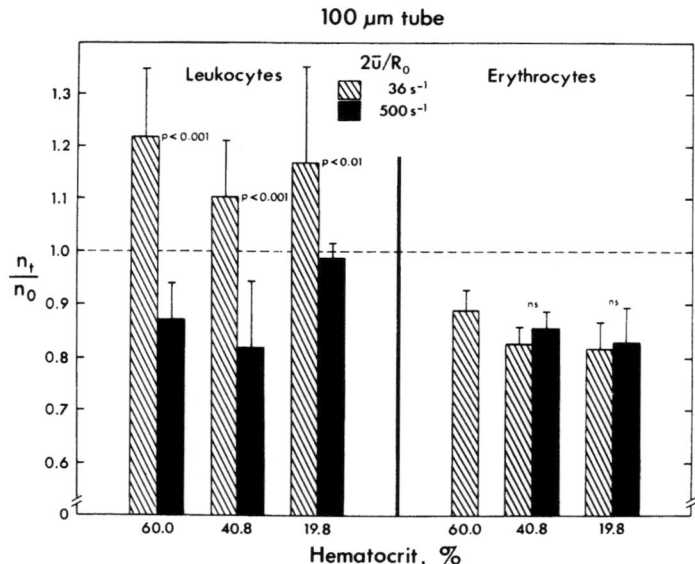

Figure 11. Bar graphs of the number concentration of leukocytes and red cells in whole blood in the 100 μm tube relative to that in the infusing reservoir, at the equivalent Poiseuille flow mean shear rates $2\overline{U}/R_o$ = 500 s^{-1} (hatched) and 36 s^{-1} (solid bars), at different reservoir hematocrits. (From Goldsmith and Spain)[75]

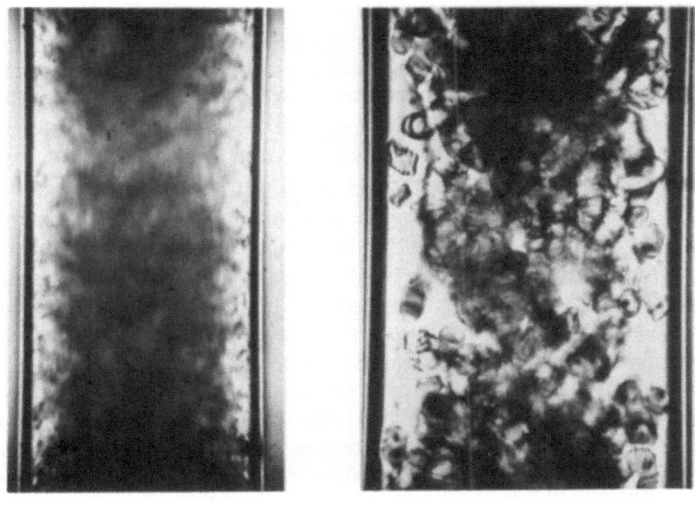

$\bar{U} = 1.54$ mm s^{-1} $\bar{U} = 0.11$ mm s^{-1}

Figure 12. Photomicrographs of the flow of blood at 20% hematocrit at high (left) and low flow rates (right) in a 100 μm diameter tube. A core of rouleaux of red cells of nonuniform width develops at low $\bar{U}$. At the tube periphery there are single rouleaux, platelets and white cells. (From Goldsmith and Spain)[77]

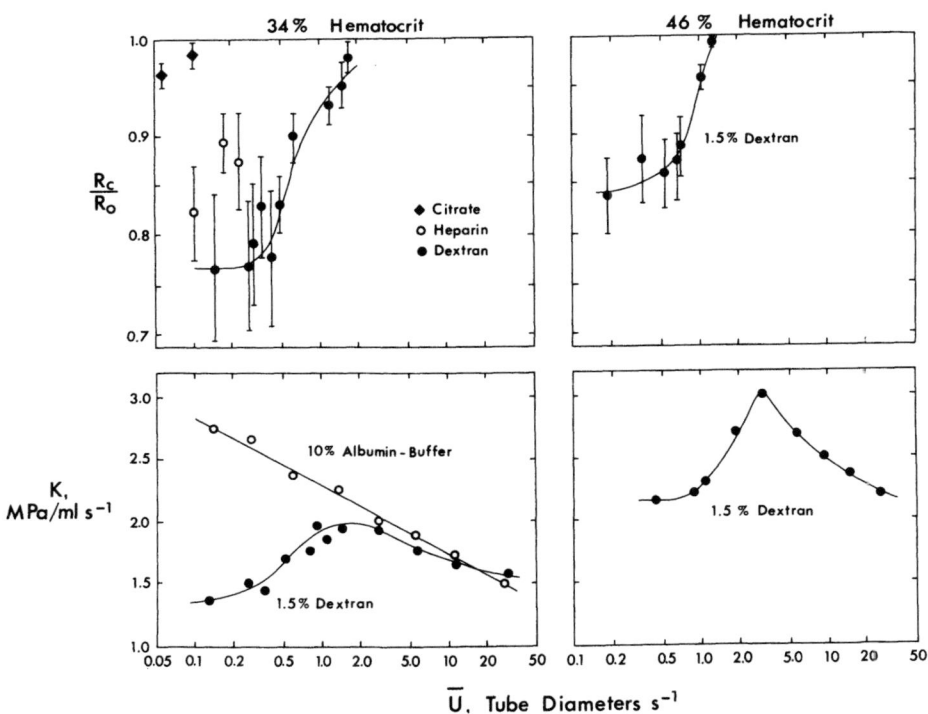

Figure 13. Hydrodynamic resistance, K (lower panels) and relative blood cell core width, R_c/R_o (upper panels; error bars = ± 1 S.D.) of washed red cells in 1.5% Dextran T-110 buffer suspensions as a function of mean blood velocity, $\bar{U}$, at 34% (left), and 46% hematocrit (right), R_o = 172 μm. Also shown are values of K obtained in 34% citrated and heparinized blood. (From Cokelet and Goldsmith)[84]

In vertically positioned tubes, enhanced red cell aggregation in the presence of Dextran 250 results in a lowered apparent viscosity at mean flow rates $\overline{U} < 2$ tube diameters s^{-1}, due to the formation of a symmetrical red blood cell core. By contrast, in horizontally positioned tubes, in which cells and aggregates sediment towards the lower tube wall, the apparent viscosity continues to increase with decreasing $\overline{U}$.[81] We have quantitated the relationship between the radius, R_c, of the red blood cell core and the hydrodynamic resistance, $K = \Delta P/\Delta Q$, as a function of decreasing $\overline{U}$, and proposed a two-phase model to predict the hydrodynamic resistance from rheological data on red blood cell suspensions and the measured R_c.[84] As illustrated in Fig. 13, the results obtained in a 344 µm diameter tube using red cells suspended in buffered T110 Dextran, show that K at first increased as $\overline{U}$ decreased from 30 to ~ 2 s^{-1}. It then decreased to a minimum at $\overline{U} < 0.5$ s^{-1}, precisely in the range when formation of the red cell core was observed. In albumin buffer, however, where there was no rouleau formation, K increased continuously with decreasing $\overline{U}$ and there was no core formation. With increasing hematocrit, the initial increase, and subsequent decrease in K was larger and, as expected, at a given $\overline{U}$, the width of the peripheral cell-depleted layer decreased (Fig. 13). With decreasing tube radius, the values of $\overline{U}$ at which core formation began, and those at which minimum core radius was reached, increased. Thus, the red cell core is able to withstand higher shear rates in the smaller tubes.

The flow patterns in horizontal tubes are somewhat different[81] as sedimentation of cells onto the lower tube wall eventually results in an almost stagnant mass of corpuscles over which plasma flows, entraining cells from the uppermost layer of the settled mass. In living vessels, this was described as sludging by Knisely[85], who thought it to be pathological. It is now accepted that rouleau formation and sedimentation also occur in healthy states, and that the above flow patterns are relevant at all times in the circulation.[76]

REFERENCES

1. C.J.K. Lin, J. Lee, and N.F. Sather, Slow motion of two spheres in a shear field, *J. Fluid Mech.* 43:35 (1970).
2. G.K Batchelor, and J.T. Green, The hydrodynamic interaction of two small freely moving spheres in a linear flow field, *J. Fluid Mech.* 56:375 (1972).
3. P.A. Arp, and S.G. Mason, The kinetics of flowing dispersions. VIII. Doublets of rigid spheres (theoretical), *J. Colloid Interface Sci.* 61:21 (1977).
4. T.G.M. van de Ven, and S.G. Mason, The microrheology of colloidal dispersions. IV. Pairs of interacting spheres in shear flow, *J. Colloid Interface Sci.* 57:517 (1976).
5. P.F. Bretherton, The motion of rigid particles in a shear flow at low Reynolds number, *J. Fluid Mech.* 14:284 (1962).
6. S.P. Tha, and H.L. Goldsmith, Interaction forces between red cells agglutinated by antibody. I. Theoretical, *Biophys. J.* 50:1109 (1986).
7. K. Takamura, H.L. Goldsmith, and S.G. Mason, The microrheology of colloidal dispersions. IX. Effects of simple and polyelectrolytes on rotation of doublets of spheres, *J. Colloid Interface Sci.* 72:385 (1979).
8. K. Takamura, H.L. Goldsmith, and S.G. Mason, The microrheology of colloidal dispersions. XI. Trajectories of orthokinetic pair-collisions of latex spheres in a simple electrolyte, *J. Colloid Interface Sci*. 82:175 (1981).
9. R.St.J. Manley, and S.G. Mason, Particle motions in sheared suspensions. II. Collisions of uniform spheres, *J. Colloid Sci.* 7:354 (1962).
10. H.L. Goldsmith, and S.G. Mason, The flow of suspensions through tubes. III. Collisions of small uniform spheres, *Proc. Roy. Soc. (London)* A282:569 (1964).
11. B.V. Derjaguin, and L.D. Landau, *Acta Physicochim. URSS* 14:633 (1941)
12. E.G. Verwey, and J.Th.G. Overbeek, "Theory of the Stability of Lyophobic Colloids," Elsevier Scient. Publ. Co., Amsterdam (1948).
13. T.G.M. van de Ven, "Colloidal Hydrodynamics," Academic Press, New York (1989).
14. G.B. Jeffery, On the motion of ellipsoidal particles immersed in a viscous fluid, *Proc. Roy. Soc. London* A 102:161 (1922).

15. K. Takamura, P. Adler, H.L. Goldsmith, and S.G. Mason, Particle motions in sheared suspensions. XXXI. Rotations of rigid and flexible dumbbells (experimental), *J. Colloid Interface Sci.* 83:516 (1981).
16. K. Takamura, H.L. Goldsmith, and S.G. Mason, The microrheology of colloidal dispersions. XI. Trajectories of orthokinetic pair-collisions of latex spheres in a cationic polyelectrolyte, *J. Colloid Interface Sci.* 82:190 (1981).
17. H. Brenner, and M.E. O'Neill, On the Stokes resistance of multiparticle systems in a linear shear field, *Chem. Eng. Sci.* 27:1421 (1972).
18. S.P. Tha, J. Shuster, and H.L. Goldsmith, Interaction forces between red cells agglutinated by antibody. II. Measurement of hydrodynamic force of breakup, *Biophys. J.* 50:1117 (1986).
19. H. L. Goldsmith, O. Coenen, and C. Timm, Measurement of the adhesion force between human red cells agglutinated by monoclonal antibody, *Biorheology* 26:564 (1989)
20. G.I. Bell, Models for the specific adhesion of cells to cells, *Science* 200:618 (1978)
21. D.F. Tees, and H.L. Goldsmith, Stochastic nature of the adhesion force between doublets of human red cells cross-linked by monoclonal antibody, 5th World Congress for Microcirculation, Louisville, KY (1991).
22. E.A. Evans, Detachment of agglutinin-bonded red blood cells, *Biophys. J.* 59:838 (1991).
23. T.G.M. van de Ven, and S.G. Mason, The microrheology of colloidal dispersions. VII. Orthokinetic doublet formation of spheres, *Colloid Polymer Sci.* 255:468 (1977).
24. M. von. Smoluchowski, Versuch einer Mathematischen Theorie der Koagulationskinetik kolloider Lösungen, *Z. Physik. Chem.* 92:129 (1917).
25. D.L. Swift, and S.K. Friedlander, The coagulation of hydrosols by Brownian motion and laminar shear flow, *J. Colloid Sci.* 19:621 (1964).
26. A.S.C. Curtis, and L.M. Hocking, Collision efficiency of equal spherical particles in a shear field, *Trans. Faraday Soc.* 66:1381 (1970).
27. G.R. Zeichner, and W.R. Schowalter, Effects of hydrodynamic and colloid forces on the coagulation of dispersions, *J. Colloid Interface Sci.* 71:237 (1979).
28. H.L. Goldsmith, O. Lichtarge, M. Tessier-Lavigne, and S. Spain, Some model experiments in hemodynamics. VI. Two-body collisions between blood cells, *Biorheology* 18:531 (1981).
29. V.A. Parsegian, and D. Gingell, Some features of physical forces between biological membranes, *J. Adhesion* 4:283 (1972).
30. G.V.F. Seaman, Electrokinetic behavior of red cells, *in*: "The Red Blood Cell,", Vol. II, D. Mac N. Surgenor, ed., Academic Press, New York (1975).
31. S. Levine, M. Levine, K.A. Sharp, and D.E. Brooks, Theory of electrokinetic behavior of erythrocytes, *Biophys. J.* 42:127 (1983).
32. D.N. Bell, S. Spain, and H.L. Goldsmith, The ADP-induced aggregation of human platelets in flow through tubes. I. Measurement of the concentration, size of single platelets and aggregates, *Biophys. J.* 56:817, (1989).
33. D.N. Bell, S. Spain, and H.L. Goldsmith, The ADP-induced aggregation of human platelets in flow through tubes. II. Effect of shear rate, donor sex and ADP concentration, *Biophys. J.* 56:817, (1989).
34. D.N. Bell, S. Spain, and H.L. Goldsmith, Extracellular Ca^{2+} accounts for the sex difference in the aggregation of human platelets in citrated platelet-rich plasma, *Thromb. Res.* 58:47 (1990).
35. D.N. Bell, S. Spain, and H.L. Goldsmith, The effect of red cells on the ADP-induced aggregation of human platelets in flow through tubes, *Thromb. Haemost.* 63:112 (1990).
36. H.L. Goldsmith, and V.T. Turitto, Rheological aspects of thrombosis and haemostasis. Basic principles and applications, *Thromb. Haemost.* 55:415 (1986).
37. D.N. Bell, Physical factors governing the aggregation of human platelets in flow through tubes. Ph.D. Thesis, McGill University (1988).
38. D.N. Bell, and H.L. Goldsmith, Platelet aggregation in Poiseuille flow. II. Effects of shear rate, *Microvasc. Res.* 27:316 (1984).

39. M.M. Frojmovic, K.A. Longmire, and T.G.M. van de Ven, Long-range interactions in mammalian platelet aggregation. II. The role of platelet pseudopod number and length, *Biophys. J.* 58:309 (1990).
40. L. Leung, and R. Nachman, Molecular mechanisms of platelet aggregation, *Ann. Rev. Med.* 37:179 (1986).
41. A.T. Nurden, Platelet membrane glycoproteins and their clinical aspects, *in*: "Thrombosis and Haemostasis," M. Verstraete, J. Vermylen, R. Lijnen, and J. Arnout, eds., Leuven University Press, Leuven, Belgium (1987).
42. E.I. Peerschke, M.B. Zucker, R.A. Grant, J.J. Egan, and M.M. Johnson, Correlation between fibrinogen binding for human platelet and platelet aggregability, *Blood* 55:841 (1980).
43. G.A. Margerie, G.S. Edgington, and E.F. Plow, Interaction of fibrinogen with its receptor as part of a multistep reaction in ADP-induced platelet aggregation, *J. Biol. Chem.* 255:154 (1980).
44. M. Kloczewiak, S. Timmons, T.J. Lukas, and J. Hawiger, Platelet receptor recognition site on human fibrinogen. Synthesis and structure-function relationship of peptides corresponding to the carboxy-terminal segment of the γ chain. *Biochem.* 23:1767 (1984).
45. S.A. Santoro, and W.J. Lawing, Competition for related but nonidentical binding sites on the glycoprotein IIb-IIIa complex by peptides derived from platelet adhesive proteins, *Cell* 48:867 (1987).
46. A.S. Aasch, L.K. Leung, M.J. Polley, and R.L. Nachman, Platelet membrane topography: colocalization of thrombospondin and fibrinogen with glycoprotein IIb-IIIa complex, *Blood* 66:926 (1985).
47. P.J. Newman, R.P. McEver, M.P. Doers, and T.S. Kunicki, Synergistic action of two murine monoclonal antibodies that inhibit ADP-induced platelet aggregation without blocking fibrinogen binding, *Blood* 69:668 (1987).
48. B.S. Coller, Biochemical and electrostatic considerations in primary platelet aggregation, *Ann. N.Y. Acad. Sci.* 416:693 (1983).
49. V.T. Turitto, A.M. Benis, and E.F. Leonard, Platelet diffusion in flowing blood, *Ind. Eng. Chem. Fundam.* 11:216 (1972).
50. H.L. Goldsmith, Red cell motions and wall interactions in tube flow, *Fed Proc.* 30:1578 (1588).
51. H.L. Goldsmith, and J. Marlow, Flow behavior of erythrocytes. II. Particle motions in sheared suspensions of ghost cells, *J. Colloid Interface Sci.* 71:383 (1979).
52. H.L. Goldsmith and S.G. Mason, The microrheology of dispersions, *in*: "Rheology: Theory and Applications," Volume IV, F.R. Eirich, ed., Academic Press, New York (1967).
53. G.J. Tangelder, H.C. Teirlinck, D.W. Slaaf, and R.S. Reneman, Distribution of blood platelets in flowing arterioles, *Am. J. Physiol. (Heart Circ. Physiol.)* 248:H318 (1985).
54. H.J. Reimers, S.P. Sutera, and J.H. Joist, Potentiation by red cells of shear-induced platelet aggregation: relative importance of chemical and physical mechanisms, *Blood* 64:1200 (1984).
55. V.T. Turitto, and H. Baumgärtner, Platelet interaction with subendothelium in a perfusion system: physical role of red cells, *Microvasc. Res.* 9:335 (1975).
56. V.T. Turitto, and H.J. Weiss, Red cells: their dual role in thrombus formation, *Science* 207:541 (1980).
57. E.F. Grabowski, L.I. Friedman, and E.F. Leonard, Effects of shear rate on the diffusion and adhesion of blood platelets to a foreign surface, *Ind. Eng. Chem. Fund* 11:224 (1972).
58. I.A. Feuerstein, B.M. Brophy, and J.L. Brash, Platelet transport and adhesion to reconstituted collagen and artificial surfaces, *Trans. Am. Soc. Artif. Intern. Organs* 21:427 (1975).
59. T. Karino and H.L. Goldsmith, Rheological factors in thrombosis and haemostasis, *in*: "Haemostasis and Thrombosis," A.L. Bloom and D.P. Thomas, eds., Churchill Livingstone, London, England (1986).
60. V.T. Turitto, Viscosity, transport and thrombogenesis, *in*: "Progress in Hemostasis and Thrombosis," Volume 6, T.H. Spaet, ed., Grune & Stratton, New York (1982).
61. N.H. Wang, and K.H. Keller, Solute transport induced by erythrocyte motions in shear flow, *Trans. Am. Soc. Artif. Intern. Organs* 25:14 (1979).

62. P.A. Aarts, P.A. Bolhuis, K.S. Sakariassen, R.M. Heethar, and J.J. Sixma, Red blood cell size is important for adhesion of blood platelets to artery subendothelium, *Blood* 62:214 (1983).
63. A.L. Zydney, and CK. Colton, Augmented solute transport in the shear flow of a concentrated suspension, *Physico Chem. Hydrodyn.* 10:77 (1988).
64. P.A. Aarts, S.A.T. van den Broek, G.W. Prins, G.D.C. Kuiken, J.J. Sixma, and R.M. Heethar, Blood platelets are concentrated near the wall and red cells in the center in flowing blood, *Arteriosclerosis* 8:819 (1988).
65. K.D. Sparks, Platelet concentration profiles in blood flow through capillary tubes, M.Sc. thesis, University of Miami, Coral Gables, FL (1983).
66. A.W. Tilles, and E.C. Eckstein, The near-wall excess of platelet-sized particles in blood flow: Its dependence on hematocrit and wall shear rate, *Microvasc. Res.* 33:211 (1987).
67. E.C. Eckstein, A.W. Tilles, and F.J. Millero, Conditions for the occurrence of large near-wall excess of small particles during blood flow, *Microvasc. Res.* 36:31 (1988).
68. D.L. Bilsker, C.M. Waters, J.S. Kippenham, and E.C. Eckstein, A freeze capture method for the study of platelet-sized particle distribution, *Biorheology* 26:1031 (1989).
69. E.C. Eckstein, and F. Belgacem, Models of platelet transport in flowing blood with drift and diffusion terms, *Biophys. J.* 60:53 (1990).
70. R.H. Phibbs, Orientation and distribution of erythrocytes in blood flowing through medium-sized arteries, *in:* "Hemorheology: Proceedings of the 1st International Conference," A.L. Copley, ed., Pergamon Press, New York (1967).
71. G. Bugliarello, and J. Sevilla, Velocity distributions and other characteristics of steady and pulsatile flow in fine glass tubes, *Biorheology* 7:85 (1970).
72. F. Seitz, On the theory of vacancy diffusion in alloys, *Phys. Rev.* 74:1513 (1948).
73. G. Vejlens, The distribution of leukocytes in the vascular system, *Acta Pathol. Microbiol. Scand. Suppl.* 33:11 (1938).
74. V. Nobis, A.R. Pries and P. Gaehtgens, Rheological mechanisms contributing to WBC-margination, *in:* "White Blood Cells: Morphology and Rheology Related to Function", U. Bagge, G.V.R. Born, and P. Gaehtgens, eds., Martinus Nijhoff, The Hague (1982).
75. H.L. Goldsmith, and S. Spain, Margination of leucocytes in blood flow through small tubes, *Microvasc. Res.* 28:204 (1984).
76. H.L. Goldsmith, G.R. Cokelet, and P. Gaehtgens, Robin Fåhraeus, 15-10-1888 to 18-08-1968: The evolution of his concepts in Cardiovascular Physiology. *Am. J. Physiol.* 257:(*Heart Circ. Physiol.* 26) H1005 (1989).
77. H.L. Goldsmith and S. Spain, Radial distribution of white cells in tube flow, *in:* "White Cell Mechanics: Basic Science and Clinical Aspects," H.J. Meiselman, M. Lichtman, and P.L. LaCelle, eds., Alan R. Liss, New York (1984).
78. R. Fåhraeus, The influence of the rouleau formation on the erythrocytes on the rheology of the blood, *Acta Med. Scand.* 161:151 (1958).
79. H.J. Meiselman, Some physical and rheological properties of human blood, D.Sc. thesis, Massachusetts Institute of Technology, Cambridge, MA (1964).
80. A.A. Palmer, and H.J. Jedrzejczyk, The influence of rouleaux on the resistance to flow through capillary channels at various shear rates, *Biorheology* 12:265 (1975).
81. W. Reinke, P. Gaehtgens, and P.C. Johnson, Blood viscosity in small tubes: Effect of shear rate, aggregation and sedimentation, *Am. J. Physiol.* 253(*Heart Circ. Physiol.* 22):H540 (1987).
82. B. Shizgal, H.L. Goldsmith, and S.G. Mason, The flow of suspensions through tubes. IV. Oscillatory flow of rigid spheres, *Can. J. Chem. Eng.* 43:97 (1965).
83. A. Karnis, H.L. Goldsmith, and S.G. Mason, The flow of suspensions through tubes. V. Inertial effects. *Can. J. Chem. Eng.* 44:181 (1966).
84. G.R. Cokelet, and H.L. Goldsmith, Decreased hydrodynamic resistance in the two-phase flow of blood through small vertical tubes at low flow rate, *Circ. Res.* 68:1 (1991).
85. M.H. Knisely, Intravascular erythrocyte aggregation (blood sludge), *in::* "Handbook of Physiology," Section 2: Circulation, Vol. III, E.M. Renkin and C.C. Michel, eds., American Physiological Society, Bethesda, MD (1965).

MODEL STUDIES OF THE RHEOLOGY OF BLOOD IN MICROVESSELS

Michael R. T. Yen

Department of Biomedical Engineering
Memphis State University
Memphis, TN 38152

INTRODUCTION

In the microcirculation, blood cannot be treated as a homogeneous fluid, but rather as a suspension. The individual cellular elements influence the hemodynamics. In this chapter, we shall study the behavior of red cells in microvessels. We will focus on the general features of cell-vessel interaction and their effect on the apparent viscosity of blood. Model experiments are used in these studies. In our models, the plasma is simulated by a silicone fluid, the red cells are simulated by gelatin pellets. With the model approach, apparent viscosity of blood in pulmonary capillaries is obtained. The Fahraeus-Lindqvist effect, Inversion of Fahraeus-Lindqvist effect, velocity distribution in microvessels, hematocrit in very narrow tubes, etc. are investigated.

APPARENT VISCOSITY AND RELATIVE VISCOSITY

The need to consider cell-vessel interaction in microvessels makes the rheology of blood in microvessel very different from that of larger vessels. To understand these topics let us consider two terms namely the apparent viscosity and relative viscosity. Suppose a certain fluid flows through a circular cylindrical tube. If the fluid is Newtonian and the flow is laminar, then the pressure-flow relationship can be described by Poiseuille law.

$$\frac{\Delta p}{\Delta L} = \frac{8\mu}{\pi a^4} \dot{Q} \qquad (1)$$

where $\Delta p/\Delta L$ is the pressure gradient between two points separated by a distance ΔL, μ is the viscosity of the fluid, a is the radius of the tube and $\dot{Q}$ is the flow rate. If the fluid is non-Newtonian (i.e. blood) this equation does not apply; but we still measure $\Delta P/\Delta L$ and $\dot{Q}$ and use the same equation to calculate a coefficient μ

$$\mu = \frac{\pi a^4}{8} \frac{1}{\dot{Q}} \frac{\Delta p}{\Delta L} \qquad (2)$$

The μ so computed is defined as the apparent coefficient of viscosity for the circular cylindrical tube μ_{app}. If μ_{app} is normalized with respect to μ_o, the plasma viscosity which is Newtonian, it is then called relative viscosity μ_r. The concept of apparent viscosity can be extended to any flow regime, including turbulent flow, as long as we can compute it from a formula that is known to work for a homogenous Newtonian fluid. The concept of relative viscosity can be extended to any flow system, even we do not know its structural geometry and elasticity, as long as flow and pressure can be measured. μ_{app} and μ_r are not intrinsic properties of blood; they are properties of blood and blood vessel interaction and depend on the data reduction procedure. If a vessel system has a geometry such that the theoretical problem for homogeneous fluid flow has not been solved, then we cannot derive an apparent viscosity for flow in such a system. But we can determine a relative viscosity if we are able to perform flow experiments in the system with both blood and a homogeneous fluid.

Apparent Viscosity of Blood in Pulmonary Capillaries

As an example, let us consider the flow of blood in the capillary blood vessels in the lung. The apparent viscosity of blood in capillary blood vessels depends on hematocrit, and on the size and shape of the blood vessels relative to those of the red blood cells. The cause of this variation in apparent viscosity is the local disturbances to flow caused by the red cells. In this chapter, the apparent viscosity is studied by testing a macroscopic model of the pulmonary alveolar sheet with elastic pellets simulate the red blood cells.

Our model is idealized so that the sheet structure is uniform everywhere. Even in such an idealized model, the red cell distribution in flow is seen to be nonuniform, or "patchy." This "patchiness" is a basic feature of the stochastic character of the particulate flow in channels of complex geometry, due to the randomness of the dispersal of particles. This may be thought of as a consequence of the lack of structural forces imposing a regular positioning of the particles. As a consequence of this patchiness, it becomes necessary to use local average values of hematocrit, flow, and pressure gradient to obtain the apparent viscosity coefficient.

Model

The morphology of the interalveolar septa or alveolar sheet has been studied extensively. From a hemodynamic point of view, the sheet is best regarded as composed of two membranes (consisting of endothelial cells, epithelial cells, and interstitial space) which are supported by a doubly periodic array of "posts." Data on the dimensions of posts, interpostal distances, and vascular space-tissue ratio are given by Sobin et al.[24]. The thickness of the sheet depends on the transmural pressure. When the transmural pressure is positive the thickness increases linearly with increasing pressure. When the transmural pressure is negative, the sheet collapses and the thickness becomes zero[25]. From these data a model as shown in Fig. 1 was constructed. The dimensions of the model are given in Table 1.

Table 1. Dimensions of the model.

ϵ	=	post diameter, 0.643 cm
w	=	width of the sheet, 9.042 cm
h	=	thickness of the sheet 2.26 cm, 1.85 cm, 1.59 cm, 1.36 cm, and 0.83 cm
a	=	interpostal distance in vertical direction, 1.91 cm
a_1	=	interpostal distance in horizontal direction, 3.81 cm
S	=	solidity ratio = $1 - \pi\epsilon^2/(2aa_1)$
μ	=	viscosity coefficient of fluid, 50 poise
r	=	radius of the open reservoir, 7.62 cm
Δ	=	distance between the two pressure taps, 7.62 cm
D_c	=	diameter of red cell, 1.079 cm
t_c	=	thickness of red cell, 0.318 cm

Number of rows and columns of posts 11 x 5, staggered

Test Apparatus

Figure 1 is a schematic diagram of the apparatus used in the experiment. The alveolar model was held upright. For the purpose of recording the pressure gradient, two pressure taps were connected to a differential pressure transducer (Sanborn PT 267 BC) by Tygon tubings, between two locations A, B shown in Fig. 1. A

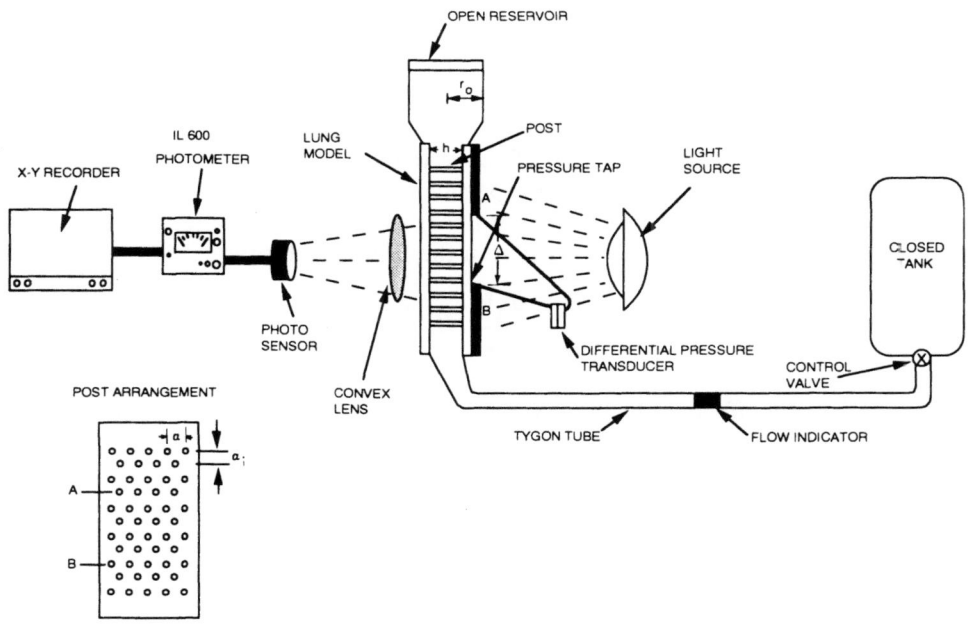

Figure 1. Schematic arrangement of test apparatus.

photosensing device was used to determine the particle concentration in the flow. The device consisted of a light source, a condensing lens and a photosensor arranged in a manner as shown in Fig. 1. A beam of light was directed on that portion of the model which lay between the pressure taps, condensed by the convex lens, focused on the photosensor, and recorded by a photometer and an X-Y plotter. The hematocrit was determined by the light absorbed by the gelatin pellets.

Before each experiment, the gelatin particles were mixed into the silicone fluid in the open reservoir at the top of the model. The mixture was stirred carefully to obtain a uniform suspension. During each test the mixture was sucked into the model at a constant pressure head of 38 cm Hg. Simultaneous readings of the pressure difference, particle concentration, and mean flow velocity were recorded.

Data Reduction

To reduce the data into an expression of the coefficient of apparent viscosity, a theoretical framework is needed. The basic theory is presented by Fung and Sobin[7]. With dimensional analysis it was shown that the pressure gradient in an alveolar sheet at small Reynolds number can be presented in the form

$$\frac{h^2}{\mu V} \nabla p = F\left(\frac{D_c}{h}, \frac{\mu_o V}{E_c h}, N_R, \frac{h}{2}\sqrt{\frac{\omega \rho}{\mu_o}}, H, \frac{w}{h}, \frac{h}{\epsilon}, \frac{\epsilon}{a}, \theta, VSTR\right) \quad (3)$$

where p is pressure, μ is the apparent coefficient of viscosity, μ_o is the viscosity of plasma, V is the mean velocity of flow, h is the sheet thickness, F is a function defined by ratio of red cell diameter D_c to sheet thickness h, the cell membrane strain parameter $\mu_o V/E_c h$, Reynolds number N_R, Womersely number $h/2 \sqrt{\omega \rho/\mu_o}$, hematocrit H, ratio of width of the sheet w to thickness h, the sheet thickness-to-postal-diameter ratio h/ϵ, the ratio of post diameter to interpostal distance ϵ/a, the cell orientation θ and the vascular solidity ratio. In capillary blood, the Reynolds number and Womerseley number are much smaller than 1, and their effects may be neglected. The cell membrane strain parameter is the ratio of the typical shear stress in the fluid $\mu_o V/h$ to the elasticity modulus of the cell membrane E_c. In general, in narrow circular cylindrical tubes, the effects of these parameters make the pressure-flow relationship nonlinear. A model experiment on alveolar sheet by the author[28] shows that the pressure flow relationship is linear when D_c/h is smaller than 0.8. This is fortunate. Equation (3) can be written in the following form.

$$\nabla p = -\frac{\mu_o V}{h^2} F\left(\frac{D_c}{h}, \frac{\mu_o V}{E_c h}, H, \frac{w}{h}, \frac{h}{\epsilon}, \frac{\epsilon}{a}, \theta, VSTR\right) \quad (4)$$

In the studies, the effects of D_c/h, $\mu_o V/E_c h$, H, w/h are isolated separately from the rest of the parameters.

$$\nabla p = -\frac{\mu_o V}{h^2} F'\left(\frac{D_c}{h}, \frac{\mu_o V}{E_c h}, H\right) k\left(\frac{w}{h}\right) f\left(\frac{h}{\epsilon}, \frac{\epsilon}{a}, \theta, VSTR\right) \quad (5)$$

This is further abbreviated to

$$\nabla p = -\frac{V}{h^2}\mu k f \qquad (6)$$

where μ stands for the apparent viscosity

$$\mu = \mu_o \, F'\left(\frac{D_c}{h}, \frac{\mu_o V}{E_c h}, H\right) \qquad (7)$$

The function k is a dimensionless factor. When

$$\frac{h}{w} < 0.2, \quad k = 12\left(1 - 0.63\frac{h}{w}\right), \qquad (8)$$

hence for all practical purposes k can be replaced by 12. The function f is called the geometric friction factor. The function f has been analyzed by Lee[17] for slow flow of a homogeneous viscous fluid. He has presented a theoretical formula for f. An effort was made by Yen[28] to measure f and to compare it with the theoretical result. The comparison of theoretical and experimental values of f is shown in Fig. 2. The agreement is gratifying for values of h/ε up to about 4. For higher values of h/ε the theoretical values gradually fall below the experimental values.

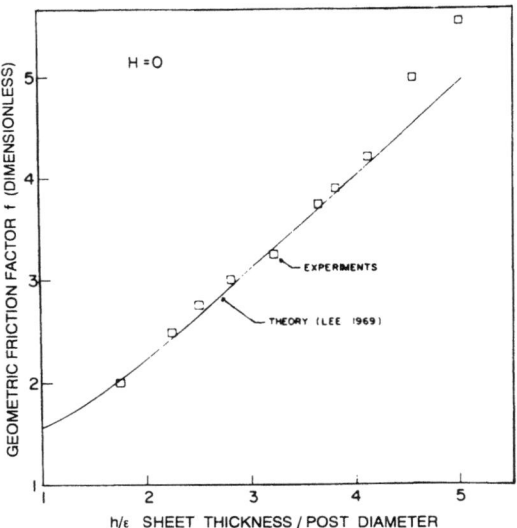

Figure 2. Comparison of theoretical and experimental "geometric friction factor."

Yen and Fung[28] experimented on a scale model of the pulmonary alveolar sheet, with the red blood cells simulated by soft gelatin pellets and with the plasma simulated by a silicone fluid. Their results show that the pressure flow relationship is quite linear and that h/ε<4 the relative viscosity of blood with respect to plasma depends on the hematocrit H in the following manner

$$\mu_r = 1 + aH + bH^2 \qquad (9)$$

Thus the apparent viscosity of blood in pulmonary capillaries is

$$\mu_{app} = \mu_o \left(1 + aH + bH^2\right) \tag{10}$$

The values of the constants a and b in equation 9 and 10 are listed in Fig. 3.

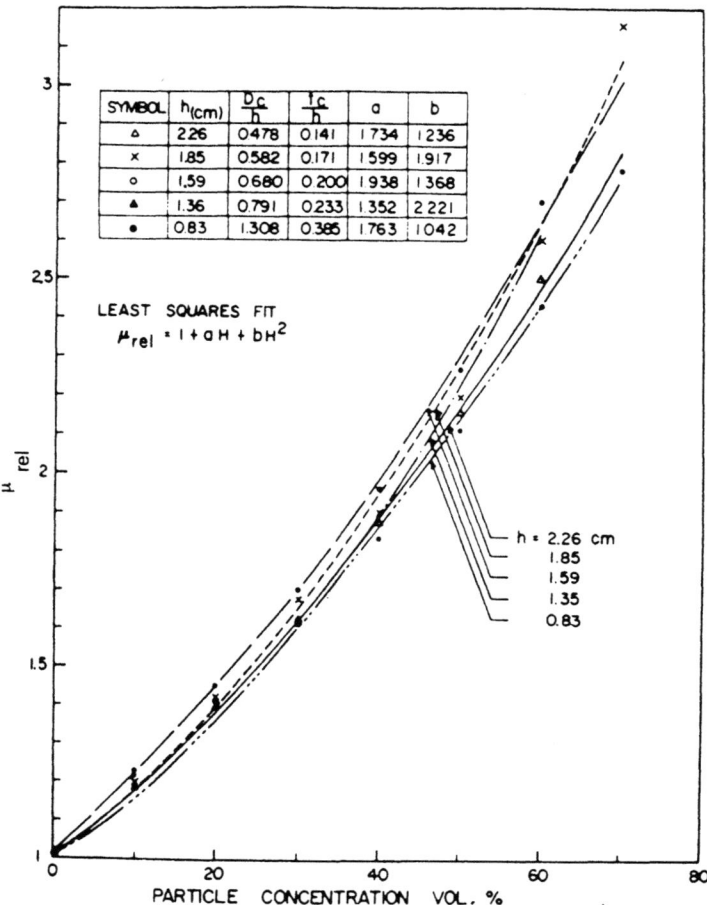

Figure 3. Plot of relative viscosity vs. particle concentration.

THE FAHRAEUS-LINDQVIST EFFECT

As we try to quantitatively understand and describe blood in the very small tubes or vessels it is essential to know the red cell concentration in the tube. This is essential because the hematocrit level of blood has a very strong influence on the flow properties of blood under all flow conditions.

Direct experimental data on the relationship between the hematocrit of blood flowing in a tube and the hematocrit of blood in the reservoir or the vessel feeding the tube have been done by some investigators.

In 1929 Fahraeus[4] in Sweden pointed out that blood of constant hematocrit is allowed to flow from a feed reservoir into a small circular cylindrical tube, the hematocrit in the tube decreases as the tube diameter decreases. Since then the observation has come to be as Fahraeus effect. Fahraeus used dog's blood in tubes down to 50 μm diameter.

Fig. 4 shows the Fahraeus equipment. It consisted of a large diameter glass tube which contained a midsection where the inside diameter was reduced to the capillary dimension in a smooth transition. The tube was horizontal and rotated at a constant rate to prevent sedimentation. The larger portion of the tube served as a reservoir.

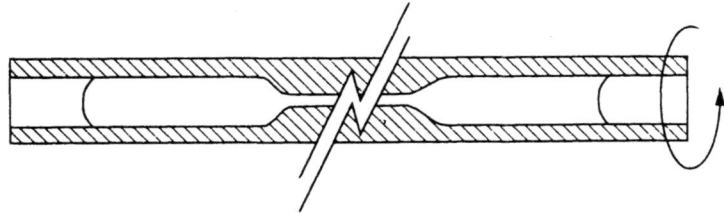

Figure 4. Fahraeus-Lindqvist equipment[4].

Barbee and Cokelet in 1971[1] did extensive series of experiments on human blood. Fig. 5 shows their experiment. The red cell suspension was placed in the right

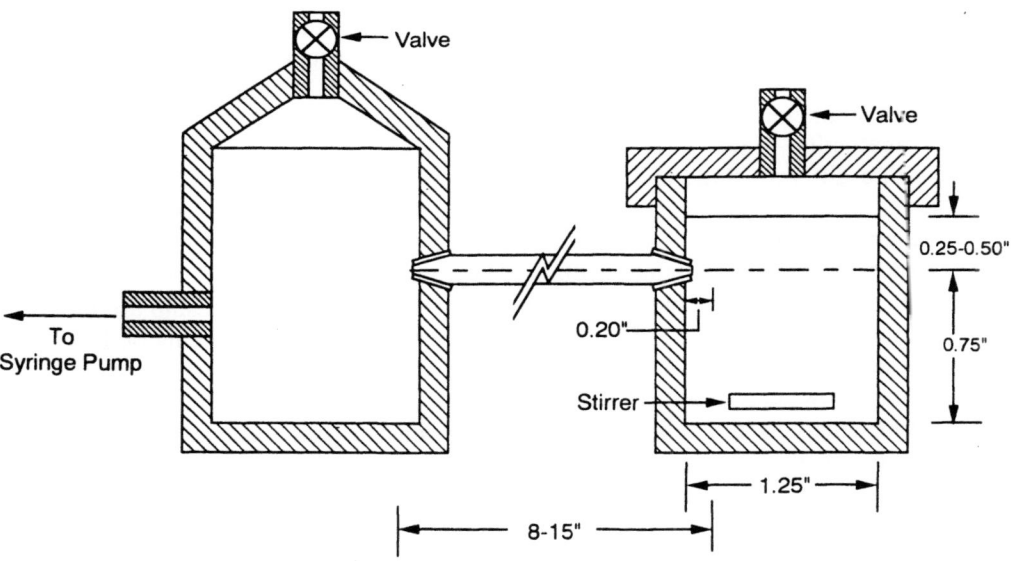

Figure 5. Schematic sketch of Barbee and Cokelet's experiment. Bypass line and lines to pressure transducers not shown[1].

reservoir. The left reservoir was completely filled with isotonic saline. During most experimental runs, the magnetic stirrer bar rotated at a rate of 150 rpm to prevent sedimentation. The pressure was constantly monitored by a pressure transducer. Entrance is abrupt. Their result (Fig. 6) shows that the trend of Fahraeus effect continues down to 29 μm. For blood in cylindrical vessels, the apparent viscosity decreases with decreasing blood vessel diameter. This was pointed out by Fahraeus and Lindqvist and is called Fahraeus-Lindqvist effect[5].

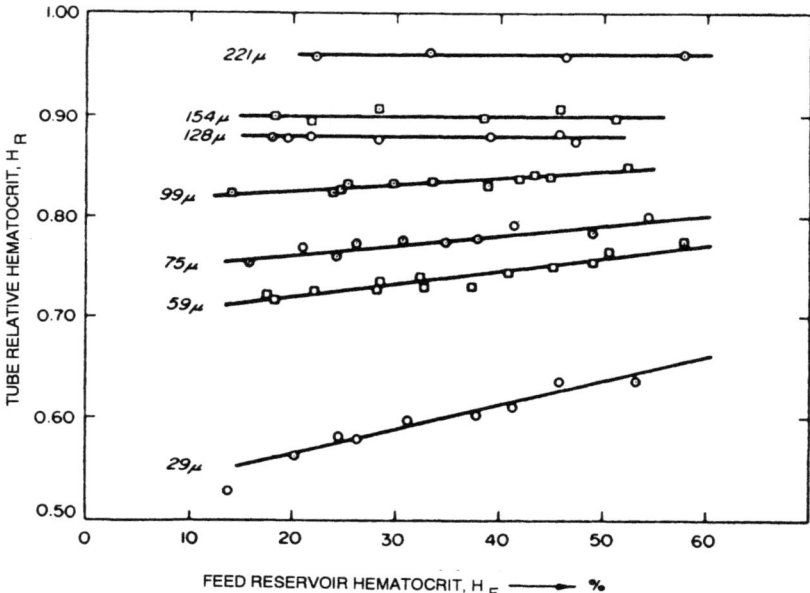

Figure 6. The Fahraeus effect. Figure reprinted from Barbee and Cokelet[1].

Barbee and Cokelet showed that (Fig. 7) the Fahraeus effect can be used to explain the Fahraeus-Lindqvist effect. The experimental data on the wall stress are plotted vs $\overline{U}$, the bulk average velocity of flow divided by the tube diameter. These relationships are shown in Fig. 7 for different hematocrits. Each of the level curves can be fitted with a power law curve with a different slope and intercept depending on the hematocrits.

The top curve represents data obtained with an 811 μm tube with feed reservoir hematocrit $H_F=0.559$. On the next curve, the circles represent experimental data obtained with a 29 μm tube with $H_F=0.559$. These two sets of data should coincide if the shear stress-shear rate relationship are the same in the two tubes. But they do not agree, because the hematocrit in these two tubes are different. The true tube hematocrit is $H_T=0.358$. If we obtain flow data in an 811 μm tube with $H_F=0.358$, we find that these are represented by the solid curve which happens to pass through the circle. Therefore the Fahraeus-Lindqvist effect appears to be due entirely to the Fahraeus effect for tubes larger than 29 μm.

INVERSION OF FAHRAEUS-LINDQVIST EFFECT

Two Models

Two types of model experiments were performed, using two different entry conditions (1) The entry condition at the tube was such that the flow transition was smooth. The cone geometry was chosen arbitrarily in those experiments, and may differ from the real arteriole-capillary blood vessel. (2) An abrupt entry section, obtained by a sharply cut circular cylindrical tube. To simulate the arteriole-capillary blood vessel junction, the capillary tube

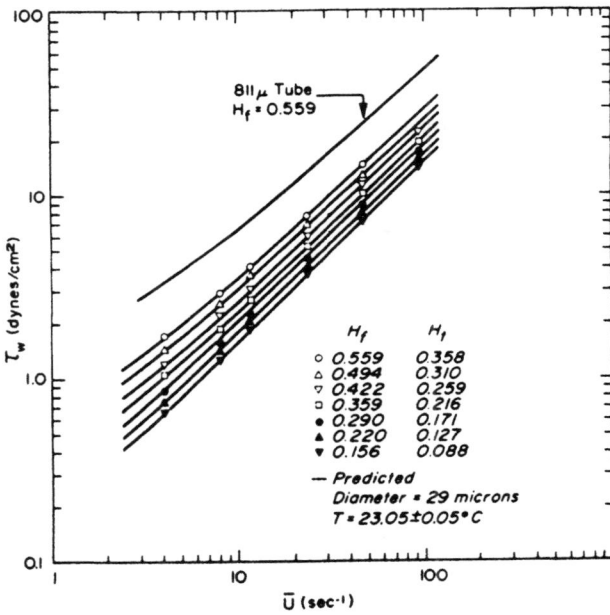

Figure 7. The flow behavior of blood in a 29μm diameter tube. The symbols are the actual flow data, recorded as the wall shear stress, τ_w, and the bulk average velocity divided by the tube diameter, $\bar{U}$. The solid curves through the points represent the behavior of the blood predicted from the data obtained in an 811μm diameter tube when the average tube hematocrit is the same as that experimentally found in the 29μm tube. In an 811-micron tube, H_F, the feed reservoir hematocrit, and H_T, the average tube hematocrit, are equal. From Barbee and Cokelet[1].

was placed perpendicular to the direction of the main flow in a larger vessel.

Model I. Smooth cone entry: The model consisted of a reservoir, an entry cone, a circular cylindrical tube which simulated a capillary blood vessel, and a draining system (Fig. 8). In each experiment, the pellets' suspension was carefully mixed and stirred uniform and then poured into the reservoir. When the valves A, B, C were open, the mixture flowed into the cylindrical tube. The flow rate was controlled by the valve C which was connected to a vacuum pump. The flow could be stopped at any time by closing valve A, B, or C. The rate of flow was measured by measuring the change of position of a flow indicator in the Tygon tubing (a plug in the tube). The pressure gradient in the test chamber was measured by a differential pressure transducer (Sanborn, PT 267BC), which was connected to two taps drilled into the cylindrical tube. The signal from the pressure transducer was amplified by a Hewlett-Packard amplifier model 311A and recorded on a Mosely 2D-2A X-Y recorder.

For the determination of particle concentration in the tube, the valves A and B were swiftly closed to stop the flow after it had reached steady state under a constant pressure head, then the number of particles presented in the tube was counted, and the volume of the pellets computed.

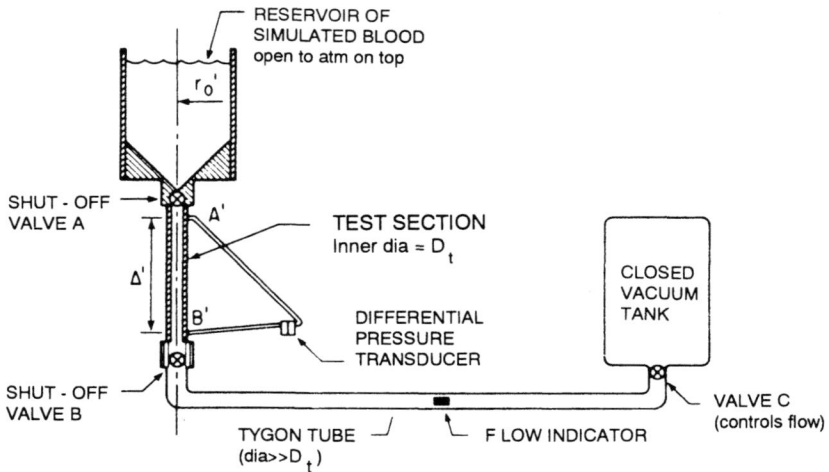

Figure 8. Schematic diagram of test Model I.

Continuous motion pictures were taken of the flow in order to record the deformation of the pellets. A set of reflection mirrors was used so that simultaneous views in two perpendicular directions can be obtained from the motion pictures.

Model II. Flat-ended tube perpendicular to external flow: This model is shown in Fig. 9. The reservoir and the inner rotating cylinder were made

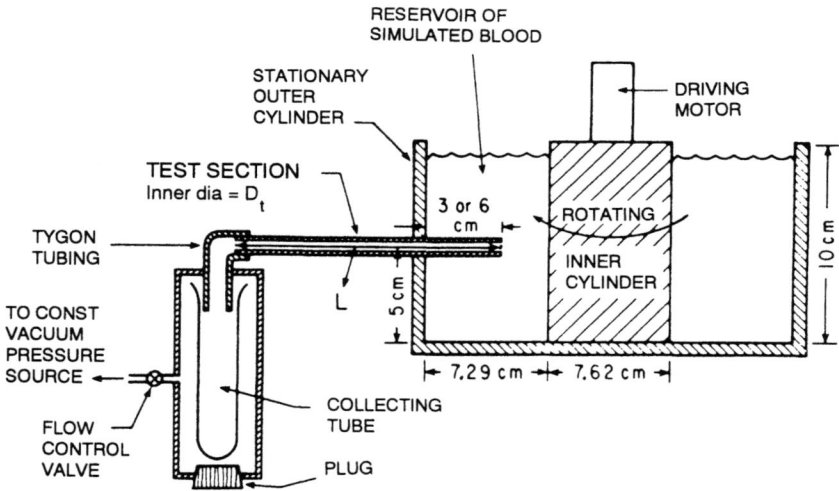

Figure 9. Schematic diagram of test Model II.

of Lucite. The test tubes which simulated the capillary blood vessels were made of steel.

The ends of the tubes were sharply cut. One end of the tube was inserted into the reservoir, which was a cylindrical tank containing a rotating inner core which imparted a steady shear flow to the fluid. This was considered to be the mainstream flow from which the test tube drains the mixture of the gelatin pellets and the silicone fluid. The axis of the test tube was perpendicular to the streamlines of the flow in the reservoir. The flow in the test tube was controlled by a vacuum suction applied through a valve to the collecting chamber. The outflow from the test tube was passed through a short segment of Tygon tubing to a "collecting tube." The steady rate of flow was computed by dividing the volume of the suspension collected with the time interval of collection. The flow rate divided by the cross-sectional area of the vessel yielded the mean flow velocity V_T in the test section.

In this test it was necessary to know the velocity of flow in the large tank. This was obtained by taking motion pictures of the particles in the tank from above. To see the motion clearly, a few pellets were made of white color to distinguish them from the rest of the pellets which were black. The velocity distribution was computed from the position changes of these white pellets.

The experiment consisted in collecting a quantity of fluid after a steady-state flow was established for a given suction pressure and for a given tangential velocity of flow in the reservoir at the entry section of the test tube. The fluid collected in the collecting tube was used to determine the discharge hematocrit, that collected in the test tube was used to determine the tube hematocrit. For the discharge hematocrit measurement the collected sample was centrifuged for 1 h at 3000 g, and the packed particle volume and the total volume were measured and their ratio measurement computed. For the tube hematocrit measurement the flow in the capillary tube was abruptly stopped by simply removing the suction pressure clamping the Tygon tubing and plugging off one end of the tube; then the fluid was blown out, and the number of particles was counted manually. The ratio of the computed particle volume to the total volume was taken as the tube hematocrit.

Results

Model I: In Fig. 10A the ratio of hematocrit in the "capillary" tube to that in the reservoir is plotted against the ratio of particle to tube diameters. The solid curves show the mean. The flags represent the bounds of standard errors. Results for four reservoir hematocrits ranging from 25 to 55% are shown. The data indicate that the ratio of the hematocrits in the tube to that in the reservoir (H_T/H_F) increases with increasing ratios of cell-to-tube diameter (D_c/D_t); i.e. for a given cell diameter, the tube hematocrit increases when the tube diameter decreases. This is a reversal of the Fahraeus effect. It is more evident at lower reservoir concentration. The ratio H_T/H_F increases to 1.07±0.05 when $D_c/D_t=1.13$ if H_F is 25% and to 0.98±0.03 if H_F is 45%.

Fig. 10B shows the variation of H_T/H_F as a function of H_F for fixed values of D_c/D_t. It shows a tendency toward increased relative hematocrit in the capillary tube at lower reservoir hematocrits.

Model II: Fig. 11 shows the ratio of the discharge hematocrit H_D to the reservoir hematocrit H_F as a function of the ratio of the mainstream velocity at the

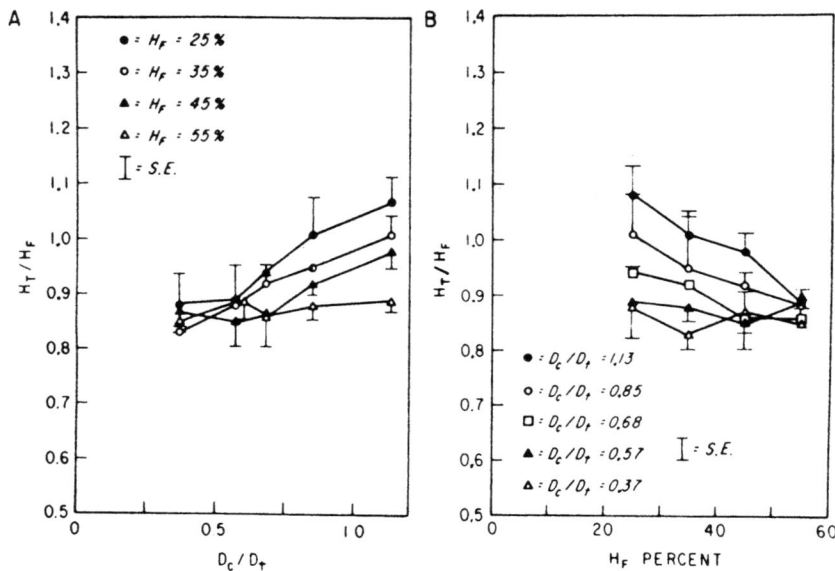

Figure 10. Experimental results from model I, showing mean and standard errors of the mean. A) Plot of ratio of tube hematocrit (H_T) to feed hematocrit (H_v) as a function of ratio of cell diameter (D_c) to tube diameter (D_t). B) Tube hematocrit-to-feed hematocrit ratio (H_t/H_v) as function of feed hematocrit (H_v).

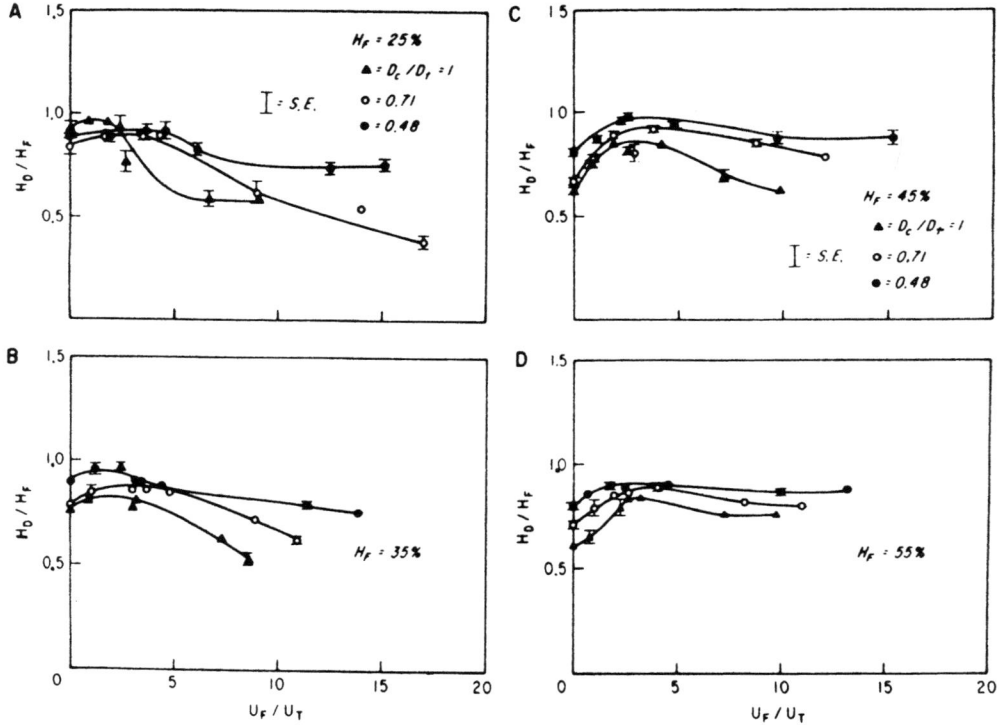

Figure 11. Experimental results from model II, showing mean and standard errors of the mean. A) Relative discharge hematocrit as function of ratio of tangential velocity in the reservoir at tube entrance and mean tube flow velocity for feed hematocrit H_F=25%. B) H_v=35%. C) H_v=45%. D) H_v=55%.

entrance section of the tube V_F to the mean velocity of flow in the tube V_T for four different values of H_F and three different tube sizes (D_c/D_E=0.48, 0.71, and 1.0). The curves are seen to be bell shaped. From a certain initial value of H_D/H_F the exit hematocrit rises to a peak value at a ratio of V_F/V_T in the range 1-4. For higher velocity ratios (V_F/V_T, 4-17) the discharge hematocrit declines.

In Fig. 12 the ratio of the hematocrit in the capillary tube to that in the reservoir is plotted against the ratio of the mainstream velocity in the reservoir at the tube entrance to the mean velocity of flow in the tube for a reservoir hematocrit of 25% in Model II, with three values of the particle-to-tube diameters. A comparison of Figs. 12 and 11A shows that the capillary tube hematocrit and discharge hematocrit are not equal.

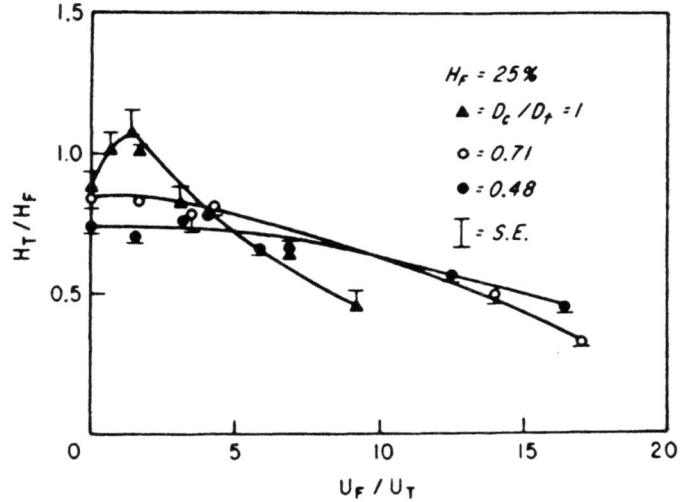

Figure 12. Relative tube hematocrit in model II as function of ratio of tangential velocity in the reservoir at tube entrance to mean tube flow velocity for feed hematocrit H_v=25%.

Figures 11 and 12 show an initial rise of discharge hematocrit and tube hematocrit when the mainstream velocity at the tube entrance increases. However, these hematocrits decrease with further increase of the mainstream velocity. Thus it is clear that the flow condition at the entrance of the tube influences the entry of particles in the tube. These figures show that the lower the reservoir hematocrit, the greater is the effect of relative velocity at the entry section on the ratio of discharge hematocrit to reservoir hematocrit. At a given velocity, the smaller the tube size, the smaller is the relative discharge hematocrit.

These experimental results show that the seemingly simple motion of suspended flexible particles from a reservoir into a small tube is a complicated phenomenon for which a simplified generalization is likely to be wrong. Our results show that the Fahraeus effect does not continue indefinitely. As the ratio of cell diameter to tube diameter D_c/D_t approaches 1 or exceeds 1, the effect is reversed. Furthermore, the hematocrit in the tube can be greatly influenced by the flow condition just outside the entrance section of the tube. This suggests that in an arteriole-capillary junction, the ratio of the hematocrit in the capillary to that in the arteriole can be influenced by the rate of flow in the arteriole.

Application of our results to in vivo blood flow must be based on the principles of model testing, which requires both geometric and dynamic similarity. In geometry, our gelatin pellets do resemble red blood cells but the entry sections of the simulated capillary blood vessels are idealized and cannot claim any similarity to the in vivo junctions. The dynamic similarity requires the simulation of (a) the ratio of the inertia force to viscous force, and (b) the red cell flexibility. The former, expressed in terms of the Reynolds number, is simulated. The Reynolds number of blood flow in the capillary blood vessels, based on the diameter of the vessel and the viscosity of the plasma, and a flow velocity of the order of 1 mm/s, is about 10^{-2}-10^{-4}. Our models have the same range. The flexibility of real red blood cells, however, is not known. We make our gelatin pellets as soft as practical, but probably still very much stiffer than red blood cells. Lacking a precise knowledge of the prototype, we do not know how to simulate it. Nevertheless, the existence of an inversion of the Fahraeus effect in model tests should be of interest to physiology, because it serves as a warning that the widely believed Fahraeus-Lindqvist effect - that the apparent viscosity of blood in blood vessels decreases monotonically with decreasing vessel diameter - does not remain true when the vessel diameter becomes so small as to approach the cell diameter. For those organs in which the hematocrit distribution has physiological significance, it would be worthwhile to measure the hematocrit in arterioles, capillaries, and venules.

VELOCITY DISTRIBUTION IN MICROVESSELS

Usually we wish to know the flow rate of whole blood. Thus if red cell velocity is measured, we should know the ratio of the average red cell velocity to the mean velocity of whole blood. In a large blood vessel this ratio approaches 1, since the cells, small compared to the tube, will be convected with the plasma. In a microvessel, the ratio will be larger than 1 because the cells tend to concentrate in the center of the tube where the velocity is higher than the average. If the vessel is so very narrow that the red cells effectively plug up the tube, then the particle velocity and the whole blood velocity tend to equal again, because the "leak back" mechanics of the plasma in the narrow gap between the red cell and the tube wall is small.

To clarify this ratio, a model experiment was done by Yen and Fung[30] with gelatin pellets simulating red cells and silicone fluid simulating plasma. The results are shown in Fig. 13. The ratio of particle velocity to mean flow velocity is plotted against the particle velocity for different values of D_c/D_t and hematocrit. The dotted line in the figure represents the mean value of all data for each value of D_c/D_t. It is seen that the particle-to-mean velocity ratio is independent of the flow rate, but dependent on D_c/D_t. When the hematocrit is 10%, the velocity ratio is 1.21 when $D_c/D_t=1$. It increases to 1.47 when $D_c/D_t=0.67$ and to 1.49 when $D_c/D_t=0.5$. The scatter of the data is seen to be very large reflecting the fact that the exact value of the velocity ratio v/V depends on the incidental factor of particle configuration

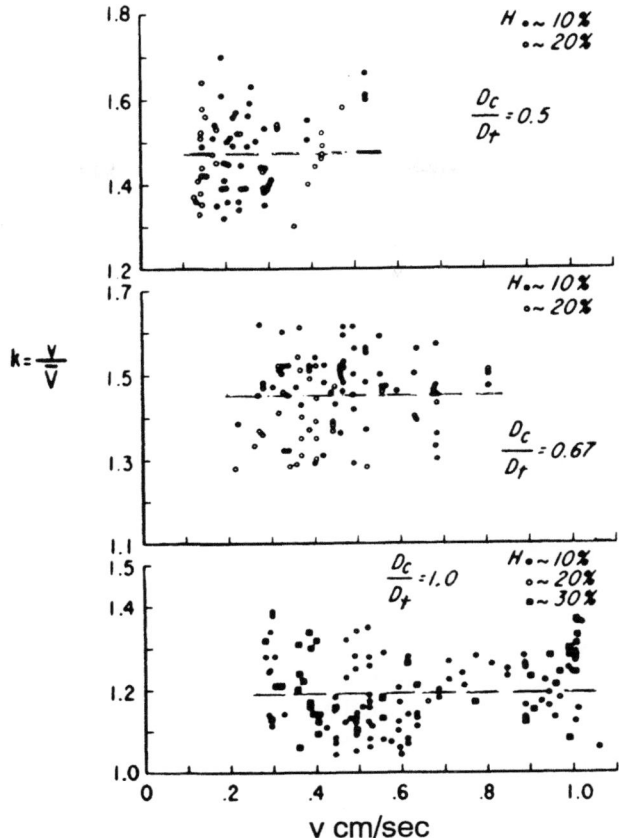

Figure 13. Experimental results of test model for $D_c/D_t = 0.5$, 0.67, and 1.0. The ratio k between particle velocity v and mean flow velocity V is plotted against v for different feed-tube hematocrits H_t. Dotted lines are mean values of k computed for combined values of H_t for each value of D_c/D_t.

relative to the tube. The red cells deform severely in such a flow, and can assume all kinds of configuration with their axes parallel to the cylinder axis or perpendicular to it or at some angles in between.

THE VELOCITY-HEMATOCRIT RELATIONSHIP

Blood flow in capillary blood vessels is often seen to be unsteady. At one instant, one may see a capillary without any red blood; in the next instant he may see the same capillary packed with flowing red cells. The capillaries without red cells are often thought to be closed by precapillary sphincters. Another cause of flow fluctuation is the basic particulate nature of the blood. To clarify this, let us consider a narrow capillary blood vessel with a diameter comparable to that of the red blood cells. Let this blood vessel bifurcate into two daughter branches at a certain point. At the bifurcation point, a red blood cell flowing down the capillary must flow into one of the two daughter branches. The forces that determine into which branch the red cell must go are the resultants of the pressure and the shear stress in the fluid. Both of these resultants tend to pull the cell into the foster channel. In the simplest case of rigid spherical balls flowing down a circular cylindrical tube that branches, this can be analyzed and demonstrated very easily. Using balls of diameter about 90% of the diameter of the tube, Fung[10] has shown that the faster branch gets all the balls, the slower branch gets none. The outcome is less certain if the ratio of ball diameter to tube diameter is smaller or if the balls are not spherical, or if they are deformable rather than rigid. Red blood cells are not spherical and are very flexible. Hence they demand a more careful analysis.

Two types of experiments may be done to resolve the problem quantitatively: a) model experiments and b) observation in vivo. A model experiment is simple and accurate, but is not real microcirculation. It can be used to demonstrate a principle and to obtain accurate data to assess theoretical calculations. In vivo observations in some animals would be desirable, but there are difficulties. 1) A very accurate method of determining the blood vessel diameter in vivo is not available. 2) The whole blood contains white blood cells which though much fewer in number than red blood cells, do exercise considerable influence on the red cell distribution, because some of them are much bigger than the red cells and are almost spherical in shape and hence would move into the faster branch much more readily. These white cells often carry a string of closely packed red cells behind them. 3) The only method available for measuring the velocity of flow in capillaries is by measuring the velocity of red cells. Since the velocity in a capillary vessel is not uniform, certain corrections are needed. These difficulties are not so serious as to completely invalidate in vivo observations, but they do complicate the problem.

A model approach is adopted by us. In our model, the plasma is simulated by silicone fluid, the red blood cells are simulated by gelatin pellets, and the blood vessel is simulated by Lucite tubes. The dimensions of the model are designed according to the principle of kinematic and dynamic similarity of flow. The entire model is large enough to permit visual observation of flow details. Instrumentation is simple and measurements are quite accurate.

A schematic diagram of the experimental apparatus is given in Fig. 14. The apparatus consists of a closed reservoir of simulated blood with an inverted T tube of Lucite attached as shown. The tube simulates a bifurcating capillary blood vessel. All branches are circular cylinders and of the same diameter. A shut-off valve is positioned between the reservoir and the testing tube to stop the flow when desired. In the experiments, three different sizes of Lucite testing tubes were used. A source of constant pressure, connected to the top of the reservoir, drove the fluid through the test tube. The ends of the tubes, as well as the junction of the T (the bifurcation point), were carefully machined. The radius of curvature of the edge of junction (the corner of the branches at bifurcation) was measured in a comparator after cutting open a test tube. The radius of curvature was found to be 0.03-0.05 mm or about 1-1.5% of the simulated red cell diameter, 3-5% of the cell thickness. A motor-driven stirrer kept the suspension uniformly mixed during the experiment.

The hematocrit and velocity relationship in bifurcation branches was determined by allowing a gelatin pellet suspension to flow through the T tube. The experiment consisted of collecting quantities of fluid from the two branches in a steady-state condition. Flow can be stopped by closing the shut-off valve after each test.

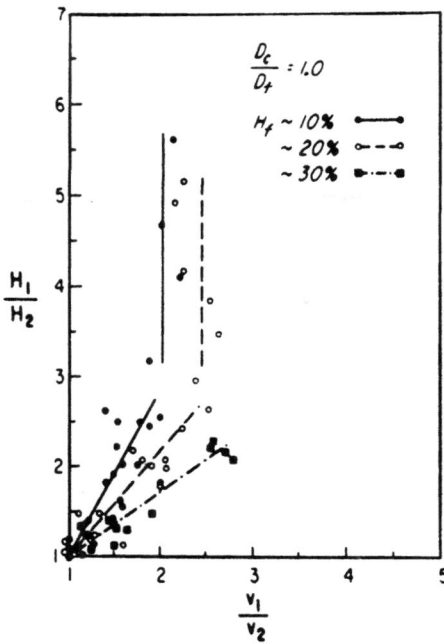

Figure 14. Experimental results of test model for $D_c/D_t = 1.0$. H_1/H_2 is plotted against v_1/v_2 for three different feed-tube hematocrits H_f.

Velocities in the two branches by cutting off a segment of one branch at a time. In each experiment, the suspension of pellets in the reservoir was carefully mixed and stirred to a homogeneous concentration. The driving pressure was kept at a constant value. The pellet velocities were obtained by measuring the time required for a pellet to travel a known distance. The mean velocity of flow in the tube was computed from the volume discharge collected in a known interval of time. The pellet velocities and the mean velocities of flow were measured simultaneously in two tubes (branch 1 and branch 2). The discharge hematocrit was measured by centrifuging the collected sample, and obtaining the ratio of the volume of the packed particles to the total volume.

The hematocrit in each tube was determined from their respective discharge hematocrit using the following relationship[26].

$$\frac{discharge\ hematocrit}{tube\ hematocrit} = \frac{mean\ speed\ of\ pellets}{mean\ speed\ of\ tube\ flow} \qquad (11)$$

Hematocrit (H_f) of the feeding tube was calculated as follows. Let V_f be the mean flow velocity in the feeding tube, V_1 be that in branch 1, V_2 be that in branch 2; and let v_f be the particle velocity in the feeding tube, v_1 be that in branch 1, v_2 be that in branch 2. Without loss of generality, we may let $v_1 \geq v_2$, and $V_1 \geq V_2$.

Then, by the principle of conservation of mass of the mixtures and of the pellets, we have, since the cross-sectional area of all branches are the same,

$$\bar{V}_f = \bar{V}_1 + \bar{V}_2 \qquad (12)$$

$$H_f v_f = H_1 v_1 + H_2 v_2 \tag{13}$$

The mean flow velocities $\overline{V}_f$, $\overline{V}_1$, $\overline{V}_2$ are related to the particle velocities v_f, v_1, v_2. These velocities were measured and the ratios $v_1/\overline{V}_1$, $v_2/\overline{V}_2$ were computed. The results are shown in Fig. 13.

It is seen that the ratio of particle velocity to mean flow velocity depends strongly on the ratio of cell diameter to tube diameter D_c/D_t, but does not vary significantly with hematocrit. Nor does it vary with flow rate. Hence for each value of D_c/D_t, we can substitute Eq. (12) by

$$v_f = v_1 + v_2 \tag{14}$$

On substituting Eq. (13) into Eq. (14), we obtain the feed tube hematocrit H_f

$$H_f = \frac{H_1 v_1 + H_2 v_2}{v_1 + v_2} \tag{15}$$

Figures 14-16 show the experimental results. In these figures, the ratio H_1/H_2 is plotted against the ratio v_1/v_2 for different feed-tube hematocrits H_f. The symbols

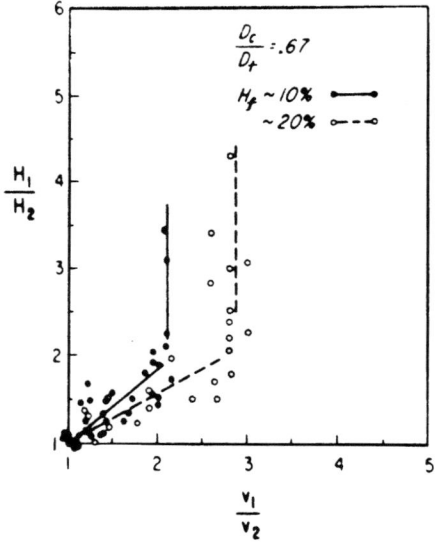

Figure 15. Experimental results of test model for $D_c/D_t = 0.67$. H_1/H_2 is plotted against v_1/v_2 for two different feed-tube hematocrits H_f.

v_1, v_2, H_1, H_2 are the particle velocities and tube hematocrits in the branches 1 and 2, respectively. It is seen that for narrow capillaries (with the cell diameter/tube diameter = D_c/D_t = 1.0, 0.67 and 0.5) the branch with faster flow (branch 1) will have more cells.

In general, for velocity ratios sufficiently smaller than a critical value, the hematocrit ratio can be expressed by a linear relationship given by

$$\frac{H_1}{H_2} - 1 = a\left(\frac{v_1}{v_2} - 1\right) \tag{16}$$

The dimensionless constant, a, depends on a number of factors, the most important of which are i) the ratio of cell diameter to tube diameter, ii) the shape and rigidity of the pellets, and iii) the hematocrit in the feeding tube.

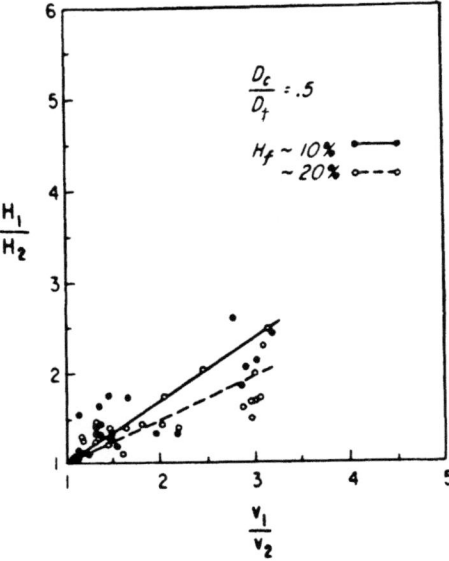

Figure 16. Experimental results of test model for $D_c/D_t = 0.5$. H_1/H_2 is plotted against v_1/v_2 for two different feed-tube hematocrits H_f.

Figs. 14 and 15 also show that for velocity ratios beyond a critical value nearly all the cells flow into the faster branch. The smaller the feeding tube hematocrit is, the smaller is the critical velocity ratio at which this phenomenon occurred. The critical velocity lies in the range of 2-3.0 when $D_c/D_t = 1.0$, with exact value depending on the feed hematocrit. The critical velocity ratio becomes higher when D_c/D_t decreases.

The result of the experiment also has other application. It is quite popular today to use radioactive microspheres in blood flow to measure the distribution of blood flow in different parts of organs. Now the distribution of microspheres into bifurcating vessels of relevant sizes must be subjected to exactly the same kind of fluid mechanical forces as the red blood cells in capillaries. Hence microspheres of 50 μm diameter will be controlled by the velocity distribution in blood vessels of diameters in the neighborhood of 50-100 μm. If the measurement of radiation is proportional to the total number of microspheres per unit volume, then it will indicate the distribution of blood flow only if the numerical concentration of microspheres per unit volume of blood is uniform; and this happens only if the velocity of flow at points of bifurcation into different branches is equal. If the velocities are unequal, then the numerical concentration of the microspheres will be different in different branches.

In particular, if the velocity ratio exceeds the critical value, which for spheres lies in the range 1-2 if the diameters of the spheres approach that of the vessels, then the distribution of number of spheres will be entirely out of proportion to the ratio of the flows.

These remarks point to a consideration that should be given to the microsphere methods, but its importance depends on the size of the spheres as well as the vascular structure of the specific organs. It also depends on how precise the measurements are intended to be. If the field of view for each measurement is so large that all the variations of concentration of the spheres are averaged out, then our remark has no importance. If the field of view is made smaller and smaller so that a more detailed picture of the concentration distributed is measured, then our remark may become important. By the same token, the larger the spheres, the larger will be the vessels in which the velocities will influence concentration; hence the larger will be the volumes in which local variations of concentration occur to vitiate the simple interpretation of radiation measurements.

Theoretical Analysis of the Experiment

An approximate analysis of the forces that act on a particle at a bifurcation point will clarify the theme of this experiment. We shall show that if the mean velocities of flow in the two daughter branches are $\overline{V}_1$ and $\overline{V}_2$, with $\overline{V}_1 > \overline{V}_2$, then the resultant forces due to pressure and shear stress are both proportional to $\overline{V}_1 - \overline{V}_2$, and act in the direction of the daughter tube with the higher velocity, $\overline{V}_1$. The difference between a rigid spherical particle and a flexible disk will be seen to be significant. The effect of the angle of branching will be shown to be rather minor. Finally, the question of how well the flexibility of the pellets used in the experiment simulates the flexibility of red blood cells in microcirculation is answered.

First, consider disk-shaped flexible pellets. Such a pellet can assume all kinds of orientation in the tube. Let us consider two configurations, one in which the plane of the disk is perpendicular to the axis of the tube, and another in which the plane of the disk is parallel to the axis of the tube. Stated in terms of the angle between the axes of symmetry of the disk and the tube, the angle is zero in the first case and 90° in the second. All other orientations are between these two. For conciseness, we say that the first case is plugging, the second is edge-on.

Consider the plugging case. Since the local velocity of flow must vanish on the wall of the tube and is maximal at the center, the flexible pellet must deform. Consistent with the principle of minimum potential energy, the deformation of a thin disk will be planar, i.e., its midplane will deform into a developable cylindrical surface, which requires no membrane strain in the midplane of the disk. All the energy of deformation is used in bending, and not in stretching the midplane. (If h represents the thickness of the disk, then the bending energy is proportional to h^3, whereas stretching energy is proportional to h. When $h \to 0$, it becomes much easier to bend than to stretch.) Let us sketch such a deformed pellet in a branching tube as shown in Fig. 17, A or B. Let us consider three sections of the tubes as indicated by dotted lines in the figure: one in the mother tube and two in the daughter tubes, all at a distance L from the point of bifurcation. In a flow of a homogeneous viscous fluid before the arrival of the pellet, let the pressures at these sections be P_0, P_1, P_2, respectively. Obviously $P_0 > P_2 > P_1$ on account of the assumption $\overline{V}_1 > \overline{V}_2$. When the pellet arrives at the junction, let the pressures acting on the pellet surface facing the

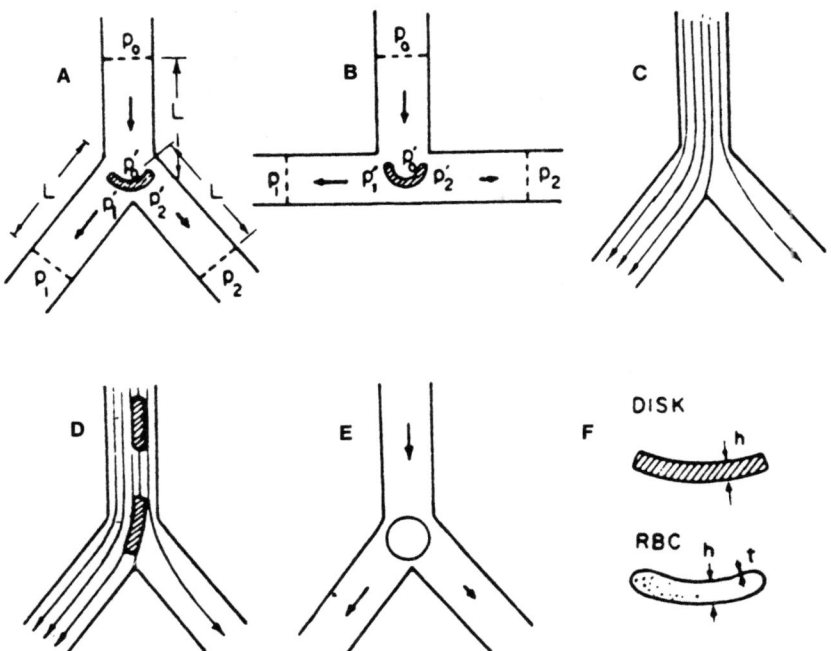

Figure 17. Illustration of the features of modeling. Flow is directed downward. Velocity in left branch (tube 1) is faster than that in right branch (tube 2). A: a plugging pellet at the fork. B: similar to A, but with branching angle equal to 90°. C: streamline pattern in the case $V_1=2V_2$, as seen in the middle section. D: pellets flowing edge on, and astride of the dividing stream surface. E: flow of a rigid spherical particle at the fork. F: cross sections of a pellet in the model and a red blood cell. Pellet is a solid disk of homogeneous elastic material. Red blood cell is considered to be a thin shell filled with a liquid.

three branches be denoted by P_0', P_1', P_2', respectively, as indicated in the figure. Now, the values of P_0', P_1', P_2' depend on how tightly the pellet plugs up the flow. If the flow is momentarily plugged completely, then we have

$$P_0' - P_0 \qquad P_1' - P_1, \qquad P_2' - P_2 \qquad (17)$$

If the flow can leak past the pellet then $P_0'<P_0$, $P_1'>P_1$, $P_2'>P_2$. But they are related, $P_1'<P_2'<P_0'$; and $P_0'=P_1'=P_2'$ only in the limiting case of no pellet. The force that acts on the pellet that pulls the pellet into branch 1 is equal to the difference of pressures acting on the two sides of the pellet multiplied by the projected area in branch 1; i.e., $(P_0'-P_1')\pi D_c^2/8$. Here D_c is the diameter of the pellet. The resultant force due to the pressure difference is F_p

$$F_p - \left(P_0'-P_1'\right)\frac{\pi D_c^2}{8} - \left(P_0'-P_2'\right)\frac{\pi D_c^2}{8} - \left(P_2'-P_1'\right)\frac{\pi D_c^2}{8} \qquad (18)$$

and is directed into branch 1 because $P_1'<P_2'$. We can convert this expression into the mean velocities of flow as follows. Assuming Poiseuille flow, we have

$$\bar{V}_1 = \frac{D_t^2}{32\mu} \frac{(P_0 - P_1)}{2L} \qquad \bar{V}_2 = \frac{D_t^2}{32\mu} \frac{(P_0 - P_2)}{2L} \qquad (19)$$

where μ is the apparent viscosity of the fluid. If Eq. 17 is applicable (flow plugged momentarily), then a combination of Eqs. 18 and 19 yields

$$F_p = 8\mu L \pi \frac{D_c^2}{D_t^2} (\bar{V}_1 - \bar{V}_2) \qquad (20)$$

In general, Eq. 17 is not applicable, and the best we can say is that

$$F_p = const \cdot \frac{D_c^2}{D_t^2} (\bar{V}_1 - \bar{V}_2) \qquad (21)$$

This shows that the resultant force is proportional to the velocity difference of the two branches, and to the square of the particle-to-tube diameter ratio.

Next, consider the edge-on case. Here it is necessary to consider the shear stress and velocity gradient, and hence the velocity distribution. The easiest way to understand the velocity distribution in a branching flow is to look at the streamline pattern. A streamline is a line whose tangent is parallel to the local velocity vector. A sketch of the streamline of a flow of a homogeneous fluid in a branching tube is shown in Fig. 17C. We can quantitate the spacing of these lines as follows. Consider an axisymmetric flow in a circular cylinder. The volume rate of flow in a circular ring of radius r and a small thickness Δr is equal to the area of the ring $2\pi r \Delta r$ multiplied by the velocity v at r

$$\Delta Q = 2\pi r \cdot v \cdot \Delta r \qquad (22)$$

We define a stream function $\psi(r)$, which is the value assigned to a streamline at radius r by the requirement that the difference between the stream functions is equal to the flow between the streamlines

$$\Delta Q = \psi(r+\Delta r) - \psi(r) = \Delta \psi = \frac{\partial \psi}{\partial r} \Delta r \qquad (23)$$

A combination of Eqs. 22 and 23 yields

$$v = \frac{1}{2\pi r} \frac{\partial \psi}{\partial r} \qquad (24)$$

Now we can assign $\psi = 0$ at the center and $\psi = Q$ at the tube radius and space the lines $\psi = const$ in such a way that Eqs. 22-24 are satisfied. These are shown in Fig. 17C for the case in which $V_1 = 2V_2$ for the flow of a homogeneous fluid. Note that there is a dividing streamline that stops at the stagnation point of the flow, located at the junction of the daughter tubes. The surface formed by all the dividing streamlines is called the dividing stream surface. All fluid on the left of the dividing stream surface goes into tube 1, all on the right into tube 2.

Now, let there be edge-on pellets in the tube. The edge-on configuration of the pellets will not cause large disturbances in the streamline pattern. If a pellet lies entirely to the left of the dividing stream surface it will flow into tube 1. If it lies to the right, it flows into tube 2. The borderline cases, in which the pellets lie astride the dividing stream surface, as shown in Fig. 17D, are the ones that command attention.

The flow around an edge-on pellet near the bifurcation point can be seen from the streamlines shown in Fig. 17D. The streamlines are crowded on the left side of the pellet, indicating that the flow velocity is high there. On the right the streamlines are rarified. This feature is accentuated especially in the neighborhood of the bifurcation point. It follows that the velocity is higher on the left side of the pellet than that on the right side, because Eq. 24 shows that v is proportional to the density of the streamlines. Since the velocity vanishes on the tube wall, the velocity gradient, i.e., the shear gradient, is higher on the left. Therefore, according to the Newtonian law of viscosity the shear stress acting on the left side of the pellet surface is larger than that on the right. The resultant shear force is obtained by integrating the shear stress over the area. Now the shear gradients on the two sides can be assumed to be proportional to V_1/D_t and V_2/D_t. The resultant shear force is therefore

$$F_s = const \cdot \frac{D_c^2}{D_t} (\bar{V}_1 - \bar{V}_2) \qquad (25)$$

and is directed toward tube 1.

Pellets in other configurations will be acted on by stress resultants similar to those given by Eqs. 21 and 25, though different in magnitude. Thus, it is seen that the net pressure and shear forces tend to pull pellets to the faster side, more definitely for pellets in the plugging configuration, less for edge-on ones. The magnitude of the force depends on D_c/D_t. The ratio of the number of cells in edge-on configuration to that in plugging configuration varies with D_c/D_t. If $D_c/D_t > 1$, practically all cells will be edge-on. The fraction of all edge-on pellets to be astride the dividing stream surface increases as D_c/D_t increases and as $\bar{V}_1/\bar{V}_2$ increases.

If the pellets were rigid spheres, as shown in Fig. 17E, the resultant shear and resultant pressure forces would both help to pull the pellets into tube 1. If $D_c/D_t \to 1$, the critical velocity ratio of the rigid spheres also tends to 1. The critical velocity ratio increases as D_c/D_t decreases. For flexible disks, it is expected that the critical velocity ratio tends to increase as the pellets become more flexible. These expectations are verified by the experimental results.

Finally, we must answer the question how flexible are the pellets used in our experiments compared with the red blood cells in microcirculation. According to Ref. 28, the gelatin used in our experiments has a shear modulus of 7.7×10^3 dyn/cm^2 and a Young's modulus of $E = 2.31 \times 10^4$ dyn/cm^2. On the other hand, the Young's modulus of the red cell membrane is not so well known. The literature has been reviewed by Skalak et al[22] who estimated the Young's modulus of the red cell membrane to be 10^4 dyn/cm^2. Hence the elasticity of the gelatin and the red cell membrane are quite similar.

To determine a dimensionless parameter that serves to compare the effect of flexibility of the pellets in the bifurcating flow, we refer again to Fig. 17. The important deformation of the pellet is bending. Suppose a pellet is bent to a curvature 1/R, R being the radius of curvature. The dimensionless parameter that determines the shape of the pellet is the product of the cell diameter and the curvature: D_c/R. Now for a disk of thickness h, diameter D_c, and Young's modulus

E, the curvature is

$$\frac{1}{R} = \frac{bending\ moment}{ED_c h^3/12} \tag{26}$$

The bending moment is proportional to the shear stress x area x moment arm, and the shear stress is equal to the coefficient of viscosity of plasma, μ, times the shear gradient. Thus

$$bending\ moment = \mu\left(\frac{V}{D_t}\right)\frac{\pi D_c^2}{8}\left(\frac{D_c}{2}\right) \tag{27}$$

Hence, the flexibility for the model is

$$\left(\frac{D_c}{R}\right)_{model} = \left(D_c \frac{\pi\mu V D_c^3}{16 D_t} \frac{12}{ED_c h^3}\right)_{model} = \left(\frac{3\pi\mu V D_c^3}{4ED_t h^3}\right)_{model} \tag{28}$$

For the red blood cell, when it is considered as a disk-shaped thin shell of uniform thickness h (RBC thickness) with wall thickness t (cell membrane thickness), see Fig. 17F, the curvature is

$$\frac{1}{R} = \frac{bending\ moment}{ED_c h^2 t/4} \tag{29}$$

Hence the flexibility parameter for the red blood cell is

$$\left(\frac{D_c}{R}\right)_{RBC} = \left(\frac{\pi\mu V D_c^3}{4ED_t h^2 t}\right)_{RBC} \tag{30}$$

For dynamic similarity between the model and the prototype, we must have

$$\left(\frac{D_c}{R}\right)_{model} = \left(\frac{D_c}{R}\right)_{RBC} \tag{31}$$

i.e.

$$\left(\frac{3\pi\mu V D_c^3}{4ED_t h^3}\right)_{model} = \left(\frac{\pi\mu V D_c^3}{4ED_t h^2 t}\right)_{RBC} \tag{32}$$

Note that the velocity of flow enters into this equation. Typical values of our experimental model are $\mu=100$ poise, $V=1$ cm/s, $D_c=3.2$ mm, $h=1$ mm, $D_t=4$ mm, $E=2.31 \times 10^4$ dyn/cm^2. Typical values for RBC in capillaries are $\mu=0.01$ poise, $V=0.1$ cm/s, $D_c=8$ μm, $h=1.5$ μm, $D_t=10$ μm, $t=50$Å, $E=10^4$ dyn/cm^2. Hence, typically,

$$\left(\frac{D_c}{R}\right)_{RBC} - 3.6 \quad (V \text{ in } mm/s) \qquad (33)$$

$$\left(\frac{D_c}{R}\right)_{model} - 0.8 \quad (V \text{ in } cm/s) \qquad (34)$$

Thus, to simulate a capillary blood flow at 1 mm/s the flow velocity in the model should be 4.3 cm/s.

The simulation of the flexibility parameter of the red blood cell is quite good in our model according to our present understanding of the red cell membrane.

The theory presented above is simple because many details are left in a qualitative state. For future refinement, the probability of finding red cells in the edge-on and plugging or other configurations must be analyzed, and the resultant pressure and shear forces must be more accurately assessed.

REFERENCES

1. Barbee JJ, Cokelet GR: The Fahraeus effect, Microvas. Res., 3:1-21, 1971.
2. Chien S, Dellenback RJ, Usami S, Gregersen MI: Plasma trapping in hematocrit determination. Differences among animal species, Proc. Soc. Exptl. Biol. Med., 119:1155-1158, 1965.
3. Dientenfass L: Inversion of the Fahraeus Lindqvist phenomenon in blood flow through capillaries of diminishing radius, Nature, 215:1099-1100, 1967.
4. Fahraeus R: The suspension stability of the blood, Physiol. Rev., 9:241-274, 1929.
5. Fahraeus R, Lindqvist T: The viscosity of the blood in narrow capillary tubes, Am. J. Physiol., 96:562-568, 1931.
6. Fung YC: Blood flow in the capillary bed, J. Biomech., 2:353-373, 1969.
7. Fung YC, Sobin SS: Theory of sheet flow in the lung alveoli, J. Appl. Physiol., 26:472-488, 1969.
8. Fung YC, Sobin SS: Pulmonary alveolar blood flow, Circ. Res., 30:470-490, 1972.
9. Fung YC: Stochastic flow in capillary blood vessels, Microvascular Res., 5:34-38, 1973.
10. Fung YC: Interaction of blood cells with vessel walls in microcirculation, Thrombosis Research, 8(Suppl. II):315-327, 1976.
11. Fung YC: Biomechanics: Mechanical Properties of Living Tissues, Springer-Verlag, New York, 1981.
12. Fung YC: Biodynamics: Circulation, Springer-Verlag, New York, 1984.
13. Fung YC: Biomechanics: Motion, Flow, Stress, and Growth, Springer-Verlag, New York, 1990.
14. Gaehtgens, P: Flow of blood through narrow capillaries: Rheological mechanisms determining capillary hematocrit and apparent viscosity, Biorheology J., 17:183-189, 1980.
15. Goldsmith HL: Deformation of human red cells in tube flow, Biorheology, 7:235-242, 1971.

16. Gregersen MI, Bryant CA, Hammerle WE, Usami S, Chien S: Flow characteristics of human erythrocytes through polycarbonate sieves, Science, 157:825-827, 1967.
17. Lee JS: Slow viscous flow in a lung alveoli model: J. Biomech., 2:187-198, 1969.
18. Lee JS, Fung YC: Modeling experiments of a single red blood cell moving in a capillary blood vessel, Microvascular Res., 1:221-243, 1969.
19. Lew HS, Fung YC: Plug effect of erythrocytes in capillary blood vessels, Biophys. J., 10:80-99, 1970.
20. Lipowsky HH, Usami S, Chien S: In vivo measurement of "apparent viscosity" and microvessel hematocrit in the mesentery of the cat, Microvascular Res., 19:297-319, 1980.
21. Schmid-Schoenbein GW, Skalak R, Usami S, Chien S: Cell distribution in capillary networks, Microvascular Res., 19:18-44, 1980.
22. Sheshadri V, Sutera SP: Concentration changes of suspensions of rigid spheres flowing through tubes, J. Colloid Interface Sci., 27:101-110, 1968.
23. Skalak RA, Tozeren A, Zarda RP, Chien S: Strain energy function of red blood cell membrane, Biophys. J., 13:245-246, 1973.
24. Sobin SS, Tremer HM, Fung YC: Morphometric basis of the sheet-flow concept of the pulmonary alveolar microcirculation in the cat, Circ. Res., 26:397-414, 1970.
25. Sobin SS, Fung YC, Tremer H, Rosenquist TH: Elasticity of the pulmonary intervalveolar microvascular sheet in the cat, Circ. Res., 30:440-450, 1972.
26. Sutera SR, Seshadri V, Croce PA, Hochmuth RM: Capillary blood flow II. Deformable model cells in tube flow, Microvascular Res., 2:420-433, 1970.
27. Svanes J, Zweifach RW: Variations in small blood vessel hematocrits produced in hypothermic rats by micro-occlusion, Microvasc. Res., 1:210-221, 1969.
28. Yen RT, Fung YC: Model experiments on apparent blood viscosity and hematocrit in pulmonary alveoli, J. Appl. Physiol, 35:510-517, 1973.
29. Yen RT, Fung YC: Inversion of Fahraeus effect and effect of mainstream flow on capillary hematocrit, J. Appl. Physiol., 42(4):578-586, 1977.
30. Yen RT, Fung YC: Effect of velocity distribution on red cell distribution in capillary blood vessels, Am. J. Physiol., 235(2):H251-H257, 1978.

FLUID DYNAMICS AND THROMBOSIS

Steven M. Slack, Winnie Cui and Vincent T. Turitto

Department of Biomedical Engineering
Memphis State University, Memphis, TN 38152

INTRODUCTION

Maintenance of blood fluidity and, conversely, arrest of bleeding at sites of vascular injury, depend on complex interactions between clotting factors, platelets, endothelial cells, the subendothelial matrix and proteins of the fibrinolytic pathway. Products generated by one system, e.g., the coagulation cascade, invariably either enhance or inhibit reactions in other pathways. Cleavage of high molecular weight kininogen, for example, amplifies the intrinsic pathway of coagulation and simultaneously releases bradykinin, a peptide with strong vasoconstrictive properties[1]. Endothelial cells secrete prostacyclin (PGI_2), a potent inhibitor of platelet activity, and both tissue plasminogen activator (t-PA) and t-PA inhibitor, components that play central roles in fibrinolysis[2]. In addition, endothelial cells express thrombomodulin, a protein that plays a role in the thrombin-dependent activation of protein C and which ultimately results in the inactivation of Factors Va and VIIIa and release of plasminogen activators[3]. Moreover, thrombin generated through the coagulation cascade converts fibrinogen to fibrin, which serves to stabilize developing hemostatic plugs but which also functions as a platelet activating agent.

Our knowledge of the interplay between components of the hemostatic system arises largely from detailed studies of isolated components in in vitro experiments. An aspect that enables the hemostatic or thrombotic process to proceed efficiently and rapidly in vivo, which has received much less attention, is flow and the consequent transport of material from one site to another as a result of flow. Platelets and blood-borne substances must travel from the bloodstream to the injured vascular surface if they are to adhere and initiate plug formation; factors secreted from endothelial cells must diffuse away from the vascular wall if they are to interact with their receptors; plasminogen must be present near the surface if it is to be incorporated into a developing clot and subsequently activated to a form capable of degrading fibrin. Thus, blood flow is crucial for efficient hemostasis and strongly influences thrombotic events occurring normally and in pathological disorders such as atherosclerosis.

Despite the apparent recognition of flow as a primary contributor to thrombotic events, the properties of flowing systems are not broadly appreciated. Even less recognized is the fact that flow can affect proteins and cells in more subtle ways than by simply augmenting their transport from one region to another. The intentions of this article are to 1) introduce the basic concepts of fluid dynamics essential for an understanding of the multi-faceted role of flow in surface-mediated thrombosis; 2) examine current hypotheses regarding the mechanism(s) by which flow alters the activity of blood and vascular factors; 3) briefly review the extensive evidence demonstrating that selected hemostatic defects are intrinsically dependent on flow; and 4) discuss how flow through regions of stenosed vessels, such as are present at sites of atherosclerotic lesions or during transient ischemic attacks, may contribute to the severity of the disorder.

FLUID DYNAMICS - BASIC CONCEPTS

Thrombosis at sites of vascular injury involves complex interactions between blood-borne elements, e.g., platelets and coagulation proteins, and constituents of the vascular wall. Blood flow constitutes an essential element of thrombotic processes by virtue of its ability to augment mass transport beyond that resulting from diffusion alone. Prior to examining the details of protein and platelet transport in flowing blood, a brief discussion of fluid dynamics will be presented. More thorough treatments of blood rheology and fluid dynamics are available in the literature[4-8].

Blood flow through the vasculature is frequently modeled as that of a constant density, constant viscosity fluid moving steadily through a rigid tube. In reality, blood vessels are compliant and blood is a complex suspension of cells and proteins whose viscosity depends strongly on hematocrit and flow rate. Under many conditions, however, those simplifying assumptions are not unreasonable and analytical solution of the appropriate fluid dynamic equations becomes possible. The results of such an analysis indicate that the blood velocity profile through a vessel is parabolic, with velocity being maximal at the centerline and falling to zero at the vessel wall. Actual measurements indicate that the profile is somewhat flattened out in the central portion of the vessel, primarily because of the presence of red blood cells[7]. In addition, the red blood cells and platelets are not distributed uniformly across the vessel, the former being present primarily in the central core of the vessel and the latter being excluded to the region nearer the wall. Careful measurements of radial concentration profiles indicate that platelets, in the absence of any surface reaction, exhibit a 5-10 fold increased concentration within 10 mm of a tube wall, whereas red cells have a decreased concentration in this zone[9,10]. This cellular distribution forms rapidly upon entering the tube and is a function of red cell concentration and wall shear rate. The occurrence of this gradient has been attributed to a "drift" force being imposed on the platelets which is dependent on radial position, but the nature and origin of this force remain obscure[10].

Two parameters used to characterize blood flow are the shear rate (γ) and the shear stress (τ). Shear rate is, by definition, the negative of the radial velocity gradient (the slope of the velocity profile at a given radial position) and reflects how rapidly the fluid velocity changes as one moves in a radial direction. The shear rate is zero at the centerline and maximal at the vessel wall. Thus, the difference in velocity of adjacent blood elements increases as one approaches the vessel wall. The shear stress is, for a Newtonian fluid, i.e., one whose viscosity is independent of shear rate, the product of shear rate and viscosity and reflects the force per unit area acting in a direction tangential to the vessel wall. At shear rates greater than ~ 200 s^{-1}, the viscosity of blood is relatively constant (3–4 cP); at lower shear rates, however, the viscosity increases, an effect due to red blood cell crosslinking by fibrinogen (Rouleaux formation)[7]. Values for shear rate range from 50 – 3000 s^{-1} in the human body (Table 1). In a partially occluded blood vessel, however, the shear rate, and

Table 1. Various fluid dynamic parameters in the human vasculature.[1,2]

Vessel	Diameter (cm)	Volumetric flow rate, Q (cm³/sec)	Reynolds Number, Re	Wall Shear Rate γ_w (sec^{-1})	Wall Shear Stress τ_w (dyne/cm²)
Ascending aorta	2.3-4.5	350	3200-6000	50-300	2-10
Femoral artery	0.5	4	280	30	10
Common carotid	0.6	5	330	25	9
Carotid sinus	0.5	3	240	240	8
External carotid	0.4	2	180	330	12
Small arteries	0.03	4 x 10^{-3}	2	1300	50
Arterioles	3 x 10^{-3}	3 x 10^{-6}	4 x 10^{-2}	1600	60
Capillaries	6 x 10^{-4}	0.8 - 6 x 10^{-8}	0.4-3 x 10^{-3}	250-2000	-

[1] Modified from Reference 7.
[2] Numbers are time-averaged values.

consequently the shear stress, may be an order of magnitude greater. This occurs because shear rate varies inversely with the third power of vessel diameter, at least in tubular flow with rigid walls. Thus, reduction of vessel diameter by half would result in an 8-fold increase in shear rate. A similar dependence of shear values on vessel dimensions exists in irregular geometries, such as may be found in a stenosed artery, and is discussed in a later section.

The processes responsible for transporting platelets and proteins from the bloodstream to the damaged vessel are diffusion and convection. The former is a random process and results from the inherent thermal energy possessed by the cell or protein. Diffusion coefficients vary among proteins but generally range from $10^{-6} - 10^{-7}$ cm^{2}s^{-1} (Table 2). Blood cells are much larger than proteins and hence have

Table 2. Experimental and theoretical values of diffusivities for various plasma proteins and blood cells.

Protein or Cell	Molecular Weight (Daltons)	Diffusivity (D_{20w}) x 10^{-7} cm^{2}s^{-1}
Albumin	66,500	6.1
α_1-Antitrypsin	51,000	5.2
α_2-Macroglobulin	725,000	2.41
Transferrin	79,600	5.0
Fibrinogen	340,000	1.97
Immunoglobulin G	150,000	4.0
Platelet[1]	-	0.0158
Red Blood Cell[1]	-	0.00675

[1]Diffusivity calculated from the Stokes-Einstein relationship assuming a plasma viscosity of 1.2 cP at 37°C.

smaller diffusion coefficients. Based on the Stokes-Einstein equation, which relates diffusion coefficients to particle size, temperature and fluid viscosity, platelet diffusivities of 10^{-9} cm^{2}s^{-1} are expected[4]. It has been determined, however, that platelet diffusivities are 2 - 3 orders of magnitude larger than predicted from classical theory; the increased effective diffusivity exhibited by platelets, which has been observed to be a function of shear rate, results from red blood cell collisions and deformations which augment platelet transport through the blood[7]. The enhancement is presumably additive to the classical Brownian motion and is maximal for cells and other large macromolecules, for whom diffusion coefficients are normally quite small. Both the enhanced diffusion and concentration gradient contribute to the ability of platelets to be transported to solid surfaces.

The net transport of mass within a fluid is accomplished by diffusion due to Brownian motion and by convection, which arises from the motion of the bulk fluid. In most cases, where only the material near the vessel wall can reach the the surface, convective transport is represented by the magnitude of the wall shear rate. In convective transport proteins and platelets are carried down the length of the tube at velocities dependent on their radial position. Their paths are, in streamline flow, parallel to the surface of the tube. In the absence of flow, mass transport depends solely on diffusion. Because diffusion (or Brownian motion) is a random process, the time required for the necessary clotting factors and platelets to reach the vascular wall would be exceedingly long. As an example, the root mean square distance traveled by a platelet at 37°C in one second is ~ 0.6 μm, using a platelet diffusion coefficient of 1.58 x 10^{-9} cm^2/sec[7]. This means that if we photographed a single platelet in a static, dilute suspension, waited 1 second, and photographed the platelet again, and repeated this process an infinite number of times, the average distance traversed by the platelet per observation would be 0.6 μm. In contrast, the distance traveled by a platelet, located 5 μm from the wall of a vessel of diameter 0.3 mm, after 1 second at a volumetric flow rate of 3.5×10^{-3} cm^3/sec (all typical values), is almost 6500 μm. This example serves to emphasize that transport by flow (convection) dominates that by diffusion. Nevertheless, as proteins and cells approach the vascular wall, their velocity decreases very rapidly and becomes zero at the surface. Thus, near the wall, diffusion assumes a greater importance.

DIRECT ACTIVATION OF BLOOD AND VASCULAR ELEMENTS BY FLOW

Although flow serves to enhance the transport of materials from the bulk fluid to the surface of blood vessels, the functional behavior of certain cells and proteins is affected by flow or, more precisely, by shear stresses generated as a consequence of flow. Such behavior is non-classical in the sense that the kinetics of small molecular reactions are generally not dependent on the local fluid dynamics. In an early study of platelet responses to shear stresses, Brown et al.[11], using a rotational viscometer, showed that shear stresses as low as 50 dyne/cm^2 induced release of ATP, ADP, and serotonin from platelet dense granules and caused platelet aggregation. Exposure to a shear stress of 100 dyne/cm^2 resulted in platelet lysis and higher levels of shear stress caused platelet fragmentation. Because large shear stresses (> 50 dyne/cm^2) can possibly exist in the vasculature (for instance in prosthetic heart valves or in partially occluded arteries), this study clearly suggests that platelet damage induced by shear stress may be a mechanism by which platelet activation and subsequent thrombus formation proceed.

In more recent studies by Moake et al.[12,13] and Ikeda and his colleagues[14], human platelets, while shearing in a viscometer, were shown to aggregate in the absence of exogenous agonists. Relatively low shear stresses (12 dyne/cm^2) induced reversible aggregate formation, a process apparently mediated by the binding of fibrinogen to the platelet membrane receptor GP IIb-IIIa since platelet rich plasma from subjects with afibrinogenemia or Glanzmann's thrombasthenia did not exhibit aggregation whereas that from subjects with Bernard-Soulier syndrome and severe von Willebrand disease did. Exposure of platelets to larger shear stresses (108 dyne/cm^2) resulted in irreversible aggregates, a process that appeared to depend on von Willebrand Factor (vWF) interactions with the platelet membrane glycoproteins GP Ib-IX and GP IIb-IIIa. Nevertheless, although these studies indicate that shear stresses can be considered a platelet agonist, the mechanism by which such forces induce platelet aggregation is not clear.

One possible explanation is that shear stresses deform the platelet membrane receptors, exposing otherwise cryptic binding regions to adhesive ligands such as fibrinogen and vWF. The ability of shear forces to degrade proteins in solution is well-known; in fact, Tirrell[15] has concluded that physiological shear stresses can and do alter the structure (and therefore function) of numerous plasma proteins. Further, Charm and Wong[16] demonstrated that shearing of plasma in a rheogoniometer resulted in significant losses in the clottability of fibrinogen, a phenomenon they suggested might be responsible for the normal turnover of fibrinogen in vivo.

Endothelial cell function has also been shown to be dependent on flow conditions.

The concept that endothelial cells simply provide a physical barrier which prevents direct contact of blood with procoagulant elements within the subendothelial matrix has changed considerably in recent years. For example, it is now known that endothelial cells actively secrete both procoagulant and anticoagulant compounds[2]. This realization, coupled with the recent interest in endothelial seeding of prosthetic materials as a means of fabricating more blood-compatible vascular grafts, has spawned studies of biochemical and physical factors influencing endothelial cell function. Of particular interest is the ability of shear stresses of physiologic magnitude to alter endothelial cell behavior with respect to its role in hemostasis. Moreover, studies examining the effect of larger stresses, such as might be generated at sites of atherosclerotic lesions, on the integrity of endothelial cells may suggest therapeutic strategies for minimizing thrombus deposition following plaque disruption or for preventing restenosis of blood vessels following angioplasty.

Recent evidence indicates that endothelial cells can synthesize prostacyclin (PGI_2), a potent naturally occurring inhibitor of platelet aggregation, tissue-plasminogen activator (t-PA), t-PA inhibitor, and platelet-derived growth factor (PDGF). That physiologic shear stresses can regulate the synthesis and expression of these hemostatic and mitogenic factors by endothelial cells has been amply demonstrated. For instance, Nollert et al.[17], studying the uptake and metabolism of arachidonic acid by human umbilical vein endothelial cells in the absence or presence of flow, reported that while uptake of arachidonic acid did not depend on flow, incorporation of arachidonic acid into diacylglycerol and phosphatidylinositol was increased in those cells subjected to shear. Moreover, preferential metabolism of arachidonic acid into PGI_2 was observed in cells exposed to flow, a process which may reflect an attempt by endothelial cells to mitigate platelet aggregation in high shear regions such as those present in partially stenosed vessels.

Work by Diamond and colleagues demonstrated that shear stresses of 15 and 25 $dyne/cm^2$ but not 4 $dyne/cm^2$ increased secretion of t-PA by endothelial cells 2-3 fold over the basal secretion rate[18]. In contrast, the release of t-PA inhibitor by endothelial cells was unaffected over the same range of shear stresses. In addition, several groups have reported that endothelial cells secrete, in response to physiologic shear stress, vasodilators with properties similar to those of nitric oxide[19-21]. Again, these observations suggest that endothelial cells, in response to shear stresses generated by flow, attempt to alleviate the forces acting upon them by releasing factors that will augment clot dissolution (t-PA) or dilate the narrowed vessel (nitric oxide).

In contrast to those studies, Hsieh et al.[22] have reported that endothelial cells subjected to physiological shear stresses (16 $dyne/cm^2$) synthesize elevated levels of platelet-derived growth factor (PDGF) mRNA. Since PDGF is a potent smooth muscle cell mitogen[23], this result suggests a mechanism by which smooth muscle cells proliferate at sites of vascular trauma, such as those resulting from angioplasty, and cause a reduction in lumen diameter.

As will be discussed more thoroughly in the next section, atherosclerotic lesions that extend into the lumen of an artery can disrupt the normal velocity profile, generating regions of both high and low shear stresses. It has been observed that platelet deposition is greatest at the apex of the occluding thrombus, where shear stresses are highest[24], whereas fibrin formation appears to be enhanced in recirculation zones, where stresses are the smallest[25]. Recently, the ability of flow to modulate the activity of coagulation factors has been examined, with emphasis on the conversion of Factor X to Xa by the tissue factor:Factor VIIa:phospholipid complex[26]. In these experiments, the interior surface of very narrow glass capillary tubes is coated with phospholipid and tissue factor. A solution of Factor X and Factor VIIa is then perfused through the tubes at controlled shear rates, the effluent collected, and the amount of Factor Xa quantitated with a chromogenic assay. Interestingly, the generation of Factor Xa increased with increasing shear rate (and shear stress). Based on Michaelis-Menten kinetics and their preliminary data, the authors determined that the maximum reaction velocity (V_{max}) increased with increasing shear rate and suggested that shear stresses somehow alter the affinity of Factor X for the tissue factor:Factor VIIa:phospholipid complex. Because Factor Xa, in combination with Factor Va and phospholipid (the prothrombinase complex), converts prothrombin to thrombin, which in turn cleaves fibrinogen into fibrin monomers, these data suggest that fibrin deposition would be increased in regions of high shear. This conflicts with the experimental observation alluded to above that more fibrin is found in low-shear recirculation zones. One explanation may be that increased shear stresses reduce the activity of the prothrombinase complex, resulting in less thrombin formation, but the exact mechanism is currently unknown.

BLOOD FLOW AND PLATELET-SURFACE INTERACTIONS

The exposure of surfaces under well-defined flow conditions has permitted the characterization of the resultant platelet adhesion and thrombus growth on the surface in terms of basic fluid dynamic and mass transport parameters and represents an in vitro simulation of in vivo events. As indicated, flow determines the rate of transport of both cells and proteins to and away from the surface and imposes fluid forces that could affect the removal of material deposited at the surface. These forces may also lead to physical activation of secretory processes[11,27]. Fluid dynamic considerations have become increasingly important in studying platelet reactivity. Platelet adhesion and thrombus formation on vascular surfaces have been shown to increase significantly with increasing shear rate[28-32]. Also, a fluid dynamic basis for the observed differences in venous, arterial, and microcirculatory thrombosis and for the shear rate-dependent platelet adhesion defect in von Willebrand's disease has been demonstrated[29-32].

Factor VIII/von Willebrand Factor

Factor VIII/von Willebrand factor is a bimolecular complex composed of Factor VIII, the clotting protein, and von Willebrand factor (vWF), a protein found in reduced quantities in the plasma, the α-granules of platelets, and the blood vessels of patients with severe von Willebrand's disease. The activity of vWF is associated with platelet retention in glass bead columns, ristocetin-induced platelet agglutination, and skin bleeding time, but is not related to clotting ability (Factor VIII:C)[33]. Patients with von Willebrand's disease exhibit abnormal adherence of their platelets to subendothelium and to native collagen fibrils in flowing blood[33-35]; the magnitude of the defect in platelet adhesion increases with the local shear rate[29]. Treatment of normal blood with antibodies to vWF prodeuces a similar flow (shear)-dependent defect in adhesion to subendothelium[36]. Similar findings have been observed in pigs with von Willebrand's disease[37]. Addition of purified fractions of vWF partially corrects the adhesion defect on subendothelium[38]. It has also been demonstrated that platelet adhesion to subendothelium depends on the amount of plasma vWF bound to the subendothelial surface[39,40]. The mechanism by which vWF influences platelet adhesion is not clear at present but apparently involves the ability of platelets to attach initially to the subendothelium[41], perhaps through its interaction with collagen Type VI in the vessel wall[42], although the ability of vWF to promote platelet spreading on subendothelium has also been suggested[40,43].

Von Willebrand factor consists of a series of multimers of molecular weights ranging from 0.8 to 20 x 10^6 daltons[44-46]. The absence of the higher molecular weight fragments in plasma characterizes type II von Willebrand's disease and these patients also exhibit a defect in the ability of their platelets to attach to subendothelium[29,38]. The extent to which the adhesion defect in type II von Willebrand's disease may be related to the absence of higher weight multimers is not clear, since levels of vWF are also quantitatively reduced in this disorder. Experiments conducted with differing size distributions of vWF multimers suggest that multimer size may not be the only consideration in platelet adhesion[47] and that surface configuration or orientation may be important in determining its functional behavior.

Platelets from patients with von Willebrand disease have normal quantities of glycoproteins in their membrane fractions[48], whereas patients with Bernard-Soulier syndrome have normal levels of vWF in their plasma. The presence of an adhesion defect in both of these disorders suggests the interesting possibility that the glycoprotein fraction (GP Ib-IX) of the platelet membrane may combine with vWF that is bound to the vessel surface or with endogenous vessel wall vWF that is present in the subendothelium to produce normal platelet adhesion. In this regard, the site on the vWF molecule that binds to GP Ib, as well as the vWF binding region on GP Ib, have been identified[49-51].

The platelet-platelet events following platelet adhesion have also been demonstrated to depend on vWF[37,52]. Indeed, the size and extent of platelet thrombi at high shear rates (1300 s^{-1}), as well as shear-induced irreversible platelet aggregation[13,14,53,54], have been shown to be mediated by vWF rather than fibrinogen. The apparent preference for vWF under high shear forces may be related to the multimeric nature of the protein in that the multiple, closely spaced binding sites offered by vWF may be more suited to resisting shear removal forces in flowing blood.

BLOOD FLOW AND THROMBOSIS IN VASCULAR PATHOLOGY

It has been known for many years that proliferative lesions and atherosclerotic plaque develop at sites of altered blood flow, such as are characteristic of vessel bifurcations[55,56]. Thus, blood flow may play a role in the early stages of atherogenesis. The extent to which thrombosis, as characterized by the abnormal interaction of platelets or coagulation proteins with the vessel wall, is involved in the development of the atherosclerotic lesion is still controversial, although the discovery that platelets contain platelet-derived growth factor (PDGF), which is a potent mitogen for vascular smooth muscle cells[57], suggests that intimal thickening at sites of atherosclerotic lesions may result from platelet adhesion, activation and subsequent release of PDGF. However, it has become abundantly clear from cardiovascular studies that the event precipitating both heart attack and stroke in man is related to thrombosis in the vicinity of the stenotic lesion[56]. Narrowed vascular lumens produced by the slowly progressing disease of atherosclerosis can be rapidly occluded by the deposition of platelets and fibrin. Severe narrowing of the lumen alters the natural streamline blood flow in the vessel and potentially leads to order of magnitude changes in the forces imposed on the vessel wall.

The fluid dynamic conditions in the vicinity of a narrowed artery are extremely complex due to the irregular nature of the lesion. Moreover, lesion geometry has not been adequately characterized and thus a true understanding of the local fluid dynamics is not attainable at present. However, typical (limiting) profiles may be developed in order to understand better the nature of flow in and around stenoses and its potential influence on subsequent blood-surface interactions. Idealized models of stenoses are also needed for the investigation of blood-vessel interactions related to such lesions. One such model has been developed by Badimon and colleagues[58]; the authors report markedly enhanced platelet deposition at the apex of the stenosis produced by a vessel wall obstructing the lumen cross-sectional area (% reductions in diameter were 0, 35, 55, and 80%). We have used a two-dimensional model to determine the local shear conditions for this model, using a Reynolds number typical of flow in the perfusion device and both a sinusoidal and step geometry to represent the stenosis. In this model, the peak wall shear rates are increased by 1-2 orders of magnitude compared with planar flow (0 % stenosis). The shear profiles are distinctly different in the two geometries, emphasizing the importance of lesion geometry in determining flow and potentially thrombotic responses.

CONCLUSIONS

Blood flow plays a multi-facted role in thrombosis. Convective transport of the necessary clotting factors and platelets from the bulk fluid to the vessel surface ensures the rapid interaction of these elements with the exposed subendothelial matric and subsequent arrest of bleeding. Furthermore, blood flow serves to remove activated products from the area, allowing for their inactivation by inhibitors. Fluid flow also appears to modulate thrombosis by directly affecting the cells involved in the process. Thus, shear stresses can activate platelets, resulting in aggregate growth, and can regulate the generation and expression by endothelial cells of prothrombotic and antithrombotic agents. In addition, flow appears to influence the coagulation cascade, although this phenomenon has not been studied as extensively. In partially occluded blood vessels, in which inordinately large shear stresses are present, flow may induce platelet activation and thrombus formation, resulting in total vessel occlusion and ischemia. Alternatively, PDGF released from shear-activated platelets, which has been shown to induce smooth muscle cell proliferation, may contribute to intimal thickening and subsequent vessel narrowing. Finally, shear stresses may cause plaque rupture at sites of atherosclerotic lesions, thereby exposing prothrombotic elements to blood. Further studies under known flow conditions will be necessary to sort out the dominant mechanism(s) by which flow influences thrombosis.

REFERENCES

1. Griffin JH, Cochrane CG: Recent advances in the understanding of contact activation reactions, Sem. Thromb. Hemostas., V:254–272, 1979.
2. Gimbrone, MA: Vacular endothelium: Nature's blood-compatible container, Ann. N.Y. Acad. Sci., 516:5–11, 1987.

3. Esmon CT: The protein C anticoagulant pathway, Arteriosclerosis and Thrombosis, 12:135-145, 1992.
4. Bird RB, Stewart WE, Lightfoot EN: Transport Phenomena, Wiley, New York, 1960.
5. Whitmore RL: Rheology of The Circulation, Pergamon Press, New York, 1968.
6. Middleman S: Transport Phenomena in the Cardiovascular System, John Wiley & Sons, New York, 1972.
7. Goldsmith HL, Turitto VT: Rheological aspects of thrombosis and haemostasis: Basic principles and applications, Thromb. Haemostas., 55:415-435, 1986.
8. Lowe GDO: Nature and clinical importance of blood rheology, In: Clinical Blood Rheology (Ed., Lowe GDO), CRC Press, Boca Raton, 1-10, 1988.
9. Eckstein EC, Tilles AW, Millero FJ: Conditions for the occurrence of large near-wall excesses of small particles during blood flow, Microvasc. Res., 36:31-39, 1988.
10. Eckstein EC, Belgacem F: Model of platelet transport in flowing blood with drift and diffusion terms, Biophys. J., 60:53-69, 1991.
11. Brown CH, et al.: Morphological, biochemical, and functional changes in human platelets subjected to shear stress, J. Lab. Clin. Med., 86:462-471, 1975.
12. Moake JL, et al.: Involvement of large plasma von Willebrand factor (vWF) multimers and unusually large vWF forms derived from endothelial cells in shear stress-induced platelet aggregation, J. Clin. Invest., 78:1456-1461, 1986.
13. Moake JL, et al.: Shear-induced platelet aggregation can be mediated by vWF released by platelets, as well as exogenous large or unusually large vWF multimers, requires adenosine diphosphate, and is resistant to aspirin, Blood, 71:1366-1374, 1988.
14. Ikeda Y, et al.: The role of von Willebrand factor and fibrinogen in platelet aggregation under varying shear stress, J. Clin. Invest., 87:1234-1240, 1991.
15. Tirrell MV: Stress-induced structure and activity changes in biologically active proteins, J. Bioeng., 2:183-193, 1978.
16. Charm SE, Wong BL: Shear degradation of fibrinogen in the circulation, Science, 170:466-468, 1970.
17. Nollert MU, et al.: The effect of shear stress on the uptake and metabolism of arachidonic acid by human endothelial cells, Biochim. Biophys. Acta, 1005:72-78, 1989.
18. Diamond SL, Eskin SG, McIntire LV: Fluid flow stimulates tissue plasminogen activator secretion by cultured human endothelial cells, Science, 243:1483-1485, 1989.
19. Tesfamariam B, Cohen RA: Inhibition of adrenergic vasoconstriction by endothelial cell shear stress, Circ. Res., 63:720-725, 1988.
20. Cooke JP, et al.: Flow stimulates endothelial cells to release a nitrovasodilator that is potentiated by reduced thiol, Am. J. Physiol., 259:H804-H812, 1990.
21. Buga GM, et al.: Shear stress-induced release of nitric oxide from endothelial cells grown on beads, Hypertension, 17:187-193, 1991.
22. Hsieh HJ, Li NQ, Frangos JA: Shear stress increases endothelial platelet-derived growth factor mRNA levels, Am. J. Physiol., 260:H642-H646, 1991.
23. Jawien A, et al.: Platelet-derived growth factor promotes smooth muscle migration and intimal thickening in a rat model of balloon angioplasty, J. Clin. Invest., 89:507-511, 1992.

24. Lassila R, et al.: Dynamic monitoring of platelet deposition on severely damaged vessel wall in flowing blood, Arteriosclerosis, 10:306-315, 1990.
25. Weiss HJ, Turitto VT, Baumgartner HR: Role of shear rate and platelets in promoting fibrin formation on rabbit subendothelium., J. Clin. Invest., 78:1072-1082, 1986.
26. Gemmell CH, Nemerson Y, Turitto VT: The effects of shear rate on the enzymatic activity of the tissue factor-factor VIIa complex, Microvasc. Res., 40:327-340, 1990.
27. Goldsmith HL, Yu SS, Marlow J: Fluid mechanical stress and the platelet, Thromb. Diath. Haemorrh., 34:32, 1975.
28. Baumgartner HR: The role of blood flow in platelet adhesion, fibrin deposition, and formation of mural thrombi, Microvasc. Res., 5:167, 1973.
29. Weiss HJ, Turitto VT, Baumgartner HR: Effect of shear rate on platelet interaction with subendothelium in citrated and native blood. I. Shear-rate dependent decrease of adhesion in von Willebrand's disease and the Bernard-Soulier syndrome, J. Lab. Clin. Med., 92:750, 1978.
30. Baumgartner HR, Turitto VT, Weiss HJ: Effects of shear rate on platelet interaction with subendothelium in citrated and native blood. II. Relationships among platelet adhesion, thrombus dimensions, and fibrin formation, J. Lab. Clin. Med., 95:208, 1980.
31. Baumgartner HR, Sakariassen K: Factors controlling thrombus formation on arterial lesions, Ann. NY Acad. Sci., 454:162, 1985.
32. Turitto VT, Baumgartner HR: Platelet interaction with subendothelium in flowing rabbit blood: Effect of blood shear rate, Microvasc. Res., 17:38-54, 1979.
33. Furkin BG, Howard MA: On von Willebrand's disease, Br. J. Haemotol., 32:151, 1976.
34. Baumgartner HR, Tschopp TB, Weiss HJ: Platelet interaction with collagen fibrils in flowing blood. II. Impaired adhesion-aggregation in bleeding disorders- a comparison with subendothelium, Thromb. Haemostas., 37:17, 1977.
35. Tschopp TB, Weiss HJ, Baumgartner HR: Decreased adhesion of platelets to subendothelium in von Willebrand's disease, J. Lab. Clin. Med., 83:296, 1974.
36. Baumgartner HR, Tschopp TB, Meyer D: Shear rate-dependent inhibition of platelet adhesion and aggregation on collagenous surfaces by antibodies to human factor VIII/von Willebrand factor, Br. J. Haemotol., 44:127, 1980.
37. Badimon L, et al.: Role of von Willebrand factor in mediating platelet-vessel wall interaction at low shear rates: The importance of perfusion conditions, Blood, 73:961, 1989.
38. Weiss HJ, et al.: Correction by factor VIII of the impaired platelet adhesion to subendothelium in von Willebrand's disease, Blood, 51:267, 1978.
39. Sakariassen KS, Bolhuis PA, Sixma JJ: Human blood platelet adhesion to artery subendothelium is mediated by factor VIII-vWF bound to the subendothelium, Nature, 279:636, 1979.
40. Bolhuis PA, et al.: Binding of factor VIII-von Willebrand factor to human arterial subendothelium precedes increased platelet adhesion and enhances platelet spreading, J. Lab. Clin. Med., 97:568, 1981.
41. Turitto VT, Weiss HJ, Baumgartner HR: Decreased platelet adhesion on vessel segments in von Willebrand's disease: A defect in initial platelet attachment, J. Lab. Clin. Med., 102:551, 1983.
42. Rand JH, et al.: 150 kDa vWF binding protein extracted from human vascular subendothelium is a type VI-like collagen, Clin. Res., 36:417a.

43. Neuwenhuis HK, et al.: Deficiency of platelet membrane glycoprotein Ib associated with a decreased platelet adhesion to subendothelium: A defect in platelet spreading, Submitted.
44. Zimmerman TS, Ruggeri ZM: Von Willebrand's disease, Prog. Hemostas. Thromb., 6:203, 1982.
45. Mourik JAV, et al.: Factor VIII, a series of homologous oligomers and a complex of two proteins, Thromb. Res., 4:155, 1974.
46. Hoyer LW, Shinoff JR: Factor VIII-related protein circulates in normal human plasma as high molecular weight multimers, Blood, 55:1056, 1980.
47. Sixma JJ, et al.: Adhesion of platelets to human artery subendothelium: Effects of factor VIII-von Willebrand factor of various multimeric composition, Blood, 63:128, 1984.
48. Jenkins CSP, et al.: Platelet membrane glycoproteins implicated in ristocetin-induced aggregation: Studies of the proteins on platelets from patients with Bernard-Soulier syndrome and von Willebrand's disease, J. Clin. Invest., 57:112, 1976.
49. Roth GJ: Developing relationships: Arterial platelet adhesion, glycoprotein Ib, and leucine-rich glycoproteins, Blood, 77:5-19, 1991.
50. Fujimura Y, Ruggeri ZM, Zimmerman TS: Structure and function of human von Willebrand factor, In: Coagulation and Bleeding Disorders: The Role of Factor VIII and von Willebrand Factor (Ed., Zimmerman TS, Ruggeri Z), Marcel Dekker, New York, 77, 1989.
51. Girma J, et al.: Structure-function relationship of human von Willebrand factor, Blood, 70:605, 1987.
52. Turitto VT, Weiss HJ, Baumgartner HR: Platelet interaction with rabbit subendothelium in von Willebrand's disease: Altered thrombus formation distinct from defective platelet adhesion, J. Clin. Invest., 74:1730-1741, 1984.
53. Peterson DM, et al.: Shear-induced platelet aggregation requires von Willebrand factor and platelet membrane glycoproteins Ib and IIb-IIIa, Blood, 69:625-628, 1987.
54. Ikeda Y, et al.: Inportance of fibrinogen and platelet membrane glycoprotein IIb/IIIa in shear-induced platelet aggregation, Thromb. Res., 51:157-163, 1988.
55. Packham MA, Mustard JF: The role of platelets in the development and complications of atherosclerosis, Semin. Hematol., 23:8-26, 1986.
56. Woolf N, Davies MJ: Interrelationship between atherosclerosis and thrombosis, In: Thrombosis in Cardiovascular Disorders (Ed., Fuster V, Verstraete M), W.B. Saunders, Philadelphia, 41-78, 1992.
57. Ross R, et al.: A platelet dependent serum factor that stimulates the proliferation of arterial smooth muscle cells in vitro, Proc. Natl. Acad. Sci. USA, 71:1207-1210, 1974.
58. Badimon L, Badimon JJ: Mechanisms of arterial thrombosis in nonparallel streamlines: Platelet thrombi grow on the apex of stenotic severely injured vessel wall, J. Clin. Invest., 84:1134-1144, 1989.

BLOOD RHEOLOGY, BLOOD FLOW AND DISEASE

Gordon D. O. Lowe

University Department of Medicine
Royal Infirmary, Glasgow G31 2ER, U.K.

INTRODUCTION

Hemorheology is the study of the flow behavior and flow properties of blood. There is increasing interest in the contribution of blood rheology factors to not only the rare hematological hyperviscosity syndromes; but also to the much commoner problems of ischemia of the heart, brain and limbs[1-5]. The present review considers briefly the measurements of blood rheology which can be made in clinical practice and their relationships to blood flow and cardiovascular disease.

RHEOLOGICAL MEASUREMENTS IN CLINICAL PRACTICE

Recommendations for standardization in measurement of plasma and whole-blood viscosity have been published by an expert panel of the International Committee for Standardization in Hematology[6]. Capillary viscometers may be preferred to rotational viscometers in clinical practice, because of their accuracy, precision, and speed of measurement; and lesser demand for blood volume, expenditure, and expertise of laboratory personnel[4,5]. A small blood sample, anticoagulated with dipotassium EDTA and stored at 4°C, is adequate for accurate measurement of high-shear, whole blood viscosity, hematocrit and plasma viscosity for up to four days[7]. The coefficients of variation for these measurements within a blood sample are 1-2%.

Measurements of low-shear blood viscosity, other measures of red cell aggregation (e.g. erythrocyte sedimentation rate, photometry) and red cell deformability are less robust, requiring measurement within eight hours of collection of a similarly stored and anticoagulated sample in order to avoid in vitro changes. Recommendations for measurement of red cell aggregation and deformability have been published[6,8].

Recent epidemiological studies have defined the distributions of these rheological measurements in random population samples, as well as their relationships to demographic variables and to cardiovascular risk markers[7,9-15].

Plasma viscosity, measured at 37°C in a capillary viscometer, is almost twice the viscosity of water (U.K. population mean 1.33 mPa s; range 1.15-1.50 mPa s). Plasma viscosity increases with age (due to increases in fibrinogen, lipoproteins and immunoglobulins), and also with cigarette-smoking (due to increased fibrinogen), hyperlipidemia (due to lipoproteins) and with hypertension. Two studies have shown that plasma viscosity is independently associated with blood pressure[10,13]. Plasma viscosity is also increased in diabetes[16].

Whole-blood viscosity, measured at high shear rates at 37°C in a capillary viscometer, is about five times the viscosity of water (U.K. population mean 3.5 mPa s) with a wide range (2.2-4.5 mPa s) which reflects the population range of hematocrit (0.35-0.54). In men, increase in plasma viscosity with age is balanced by a slight fall in hematocrit, hence blood viscosity remains relatively constant. Women have lower hematocrit levels and hence lower blood viscosity than men, as well as a lower risk of ischemic heart disease. This sex difference is reduced in women with prevalent ischemic heart disease, women taking oral contraceptives which increase cardiovascular risk[17], and following the menopause which also increases cardiovascular risk. Blood viscosity also increases with cigarette-smoking, hyperlipidemia and hypertension (due to increases in hematocrit as well as in plasma viscosity), and in diabetics, especially if hypertensive[16].

Red cell aggregation, measured by a photometric method, is also associated with most cardiovascular risk markers, with the exception of smoking which causes dose-dependent reduction in red cell aggregation possibly due to reduction in red cell deformability which is required for aggregation[11]. Red cell deformability, measured by filtration methods, may also be reduced in diabetics[16]. Filtration of whole blood through micropore filters (diameter under 8 μm) is dominated by leukocytes, especially rigid monocytes and activated neutrophils. The major cardiovascular risk marker associated with the leukocyte count is cigarette-smoking[11].

BLOOD RHEOLOGY IN DISEASE

There is little doubt that major abnormalities in rheological factors in the rare hematological disorders which cause overt hyperviscosity syndromes play a causal role by influencing organ blood flow. These syndromes include the polycythemias (increased hematocrit), plasma hyperviscosity syndrome (increased plasma viscosity and red cell aggregation due to monoclonal or polyclonal hypergamma globulinemias), hyperleukocytic leukemias (increase in circulating immature, rigid leukocytes) and hemolytic anemias (increased red cell ridigity). Acute reduction in the rheological abnormality frequently increases organ blood flow and ameliorates clinical features in these hyperviscosity syndromes[2,4,5].

More controversial is the possible role of rheological factors in promoting common vascular disorders including ischemic heart disease, stroke and peripheral ischemia. Epidemiologically, there is increasing evidence that rheological variables are associated not only with cardiovascular risk markers (as noted above), but also with cardiovascular disease. Thus, plasma and whole-blood viscosity are associated with prevalent ischemic heart disease[14,18,19] and with prevalent peripheral arterial disease[20]. Recently, plasma viscosity was found to be a primary risk predictor for ischemic heart disease[15] and also a secondary predictor following recovery from myocardial infarction[21]. Patients with acute stroke have increased plasma and blood viscosity, which predicts acute mortality[22], as well as recurrent stroke after recovery[23]. Increases in hematocrit and fibrinogen appear partly responsible for these viscosity changes.

The changes in viscosity associated with cardiovascular disease cannot be explained by mutual associations with risk markers. They may be consequences of the arterial disease or the resulting ischemia. However they arise, it is worth considering whether they might promote arterial disease.

Promotion of arterial disease might occur via atherogenesis, thrombogenesis, or ischemia[1].

Atherosclerosis is a strikingly focal disease, being localized to bends and bifurcations: areas where separation of flow streamlines occurs result in regions of low shear stress. In such regions, the systemic increases in high-shear blood viscosity associated with cardiovascular risk markers may be magnified as local increases in blood viscosity. It is possible that such increases may favor atherogenesis by increasing interactions of adhesive platelets or leukocytes with the vessel wall[3], or by increasing infiltration of fibrinogen or lipoproteins. Experimental studies are required to investigate these hypotheses. Meanwhile, blood viscosity has been associated with the extent of coronary artery disease at angiography[24], and with the extent of peripheral arterial disease in a random population sample[20].

Thrombosis at an arterial stenosis is now recognized as the usual pivotal event which precipitates acute, spontaneous ischemia or infarction. Increases in blood viscosity, hematocrit and plasma viscosity increase platelet and leukocyte adhesion and aggregation[1-5], as well as increase shear stress on the arterial wall: such effects might favor formation of the initial, platelet- and leukocyte-rich thrombus as blood flows through an arterial stenosis. Distal to such a stenosis, flow separation again results in areas of low shear stress[3] where increases in low-shear blood viscosity may favor recirculation and interaction of activated cells and coagulation factors, leading to formation of the secondary thrombus which is rich in red cells and fibrin. Experimental studies are again required to study this hypothesis. Meanwhile, it is relevant to note two pieces of information:

(a) Patients with unstable angina (who have partially occlusive thrombi in the coronary arteries) have significantly lower levels of blood viscosity, hematocrit and fibrinogen than patients with myocardial infarction (who have fully occlusive thrombi)[25].

(b) In a population study, plasma viscosity was the variable most strongly associated with an imbalance of blood coagulation over fibrinolysis, as measured by the ratio of fibrinopeptide A (split from fibrinogen by thrombin) to fibrinopeptide $B\beta 15-42$ (split from fibrinogen by plasmin)[26]. This finding raises the possibility that increased plasma viscosity might directly affect the balance of fibrin formation and dissolution at the vessel wall, where shear stresses are highest: for example, by shear effects on tissue factor, or on tissue plasminogen activator production by the endothelium.

Increased viscosity may also promote the ischemia which arises distal to athero-thrombotic stenoses or occlusions. In acute myocardial infarction, thrombolytic therapy with streptokinase reduces mortality as effectively as tissue plasminogen activator, despite a lower rate of early lysis of coronary thrombi. One possible explanation is that the greater degree of fibrinogen reduction, and hence viscosity reduction, with streptokinase offsets this disadvantage by increasing myocardial blood flow in the ischemic myocardium[27].

In acute stroke, studies of viscosity reduction by hemodilution with dextran have been disappointing, possibly because dextran increases plasma viscosity, or because individual hemodynamic monitoring is required. More recently, a study of hemodilution with albumin and crystalloid, using hemodynamic monitoring and considering both the initial hematocrit and the hydration status of the patient, has shown benefit[28].

Viscosity elevation may also reduce blood flow distal to arterial stenoses in chronic, exercise-induced ischemia (stable angina or intermittent claudication). In the presence of a given degree of arterial stenoses in the lower limb, the risk of claudication increases significantly with increasing plasma viscosity[20]. Reductions in plasma viscosity achieved by exercise training[29], reduction in cigarette-smoking[30], or by several vasoactive drugs[31] are associated with increase in walking distance in claudicants, and viscosity reduction is a prime candidate for the mechanism of benefit. Whether or not these reductions in viscosity contribute to the beneficial effects of these interventions in stable angina is not known, but seems possible.

REFERENCES

1. Lowe GDO: Blood rheology in arterial disease, Clin. Sci., 71:137, 1986.
2. Chien S, Dormandy JA, Ernst E, Matrai A: Clinical Hemorheology, Martinus Nijhoff, Dordrecht, 1987.
3. Karino T, Goldsmith HL: Rheological factors in thrombosis and haemostasis, In: Haemostasis and Thrombosis, (Eds., Bloom AL, Thomas DP), 2nd ed., Churchill Livingstone, Edinburgh, 1987.
4. Lowe GDO: Blood Rheology and Hyperviscosity Syndromes, Bailliere's Clin. Haematol., 1:597, 1987.
5. Lowe GDO: Clinical Blood Rheology, CRC Press, Boca Raton, 1988.
6. International Committee for Standardization in Haematology (Expert Panel on Blood Rheology): Guidelines for measurement of blood viscosity and erythrocyte deformability, Clin. Hemorheol., 6:439, 1986.
7. Lowe GDO, Smith WCS, Tunstall-Pedoe H, et al.: Cardiovascular risk and haemorheology. Results from the Scottish Heart Health Study and the MONICA-Project, Glasgow, Clin. Hemorheol., 8:517, 1988.
8. International Committee for Standardization in Haematology (Expert Panel on Blood Rheology): Guidelines on selection of laboratory tests for monitoring the acute phase response, J. Clin. Pathol., 41:1203, 1988.
9. Ernst E, Koenig W, Matrai A, Keil U: Plasma viscosity and hemoglobin in the presence of cardiovascular risk factors, Clin. Hemorheol., 8:507, 1988.
10. Koenig W, Sund M, Ernst E, Keil U, Rosenthal J, Hombach V: Association between plasma viscosity and blood pressure. Results from the MONICA-Project, Augsburg, Amer. J. Hypertens., 4:529, 1991.
11. Lowe GDO, Lee AJ, Rumley A, Smith WCS, Tunstall-Pedoe H: Epidemiology of haematocrit, white cell count, red cell aggregation and fibrinogen: the Glasgow MONICA Study, Clin. Hemorheol., in press, 1992.
12. de Simone G, Devereux RB, Chien S, Alderman MH, Atlas SA, Laragh JH: Relation of blood viscosity to demographic and physiologic variables and to

cardiovascular risk factors in apparently normal adults, Circulation, 81:107, 1990.
13. Smith WCS, Lowe GDO, Lee AJ, Tunstall-Pedoe H: Rheological determinants of blood pressure in a Scottish adult population, J. Hypertens., 10:467, 1992.
14. Yarnell JWG, Sweetnam PM, Hutton RD, et al.: Plasma and blood viscosity in ischaemic heart disease: the Caerphilly studies, Clin. Hemorheol., 8:501, 1988.
15. Yarnell JWG, Baker IA, Sweetnam PM, et al.: Fibrinogen, viscosity, and white blood cell count are major risk factors for ischemic heart disease. The Caerphilly and Speedwell Heart Disease Studies, Circulation, 83:836, 1991.
16. McRury SM, Lowe GDO: Blood rheology in diabetes mellitus, Diabetic Med., 7:285, 1990.
17. Lowe GDO, Drummond MM, Forbes CD, Barbenel JC: Increased blood viscosity in young women using oral contraceptives, Am. J. Obstet. Gynecol., 137:840, 1980.
18. Baker IA, Sweetnam P, Elwood PC, et al.: Haemostatic factors associated with ischaemic heart disease in men aged 45 to 65 years. The Speedwell Study, Br. Heart J., 47:490, 1982.
19. Yarnell JWG, Sweetnam PM, Elwood PC, et al.: Haemostatic factors and ischaemic heart disease: the Caerphilly Study, Br. Heart J., 53:483, 1985.
20. Lowe GDO, Donnan PT, McColl P, et al.: Blood viscosity, fibrinogen, and activation of coagulation and leucocytes in peripheral arterial disease: the Edinburgh Artery Study, Thromb. Haemostas., 65:851, 1991.
21. Martin JF, Bath PMW, Burr ML: Influence of platelet size on outcome after myocardial infarction, Lancet, 338:1409, 1991.
22. Lowe GDO, Anderson J, Barbenel JC, Forbes CD: Prognostic importance of blood rheology in acute stroke, In: Cerebral Ischemia and Hemorheology, (Eds., Hartmann A, Kuschinsky W), Springer Verlag, Berlin, 1987.
23. Ernst E, Resch KL, Matrai A, Buhl M, Schlosser P, Paulsen HF: Impaired blood rheology: a risk factor after stroke?, J. Intern. Med., 229:457, 1991.
24. Lowe GDO, Drummond MM, Lorimer AR, et al.: Relationship between extent of coronary artery disease and blood viscosity, Br. Med. J., i:673, 1980.
25. Douglas JT, Lowe GDO, Hillis WS, Rao R, Hogg KJ, Gemmill JD: Blood and plasma hyperviscosity in acute myocardial infarction compared to unstable angina: rapid reversal by thrombolysis, Thromb. Haemostas., 62:590, 1989.
26. Lowe GDO, Wood DA, Douglas JT, et al.: Relationships of plasma viscosity, coagulation and fibrinolysis to coronary risk factors and angina, Thromb. Haemostas., 65:339, 1991.
27. Lowe GDO: Clinical and laboratory aspects of thrombolytic therapy, In: Recent Advances in Blood Coagulation 5, (Ed., Poller L), Churchill Livingstone, Edinburgh, 1991.
28. Goslinga H, Eijzenbach VJ, Heuvelmans JHA, et al.: Custom-tailored hemodilution with albumin and crystalloids in acute ischemic stroke, Stroke, 23:181, 1992.
29. Ernst E, Matrai A: Intermittent claudication, exercise and blood rheology, Circulation, 76:1110, 1987.
30. Ernst E, Matrai A: Abstention from chronic smoking normalizes blood rheology, Arteriosclerosis, 64:75, 1987.
31. Lowe GDO: Drugs in cerebral and peripheral arterial disease, Br. Med. J., 300:524, 1990.

BIOMECHANICAL ASPECTS OF TISSUE GROWTH AND ENGINEERING

Y. C. Fung

Department of AMES - Bioengineering R-012
University of California, San Diego
La Jolla, California   92093
USA

ABSTRACT

The historical development of the concept of stress modulated tissue remodeling is reviewed. Concrete new results are presented to show the following: 1) Morphological and structural remodeling of blood vessels due to change of blood pressure. 2) Remodeling of the zero-stress state of blood vessels as revealed by the change of the opening angles of the aorta due to hypertension. 3) Remodeling of mechanical properties as quantified by the change of material constants in the nonlinear constitutive equation. A unified interpretation of these remodeling processes is that a certain stress-growth law exists. I offer a simple mathematical stress-growth law for discussion.

Many of the papers presented in this book are concerned with cardiovascular prosthesis. These prosthesis when implanted into a patient must work with living tissues side by side. The prosthesis will not remodel, the living tissues will. Obviously the compatibility between the tissue and prosthesis should be engineered.

INTRODUCTION

In this chapter, we shall discuss the changes in geometry, structure, and mechanical properties that occur in the cardiovascular system as the stress in the system changes. These changes are the results of tissue remodeling. They occur when an organ is subjected to environmental changes such as hypoxia, high altitude, zero gravity, and hyperbaric conditions; or to disease, drugs, injury, surgery, healing, and rehabilitation. An understanding of tissue remodeling is important to the understanding of physiology and pathophysiology. It is also important to tissue engineering, e.g. the making of tissue substitutes with living cells, such as a skin substitute made with the patient's own keratinocytes cultured in a polymer scaffold, an artificial blood vessel covered with the patient's own endothelial cells, or a blood vessel substitute with animal tissue. The art of creating these tissue substitutes requires the knowledge of how do the tissues grow and change in a changing stress environment, and how would the substitute work in conjunction with the natural tissue in the human body.

The science of tissue remodeling and tissue engineering rests on multiple foundations: nutritional, immunological, cellular, biochemical, and biomechanical. Among all foundations there must be a biomechanical one. This is because in a living tissue or a tissue substitute, cells are contiguous and must interact mechanically with each other. Since the force of interaction of one part of a cell on another is measured by stress, the classical continuum mechanics applies. Thus, although classical continuum mechanics traditionally has not been concerned with biology, and conversely, old biology has not been bothered with continuum mechanics, yet the two subjects must now work together to solve a common problem. Biology must be concerned with mechanics because cells interact. Mechanics must encompass biology also because cells form a continuum, and interact.

Furthermore, cells can grow or resorb. So the structure formed by cells, the tissue, can also grow or resorb. The phenomenon of growth or resorption is a mass transfer phenomenon. Mass transfer inside cells and between cells depends on the strain in the cell membrane, strain in intracellular reticulum, and strain in extracellular matrix. Hence mass transfer in living tissue depends on the state of stress and strain; so do growth and resorption, and tissue remodeling.

HISTORICAL DEVELOPMENT OF THE CONCEPT OF STRESS MODULATED TISSUE REMODELING

Living organisms are endowed with a certain ability to heal when damaged. Traditionally, the art of healing is studied in surgery and medicine. Orthopedic surgeons were the first to pay attention to the role played by biomechanics in the healing of bone fracture. In 1866, G.H. Meyer[1] presented a paper on the structure of cancellous bone and demonstrated that "the spongiosa showed a well-motivated architecture which is closely connected with the statics of bone." A mathematician, C. Culmann, was in the audience. In 1867, Culmann[2] presented to Meyer a drawing of the principal stress trajectories on a curved beam similar to a human femur. The similarity between the principal stress trajectories and the trabecular lines of the cancellous bone is remarkable. In 1869, J. Wolff[3] claimed that there is a perfect mathematical correspondence between the structure of cancellous bone in the proximal end of the femur and the trajectories in Culmann's crane. In 1880, W. Roux[4] introduced the idea of "functional adaptation." A strong line of research followed Roux. Pauwels1[5], beginning in his paper in 1935 and culminating in his book of 1965, which was translated into English in 1980, turned these thoughts to precise and practical arts of surgery. Vigorous development continues. In 1980's, Carter and his associates[6,7] have published a hypothesis about the relationship between stress and calcification of the cartilage into bone. Cowin and his associates[8-10] have developed mathematical theory of Wolff's law. Fukada[11], Yasuda[12], Bassett[13],

Saltzstein and Pollack[14], and others have studied piezoelectricity of bone and developed the use of electromagnetic waves to assist healing of bone fracture. Lund[15], Becker[16], Smith[17] and others have studied the effect of electric field on the growth of cells and on the growth of amputated limb of frog. Recently, the biology of bone cells is advancing rapidly, and molecular biological transformation of muscle into bone has been announced.

Thus the motivation, the knowledge, and the art of providing a suitable stress to the bone and cartilage to promote healing and rehabilitation, is advancing rapidly.

REMODELING OF SOFT TISSUES IN RESPONSE TO STRESS CHANGES

The best known example of soft tissue remodeling due to change of stress is the hypertrophy of the heart caused by a rise in blood pressure. Another famous example was given by Cowan and Crystal[18] who showed that when one lung of a rabbit was excised, the remaining lung expanded to fill the thoracic cavity, and it grew until it weighed approximately the initial weight of both lungs.

On the other hand, animals exposed to the weightless condition of space flight have demonstrated skeletal muscle atrophy. Leg volumes of astronauts are diminished in flight. In flight vigorous daily exercise is necessary to keep astronauts in good physical fitness over a longer period of time, see references in Fung[19].

Immobilization of muscle causes atrophy. But there is a marked difference between stretched immobilized muscle versus muscle immobilized in the resting or shortened position. Fundamentally, growth is a cell biological phenomenon at molecular level. Stress and strain keep the cells in a certain specific configuration. Since growth depends on cell configuration, it depends on stress and strain.

STRESS IN BIOLOGICAL TISSUES

Since all the readers of this book are not engineers, and since the biologists', physicians' or surgeons' concept of stress and strain may be quite different from that of the engineers', I will begin this Chapter with an explanation of terminology.

Consider a little cube of material in one's body, as shown in Fig. 1(A). The cube has six faces. On the face No. 1, which is perpendicular to the X_1 coordinate axis, there acts three forces: one is perpendicular to the surface and acts in the X_1 - axis direction. Another one is parallel to the surface and acts in the direction of X_2 axis. The third one is also parallel to the surface but acts in the direction of the X_3 axis. Dividing these forces by the area of the surface, we obtain three numbers, T_{11}, T_{12}, T_{13}. T_{11} is called a *normal stress*. T_{12}, T_{13} are called *shear stresses*. T_{11}, T_{12}, T_{13} have the units of force as per unit area. In English system we measure stresses by the units of *pounds per square inch*. In the *International System of Units* (SI Units) stresses are expressed in terms of *Newtons per square meter*. One pound per square inch is equal to 6894 Newtons per square meter. One Newton per square meter is also said to be one *Pascal*. It is so named internationally in honor of Louis Pascal.

Similarly, on the surface No. 2 which is perpendicular to the X_2 axis, there act three stresses, T_{21}, T_{22}, T_{33}, as shown in Fig. 1(A). On the surface No. 3, which is perpendicular to the X_3 - axis, there are three stresses T_{31}, T_{32}, T_{33}. On the other three surfaces, perpendicular to the negative X_1-axis, negative X_2-axis, negative X_3-axis,

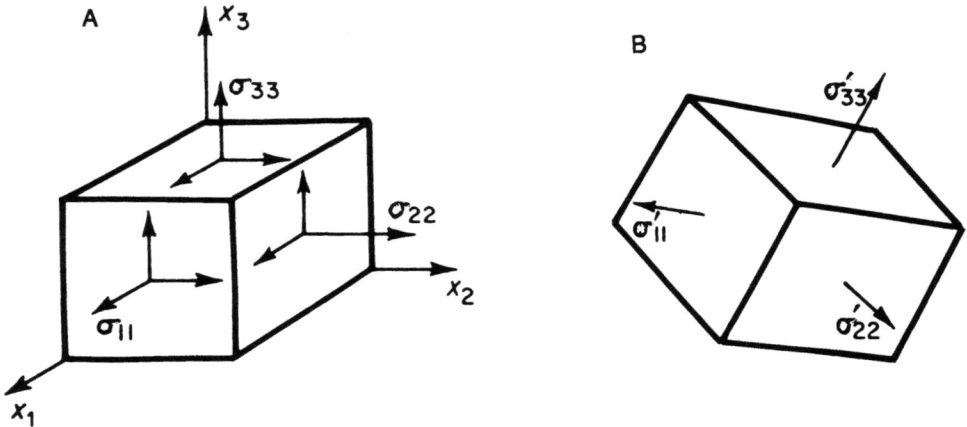

Figure 1.(A) Stress is force per unit area acting on a surface. The component perpendicular to the surface is normal stress. The components tangential to the surface is shear stress. On a small cube, each face has three stresses acting on it. (B). At any point in a body, one can find a set of principal axes. If we imagine isolating a little cube oriented in the principal direction, then on each face there acts only one stress: the normal, principal stress.

the stresses are respectively the same as those acting on the surfaces perpendicular to the positive X_1, X_2, X_3 axes. Thus altogether the stresses acting in the little cube can be listed in a table below:

Components of Stresses

	1	2	3
Surface normal to X_1	T_{11}	T_{12}	T_{13}
Surface normal to X_2	T_{21}	T_{22}	T_{23}
Surface normal to X_3	T_{31}	T_{32}	T_{33}

The whole table specifies the state of stress in the little cube. Hence at every point in a body, where such a cube can be drawn, there are nine components of stresses. Fortunately, it can be shown that the three pairs of shear stresses are always equal:

$$T_{12} = T_{21}, \qquad T_{23} = T_{32}, \qquad T_{31} = T_{13}$$

Therefore, at every point, there are six independent components of stresses. Thus stress is a quantity that needs six numbers (components) to describe it at every point.

It can be shown that at any given point in a body, one can find a certain orientation of the cube on whose surfaces all the shear stresses are zero, see Fig. 1(B). In such a so-called **principal orientation**, the stress components listed as a matrix appear as:

                    Principal Stresses

	1	2	3
Surface ⊥ principal direction 1	T_{11}	0	0
Surface ⊥ principal direction 2	0	T_{22}	0
Surface ⊥ principal direction 3	0	0	T_{33}

Then the components of stresses T_{11}, T_{22}, T_{33} are called *principal stresses*. To describe the state of stress at any point in a body by the principal stresses, one must specify also the principal directions. Hence one still need six numbers to describe the stress.

Conclusion: Stress is a quantity that needs six numbers to specify it. The six numbers can be the six components of the stress, or the three principal stresses together with three angles that describe the orientation of the principal axes.

In the rest of this chapter, the word stress will be used in strict accordance with the above definition.

MORPHOLOGICAL AND STRUCTURAL REMODELING OF BLOOD VESSELS DUE TO CHANGE OF BLOOD PRESSURE

The pressure of circulating blood oscillates in every pulse wave. It varies from place to place as blood circulates. What normally referred to as systemic blood pressure is the difference of the pressure of blood in the aorta at the aortic valve and that in the right atrium. This is the pressure that drives the entire systemic circulation. The corresponding driving pressure for the pulmonary circulation is the difference between the pressure in the pulmonary artery at the pulmonic valve and the pressure in the left atrium. Both the systemic and pulmonary circulations are characterized by the systolic and diastolic pressures. When these pressures change, the blood pressure in every vessel in the body changes. When blood pressure changes, the stress in the blood vessel wall changes.

To explain the stress state in blood vessels, let us first refer to a sketch of the heart, the aorta, and the pulmonary arteries in Fig. 2. To the right hand side, on top is a sketch of a magnified view of a short segment of the pulmonary artery *in vivo* (at normal pulmonary blood pressure). Note the *tethering of lung parenchyma* (tissue of alveolar walls) on the outside of the pulmonary artery in the lung. Next to this *in vivo* segment shows the same segment at the *no-load* condition (blood pressure, airway pressure, and tissue stress all reduced to zero). At the same magnification the diameter of the pulmonary artery is much reduced by the removal of the load. But the no-load condition is not without stress. The stress remaining in the vessel wall at the no-load condition is called the *residual stress*. To remove the residual stress, a cut is made in the radial direction. The vessel opens up. It can be shown that further cutting in various ways will not cause any further measurable deformation. Hence we call the opened up segment a vessel wall at *zero-stress state*. The stresses in the vessel wall at the *in vivo* and no-load states are sketched in an infinitesimal element in the next row.

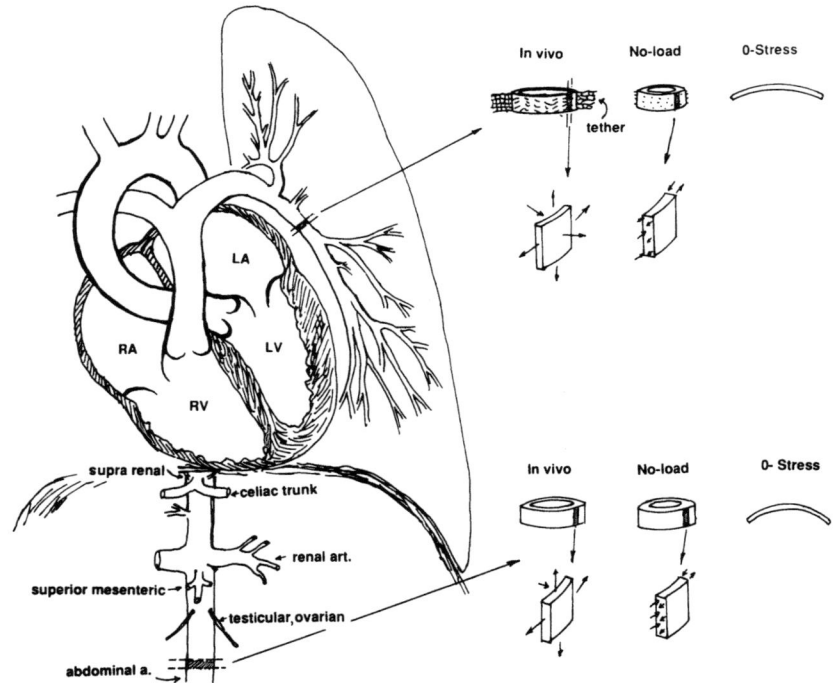

Figure 2. A sketch of the heart, the aorta, and the pulmonary arteries, and the stresses in them. See text for details.

At the lower right hand side are similar sketches of a short segment of the aorta at the *in vivo*, no-load, and zero-stress states. Below them in the last row are shown the stresses in the aortic wall at these states. The aorta differs from the pulmonary artery in several respects. First, it is of thicker wall and subjected to a higher blood pressure. Second, there is little tethering on the adventitia of the aorta, while the lung parenchyma tethers fully on the pulmonary artery. The external pressure on the abdominal aorta is the abdominal pressure. The external load acting on the outside wall of the pulmonary arteries in the lung consists of the gas pressure in alveoli and the tissue stress. The opening angles of the zero-stress state of both vessels vary with location.

As it is sketched in Fig. 2, at the *in vivo* condition at normal blood pressure, the circumferential stress is usually tensile and is the largest stress component in the vessel wall, the longitudinal stress components exists because the vessel is normally stretched in the axial direction. The radial stress component is compressive at the inner wall where it is equal to the blood pressure, and it gradually decreases to zero (atmospheric) at the outer wall.

The systemic blood pressure can be changed in a number of ways, by drugs, by high salt diet, by constricting flow of blood to kidney, etc. If the aorta is constricted severely by a stenosis above the renal arteries (Fig. 2), the aorta above the stenosis will become hypertensive, the whole upper body supplied by the upper aorta will become hypertensive, whereas the aorta below the stenosis will become hypotensive at first, but the reduced blood flow to the kidney will cause the kidney to secrete more of the enzyme renin into the blood stream and raise the blood pressure. If the stenosis was below the kidney arteries and is sufficiently severe, then the lower body will become hypotensive. Such a constriction can be simulated by a metal clamp, and is used in experiments.

The pulmonary blood pressure can also be changed by a number of ways. A most convenient way in the laboratory is to change the oxygen concentration of the gas breathed by the animal. If the oxygen concentration is reduced from normal (i.e. hypoxic), the smooth muscle cells in the pulmonary blood vessels contract, the vessel diameters are reduced, and the pulmonary blood pressure goes up. This is the reaction human encounters when sea level persons go to high altitude.

I will show an example to demonstrate how fast blood vessels remodel themselves when the blood pressure changes. Fung and Liu[20] created high blood pressure in rat's lung by putting rats in a low oxygen chamber. The chamber's oxygen concentration is about the same as that at the Continental Divide of the Rocky Mountains in Colorado, about 12,000 ft. Nitrogen is added so that the total pressure is the same as the atmospheric pressure at sea level. When a rat enters such a chamber, its systolic blood pressure in the lung will shoot up from the normal 15 mm Hg to 22 mm Hg within minutes, and maintains in the elevated pressure of 22 mm Hg for a week, then gradually rises to 30 mm Hg in a month. (Its systemic blood pressure remains essentially unchanged in the meantime.) Under such a step rise in blood pressure in the lung, its pulmonary blood vessel remodels. To examine the change, a rat is taken out of the chamber at a scheduled time. It was anesthetized immediately by an intraperitoneal injection of pentobarbital sodium according to a procedure and dosage approved by the University, NIH and Department of Agriculture, then dissected according to an approved protocol. The specimens were fixed first in glutaraldehyde, then in osmium tetraoxide, embedded in Medcast resin, stained with toluidine blue O, and examined by light microscopy.

Figure 3 shows how fast the remodeling proceeds. In this figure, the photographs in each row refers to a segment of the pulmonary artery as indicated by the leader line. The first photograph of the top row shows the cross-section of the arterial wall of the normal three month old rats. The specimen was fixed at the no-load condition. In the figure, the endothelium is facing upward. The vessel lumen is on top. The endothelium is very thin, of the order of a few micrometer. The scale of 100 μm is shown at the bottom of the figure. The dark lines are elastin layers. The upper, darker half of the vessel wall is the media. The lower, lighter half of the vessel wall is the adventitia. The second photo in the first row shows the cross-section of the main pulmonary artery two hours after exposure to lower oxygen pressure. There is evidence of small fluid vesicles and some accumulation of fluid in the endothelium and media. There is a biochemical change of elastin staining in vessel wall at this time. The third photograph shows the wall structure 12 hours later. It is seen that the media is greatly thickened, while the adventitia has not changed very much. At 96 hours of exposure to hypoxia, the photograph in the fourth column shows that the adventitia has thickened to about the same thickness as the media. The next two photos show the pulmonary arterial wall structure when the rat lung is subjected to 10 and 30 days of lowered oxygen concentration. The major change in these later periods is the continued thickening of the adventitia.

The photographs of the second row show the progressive changes in the wall of a smaller pulmonary artery. The third and fourth rows are photographs of arteries of even smaller diameter. The inner diameter of the arteries in the fourth row is of the order of 100 μm, approaching the range of sizes of the arterioles. The remodeling of the vessel wall is evident in pulmonary arteries of all sizes.

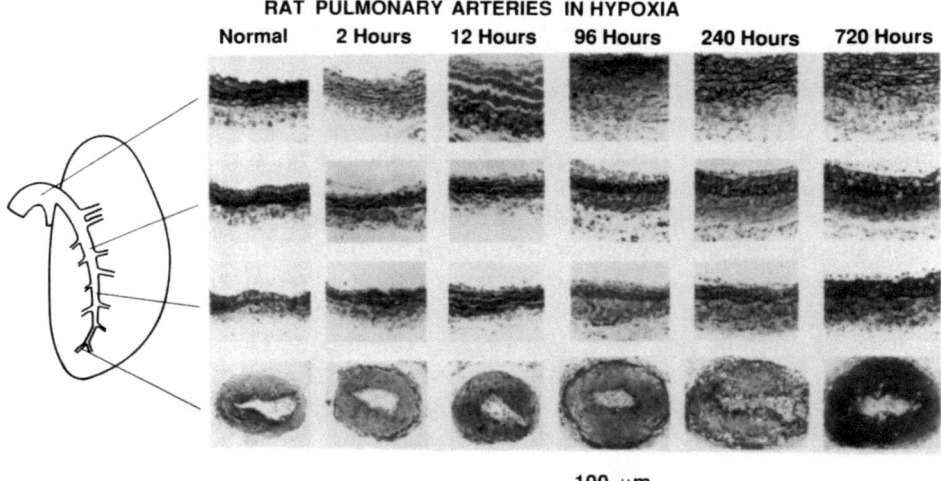

Figure 3. Photographs of histological slides from four regions of main pulmonary artery of a normal rate and hypertensive rats with different periods of hypoxia. Specimens were fixed at no-load condition. See text for details. From Fung and Liu (1990). Reproduced by Permission.

Thus we see that the active remodeling of blood vessel wall proceeds quite fast. Histological changes can be identified within hours. The maximum rate of change occurs within a day or two.

THE OPENING ANGLE OF THE ZERO-STRESS STATE OF BLOOD VESSEL

When the structure of a blood vessel changes, the mechanical property of the vessel will change. Among these properties, a fundamental one is the zero-stress state.

To analyze stress and strain in a body, it is necessary to know the zero-stress state: the geometric shape assumed by the body when the stress is zero everywhere. In the literature, virtually all publications assume that the zero-stress state of a body is the state when all external loads are removed. But is it? As we have sketched in Fig. 2, if we cut an aorta twice by cross-sections perpendicular to the longitudinal axis of the vessel, we obtain a ring. If we cut the ring radially, it will open up into a sector. By using equations of static equilibrium, we know that the stress resultants and stress moments are zero in the open sector. Whatever stress remains in the vessel wall must be locally in equilibrium. If one cut the open sector further, and can show that no additional strain results, then we can say that the sector is in zero-stress state. We did a simple experiment illustrated in Fig. 4 (from Fung and Liu[21]). Five consecutive segments (rings) 1 mm long each were cut from a rat aorta. The first four segments were then cut radially and successively at four positions around the circumference as indicated in Fig. 5, namely, inside, outside, anterior, posterior; designated as I, O, A, P, respectively, approximately 90 degrees apart successively. The fifth segment was cut in all four positions, resulting in four pieces designated *a, b, c, d*. The open sectors of the first four rings are shown in the upper row on the right. When the *four pieces of the fifth ring were reassembled in the order of abcd, bcda, cdab, bcda*, with tangents matched at successive ends, we obtain four configurations shown in the lower row of Fig. 4. They resemble the shape of the four cut segments of the first row quite well. This tells us that the arterial wall is not axisymmetric, that different parts of the circumference are different, and that one cut is almost as good as four cuts in relieving the residual stress.

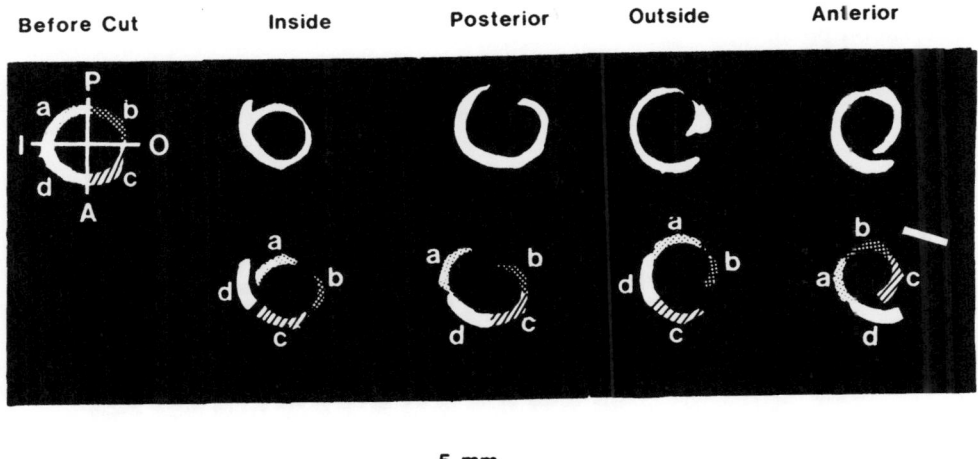

Figure 4. Test of the hypothesis that one cut is sufficient to reduce an arterial segment at no load to the zero-stress state. See text. From Fung and Liu (1989).

Having been assured that the open sector represents zero-stress state of a blood vessel, we conclude that the zero-stress state of an artery is not a tube. It is a series of open sectors. To characterize the open sectors, we define an *opening angle* as the angle subtended by two radii drawn from the mid-point of the inner wall (endothelium) to the tips of the inner wall of the open sections (see Fig. 5). Although a full description of the geometry of the sector can only be done with a photograph or its equivalent, the opening angle does provide a simple numerical index.

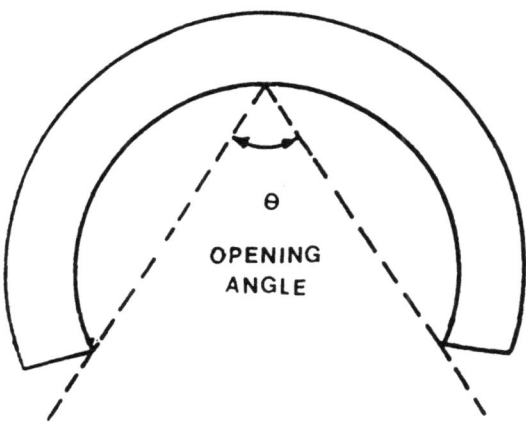

Figure 5. Definition of the opening angle.

The photographs in the first column of Fig. 6 show a more complete picture of the zero-stress state of a normal young rat aorta[21]. The entire aorta was cut successively into many segments of approximately one diameter long. Each segment was then cut radially at the "inside" position indicated in Fig. 5. It was found that the opening angle varied along the rat aorta: it was about 160° in the ascending aorta, 90° in the arch, 60° in the thoracic region, 5° at the diaphragm level, 80° toward the iliac bifurcation point.

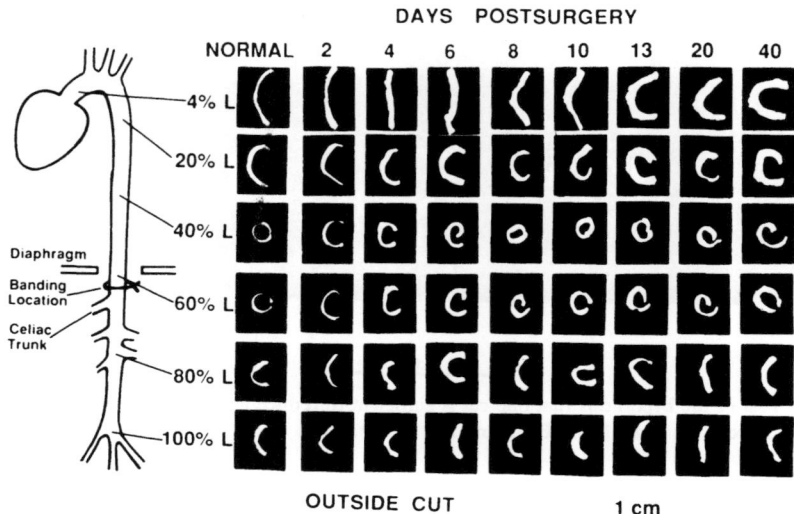

Figure 6. Photographs of the cross sections of short segments of the rat aorta cut radially. The first column shows the zero-stress state of the aorta of normal rat. The rest shows the change of zero-stress state due to vessel remodeling after a sudden onset of hypertension. The photos are arranged according to days after surgery from left to right, and according to location on the aorta from top to bottom, expressed as distance from the heart in percentage of the total length. The location of the metal clip used to induce hypertension is shown in the sketch at left. The arcs of the blood vessel wall do not appear smooth because of some tissue attached to the wall. In reading these photographs, one should mentally delete these tethered tissues. From Fung and Liu (1989).

Following the common iliac artery down a leg of the rat, we found that the opening angle was in the 100° level in the iliac artery, dropped down in the popliteal artery region to 50°, then rose again to the 100° level in the tibial artery[22]. In the medial plantar artery of the rat, the micro arterial vessel 50 mm diameter had an opening angle of the order of 100°[22].

There are similar spatial variations of opening angles in the aorta of the pig and dog[23], although there were quantitative differences. We also found significant opening angles in pulmonary arteries[20], systemic and pulmonary veins[24], and trachea[25]. Thus we conclude that the zero-stress state of blood vessels and trachea are sectors whose opening angles vary with the locations on the vessel, and with animal species.

REMODELING THE ZERO-STRESS STATE OF BLOOD VESSEL: CHANGE OF THE OPENING ANGLE OF AORTA DUE TO HYPERTENSION

We created hypertension in rats by constricting the abdominal aorta with a metal clip placed right above the celiac trunk (see Figs. 2 and 6). The clip severely constricted the aorta locally, reduced the cross-sectional area of the lumen by 97%, with only about 3% of the normal area remaining[21,22]. This caused a 20% step-increase of blood pressure in upper body, and a 55% step-decrease of blood pressure in the lower body immediately following the surgery. Later, the blood pressure increased gradually, following a course shown in Fig. 7. It is seen that in the upper body the blood pressure rose rapidly at first, then more gradually, tending to an asymptote at about 75% above normal. In the lower body, the blood pressure rose to normal value in about four days, then gradually increased further to an asymptotic value of 25% above normal. Parallel with this change of blood pressure, the

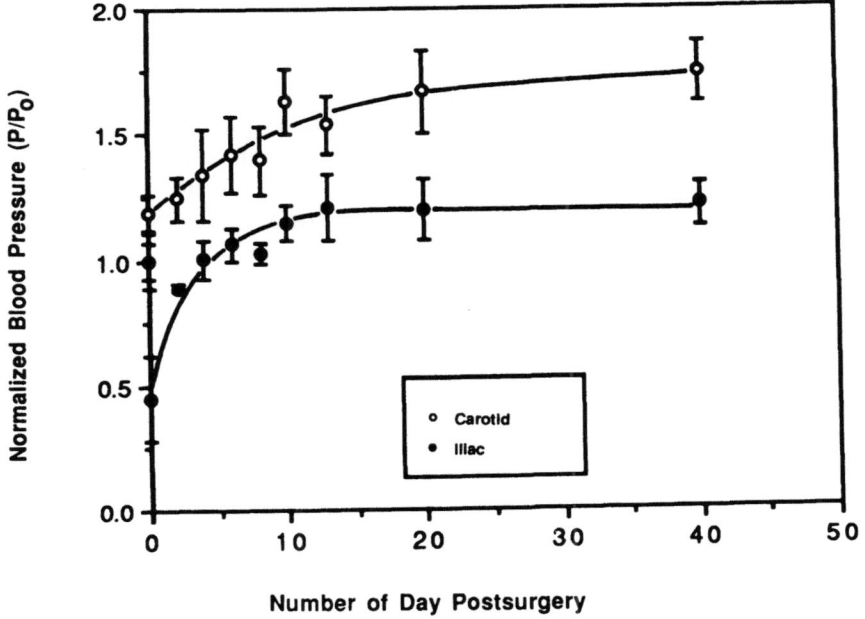

Figure 7. The course of change of blood pressure (normalized with respect to that before surgery) after banding the aorta. From Fung and Liu (1989).

zero-stress state of the aorta changed. The changes are illustrated in Fig. 6 in which the location of any section on the aorta is indicated by the percentage distance of that section to the aortic valve measured along the aorta divided by the total length of the aorta. Successive columns of Fig. 6 show the zero-stress configurations of the rat aorta at 0, 2, 4, ..., 40 days after surgery. Successive rows refer to successive locations on the aorta.

Figure 6 shows that following a sudden increase in blood pressure, the opening angles increased, gradually at first, peaked in two to four days, then decreased gradually to an asymptotic value. Variation with the location of the section on the aorta was great. The maximum change of the opening angle occurred in the ascending aorta, where the total swing of the opening angle was as large as 88°.

Thus we found that the blood vessel changed its opening angle in a few days following the blood pressure change. We found similar changes in pulmonary arteries after the onset of pulmonary hypertension by exposing rats to hypoxic gas containing 10% oxygen, 90% nitrogen, at atmospheric pressure[20].

Since opening angle changes reflect structural changes, we conclude that blood vessels remodel significantly with modest blood pressure changes.

REMODELING OF MECHANICAL PROPERTIES

The study of the mechanical properties of blood vessels has a long history, a large literature, and some specialized terminology. The reader is referred to an extensive list of references in Fung[26] and Fung[19]. In the following, I chose just one approach with which some data on the effect of vessel wall remodeling have been obtained.

On the basis of the fact that blood vessels are viscoelastic, and that their viscoelasticity is very special (but is common with most living soft tissues), in the sense that in a cyclic loading and unloading process the stress-strain curve and the hysteresis loop (i.e, the energy dissipation per cycle of loading and unloading divided by the strain energy of loading) are not very sensitive to the strain rate (i.e., the frequency of loading-unloading cycling). Typically, the variation of the stress at a given strain is no more than a factor of two or three when the strain rate varies over a range of one million. The size of the hysteresis loop also varies no more than a factor of two or three when the strain rate varies a million fold. Hence, even though the viscoelastic mechanism is very complex, overall we can regard the vessel wall tissue as *pseudo-elastic*, which means that in a periodic process (such as in regular heart beat) the vessel wall is elastic in the loading process with a certain set of elastic constants, and is also elastic in the unloading process with a different set of elastic constants[26,27]. Further, we assume that a *pseudo-elastic strain energy function* exists, denoted by the symbol $\rho_o W$, and expressed as a function of the nine components of strain, E_{ij}, (i = 1, 2, 3, j = 1, 2, 3), which is symmetric with respect to E_{ij} and E_{ji}, so that the stress components can be derived by a differentiation[28]:

$$S_{ij} = \frac{\partial \rho_o W}{\partial E_{ij}} \tag{1}$$

Here ρ_o is the density of the material at the zero-stress state, W is the strain energy per unit mass, $\rho_o W$ is the strain-energy per unit volume, E_{ij} are strains measured with respect to the material configuration at zero-stress state. The physical meaning of this statement is that in isothermal process $\rho_o W$ can be identified with Helmholtz's free energy function, and in adiabatic case $\rho_o W$ can be identified with Gibbs' thermodynamic potential[26].

With the above point of view, our problem is to determine the pseudo-elastic strain energy function $\rho_o W$. Here there are two approaches. One is to regard the blood vessel wall as an incompressible material and derive $\rho_o W$ as a function of E_{ij} in *three dimensions*[29]. The other is to assume the blood vessel as a cylindrical body with axisymmetry in mechanical properties and limit ourselves to axisymmetric loading and deformation. Then we are only concerned with two strain components: the circumferential strain, E_{11}, and the longitudinal strain, E_{22}. These strains may be different in different layers of the blood vessel: endothelium, media, and adventitia; and in each layer vary with the radial coordinate r. The radial strain is easily computed from the condition of incompressibility:

$$\lambda_1 \lambda_2 \lambda_3 = 1, \quad \lambda_i = \sqrt{\frac{E_{ii}^2 - 1}{2}} \quad (i \text{ not summed}). \tag{2}$$

Then $\rho_o W$ is derived as a function of E_{11}, E_{22}, and Eq. 1 is used with i, j limited to 1, 2.

This second approach may be called a *two-dimensional approach*. It can be applied to every layer of the blood vessel wall, so that the elastic constants are determined separately for the media and the adventitial layers. But in the literature available so far the vessel wall is treated as a homogeneous material without layering in mechanical property. In two dimensions, the material is of course compressible.

For analytical representation of $\rho_o W$ for arteries in the two dimensional approach, a polynomial form has been used by Patel and Vaishnav[30,31], a logarithmic form has been used by Hayashi et al[32], and an exponential form has been used by Fung et al[24,25,27,28]. Our experience with the exponential form has been satisfactory (cf. discussion in Fung et al[33]), and a simple computing program for the identification of material constants with experimental curves exists Hence we have used it to study the blood vessel remodeling problem. We assume the following form for $\rho_o W$:

$$\rho_o W = C \exp [a_1 E_{11}^2 + a_2 E_{22}^2 + 2 a_4 E_{11} E_{22}]$$

where C, A_1, a_2, a_4 are material constants, E_{11} is the circumferential strain, E_{22} is the longitudinal strain, both referred to the zero-stress state.

Experiments have been done on rat arteries during the course of development of diabetes after a single injection of streptozocin. When the vessel wall is treated as one homogeneous material, the results are presented in Table 1, from Liu and Fung[34]. It is seen that the material constants change with the development of diabetes. In Liu and Fung[34], the corresponding remodeling of the zero-stress state is shown.

Table 1. Coefficients C, a_1, a_2 and a_4 of the stress–strain relationship of the thoracic aorta of 20- day diabetic and normal rats.

Group	C (N/cm²)	a_1	a_2	a_4*
Normal rats				
1	13.69	0.99	2.11	0.0036
2	9.19	1.55	2.88	0.0036
3	16.19	0.77	1.81	0.0036
4	9.78	0.85	3.94	0.0036
Mean±SD	12.21±3.32	1.04±0.35	2.69±0.95	0.0036
20-day Diabetic rats				
1	8.59	2.76	4.30	0.0036
2	28.95	0.51	1.93	0.0036
3	12.34	1.35	4.06	0.0036
4	11.40	1.49	3.47	0.0036
Mean±SD	15.32±9.22	1.53±0.92	3.44±1.07	0.0036

* a_4 was fixed as the mean value from the normal rats.

Morphologicaly remodeling in diabetes is also profound, but is yet unpublished.

Similar data have been obtained from rat arteries in the course of remodeling following the initiation of hypertension and hypotension, but yet unpublished.

A UNIFIED INTERPRETATION OF THE MORPHOLOGICAL, STRUCTURAL, ZERO-STRESS STATE, AND MECHANICAL PROPERTIES REMODELING

The first thing all this remodeling tells us is that living tissues remodel themselves in response to stress changes. In a blood vessel, the stress changes in response to blood pressure changes in nonuniformly distributed in the vessel wall, hence one can expect that the remodeling process is nonuniform in the blood vessel wall. Fung[35] has proposed that the effect of the residual stress in blood vessels is to make the stress distribution more uniform in the vessel wall at normal blood pressure. Recently, this proposition has been verified by Fung and Liu[36] by measuring the strains in vessel wall directly. The results are in good agreement with an earlier theoretical calculation by Chuong and Fung[29].

In a series of papers since 1987, Hayashi and Takamizawa[37], Takamizawa and Hayashi[32,33,38,39] used the "uniform strain" hypothesis for blood vessels at homeostasis and computed the residual strain and zero-stress state on the basis of that hypothesis. Such a uniform stress or uniform strain hypothesis at homeostasis would simplify the analysis of homeostatic stress and strain in blood vessels. It is very convenient. But it is a phenomenological statement. The empirical basis of this phenomenological hypothesis lies in the observations mentioned above. In 1983 I felt that the emperical basis was not sufficiently strong and there was a need to search for a broader foundation and a firmer empirical base. This need still exists today. We would like to find a higher principle from which this hypothesis can be derived. I believe that such a higher principle is the stress-growth law. A suggestion is given below.

SPECULATION ON STRESS-GROWTH OR STRAIN-GROWTH RELATIONSHIP

Tissue growth can be affected by many factors: nutrition, growth factors, physical and chemical environment, diseases, as well as stress and strain. If other things were equal, then a stress-growth law (or a strain-growth law) may exist. Since a blood vessel is a composite material made of cells and extracellular matrix containing collagen, elastin and other substances, and each substance may have a stress-strain-growth law, there may be as many laws as there are materials.

At the present time, such a stress-growth law is unknown for blood vessels. Hence the most we can do is to speculate. I'd like to present a possible form of such a law. Referring to Fig. 8, let the solid curve represent a relationship between the rate of growth of the mass of a material M and the stress or strain acting in the material, s. The symbol s may represent a component of the stress or strain tensor, or a stress or strain invariant: exactly what it is would have to be determined later. Let a represent a homeostatic stress, at which the tissue can maintain a steady-state. M, the rate of growth of a material in a tissue, is positive when the stress or strain s exceeds a. M is negative when s is less than a. The homeostatic condition of blood vessel wall at normal blood pressure is represented by the point a. The rate of growth M, however, cannot increase indefinitely with increasing s. A well-known phenomenon of bone resorption under excessive stress or strain suggests that another homeostatic stress or strain b exists beyond which resorption occurs. Similarly, the negative rate when the stress or strain is less than a cannot be unbounded in the whole range of $s < a$. Suppose that resorption stops when $s = c$, where $c < a$; then c is another homeostatic stress or strain. If the rate of growth M is a continuous function of the stress or strain s, and this function has zeros at a, b, c, and if the trend of change of M at a, b, c, is as discussed above, then I would like to propose the following simple relation (Fung[19]):

$$M = C(s-a)^{k_1}(b-s)^{k_2}(s-c)^{k_3} \qquad (3)$$

in which M is the rate of increase of the volume of the tissue, s is the stress or strain, a, b, c are constants having the units of stress (N/m²), C is a constant with units (m⁵ N⁻¹ s⁻¹), k_1, k_2, k_3, are dimensionless numbers. When M is plotted against s, Eq. (1) is illustrated in Fig. 8. When s lies between a and b or 0 and c, the tissue grows. When s exceeds b or falls between a and c, the tissue resorbs. As s tends to zero, I assume M to be positive as in cell culture in petrie dishes. The exponents k_1, k_2, k_3 determine how fast the growth rate changes when s deviates from the homeostatic state. If $k_1 > 1$, the slope of the growth curve is zero at $s = a$, then small deviation has little influence. The slope of the growth curves is infinite at a if $k_1 < 1$. $k_1 = 1$ signals a finite slope.

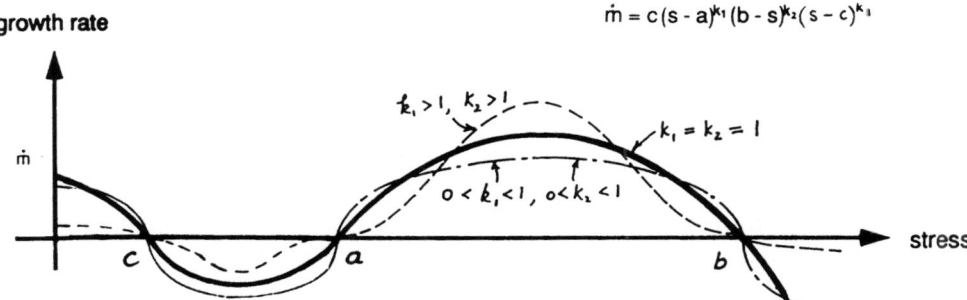

Figure 8. The author's proposed stress-growth law. From Fung (1990, p. 530).

I am using this theoretical proposal as an experimental hypothesis. I hope to find out experimentally whether Eq. (1) is true or false, whether it can or cannot be modified to obtain better results.

The exact definition of s is a subject to be studied. It could be any one of the strain or stress invariants, e.g., the maximum principal stress or strain, the maximum shear stress or strain, or the von Mises or octahedral stress or strain. It is unlikely to be the first invariant (sum of the three normal stresses or strains) unless the material is compressible.

RELEVANCE TO TISSUE ENGINEERING

Tissue engineering is a field dedicated to the engineering of living tissues[40,41]. The field of tissue engineering lies between the field of genes and cells and molecular biology on the one hand, and that of organ physiology and holistic medicine on the other hand. To master tissue engineering one must know how the health of tissues is maintained, improved, or degenerated. To understand artificial tissues containing live cells, one must know more about the behavior of natural cells in association with artificial materials. If we talk about tissue engineering of arteries, we should at least know how natural arteries behave. What we discussed above is elementary and fundamental for tissue engineering. To create and control a tissue, we must know all the relevant stress-growth laws. We have much to learn. The task is big. That's why I offer the speculative Eq. (1) as a starter, to invite further thinking and experimentation.

CONCLUSION

The ability of a living tissue to remodel itself under changed stress condition is an ability unmatched by engineering structures. This ability has obvious effects on health and disease, and worthy of a thorough understanding. Presently, we are at the beginning of a long road of exploration. Many of the papers presented in this book are concerned with cardiovascular prosthesis. These prosthesis when implanted into a patient must work with living tissues side by side. The prosthesis will not remodel, the living tissues will. Obviously the compatibility between the tissue and prosthesis should be engineered.

REFERENCES

1. Meyer GH: Die Architektur der spondiosa, Archiv. fur Anatomie, Physiologie, und wissenschaftliche Medizin (Reichert und wissenschafliche Medizin, Reichert und Du Bois-Reymonds Archiv), 34:615-625, 1867.
2. Culmann C: Die graphische Statik, Meyer und Zeller, Zurich, 1866.
3. Wolff J: Uber die innere Architektur der spongiosen Substanz. Zentralblatt fur die medizinische Wissenschaft, VI. Jahrgang, 223-234, 1869.
4. Roux W: Gesammelte Abhandlungen uber die entwicklungs mechanik der Organismen, W. Engelmann, Leipzig, 1895.
5. Pauwels F: Biomechanics of the Locomotor Apparatus, German ed. 1965. English trans. by Maqnet P, Furlong R, Springer-Verlag, Berlin, New York, 1980.
6. Carter DR, Fyhrie DP, Whalen RT: Trabecular bone density and loading history: Regulatrion of connective tissue biology by mechanical energy, J. Biomech., 20:785-794, 1987.
7. Carter DR, Wong M: Mechanical stresses and endochondral ossification in the chondroepiphysis, J. Orthop. Res., 6:148-154, 1988.
8. Cowin SC: Modeling ot eh stress adaptation process in bone, Cal. Tissue Int., 36:(Suppl.) S99-S104, 1984.
9. Cowin SC: Wolff's law of trabecular architecture at remodeling equilibrium, J. Biomech. Eng., 108:83-88, 1986.
10. Cowin SC, Hart RT, Balser JR, Kohn DH: Functional adaptation in long bones: establishing in vivo values for surface remodeling rate coefficients, J. Biomech., 12:269, 1985.
11. Fukada E: Piezoelectric properties of biological macromolecules, Adv. Biophys., 6:121, 1974.

12. Yasuda I: Mechanical and electrical callus, Ann. N.Y. Acad. Sci., 238:457-465, 1974.
13. Bassett CAL: Pulsing electromagnetic fields: A new approach to surgical problems, In: Metabolic Surgery (Eds., Buchwald H, Varco RL), Grune and Stratton, New York, 255-306, 1978.
14. Satzstein RA, Pollack SR: Electromechanical potentials in cortical bone. II. Experimental analysis, J. Biomech., 20:271, 1987.
15. Lund EJ: Experimental control of organic polarity by the electric current. I. Effects of the electric current of regenerating internodes of obelia commisuralis, J. Exper. Zool., 34:471, 1921.
16. Becker RO: The bioelectric factors in amphibian limb regeneration, J. Bone Joint Surg., 43A:6431, 1961.
17. Smith SD: Effects of electrode placement on stimulation of adult frog limb regeneration, Ann. N.Y. Acad. Sci., 238:500, 1974.
18. Cowan MJ, Crystal RG: Lung growth after unilateral pneumonectomy: Quantitation of collagen synthesis and content, Am. Rev. Respir. Disease, 111:267-276, 1975.
19. Fung YC: Biomechanics: Motion, Flow, Stress, and Growth, Springer-Verlag, New York, 1990.
20. Fung YC, Liu SQ: Changes of zero-stress state of rat pulmonary arteries in hypoxic hypertension, J. Appl. Physiol., 70(6):2455-k2470, 1991.
21. Fung YC, Liu SQ: Change of residual strains in arteries due to hypertrophy caused by aortic constriction, Circ. Res., 65:1340-1349, 1989.
22. Liu SQ, Fung YC: Relationship between hypertension, hypertrophy, and opening angle of zero-stress state of arteries following aortic constriction, J. Biomech. Eng., 111:325-335, 1989.
23. Han HC, Fung YC: Species dependence on the zero-stress state of aorta: pig vs rat, J. Biomech. Eng., 113:446-451, 1991a.
24. Xie JP, Yang RF, Liu SQ, Fung YC: The zero-stress state of rat vena cava, J. Biomech. Eng., 113:36-41, 1991.
25. Han HC, Fung YC: Residual strains in porcine and canine trachea, J. Biomech., 24:307-315, 1991b.
26. Fung YC: Biomechanics: Mechanical Properties of Living Tissues, Springer-Verlag, New York, 1981.
27. Fung YC: Stress-strain-history relations of soft tissues in simple elongation, In: Biomechanics: Its Foundations and Objectives (Ed., Fung YC), Prentice-Hall, Englewood Cliffs, New Jersey, 181-208, 1971.
28. Fung YC: Biorheology of soft tissues, Biorheology, 10:139-155, 1973.
29. Chuong CJ, Fung YC: Three-dimensional stress distribution in arteries, J. Biomech. Eng., 105:268-274, 1983.
30. Patel DJ, Vaishnav RN: Basic Hemodynamics and its Role in Disease Process, University Park Press, Baltimore, Maryland, 1980.
31. Patel DJ, Vaishnav RN: In: Cardiovascular Fluid Dynamics (Ed., Bergel DH), Academic, New York, 2:1-64, 1972.
32. Hayashi K, Handa H, Mori K, Moritake K: J. Soc. Material Sci. (Japan), 20:1001-1011, 1971.
33. Fung YC, Fronek K, Patitucci P: Pseudoelasticity of arteries and the choice of its mathematical expression, Am. J. Physiol., 237:H620-H631, 1979.
34. Liu SQ, Fung YC: Influence of streptozocin-diabetes on zero-stress states of rat pulmonary and systemic arteries, Diabetes, 41:136-146, 1992.

35. Fung YC, Fukada E, Wang JJ: Biomechanics in China, Japan, and U.S.A., Proc. of an intern. conf. held in Wuhan, China, Science Press, Beijing, China, 1-13, May 1983.
36. Fung YC, Liu SQ: Strain distribution in small blood vessels with zero-stress state taken into consideration, Am. J. Physiol. Heart and Circulation, 262(2):H544-H552, 1992.
37. Hayashi K, Takamizawa K: Stress and strain distributions in residual stresses in arterial walls, In: Progress and New Directions of Biomechanics (Eds., Fung YC, Hayashi K, Seguchi Y), MITA Press, Tokyo, Japan, 185-192, 1987.
38. Takamizawa K, Hayashi K: Strain energy density function and uniform strain hypothesis for arterial mechanics, J. Biomech., 20:7-17, 1987.
39. Takamizawa K, Hayashi K: Uniform strain hypothesis and thin-walled theory in arterial mechanics, Biorheology, 25:555-565, 1988.
40. Fung YC: Cellular growth in soft tissues affected by stress level in service, In: Tissue Engineering (Eds., Skalak R, Fox DF), Alan Liss, New York, 1988.
41. Skalak R, Fox DF: Tissue Engineering, Alan Liss, New York, 1988.

FLOW AND VASCULAR GEOMETRY

Harry L. Goldsmith and Takeshi Karino

McGill University Medical Clinic
Montreal General Hospital
Montreal, Quebec
Canada

INTRODUCTION

Fluid mechanical factors play an important role in the localization of sites of atherosclerosis, the focal deposition of platelets resulting in thrombosis, and the formation of aneurysms in the human circulation. The localization is confined mainly to regions of geometrical irregularity where vessels branch, curve and change diameter and where blood is subjected to sudden changes in velocity and/or direction. In such regions, flow is disturbed and separation of streamlines from the wall, with formation of eddies, is likely to occur. We shall describe the flow patterns and fluid mechanical stresses at these sites and consider their possible involvement in the genesis of the above mentioned vascular diseases. However, in order to understand the relationship between vessel geometry and the observed flow patterns, it is first necessary to deal with some aspects of the mechanics of flow in branching, expanding and curved vessels. Such a treatment will also serve to dispel the notion, common among physicians and surgeons, that the formation of eddies at sites of disturbed flow represents turbulent flow.

BASIC FLUID MECHANICAL PRINCIPLES

Unsteady and Disturbed Flows in the Circulation

In general, within the circulation there is not a fully developed viscous, or Poiseuille flow with a parabolic profile. The latter, shown in Fig. 1, results in a distribution of fluid velocity in the axial direction, u(r), given by:

$$u(r) = \frac{2Q}{\pi R^4} (R^2 - r^2) \qquad (1)$$

and velocity gradient, or shear rate, G, given by:

$$G(r) = -\frac{du}{dr} = -\frac{4Q}{\pi R^4} r \qquad (2)$$

with a value at the wall:

$$G(R) = \frac{4Q}{\pi R^3} = \frac{2u(0)}{R} \tag{3}$$

Q being the volume flow rate and u(0) the centerline fluid velocity, $= 2\overline{U}$, where $\overline{U} = Q/\pi R^2$ is the mean velocity in Poiseuille flow.

As shown in Fig. 1, deviations from the parabolic profile occur in small vessels even at low flow rates, and are related to the effect of the crowding of red cells on the flow,[1-3] a phenomenon previously documented in suspensions of rigid spheres and discs[4] and in emulsions.[5,6] Moreover, since blood is not a homogeneous Newtonian liquid, the velocity profiles are affected by flow rate and hematocrit, particularly in vessels < 0.5 mm in diameter.

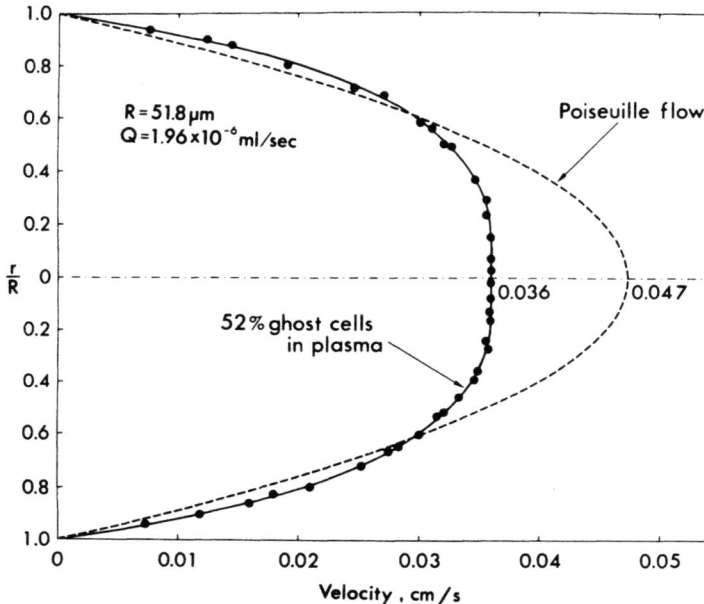

Figure 1. Distribution of velocity in a 50% ghost cell suspension flowing through a 103.6 μm diameter tube, obtained from the paths of tracer red cells, illustrating the deviation of the velocity profile from the parabolic one in Poiseuille flow. The velocity and shear rate distribution for Poiseuille flow were calculated assuming the blood to be a pure liquid flowing at the same volume flow rate as the ghost cell suspension. (From Karino and Goldsmith)[1]

In large vessels in which inertial effects are important, flow is not steady but pulsatile and exhibits kinetic energy effects as shown in Fig. 2. The flow oscillates periodically and is somewhat out of phase with the pressure gradient.[7] Also, within a given vessel, not all parts of the liquid move in phase with each other. Near the wall, where the shear rate, now dependent on both radial distance and time, is high and wall friction forces predominate, the liquid elements are almost in phase with the pressure gradient. In the center of the stream, however, the kinetic energy is high and inertia of the liquid results in its lagging increasingly behind the pressure gradient, and hence behind flow near the wall, as shown in Fig. 2.

Flow is also seriously affected locally by branching and curvature, where inertial effects result in the production of secondary flows having radial components, and sometimes disturbance created in the flow at one branch has not had time to dissipate before that due to another branch comes into play. That such effects may be particularly important with regard

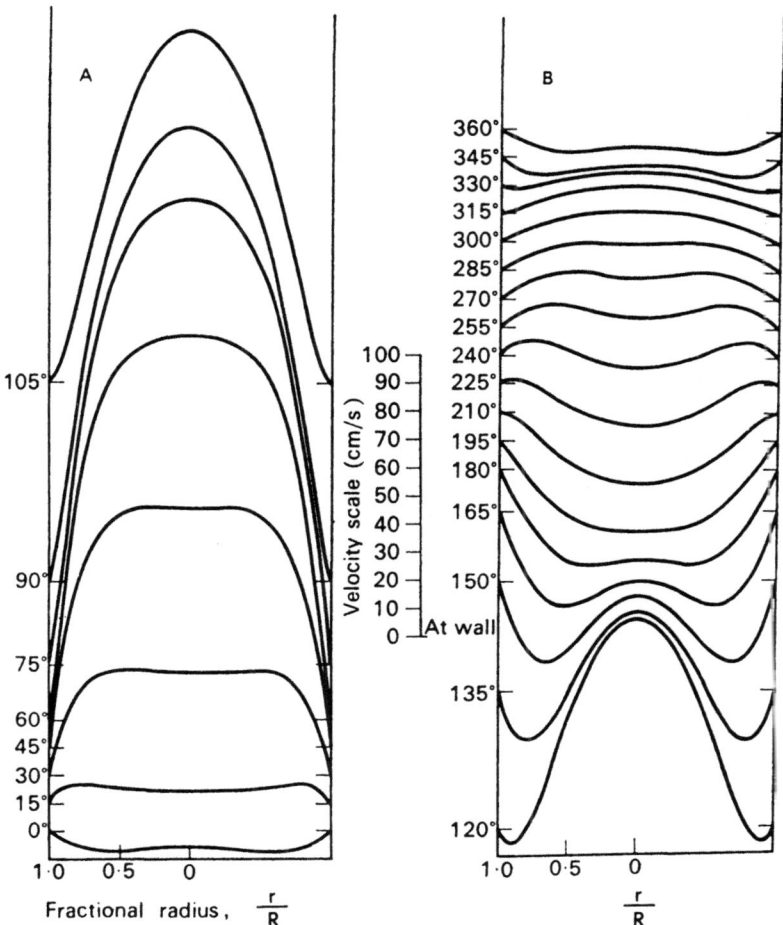

Figure 2. Pulsatile velocity profile in the femoral artery of a dog calculated from the measured pressure gradient, plotted at intervals $\omega t = 15°$ over one cycle ($\omega = 2\pi f$ is the angular velocity in radians/s where f is the frequency in cycles/s, t is time). The curves were obtained by summing together the first 4 harmonics of the flow curves together with a parabolic velocity profile representing the steady forward flow. The reversal of flow begins at $\omega t = 120°$ of the cycle at the wall while flow in the center is still moving forward. As a result, the maximum reverse flow occurs at r/R between 0.3 and 0.4. (From MacDonald's Blood Flow in Arteries)[7]

to the genesis of thrombosis and atherosclerosis, has long been suspected by investigators in the biomedical as well as the physical and engineering sciences. Rapid changes in the rate and/or direction of fluid motion have been held responsible for bringing about alterations and injury, not only to endothelium[8-10] and the media,[11] but also to the corpuscles.[12,13] This, in turn could lead to the aggregation and adhesion of the cells to the injured vessel wall. Indeed, atheromatous plaques and platelet thrombi have been observed at bifurcations and stenoses.[14-16] The question here is whether disturbed patterns of flow lead to increased, localized interactions with the vessel wall. Answers to this question have been obtained from microrheological studies, in our laboratory, of the flow behavior of blood cells in idealized models of stenoses[17,18] and T-junctions,[19,20] as well as in fixed transparent segments of natural arteries and veins.[21-27]

The term disturbed flow is used to distinguish the flow regimes encountered at stenoses, branches and curved segments of vessels from laminar and turbulent flow. Laminar flow, as seen in the example of Poiseuille flow (Fig. 1), is characterized by the steady, streamline motion of the fluid in layers parallel to the wall. In turbulent flow, which occurs beyond a critical value of the Reynolds number, the fluid elements exhibit irregular or random motion with respect to time and space.

In branching and non-uniform diameter vessels in the circulation there exists an additional flow regime that is not observed in uniform diameter straight tubes. Here, there are secondary fluid motions in directions away from that of the primary flow, and often there is separation of the streamlines from the vessel wall, with the formation of a vortex or a recirculation zone between the forward flowing mainstream and the wall. To describe this flow regime, which is neither laminar nor turbulent, we have come to use the term "disturbed flow."

Flow Parameters in the Human Arterial Tree

The velocity distributions shown in Fig. 1 were obtained under flow conditions in which the work required to shear the suspending liquid, and to cause the suspended particles to translate, rotate and deform, was done as viscous work. Here we deal with flow in large vessels in which velocities during the cardiac cycle are high (> 10 cm s^{-1}), inertial work is required to accelerate and decelerate the blood, and the tube Reynolds number is high:

$$Re_t = \frac{2R\overline{U}\rho}{\eta} \qquad (4)$$

where ρ and η are the respective fluid density and viscosity. Using equations (1) - (4) and assuming Poiseuille flow, representative values of Reynolds number, wall shear rate, G(R), may be calculated for various parts of the vascular system from the known vessel diameters and volume flow rates. These are given in Table I where it can be seen that mean linear velocities over one cycle decrease from > 800 to < 1 mm s^{-1} and the mean Re_t from 6000 to $< 10^{-3}$, in going from the heart to the microcirculation where viscous effects dominate the flow. The corresponding mean wall shear rates, and hence shear stresses, increase in going to the small vessels, reaching a maximum in the arterioles. It is evident, however, that wall shear rates at peak systole in some large arteries can be as large as those in small vessels.

Disturbed Flow Due to Sudden Changes in Flow Velocity

When blood is subjected to a change in mean velocity which occurs suddenly and is of sufficient magnitude, flow separation and the formation of an eddy will occur. The way in which such flows are generated can be illustrated using the concept of the boundary layer.

In fluid mechanics, an ideal fluid is defined as one that has no internal friction or viscosity, $\eta = 0$. As a consequence it is able to flow past a solid surface and maintain a slip velocity at the very interface. A real fluid has, to a greater or lesser degree (as in plasma), internal friction, and a true slip velocity is impossible. If the fluid has a low viscosity, and is subjected to flow at a high velocity, i.e., the Reynolds number is large, one can picture the flow of a real fluid past a solid body as being composed of two zones. The first zone is confined to a thin layer near the solid boundary where, under the influence of viscosity, frictional forces retard the fluid motion. The velocity increases from zero at the interface to the full value in the mainstream. The mainstream is the second zone, and is regarded as an ideal frictionless fluid in which there is no velocity gradient. The first zone is known as the boundary layer and its thickness can be calculated to be inversely proportional to the square root of a Reynolds number, Re_y, defined by $Re_y = \overline{U}y\rho/\eta$, where y is a characteristic distance perpendicular from the edge of the solid body. This approach is applicable strictly to gases and liquids of low kinematic viscosity ($= \eta/\rho$) flowing past streamlined bodies (e.g., an airplane wing). The analysis should not be extended to liquids subjected to laminar flow in tubes at high Reynolds numbers. The concept of the boundary layer in explaining flow separation, however, is a useful one, as the following qualitative considerations show.

TABLE I. FLOW PARAMETERS IN THE HUMAN ARTERIAL TREE

Vessel	Diameter mm	Mean Linear Velocity $\overline{U}$, mm s^{-1}			Mean Reynolds Number Re$_t$	Wall Shear Rate G(R), s^{-1} [1]		
		min.	max.	mean		min.	max.	mean
Ascending Aorta[2]	23.0 - 43.5	– 876	–	245	3210 - 6075	–	– 305	45
Femoral Artery[3]	5.0	-350	1175	188	283	-560	1885	302
Common Carotid[3]	5.9	99	388	187	332	134	526	253
Carotid Sinus[4]	5.2	85	325	156	244	130	500	240
External Carotid[4]	3.8	83	327	157	180	175	687	331
Small Arteries	0.3	–	–	50	2.3	–	–	1335
Arterioles	0.025	–	–	5	0.038	–	–	1600
Capillaries [5]	0.012	0.39	1.74	0.84	1.5 - 6.6×10^{-3}	260	1290	560

[1] Assuming Poiseuille Flow
[2] Mean systolic values from MacDonald's Blood Flow in Arteries[7]
[3] Values obtained *in vivo* by Anliker *et al.*[28]
[4] Values calculated assuming 65% flow into the internal carotid. Measured wall shear rates from *ex vivo* studies in steady flow[23] at Re$_t$ = 592 show that, at the outer wall of the sinus, there is a recirculating flow with G(R) varying from -135 to +530 s^{-1}. At the inner wall, G(R) > 2000 s^{-1}. At the inner wall of the external carotid, G(R) > 1000 s^{-1}.
[5] Values from Bollinger *et al* [29] from red cell velocities in human nailfold capillaries of large diameter. Calculations of wall shear rate would correspond to that in plasma flow. In "Bolus Flow" of a train of red cells, much higher wall shear rates would exist in the plasma layer surrounding the red cells.

Consider the sudden tubular expansion shown in Fig. 3. A liquid flows through the tube at a constant driving pressure in the axial, Z-direction, is suddenly decelerated as it enters the expansion, due to the sudden increase in cross-sectional area. Assuming the mainstream to be an ideal fluid, one can apply Bernoulli's equation relating pressure, P, to linear velocity, U, of an ideal fluid:

$$P + \rho gh + \frac{1}{2}\rho U^2 = \text{constant} \tag{5}$$

h being the height the liquid has traveled from a reference level. If the tube is horizontal (h is constant), differentiation of Eq. (5) with respect to flow distance z yields:

$$\frac{dP}{dz} + \rho U \frac{dU}{dz} = 0 \tag{6}$$

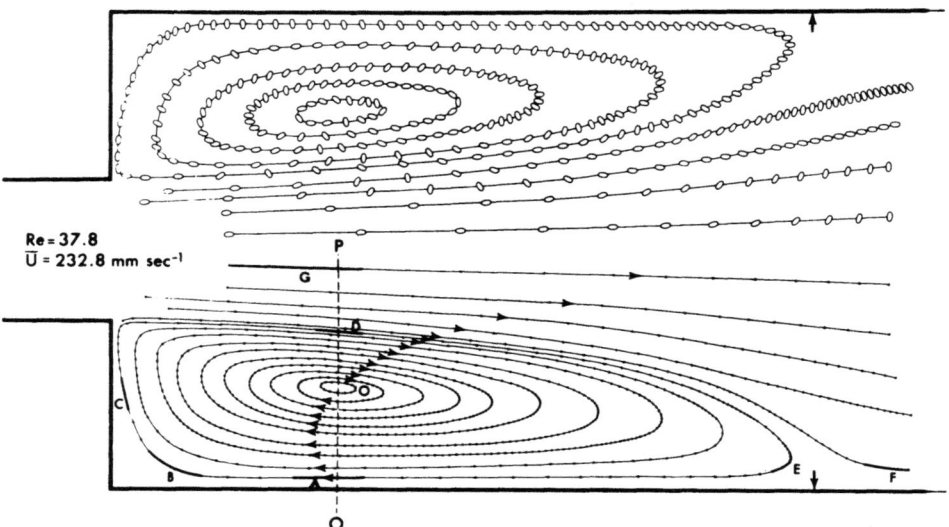

Figure 3. Flow separation as a liquid enters a sudden concentric expansion of vessel lumen from 150 to 500 μm and is rapidly decelerated, leading to a positive adverse gradient in pressure in the direction of flow. Liquid near the wall, unable to overcome the pressure gradient, is forced to a standstill and flows backward, leading to the formation of an annular vortex whose size increases with increasing inflow Reynolds number, Re, and the degree of expansion of the lumen. The mainstream is pushed away from the wall, and is separated from the vortex, from the entrance of the expansion to the reattachment point shown by the arrows. The streamlines and orientations of hardened red cells in the orbits of the annular vortex are shown in the lower and upper half, respectively. Velocities in mm s^{-1} at O: 2.4, A: 3.6, B: 3.0, C: 6.9, D: 94, E: 0.84, F: 0.95 and G: 272. (From Karino and Goldsmith)[18]

This shows that during deceleration on entry into the expansion, the gradient of flow has decreased ($dU/dz < 0$), hence the pressure gradient, dP/dz, must increase to satisfy the equation, i.e., there is a positive, adverse pressure gradient attempting to drive the fluid in the reverse direction. In the "frictionless layer" (mainstream), the kinetic energy of the fluid is sufficient to overcome this pressure gradient. However, the fluid within the boundary layer, which comes under the influence of the same pressure field, consumes its much smaller kinetic energy before having moved very far. It is eventually forced to a standstill and caused to flow backward by the external pressure. Fresh fluid arrives and experiences the same retardation, and the decelerated portion of the stream rapidly increases in volume, pushing the mainstream away from the boundary: flow separation has occurred. The fluid in the reverse flow regions coils and a vortex is formed (Fig. 3). It is evident from Eq. (6) that the decrease in U as well as the rate of deceleration, $-dU/dz$, influences the size of the effect.

Theoretically, flows through stenoses[30-33] and sudden expansions of vessel lumen[34-36] have been studied by solving the equations of motion and continuity using sophisticated computing techniques, but regarding blood as a homogeneous fluid.

Disturbed Flow Due to Sudden Changes in Direction of Flow

When there is a change in the direction of flow, as at a bend or bifurcation, the latter illustrated in Fig. 4, a fluid experiences a transverse pressure gradient across the tube forcing it to change direction. The elements of fluid with the highest kinetic energy will continue to move to the outside of the bend, while fluid with low kinetic energy will move to the inside under the action of the pressure gradient. This gives rise to a secondary cross-flow. If the fluid enters the bend with a parabolic velocity distribution, the elements having high kinetic energy are situated in the tube center surrounded by elements with low kinetic energy at the periphery. In the bend, a secondary flow consisting of two semicircles is set up. In the case of a bifurcation (Fig. 4), low velocity fluid at the top and bottom of the parent tube moves across the upper and lower regions of the daughter tube to form a double helical flow near

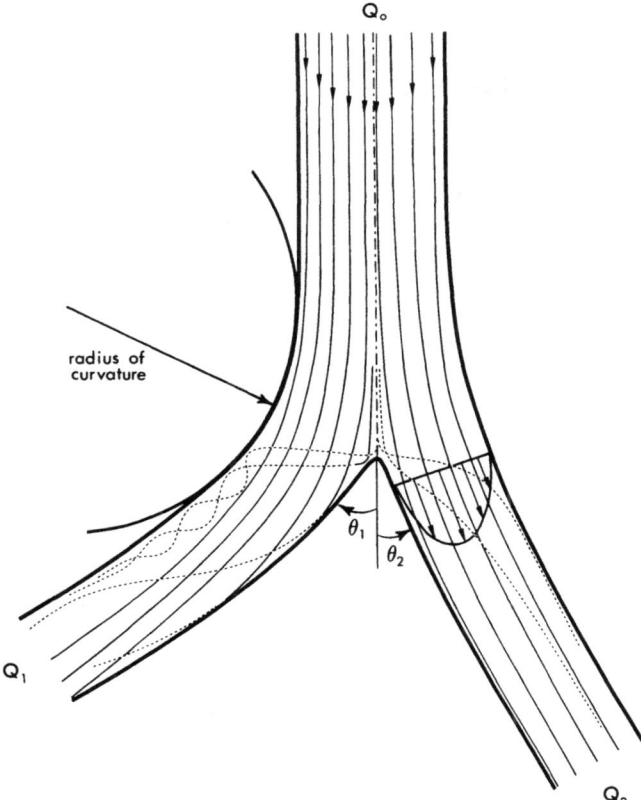

Figure 4. Flow through a bifurcation showing streamlines of the mainstream in the median plane (solid lines) which move towards the inner wall of the daughter tubes causing a secondary flow of low velocity fluid (dashed lines), which moves across the upper and lower regions at the entry of the daughter tube, to form a double helical flow near the outside wall. The large radii of curvature of the outer walls are typical of those in the arterial circulation. The extent of secondary flow increases with Reynolds number, bifurcation angle and ratio of cross-sectional area of daughter tubes to parent tube. The velocity profile in the median plane of one of the daughter tubes is also shown. (From Goldsmith and Turitto)[37]

the outside wall. High velocity fluid in the center of the parent tube moves on without being appreciably diverted. Fluid coming from the side wall of the parent tube may separate at the corner of the bifurcation and the streamlines from the top and bottom fill in the separated region. Fig. 4 also shows that the velocity profile in the diametrical plane of the daughter tube is skewed towards the flow divider. This is because relatively high velocity fluid at the center of the parent vessel is brought into close proximity of the wall of the flow divider.

The value of the Reynolds number in the parent tube at which secondary flow patterns and flow separation are first seen depends very much on the geometry of the bifurcation: the bifurcation angle, θ, the area ratio defined as the sum of the cross-sectional area of the daughter branches to the cross-sectional area of the parent branch, and the radii of curvature of the outer walls (Fig. 4).

In arterial bifurcations, θ varies from 30° to 120°, but the radii of curvature are large with gentle bends, as shown in Fig. 4, and this together with area ratios that are generally < 1.25 (i.e., there is not a marked decrease in blood velocity) minimizes the secondary flow.

Flows in bifurcations and T-junctions[38-45] and bends[36] have also been investigated by solving the equations of motion and continuity with the aid of advanced computing techniques.

MODELS OF DISTURBED FLOW IN THE CIRCULATION: SUDDEN CHANGE IN FLOW VELOCITY

Blood Cells in an Annular Vortex

We undertook an extensive study of the flow behavior of human blood cells in the annular vortex formed distal to a sudden tubular expansion of a 150-μm into a 500-μm diameter glass tube.[18] Such a flow geometry, illustrated in Fig. 3 and described above, served as a model of an arterial stenosis.

Red Blood Cells: As predicted by fluid mechanical theory,[34] when dilute suspensions of erythrocytes were subjected to steady flow through the model stenosis, a captive annular ring vortex was formed downstream of the expansion. Figure 3 shows paths and orientations of the erythrocytes in the median plane of the tube. During a single orbit, the measured particle paths and velocities, as well as the locations of the vortex center and reattachment point, were in good agreement with those predicted by the theory applicable to the fluid. Over longer periods, however, single cells and small aggregates < 20 μm in diameter migrated outward across the closed streamlines of the vortex predicted by the theory and left the vortex after describing a series of spiral orbits of continually increasing diameter until they rejoined the mainstream. In contrast, aggregates of cells > 30 μm in diameter remained trapped within the vortex, assuming equilibrium orbits or staying at the center. In pulsatile flow (a sinusoidal oscillatory flow superimposed in parallel with the steady flow), the observed phenomena were qualitatively similar to those described in steady flow. The vortex varied periodically in size and intensity; the axial location of the vortex center and reattachment point oscillated in phase with the upstream fluid velocity between maximum and minimum positions about a mean which corresponded to that measured in the absence of the component of oscillatory flow. At higher hematocrits from 15 to 45%, migration of single cells still persisted and resulted in the lowering of the hematocrit in the vortex region both in steady and pulsatile flow.[18]

The mechanism underlying the particle migration phenomenon is not a simple one, but it is likely that it is partly due to the dilution effect of the cell-poor plasma taken into the vortex from the fluid layer adjacent to the vessel wall proximal to the expansion. The mechanism for the trapping of large aggregates in the vortex was qualitatively explained by using existing fluid mechanical theories concerned with lateral particle migration near a tube wall[46] and by the operation of a mechanical wall effect.[47]

Platelet Aggregation in the Vortex: The flow behavior and interactions of human platelets in the annular vortex were studied at 37°C, using heparinized or citrated platelet-rich plasma (PRP) as well as washed platelets in Tyrodes-albumin solutions. It was demonstrated that the vortex provided favorable conditions for the spontaneous aggregation of normal human platelets through shear-induced collisions of particles while circulating in its orbits.[48] In a given suspension, the formation and growth of platelet aggregates could only be observed in a narrow range of Reynolds numbers (based on upstream linear velocity and tube diameter), which varied from suspension to suspension. Thus, in heparinized PRP, containing many sphered platelets with pseudopods and some microaggregates of two to six cells, the rate and extent of aggregation were the highest, with large elongated floating aggregates (> 100 μm in length) being observed to form in less than 1 minute and within the widest range of Re_t, based on upstream velocity and diameter, between 4.5 and 17.

From measurements of the distribution of shear rate in the vortex, it was possible to estimate the collision frequency between platelets, using two-body collision theory applicable to rigid spheres.[49,50] Assuming "platelet spheres" of volume 7.5 μm^3 and radius b = 1.2 μm, at a number concentration n = 5 × 10^5 per μl, the two body collision frequency, j_s, may be computed from:

$$j_s = \frac{32}{3} G_{rz} n b^3; \quad G_{rz} = \frac{\partial W}{\partial z} + \frac{\partial U}{\partial r} \qquad (7)$$

where G_{rz} is the component of the shearing rate of strain[18] in the vortex, r and z being the radial and axial coordinates, respectively. In the 15 μm periphery where most of the collisions occur and where flow is approximately parallel to the vessel wall, G_{rz} resembles a simple shear rate, $G(r) = dU/dr$. In the outermost orbit shown in Fig. 5, at an upstream Re = 12.2, G_{rz} varies from ~ 100 s^{-1} near the wall at B to 500 s^{-1} at H on the axial side of the vortex boundary, and Eq. (7) yields values of j_s varying from 1 to 5 collisions per cell per second. The volume in the outer 15 µm is ~ 10^{-2} µl, hence the total number of two-body collisions, $J_s = j_s \times n/2$ varies from 2.5 to 12.5 × 10^3 two-body collisions per second.

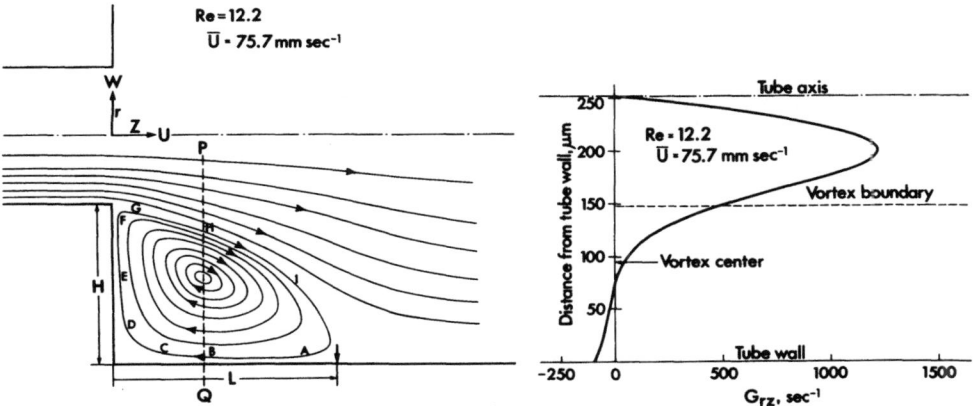

Figure 5. Left: Streamlines, radial, r, and axial, Z, coordinates of the flow field in the common median plane of the annular vortex at the sudden expansion of a 150 into 500 µm diameter tube (H = 175 µm; Re = 12.2). U and W are the respective axial and radial fluid velocities and the dashed line PQ indicates the axis of symmetry of the vortex. Right: Distribution of the shearing rate of strain along the dashed line PQ as calculated from theory[34] for PRP at 37°C. (From Karino and Goldsmith)[48]

The collision capture efficiency, α, is defined as the ratio of the collision capture frequency (number of collisions per second resulting in the formation of permanent doublets) to the collision frequency, j_s. Even if α is only 1%, at $J_s = 5 \times 10^3$ s^{-1}, 50 doublets per second would form in the vortex, and within 50 seconds, no singlets would remain. This is what is observed in heparinized plasma.

In citrated PRP and washed platelet suspensions, in which few microaggregates were seen prior to flow, the degree of aggregation was much reduced. However, when platelets in these suspensions were activated with subthreshold concentrations of ADP or thrombin, the large aggregates seen in heparinized PRP again formed. When the above suspensions were subjected to pulsatile flow in the expansion tube, there was a marked decrease in the number and size of the aggregates. Presumably, this was due to the continuously changing orbits of particles during the alternate expansion and contraction of the vortex, which shortened their residence times, as well as to the large variation in the shear rate in each cycle, beyond the range favorable for platelet aggregation.

The above results suggest that formation of platelet aggregates in vortices will be more likely to occur in the venous circulation, where the flow is steadier and the Reynolds number is lower, than in the arterial circulation.

Wall Adhesion of Platelets in the Vortex: The effects of disturbed flow on initial platelet adhesion to the vessel wall were studied using a large-scale expansion flow tube (0.92 into 3.00 mm diameter) whose inner wall was coated with collagen fibers, and suspensions of washed human platelets containing washed erythrocytes at hematocrits from 0 to 50%.[51] As illustrated in Fig. 6, platelet adhesion was localized within the vortex and downstream on either side of the reattachment point with a local minimum at the reattachment point itself. Furthermore, platelet adhesion increased, and both adhesion peaks became more pronounced, as the hematocrit increased. Surprisingly though, the adhesion peak in the vortex decreased and flattened out as the Reynolds number increased. These results are inconsistent with a diffusion-controlled platelet adhesion, which should show an increase in adhesion number density with increasing shear rate.(52,53) It appears that the particular flow pattern within the vortex is responsible for this localization. Thus, as illustrated in the upper panel of Fig. 6, only those cells carried by the curved streamlines to within one particle radius of the surface interact with the vessel wall and adhere to it on both sides of the reattachment point, which is also a stagnation point.(51,54) It follows from this that platelet adhesion onto the vessel wall, whether it be the natural endothelium or an artificial surface, will be localized wherever there is a stagnation point (or a reattachment point if it is a result of flow separation) where blood cells are carried by the flow toward the vessel wall along curved streamlines having a pronounced radial velocity component. If this mechanism operates in the circulation, a relatively higher adhesion of platelets, and hence a higher risk of thrombus formation is predictable, not only in regions of disturbed flow such as downstream of aortic or venous valves, mural thrombi, and stenoses, but also in all the branching arteries at the flow divider where there is a stagnation point.

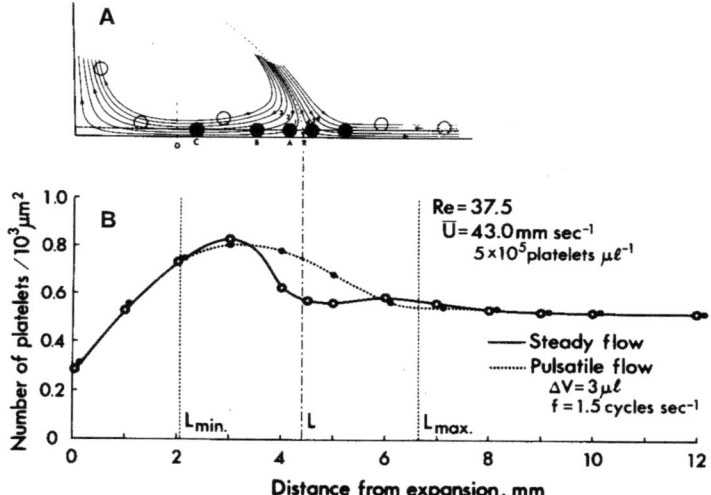

Figure 6. (A) Schematic representation of fluid streamlines near the tube wall downstream of the tubular expansion, showing the convective transport of particles in bulk flow to the vessel wall along the radially directed curved streamlines on either side of the reattachment point, R. The black circles represent the particles which are carried by the flow along the streamlines 1 - 6 within the critical distance for collision with the walls at points A to C in the vortex and the corresponding points downstream. The open circles represent particles which do not come close enough to collide with the wall. (B) Plot of the measured number density of adhering platelets obtained from an experiment carried out with a suspension of washed human platelets in Tyrodes-albumin solution containing no red cells. The figure shows the relationship between the flow pattern and the degree of platelet adhesion in, and downstream from the vortex in steady flow. (From Karino and Goldsmith)(54)

IN VIVO SUDDEN EXPANSION OF FLOW: A VENOUS VALVE

As an extension of the above described studies to natural blood vessels, one of us (T.K) developed a novel method for observing flow patterns in transparent arterial and venous segments from dogs, and arterial segments from humans, postmortem.(21) The results obtained in an isolated transparent dog saphenous vein containing a bileaflet valve[22] are shown in Fig. 7 which gives the flow patterns as observed along the common median plane of the vein and valve. There is an expansion flow with flow separation occurring at the edge of the valve leaflet, which under physiological flow conditions resulted in the formation of large paired vortices, located symmetrically on both sides of the bisector plane of the valve leaflets in each valve pocket. Particles continually entered the valve pockets from the mainstream, spending a long time in spiral orbits of decreasing diameter, while moving away from the bisector plane, and eventually left the vortex. Experiments carried out with hardened erythrocyte suspensions at 25% hematocrit showed that another smaller counter-rotating secondary vortex, driven by the large primary vortex, existed deep in each valve pocket (blank area of Fig. 7) where venous thrombi are believed to originate.[55,56] Furthermore, the erythrocyte concentration in this secondary vortex remained appreciably lower than that in the primary vortex. In such stagnant regions, fluid circulated with extremely low velocities, thus creating a very low shear field which allowed erythrocytes to form aggregates. The results suggest that in some pathological states, the valve-pocket vortices could act as automatic traps and generators of thrombi in a fashion similar to that demonstrated in the annular vortex described above.[18,48]

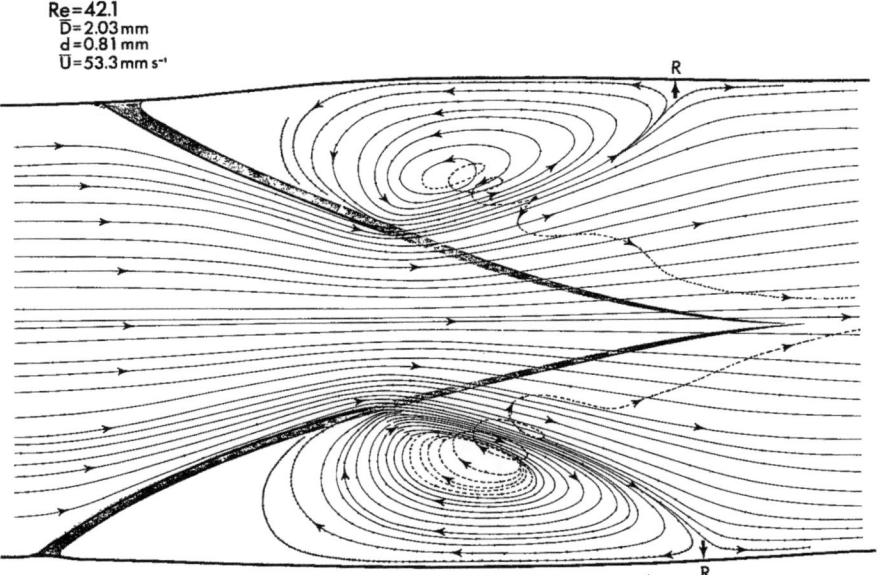

Figure 7. Detailed flow patterns in the common bisector plane of the valve leaflets in a 2 mm diameter dog saphenous vein containing a bileaflet valve and showing the formation of a spiral recirculation zone in each valve pocket. In fact, these consist of a pair of recirculation zones located symmetrically on both sides of the bisector plane. The solid lines are the paths of particles located in or close to, and the dashed lines are those located far away from, the bisector plane (the projection of the particle paths on the bisector plane). The arrows at R indicate the location of the reattachment point. (From Karino and Motomiya)[22]

MODELS OF DISTURBED FLOW IN THE CIRCULATION: SUDDEN CHANGE IN FLOW DIRECTION

Flow Patterns at Model T-Bifurcations

We also undertook an investigation of the flow patterns and distributions of fluid velocity and shear rate in glass models of T-junctions with branching angles from 30° to 150°, and side-to-main-tube diameter ratios from 0.33 to 1.0, over a wide range of inflow

Reynolds numbers, Re_o, and branch-to-parent tube flow ratios Q_1/Q_o.[19,20] Figure 8 gives the flow pattern obtained by photographing and analyzing the motions of 50 μm diameter latex spheres in the common median plane of a 90° uniform diameter T-junction when the main branch was partially occluded so that 80% of the flow left through the side branch (unlikely to occur *in vivo*). A large recirculation zone formed in the main tube (due to the sudden deceleration of fluid velocity as a portion of the flow is drawn off into the side tube) and a small, secondary recirculation zone was formed in the side tube (due to the sudden change in direction of the fluid which continues to move to the outside of the 90° corner, toward the flow divider). Particles entering the large recirculation zone described complicated orbits; some of them rejoined the flow through the main tube, others entered the side branch in a paired, spiral secondary flow with pronounced radial components, and some of these circulated through the secondary recirculation zone. When the degree of occlusion of the main tube was gradually reduced, the large recirculation zone became smaller and eventually disappeared as the flow rate ratio was reversed ($Q_1/Q_o = 0.8$), while that in the side branch grew in size.

By varying the branching angle it was shown that the critical Re_o for the formation of the main recirculation zone was lowest at 90° for all Q_1/Q_o, whereas for the side recirculation zone it decreased as the branching angle increased from 45 to 135°. When the diameter of the side tube was decreased, the main recirculation zone, which actually consisted of a pair of spiral secondary flows located symmetrically on both sides of the common median plane of the T-junction, became smaller and thinner and was confined to a thin layer adjacent to the tube wall, wrapped around the mainstream.[20]

The effect of radius of curvature of the walls at the junction was studied by comparing the critical inflow Reynolds numbers, Re_o, for the formation of recirculation zones and their sizes in the square T-junction (Fig. 8; radii of curvature < 2% of tube radius) with that in a rounded T-junction (radii of curvature ~ tube radius).[19] It was found that the recirculation zone in the side tube formed at a much lower Re_o in the square than the rounded junction, and that at a given Re_o and Q_1/Q_o, a larger main recirculation zone existed in the rounded

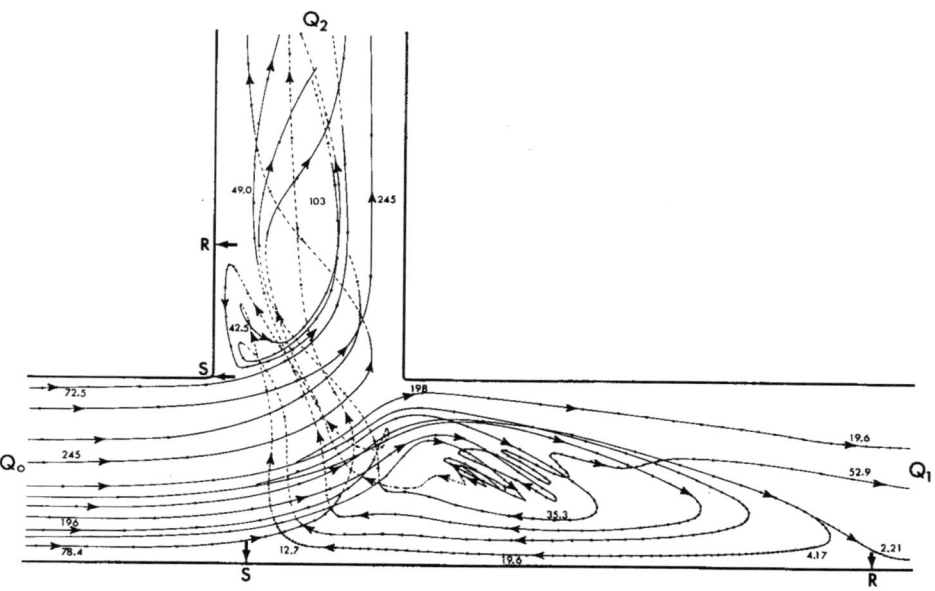

Figure 8. Flow patterns in the common median plane of a 3 mm diameter glass T-junction, traced by the paths of 50 μm polystyrene spheres in aqueous glycerol, drawn through the points measured at 22 ms intervals (numbers indicate velocities in mm s^{-1}). Suspension enters at left with a mean velocity $\overline{U}_o$ and at a Reynolds number Re_o; 80% of the flow leaves through the side tube. The very low radius of curvature at the corner opposite to, and at the flow divider results in a small corner vortex at the entry of the side tube filled with fluid from the large main vortex in the parent vessel. Solid lines represent particles traveling in, or close to the median plane, dashed lines those closer to the tube wall. Arrows at S and R indicate the respective separation and reattachment points. (From Karino *et al.*)[19]

junction. It appears that the formation of the side recirculation zone is largely affected by curvature of the wall at the bend opposite to the flow divider while that of the main recirculation zone is largely affected by the curvature at the flow divider.

IN VIVO EXAMPLES OF BIFURCATIONS AND CURVED SEGMENTS

Aortic T-Junctions

As an example of a natural arterial T-junction, we investigated the flow patterns in transparent segments of a dog abdominal aorta containing branches of the celiac, superior-mesenteric and right and left renal arteries.[24,25] The flow patterns illustrated in Fig. 9 for the aorto-celiac junction resemble those observed in the model T-junctions, but the degree of flow disturbance, even at an inflow Reynolds number as high as 609, is much less. At the geometrical flow ratio (Q_1/Q_2 = area ratio, aorta:celiac artery) flow separation occurred at $Re_o \sim 300$ well below the mean physiological $Re_o \sim 700$. However, instead of a large main standing recirculation zone as in the glass models, under physiologic flow conditions, there was only a pair of recirculation zones, confined to a thin layer close to the wall surrounding the mainstream. There was no side recirculation zone, no doubt due to the gentle curvature of the bend opposite the flow divider. This characteristic was shared by all the aortic T-junctions studied, as was the very sharp curvature of the bend at the flow divider. From the results obtained in the glass model T-junctions, this represents the optimum condition for minimizing the size of both regions of separated flow. Nevertheless, the curved streamlines of the recirculation zone and secondary flows will bring blood cells towards the vessel wall in a zone around the flow divider and in the side branch on the outer wall.

As the inflow Reynolds number was increased, the recirculation zones increased in length, and then, at $Re_O \sim 1200$, there was a sudden transition from disturbed to turbulent flow. Typical flow patterns observed at an early stage of turbulent flow at the same celiac and cranial-mesenteric (superior-mesenteric) artery junctions as those shown in Fig. 9, are illustrated in Fig. 10. It is evident that the recirculation zones of Fig. 9 have completely disappeared and the tracer particles exhibited random motion, and sometimes described small circles near the vessel wall. Flow in the side branches, however, remained non-turbulent.

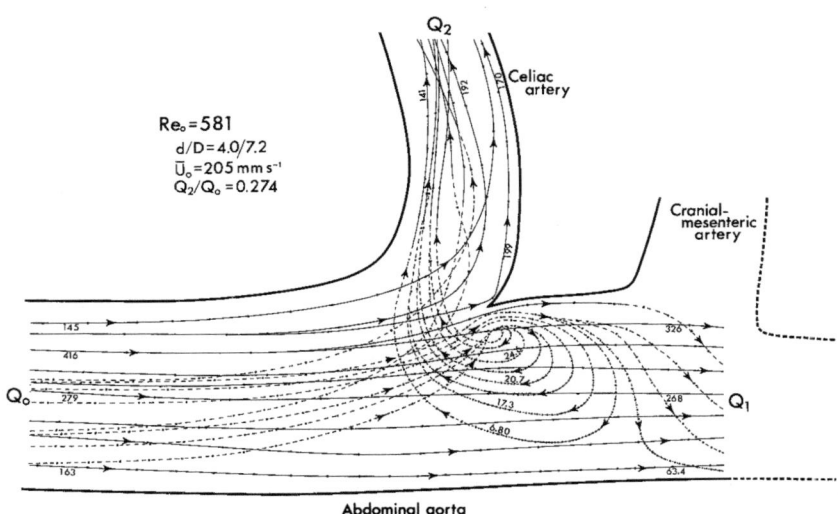

Figure 9. Flow patterns at the aortoceliac artery junction of an isolated transparent dog abdominal aorta as traced by the paths of 200 μm polystyrene spheres suspended in oil. Unlike the model T-junction of Fig. 8, curvature of the wall opposite the flow divider is very high, and that at the flow divider is very low. This minimizes the flow disturbance and results in the formation of paired thin-layered recirculation zones on both sides of the common median plane, adjacent to the vessel wall wrapped around the undisturbed mainstream. Numbers on the streamlines indicate sphere velocities in mm s^{-1}. (From Karino et al.)[25]

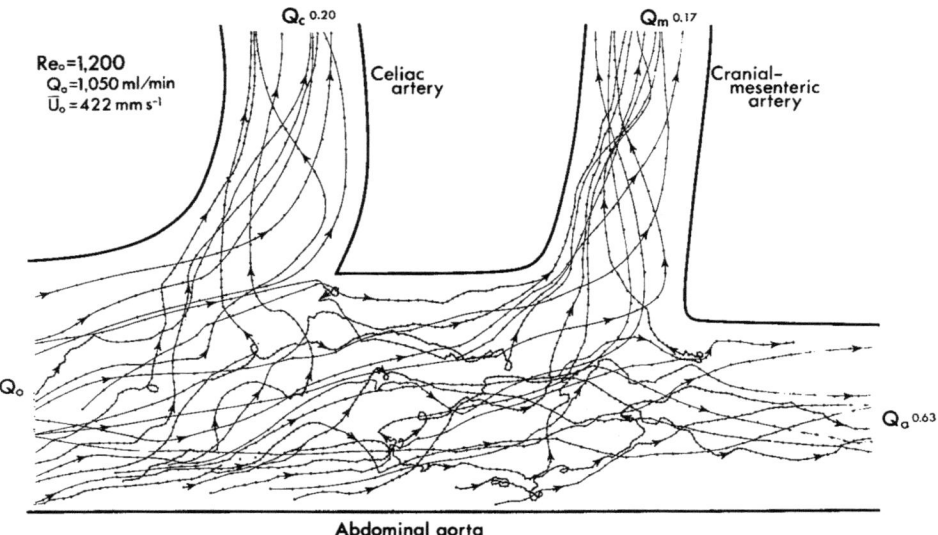

Figure 10. Particle flow behavior at the onset of turbulent flow. The figure shows the random-like paths of tracer polystyrene spheres located near the artery walls, in the same segment of the dog abdominal aorta as in Fig. 11. Note that the recirculation zones shown in Fig. 9 have completely disappeared. (From Karino et al.)[24]

The Human Carotid Bifurcation: A Unique Y-Bifurcation

The human carotid bifurcation, unlike other Y-bifurcations, exhibits a marked flow disturbance associated with a large recirculation zone located in the carotid sinus. The flow patterns were studied in detail using a transparent arterial segment prepared from a human subject post-mortem.[23] It was found that a standing recirculation zone consisting of a pair of complex spiral secondary flows, located symmetrically on both sides of the common median plane of the bifurcation, formed in the carotid sinus over a wide range of inflow Reynolds number, Re_o, and flow ratios, Q_1/Q_o (internal/common carotid). Figure 11, upper part, shows the detailed flow patterns in the carotid sinus. Particles were deflected at the flow divider and traveled laterally and very slowly along the wall above and below the common median plane, almost at right angles to, and encircling the mainstream. They then changed direction, moving back along the outer wall of the internal carotid artery at the site of the sinus, describing spiral orbits in the recirculation zone before rejoining the mainstream. Downstream from the stagnation point (R), a strong counter-rotating double helicoidal flow developed. The formation and the size of the recirculation zone were largely dependent on Q_1/Q_o as well as on Re_o. The size of the recirculation zone increased from ~ 4 mm at Re_o = 300 to a maximum of ~ 9 mm at $Re_o > 800$.

Measurements of the velocity profiles (Fig. 11, lower part) showed that these were strongly skewed towards the inner walls of the bifurcation, creating a high shear field along the vessel wall downstream from the flow divider. Due to the presence of the paired standing recirculation zones in the carotid sinus, the wall shear rate, and hence the wall shear stress, changes sign and becomes negative at the separation point (S); it becomes positive again downstream from the stagnation point, indicated by R. Thus, in the carotid sinus, there is a region where the vessel wall is stretched in opposite directions by the counter-directed wall shear stresses.

The results suggest that, under physiological conditions (mean Re_o ~ 600, Q_1/Q_o ~ 0.7),[57,58] a standing recirculation zone exists in the carotid sinus, thereby affecting local mass transfer and interactions of blood cells with the vessel wall which may lead to the incidence of thrombosis and atherosclerosis in this region.

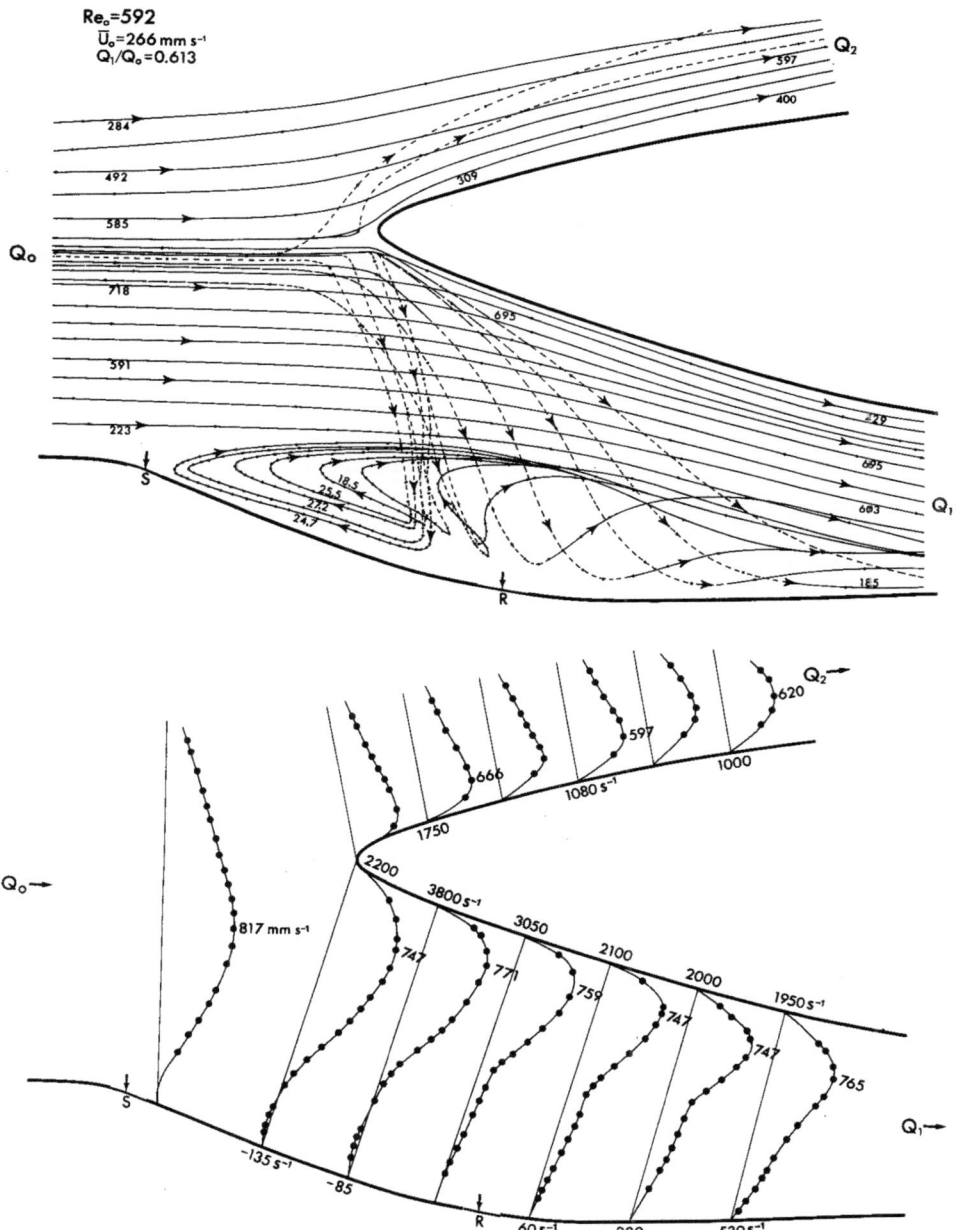

Figure 11. Upper: Flow patterns in the human carotid artery bifurcation in steady flow showing formation of a recirculation zone (paired spiral secondary flows) and a counter-rotating double helicoidal flow, both located symmetrically on either side of the common median plane in the internal carotid artery. Solid lines are paths of particles in or close to, and dashed lines paths far away from the common median plane. Arrows at S and R denote respective locations of the separation and stagnation points; numbers on the streamlines indicate the translational velocities in mm s^{-1}. Lower: Distribution of fluid velocity and shear rate in the common median plane of the bifurcation. Measured maximum velocities at each cross-section are given in mm s^{-1}. (From Motomiya and Karino)[23]

In Vivo Example of a Curved Vessel: The Aortic Arch

The aortic arch and descending aorta provide an example of markedly disturbed flow not only because of the curvature of the aorta itself, but also because of flow at very high mean Reynolds number (> 1,000), and the existence of major branches in the arch. We studied the flow patterns using isolated transparent dog arterial trees containing the whole heart, the aortic arch and descending aorta.[59] Inflow Reynolds numbers varied from 500 to 2,000, and most experiments were carried out with 15% of the aortic inflow distributed to the brachiocephalic artery and 20% to the left subclavian artery. Under these conditions, formation of complex secondary flows and eddies was observed at each branching site of the aortic arch and in the aortic sinus. Figure 11 shows tracings of particle paths during steady flow in the aortic arch of a young dog where it was found that the flow consists of three major components: quasi parallel undisturbed flow located close to the common median plane of the ascending aorta and the two daughter branches, spiral secondary flows located near the ventral periphery of the aorta and the side branches, and the mainflow to the side branches. Flow separation does occur at the inner (lower) wall of the aortic arch. However, no recirculation zone was formed downstream of the separation point. Instead, the region of separated flow was filled with the peripheral spiral secondary flows. Thus, looking down the aorta, the flow in the arch appeared as a single helicoidal flow revolving in a clockwise direction, rather than the two that had been anticipated. When the flow rates in the side branches were reduced from the control values, a recirculation zone was formed in each branch adjacent to the vessel wall opposite to the flow divider. Particles in the mainstream were deflected sideways at the flow divider and travelled laterally. Some entered the side branches, while others traveled backward along the vessel wall and, after describing multiple spiral orbits, rejoined the mainflow in the descending aorta or the side branches. In vessels having a partially opened aortic valve, it was also observed that strong eddies were formed at the entrance region of the aorta slightly downstream of the aortic sinus.

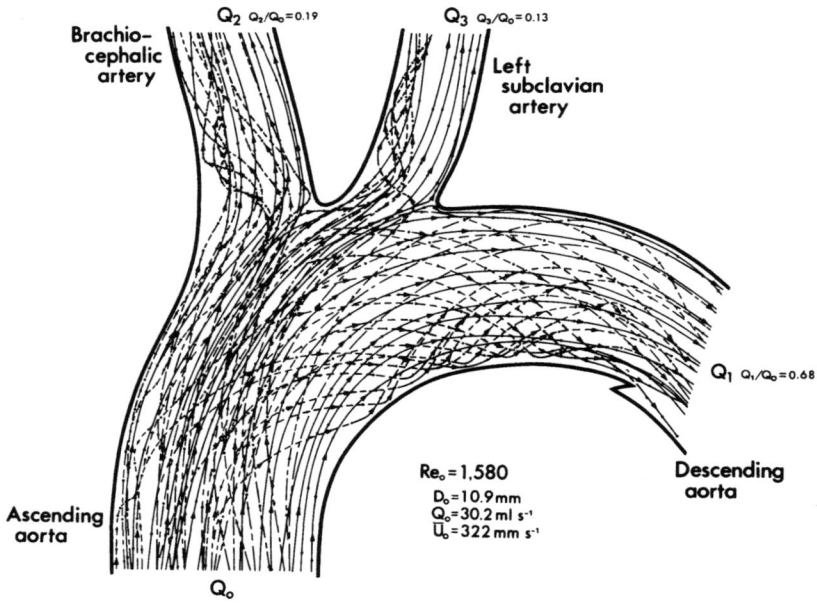

Figure 12. Disturbed flow due to vessel curvature in the dog aortic arch at an inflow Reynolds number Re_o = 1,580, with flow patterns obtained using an isolated transparent dog arterial tree containing the whole heart, aortic arch and descending aorta. Shown are fluid streamlines traced from the paths of polystyrene spheres demonstrating the existence of three major flow components: a quasi parallel undisturbed flow located close to the common median plane of the ascending aorta and two daughter branches (solid lines), a single clockwise spiral helicoidal secondary flow near the ventral periphery of the aorta and branches (dashed lines), and secondary flows generated at the flow dividers of the side branches (dashed lines). (From Sohara and Karino)[59]

The velocity distributions in the common median plane were calculated at several locations in the aorta and the side branches. In all vessels studied, the velocity distributions in the entrance region of the aorta were blunted and skewed towards the inner (lower) wall of the arch. Downstream of the branching site of the subclavian artery, the skewness of the velocity distribution reversed, and the location of maximum velocity shifted toward the outer wall of the arch.

The velocity distribution in the diametrical plane normal to the common median plane was also calculated and also found to be blunted, here skewed toward the dorsal side of the aorta. This is largely due to the fact that the axes of the side branches are located not exactly on the median plane of the aortic arch, but off center toward the dorsal wall of the aorta. In the case of older dogs, the curvatures of the bend at the apex portion of the aortic arch were much sharper than those of younger dogs, resulting in the formation of more pronounced spiral helicoidal flows in the aorta. It was also observed that behind the valve leaflets, a well defined standing vortex was formed in each aortic sinus at all the Reynolds numbers studied.

FLOW PATTERNS AND VASCULAR DISEASE

Flow Patterns in the Human Circle of Willis

Aneurysms: It is known from statistical data that the great majority of intracranial saccular aneurysms occur selectively at certain branching sites of the circle of Willis. To study the possible connection between blood flow patterns and the localization of these aneurysms in human intracranial cerebral arteries, flow patterns and velocity distributions at the major branching sites of the circle of Willis were obtained using isolated transparent segments prepared from humans postmortem.[60] Saccular aneurysms were found at the flow divider of 3 anterior communicating-anterior cerebral artery junctions having a long and relatively large diameter communicating artery, as shown in Fig. 13. In each case, the aneurysm was located around the flow divider having a large branching angle where the fluid elements from the central core of the inflow vessel, located at the leading edge of the velocity profile, directly impinged on the vessel wall around the flow divider. Incipient aneurysms were also found at the flow divider of the middle cerebral artery bifurcation and at the branching site of the anterior choroidal artery from the internal carotid artery where flow patterns similar to those described above existed. By contrast, the approaching velocity profiles at the bifurcation of the internal carotid artery and at the basilar artery, at which sites the incidence of aneurysm formation is lower, were found to be either blunted, or bipolar. In the case of the carotid bifurcation, this was due to the development of strong swirling flows in the carotid siphon. In the case of the basilar artery bifurcation, shown in Fig. 14, it was due to insufficient entrance length for the development of a parabolic profile proximal to the bifurcation. Here, the approaching velocity profile was largely dependent on the anatomical structure of the vertebral arteries. The approaching velocity profile was bipolar or blunted when the diameters of the two vertebral arteries were equal or very close to each other and formed a symmetrical Y-junction at the entrance of the basilar artery, whereas it was quasi-parabolic when the diameters of the two vertebral arteries were very different. The above findings suggested that the blunted or bipolar shaped velocity profile observed in both bifurcations of the internal carotid artery and the basilar artery may play a protective role in aneurysm formation at these two sites.

As shown in the figure for the case of approximately equal diameter daughter vessels, the flow was considerably disturbed at the Y-bifurcation and complex secondary flows were found to exist over a wide range of Re_o and flow ratios Q_1/Q_2. Particles in the mainstream were deflected sideways at the flow divider, traveled backwards along the vessel wall and entered one of the right or left superior cerebellar arteries. With increasing Re_o, the deflected particles started to describe multiple spiral orbits, creating a region of highly disturbed flow over the entire lumen of the basilar artery between the superior cerebellar arteries and the flow divider of the bifurcation.

Atherosclerotic wall thickening: Local flow patterns appear to be involved in the localization of atherosclerosis, as illustrated by studies of the middle cerebral bifurcation.[26] In each of five vessels isolated, atherosclerotic thickening of the vessel wall was found to be localized around the hips of the bifurcation. A standing recirculation zone, very similar to

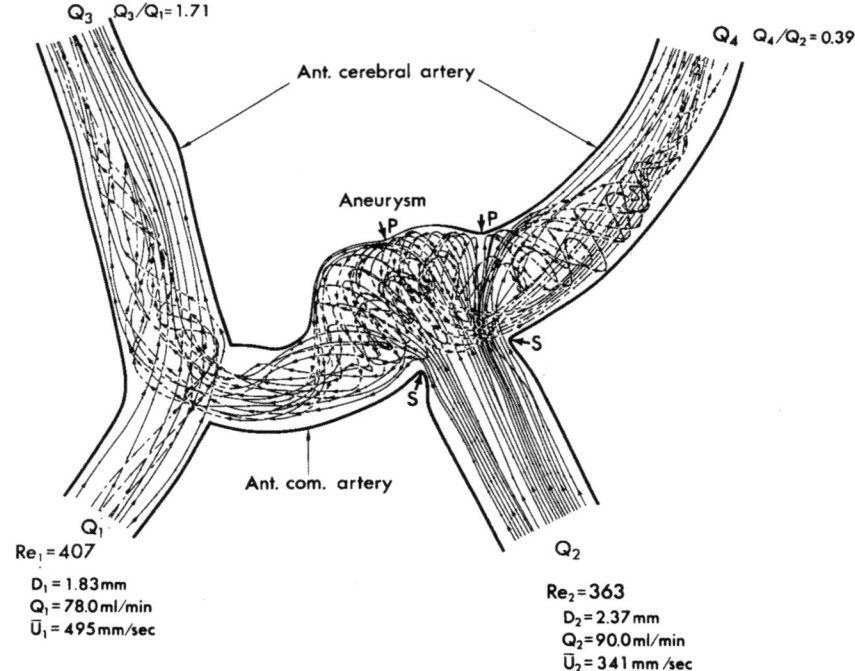

Figure 13. Flow patterns in an H-type anterior communicating artery junction with an aneurysm, showing the formation of spiral secondary flows in the aneurysmal sack which extend distally into both the anterior cerebral artery and the anterior communicating artery. (From Karino)[26]

that observed in the carotid artery bifurcation, formed along the outer wall of one or both daughter vessels (depending on the Reynolds number in the parent vessel and the flow ratios in the two daughter vessels) at the exact locations where the atherosclerotic thickening of the vessel wall occurred. Furthermore, at the normal physiological range of flow rates and flow ratios tested, there was a strong correlation between the longitudinal length of the regions of disturbed flow and that of the atherosclerotic wall thickening. Figure 15 shows the flow patterns observed in steady flow in one of the bifurcations having an almost perfectly symmetric structure and spatial arrangement of the daughter vessels. As evident from the figure, even when the flow in the parent vessel was distributed equally to the two daughter vessels, the region of disturbed flow and atherosclerotic wall thickening was much longer in the right branch than that in the left branch where the wall thickening was confined to only a very narrow area. In pulsatile flow, the complex spiral secondary flows and the recirculation zones oscillated in phase with the pulsatile flow velocity, and the locations of the stagnation and separation points situated on the outer walls of the bifurcation moved back and forth along the vessel wall. However, the general flow patterns remained the same as those observed in steady flow. This was true for all five vessels studied. A similar observation was made at an arterial bend located further downstream from the middle cerebral artery bifurcation. Here, a recirculation zone was formed along the inner wall of the bend slightly downstream from the apex at the very site of atherosclerotic wall thickening.

Flow Patterns and Atherosclerotic Lesions in Coronary Arteries

The exact anatomical locations of atherosclerotic lesions and flow patterns in the human left and right coronary arteries have been studied in detail using transparent coronary arterial trees prepared from humans postmortem.[27] It was found that atherosclerotic lesions

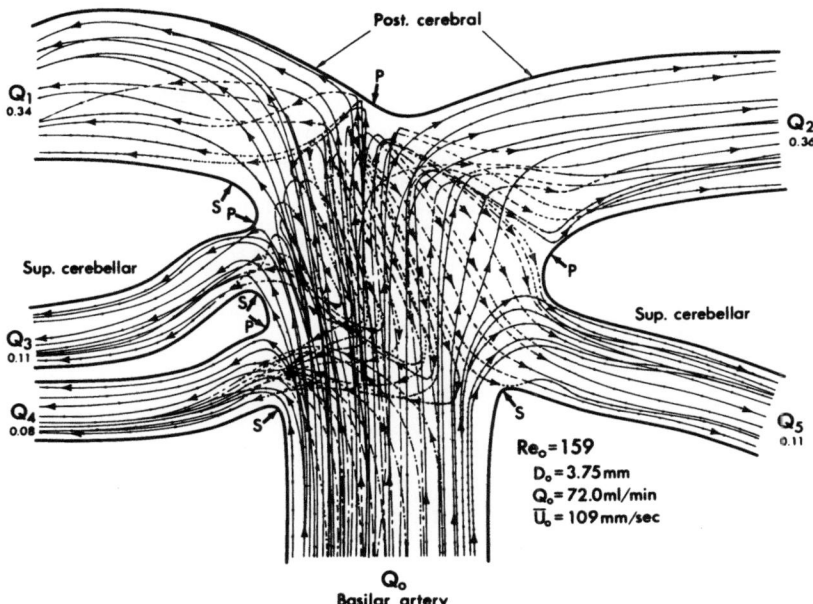

Figure 14. Flow patterns at the human basilar artery bifurcation showing the formation of complex secondary flows in the basilar artery proximal to the flow divider. The arrows at S and P indicate the respective locations of the separation and stagnation points. (From Karino)[26]

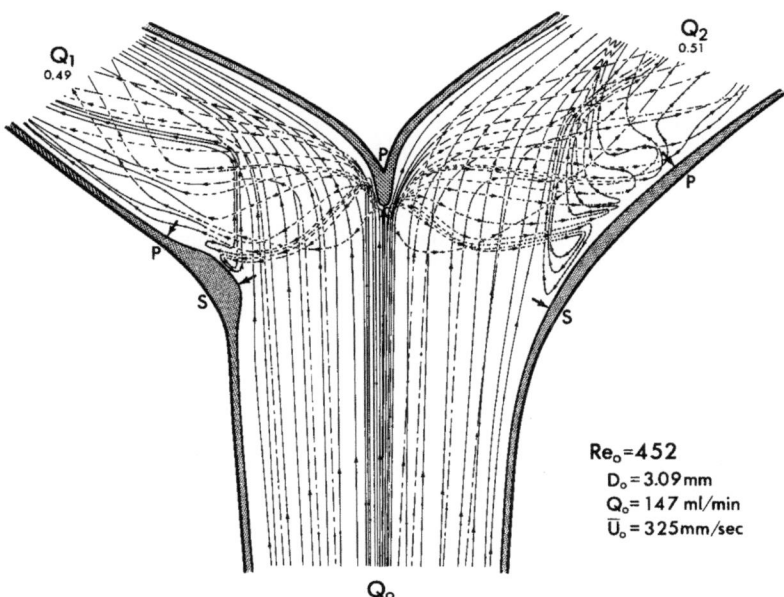

Figure 15. Detailed flow patterns at the middle cerebral artery bifurcation, showing the formation of secondary flows and recirculation zones along the outer walls of the bifurcation in steady flow, whose locations and longitudinal lengths closely match those of the atherosclerotic wall thickenings. Solid lines are the paths of particles in or close to the common median plane, dashed lines are paths far away from the common median plane. Arrows at S and P denote the respective locations of separation and stagnation points. (From Karino)[26]

develop exclusively at the outer wall (hips) of major bifurcations and T-junctions, and at the inner wall of curved segments, where flow was either slow or disturbed with the formation of slow recirculation and secondary flows and where wall shear stress was low. In no instance were atherosclerotic lesions found at the flow divider where fluid velocity and wall shear stress were high and where the formation of early atherosclerotic lesions have been observed in experimental animals fed high cholesterol diets. In the example shown in Fig. 16, that of the trifurcation of the left main coronary artery (LMCA), prepared from a 61-year old male subject, flow separation occurred at the outer wall (hips) of the bifurcation, creating regions of separated flow. These were then filled with spiral and recirculation flows formed as a result of the strong deflection of the flow from the LMCA at the obtuse angle flow

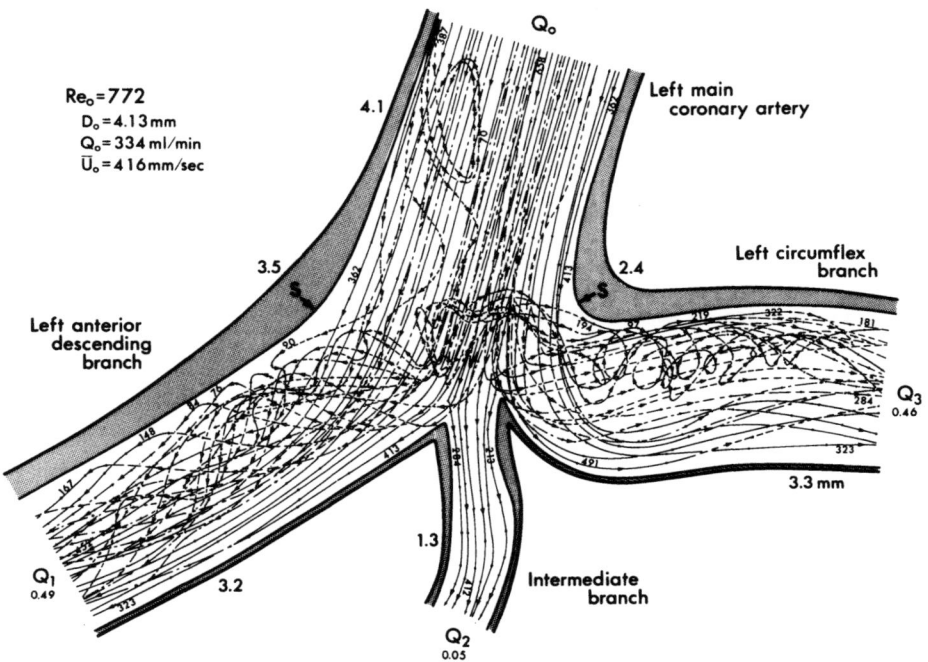

Figure 16. Flow patterns in steady flow at the trifurcation of the left main coronary artery (LMCA) in an arterial tree, prepared from a 61-year old male subject, observed normal to the common median plane of the LMCA and its two major branches, the left anterior descending and left circumflex arteries. Recirculation zones and complex secondary flows form at these branches and atherosclerotic wall thickenings were found at the hips of the trifurcation adjacent to the regions of disturbed flow. (From Asakura and Karino)[27]

divider. Adjacent to the regions of disturbed flow are the very sites where atherosclerotic lesions were localized. Thus, as shown in Fig. 18, in the left anterior descending branch (LAD), atherosclerotic lesions were found along the inner (lower lateral) wall of the gently curved segment of the proximal portion of the LAD where the measured fluid velocity and wall shear stress were relatively low. The deflection of the flow was much stronger in the lower half of the common median plane in both the LAD and the left circumflex branch resulting in the formation of spiral secondary and recirculation flows along the lower lateral wall of the trifurcation.

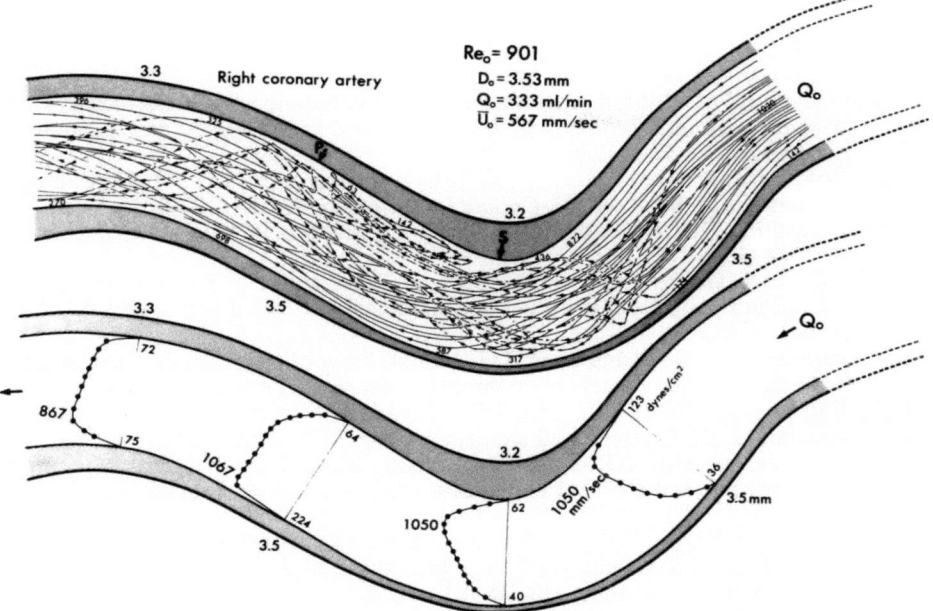

Figure 17. Flow patterns (upper) and distribution of fluid axial velocity (lower) as observed in steady flow in the common median plane (parallel to the pericardium) of an arterial segment with multiple bends located in the middle to distal portions of the right coronary artery, prepared from the same subject as for Fig. 16. Secondary flows and a recirculation zone develop adjacent to the inner wall of the middle bend. The lower portion of the figure shows changes in velocity distribution along the arterial segment as the region of high shear rate moves from the inner wall, proximal to the central sharp bend, to the outer wall at the gentle second bend. (From Asakura and Karino)[27]

In the right coronary artery, as illustrated in Fig. 17 (upper part) in an arterial segment with multiple bends, most of the atherosclerotic lesions were confined to the curved segments along the inner wall where flow was either disturbed with formation of a recirculation zone or very slow. Figure 17 shows that flow separation occurred at the inner wall of the middle and distal bends. The regions of separated flow were filled with fluid from the peripheral thin-layered secondary flow which traveled along the vessel wall, all the way from the outer wall of each bend. In the middle bend, a thin-layered standing recirculation zone was formed along the upper (pericardial side) inner wall, just distal to the apex of the sharp bend where atherosclerotic wall thickening was localized. The lower part of the figure shows the velocity distributions which are seen to drastically change within a few vessel diameters between locations proximal and distal to the sharp bend. In the proximal portion, the velocity distributions were skewed towards the inner wall (higher shear rate region), thus facilitating flow separation and formation of a recirculation zone along the inner wall distal to the apex of the bend. Just distal to the apex of the bend, the peak in the velocity distribution gradually shifted towards the outer wall of the first bend, and hence towards the inner wall of the second bend, again favoring flow separation at the inner wall of the second bend. Atherosclerotic wall thickenings were localized in an alternating manner at the inner wall of each bend, with a maximum occurring in regions of recirculation where fluid velocity and wall shear stress were low. Thus, the results here, as well as in the circle of Willis indicate that there is a strong correlation between the sites of low flow velocity (low wall shear stress) and the preferred sites for the genesis and development of atherosclerosis in man.

REFERENCES

1. T. Karino and H.L. Goldsmith, Rheological factors in thrombosis and haemostasis, *in*: "Haemostasis and Thrombosis," A.L. Bloom and D.P. Thomas, eds., Churchill Livingstone, London, England (1986).
2. H.L. Goldsmith, and J. Marlow, Flow behavior of erythrocytes. II. Concentrated suspensions of ghost cells, *J. Colloid Interface Sci.* 73:383 (1979).
3. H.L. Goldsmith, Red cell motions and wall interactions in tube flow, *Fed. Proc.* 30:1578 (1971).
4. A. Karnis, H.L. Goldsmith, and S.G. Mason, The kinetics of flowing dispersions. I. Concentrated suspensions of rigid particles, *J. Colloid Interface Sci.* 22:531 (1966).
5. F.P. Gauthier, H.L. Goldsmith, and S.G. Mason, Flow of suspensions through tubes. X. Liquid drops as models of erythrocytes, *Biorheology* 9:205 (1972).
6. E.B. Vadas, H.L. Goldsmith, and S.G. Mason, The microrheology of colloidal dispersions. III. Concentrated emulsions, *Trans. Soc. Rheol.* 20:373 (1976).
7. W.W. Nichols, and M.F. O'Rourke, "MacDonald's Blood Flow in Arteries: Theoretic, Experimental and Clinical Principles," 3rd ed., Lea & Febiger, Philadelphia (1990).
8. M.A. Reidy, and D.E. Bowyer, Scanning electron microscopy of arteries. The morphology of aortic endothelium in hemodynamically stressed areas associated with branches, *Atherosclerosis* 26:181 (1977).
9. S. Glagov, Hemodynamic risk factors: Mechanical stress, mural architecture, medial nutrition and the vulnerability of arteries to atherosclerosis, *in*: "The Pathogenesis of Atherosclerosis," R.W. Wissler and J.C. Geer, eds., Williams and Wilkins, Baltimore (1972).
10. D.L. Fry, Hemodynamic factors in atherogenesis, *in*: "Cardiovascular Diseases," P. Scheinberg, ed., Raven Press, New York (1976).
11. M.R. Roach, The effect of bifurcations and stenoses on arterial disease, *in*: "Cardiovascular Flow Dynamics and Measurements," N.H.C. Hwang and N.A. Normann, eds., University Park Press, Baltimore (1977).
12. J.F. Mustard, E.A. Murphy, H.C. Rowsell, and H.G. Downie, Factors influencing thrombus formation *in vivo*, *Am. J. Med.* 33:621 (1962).
13. J.F. Mustard, and M.A. Packham, The role of blood and platelets in atherosclerosis and the complications of atherosclerosis, *Thromb. Diathes. Haemorrh.* 33:444 (1975).
14. H.D. Geissinger, J.F. Mustard, and H.C. Rowsell, The occurrence of microthrombi on the aortic endothelium of swine, *Can. Med. Assoc. J.* 87:405 (1962).
15. J.R.A. Mitchell, and C.J. Schwartz, The relationship between myocardial lesions and coronary disease. II. A select group of patients with massive cardiac necrosis of scarring, *Brit. Heart J.* 25:1 (1963).
16. M.A. Packham, H.C. Roswell, L. Jorgensen, and J.F. Mustard, Localized protein accumulation in the wall of the aorta, *Exp. Mol. Path.* 7:214 (1967).
17. S.K. Yu, and H.L. Goldsmith, Behavior of model particles and blood cells at spherical obstructions in tube flow, *Microvasc. Res.* 6:5 (1973).
18. T. Karino, and H.L. Goldsmith, Flow behaviour of blood cells and rigid spheres in an annular vortex, *Phil. Trans. Roy. Soc. (London)* B 279:413 (1977).
19. T. Karino, H.H.M. Kwong, and H.L. Goldsmith, Particle flow behavior in models of branching vessels: I. Vortices in 90° T-junctions, *Biorheology* 16:231 (1979).
20. T. Karino, and H.L. Goldsmith, Particle flow behavior in models of branching vessels. II. Effect of branching angle and diameter ratio on flow patterns, *Biorheology* 22:87 (1985).
21. T. Karino, and M. Motomiya, Flow visualization in isolated transparent natural blood vessels, *Biorheology* 20:119 (1983).
22. T. Karino, and M. Motomiya, Flow through a venous valve and its implication in thrombus formation, *Thromb. Res.* 36:245 (1984).
23. M. Motomiya, and T. Karino, Particle flow behavior in the human carotid artery bifurcation, *Stroke* 15:50 (1984).
24. T. Karino, M. Motomiya and H.L. Goldsmith, Flow patterns in model and natural vessels, *in*: "Biologic and Synthetic Vascular Prostheses," J. Stanley, ed., Grune & Stratton, New York (1982).

25. T. Karino, M. Motomiya, and H.L. Goldsmith, Flow patterns at the major T-junctions of the dog descending aorta, *J. Biomechanics* 23:537 (1990).
26. T. Karino, Microscopic structure of disturbed flows in the arterial and venous systems, and its implication in the localization of vascular diseases, *Intern. Angiology* 5:297 (1986)
27. T. Asakura, and T. Karino, Flow patterns and spatial distribution of atherosclerotic lesions in human coronary arteries, *Circ. Res.* 66:1045 (1990).
28. M. Anliker, M. Casty, P. Friedli, R. Kubli and H. Keller, Non-invasive measurement of blood flow, in: "Cardiovascular Flow Dynamics and Measurements," N.H.C. Hwang and N.A. Normann, eds., University Park Press, Baltimore (1977).
29. A. Bollinger, P. Butti, P. Barras, H. Trachler, and N. Siegenthaler, Red blood cell velocity in nailfold capillaries of man, measured by a television microscopy technique, *Microvasc. Res.* 6:61 (1974).
30. J.H. Forrester, and E.F. Young, Flow through a converging-diverging tube and its implications in occlusive vascular disease. I. Theoretical development, *J. Biomech.* 3:297 (1970).
31. J.H. Forrester, and E.F. Young, Flow through a converging-diverging tube and its implications in occlusive vascular disease. I. Theoretical and experimental results and their implications, *J. Biomech.* 3:307 (1970).
32. J.-S. Lee, and Y.-C. Fung, Flow in non-uniform small blood vessels, *Microvasc. Res.* 3:272 (1973).
33. M.D. Deshpande, D.P. Giddens, and R.F. Mabon, Steady laminar flow through modelled vascular stenoses, *J. Biomech.* 9:165 (1976).
34. E.O. Macagno, and T.-K. Hung, Computational and experimental study of a captive annular eddy, *J. Fluid Mech.* 28:43 (1967).
35. T.-K. Hung, Vortices in pulsatile flows, in: "Proceedings of 5th International Congress of Rheology," S. Onogi, ed., University Park Press, Baltimore (1970).
36. K. Perktold, Pulsatile non-Newtonian blood flow simulation through a bifurcation with an aneurysm, *Biorheology* 26:1011 (1989).
37. H.L. Goldsmith, and V.T. Turitto, Rheological aspects of thrombosis and haemostasis: Basic principles and applications, *Thromb Haemost.* 55:415 (1986).
38. N.S. Lynn, V.G. Fox, and L.W. Ross, Computation of fluid dynamical contributions to atherosclerosis at arterial bifurcations, *Biorheology* 9:61 (1972).
39. R. Brech, and B.J. Bellehouse, Flow in branching vessels, *Cardiovasc. Res.* 7:593 (1973).
40. L.W. Ehrlich, Digital simulation of periodic flow in a bifurcation, *Computer and Fluids* 2:237 (1974).
41. K. Kandarpa, and N. Davids, Analysis of the fluid dynamic effects on atherogenesis at branching sites, *J. Biomech.* 9:735 (1976).
42. V.O. O'Brien, L.W. Ehrlich, and M.H. Friedman, Unsteady flow in a branch, *J. Fluid Mech.* 75:315 (1976).
43. D. Agonaffer, C.B. Watkins, and J.N. Cannon, Computation of steady flow in a two-dimensional arterial model, *J. Biomech.* 18:695 (1985).
44. G. Enden, M. Israeli, and U. Dinnar, A numerical simulation of the flow in a T-type bifurcation and its application to an 'end to side' fistula, *J. Biomech. Eng.* 107:321 (1985).
45. C.C.M. Rindt, A.A. van Steenhoven, A. Segal, R.S. Reneman, and J.D. Jansen, Analysis of the flow field in a 3D-model of carotid artery bifurcation, in: "Proceedings of the World Congress of Medical Physics and Biomedical Engineering," J.W. Clark, P.I. Horner, A.R. Smith, and K. Strum, eds., *Physics in Med. Biol.* 33:,375 (1988).
46. R.G. Cox, and S.K. Hsu, The lateral migration of solid particles in a laminar flow near a plane wall, *Int. J. Multiphase Flow* 3:201 (1977).
47. A. Karnis, and S.G. Mason, The flow of suspensions through tubes. VI. Meniscus effects, *J. Colloid Interface Sci.* 23:120 (1967).
48. T. Karino, and H.L. Goldsmith, Aggregation of platelets in an annular vortex distal to a tubular expansion, *Microvasc. Res.* 17:217 (1979).
49. M. von Smoluchowski, Versuch einer mathematischen Theorie der Koagulationskinetik kolloider Lösungen, *Z. Physik. Chem.* 92:129 (1917).

50. T.G.M. van de Ven, and S.G. Mason, The microrheology of colloidal dispersions. VII. Orthokinetic doublet formation of spheres, *Colloid Polymer Sci.* 255:468 (1977).
51. T. Karino, and H.L. Goldsmith, Adhesion of human platelets to collagen on the walls distal to a tubular expansion, *Microvasc. Res.* 17:238 (1979).
52. V.T. Turitto, Viscosity, transport and thrombogenesis, *in*: "Progress in Hemostasis and Thrombosis," T.H. Spaet, ed., Grune & Stratton, New York (1982).
53. H.L. Goldsmith and T. Karino, Mechanically induced thromboemboli, in: "Quantitative Cardiovascular Studies: Clinical and Research Applications," N.H.C. Hwang, D.R. Gross and D.J. Patel, eds., University Park Press, Baltimore (1978).
54. T. Karino, and H.L. Goldsmith, Role of cell-wall interactions in thrombogenesis and atherogenesis: A microrheological study, *Biorheology* 21:587 (1984).
55. L. Diener, J.L.E. Ericsson and F. Lund, The role of venous valve pockets in thrombogenesis. A postmortem study in a geriatric unit, *in*: "Atherogenesis", T. Shimamoto and F. Numano, eds., Excerpta Medica, Amsterdam (1969).
56. S. Sevitt, Pathology and pathogenesis of deep vein thrombi, *in*: "Venous Problem," J.J. Bergan and J.S.T. Yao, eds., Year Book Medical Publishers, Chicago (1978).
57. K. Kristiansen, and J. Krog, Electromagnetic studies on the blood flow through the carotid system in man, *Neurology* 12:20 (1962).
58. S. Uematsu, A. Yang, T.J. Preziosi, R. Kouba, and T.J.K. Toung, Measurement of carotid blood flow in man and its clinical application, *Stroke* 14:256 (1983).
59. Y. Sohara and T. Karino, Secondary flows in the dog aortic arch, *in*: "Fluid Control and Measurement," M. Harada, ed., Pergamon Press, Oxford (1985).
60. T. Karino, N. Kobayashi, S. Mabuchi, and S. Takeuchi, Role of hemodynamic factors in the localization of saccular aneurysms in the human circle of Willis, *Biorheology* 26:526 (1989).

EX VIVO MODELS FOR STUDYING THROMBOSIS: SPECIAL EMPHASIS ON SHEAR RATE DEPENDENT BLOOD-COLLAGEN INTERACTIONS

Kjell S. Sakariassen, Helge E. Roald, and José Aznar Salatti

Biotechnology Centre of Oslo
University of Oslo
P.O.Box 1125
0317 Oslo, Norway

INTRODUCTION

Perfusion models that allow exposure of artery subendothelium, collagen and other surfaces to non-anticoagulated blood (ex vivo models) or to anticoagulated blood (in vitro models) under well controlled experimental conditions have given much useful information about thrombotic mechanisms[1-15]. This includes a better understanding of molecular mechanisms of platelet-surface adhesion[3,6,13,16-19], platelet-platelet interaction[3,13,16,20-23], and the interplay between coagulation and platelet function in thrombus formation[4,12-14,22-27]. However, the blood flow, and in particular the shear rate, plays an unambiguous role in these events[1-3,13,14,16,17,19,21-25,28,29].

This brief essay is restricted to four ex vivo perfusion models which have been used extensively to study thrombotic mechanisms during the last two decades[3,4,7,9-11,14]. Particular emphasis is laid on shear rate dependent mechanisms of blood-collagen interactions in one of the human models[14].

EX VIVO THROMBOSIS MODELS

Four ex vivo models of thrombosis with unique perfusion chambers have been developed. These devices allow exposure of thrombogenic surfaces to flowing non-anticoagulated blood of man and animal. The first ex vivo model was developed by H.R. Baumgartner in the early seventies[4], and the other models were developed in the eighties by S.R. Hanson[9], L. Badimon[10] and by K.S. Sakariassen[14] and their colleagues. Mechanisms of thrombus formation and evaluation of anti-thrombotics have been investigated and quantified by these models under well controlled conditions of exposure time, reactivity of thrombogenic surfaces and the properties of the blood flow. However, experimental elucidation of blood flow characteristics, such as flow profile and wall shear rate at the reactive surface has so far been performed only for the perfusion chambers developed by Baumgartner[30,31], and by Sakariassen and co-workers[7]. The wall shear rate at the thrombogenic surface in the other models are theoretically derived from the dimensions of the respective blood flow channels and the actual blood flow rates (Table 1).

Table 1. Factors which determine wall shear rate in ex vivo perfusion chambers

Perfusion chamber	Flow channel dimensions (mm)		Flow rate (ml/min)	Wall shear rate (sec^{-1})	Ref
	Average effective Annular width[1]				
Annular	1.20		10	50	37
-	1.20		20	100	37
-	0.35		10	650	37
-	0.35		40	2600	49
-	0.35		50	3300	49
	Slit height	Slit width			
Parallel-plate	1.17	8.0	10	100	14
-	0.45	8.0	10	650	14
-	0.28	5.0	10	2600	14
-	0.28	5.0	20	5200	41
	Diameter				
Cylindrical	3.2		20	103	27
-	4.0		240	636	26
	Idealized diameter				
Semi-cylindrical[2]	2.0		5	106	11
-	1.0		20	3380	11

[1]Distance between subendothelial surface and inner wall of outer cylinder.
[2]The cross-section of the chamber is U-shaped.[11]

The models are briefly discussed below, with particular emphasis on perfusion chambers, blood flow properties at the thrombogenic surface and their application in man and animal.

Annular Perfusion Chamber Model

The ex vivo model with the annular perfusion chamber developed by H.R. Baumgartner[1,4] has been used in man[4,13,25] and rabbit[13,24,32,33].

Human model: Non-anticoagulated blood is drawn directly from an antecubital vein by a roller pump over arterial subendothelium which is positioned on a rod in the annular perfusion chamber. The pump is placed distally to the chamber. The chamber consists of a central rod surrounded by an outer cylinder which is connected to silastic tubing by two connecting portals. A de-endothelialized vessel segment of about 1.5 cm length from either rabbit aorta[1,4,13] or the human umbilical artery[34,35] is everted by a budprobe and subsequently mounted on the central position of the rod. Rabbit artery segments are de-endothelialized in situ by the balloon catheter technique[1,36] and human arteries in vitro by brief exposure to air[34]. The distance of the gap between the subendothelial surface and the inner wall of the outer cylinder of the chamber, which is defined as the effective annular width, and the blood flow rate determine the shear rate at the wall (Table 1). Wall shear rates in ex vivo experiments have ranged from 50 to 3300 sec^{-1}[23,37]. Blood flow characteristics, such as flow profile in the annulus and shear rate at the vessel wall, were experimentally elucidated by laser doppler velocimetry[31] and were according to theoretical predictions given by the formula:

$$\gamma = f \frac{Q}{r^3} \qquad (1)$$

where γ is the vessel wall shear rate (sec^{-1}), Q the average annular flow rate (ml/sec), r the effective annular width (cm) and f a function of the physical parameters of the system[13].

Rabbit model: The annular perfusion chamber is usually inserted into an extracorporeal shunt where blood is recirculated from a carotid artery over the subendothelial surface in the chamber and into a jugular vein. The blood flow, and thus the wall shear rate, is controlled by a roller pump placed distally to the chamber[24,32].

Parallel-plate Perfusion Chamber Model

The ex vivo model with the parallel-plate perfusion chamber developed by Sakariassen and colleagues[7,14] has been used in man[14] and dog[38].

Human model: The principles of the human ex vivo model are similar to that described for the ex vivo system with the annular perfusion chamber[4,13]. Blood is drained from an antecubital vein over a thrombogenic surface by a roller pump placed distally to the chamber. A plastic cover slip coated with various material such as either human type III collagen fibrils[14], cultured human endothelium[39,40], extracellular matrix of cultured human endothelium[40], or artificial surfaces[14] is placed in a cover slip holder which is positioned into the parallel-plate perfusion chamber and subsequently exposed to flowing blood. The chamber has two connecting portals for silastic tubing, analogous to the annular chamber. The flow slit is rectangular at the cover slip, and its dimensions, height and width, together with the blood flow rate in the flow channel determine the shear rate at the cover slip (Table 1). Blood flow profiles in the rectangular flow channel and corresponding shear rates at the cover slip were experimentally determined by laser doppler velocimetry[7], and were according to theoretical predictions given by the formula:

$$\gamma = 1.03 \frac{6Q}{ab^2} \qquad (2)$$

where γ is the shear rate (sec^{-1}) at the cover slip, Q the average blood flow rate (ml/sec), a the slit width (cm) and b the slit height (cm). Shear rates at the cover slip have been varied from 100 to 5200 sec^{-1}[14,41] (Table 1).

Dog model: Ex vivo perfusions with dogs are performed with the chamber in an extracorporeal shunt between the carotid artery and a jugular vein[38]. The blood flow in the extracorporeal shunt, and thus the shear rate at the cover slip, is controlled by a roller pump placed distally to the chamber.

Cylindrical Perfusion Chamber Model

The cylindrical ex vivo perfusion chamber model was developed by S.R. Hanson and colleagues[9] and is used in baboons (Papio Cynocephalus).

The cylindrical perfusion chamber is inserted into a chronic arteriovenous shunt of the femoral artery and vein. The walls of the circular chamber are coated with either purified type I monomeric collagen[12] or with synthetic vascular graft material such as Dacron[9]. The blood flow rate in the chamber, and thus the shear rate at the wall is controlled with a roller pump or by a clamp placed distally to the chamber. The wall shear rate is calculated from the blood flow rate in the chamber and the radius of the cylindrical chamber (Table 1) according to the formula:

$$\gamma = 4 \frac{Q}{\pi r^3} \qquad (3)$$

where γ is the shear rate (sec^{-1}) at the wall, Q the average flow rate (ml/sec) and r the radius (cm) of the circular chamber.

Calculated wall shear rates are generally in the range from 100 to 640 sec^{-1}[26,27].

Semi-Cylindrical Perfusion Chamber Model

The semi-cylindrical ex vivo perfusion chamber model used in pigs was developed by L. Badimon and colleagues[10].

The semi-cylindrical perfusion chamber is inserted into an arteriovenous shunt located between the carotid artery and the contralateral jugular vein[10]. The chamber has a cylindrical shape, but the thrombogenic surface of either pig artery subendothelium or fibrillar collagen type I is exposed to the blood flow as a planar surface which replaces a portion of the cylindrical wall. A peristaltic pump placed distally to the chamber controls the blood flow rate and the wall shear rate in the chamber. The wall shear rate is theoretically derived from the blood flow rate in the chamber and the idealized diameter of the semi-cylindrical blood flow channel (Table 1) according to the formula:

$$\gamma = 8 \frac{Q}{d} \qquad (4)$$

where γ is the shear rate (sec^{-1}) at the wall, Q the average flow rate (ml/sec) and d the diameter (cm) of the idealized cylindrical shaped flow channel[11].

Calculated wall shear rates ranging from 106 to 3380 sec^{-1} have been used in this model[10].

Methods to Quantify Thrombus Formation

Thrombus formation on the thrombogenic surfaces in these chambers has been quantified by (A) morphometry[3,13,42,43], (B) pressure differences between chamber entrance and exit as a measure of chamber occlusion by thrombotic material[22,44] and by (C) counting or gamma camera imaging of radiolabelled blood platelets[6,9,13]. Morphometry and radiolabelled platelet counting are used as end-point measurements, while pressure monitoring and radiolabelled platelet imaging are used to measure thrombus formation continuously.

Deposition of fibrinogen/fibrin on the thrombogenic surface is quantified by radiolabelled fibrinogen[12], and differentiation of deposited fibrin and fibrinogen are quantified by radioimmunoassays of plasmin digested thrombotic deposits[45].

Morphometry: Morphometric analysis of deposited thrombotic material[3,13,42,43] gives the most detailed analysis of thrombus dimensions and thrombotic mechanisms. However, the method is time-consuming and limited to end-point measurements. The method is used to evaluate thrombotic deposits from ex vivo perfusions with annular[4] and parallel-plate perfusion chambers[14].

The method relies on stained semi-thin sections of Epon embedded material[3,13,43]. The sections are cut perpendicularly to the direction of the blood flow at well defined axial positions[13,46,47]. The method of preparing such sections as specially developed to quantify blood-surface interactions and as recently reported in detail[13,43]. Two types of evaluation per section is performed, (A) standard morphometry[3,13,43] and (B) computer-assisted morphometry[42,43].

Standard morphometry, light microscopy at 1000x magnification[3,14,43]: This method quantifies the percent of the surface covered with platelets and is synonymous with percent platelet adhesion. Platelet adhesion is furthermore differentiated into percent of the surface covered with platelets which are not spread out on the surface (percent contact platelets) and percent of the surface covered with spread platelets (percent spread platelets). Fibrin deposition is quantified as the percent of the surface covered with fibrin, and is synonymous with percent fibrin deposition. Unfortunately, the method does not take into account the thickness of the fibrin meshes.

Computer-assisted morphometry, light microscopical-monitor projection at 2700x magnification[42,43]: This method quantifies such thrombus dimensions as volume and height and number of thrombi on the surface. These parameters are expressed as thrombus height (μm), thrombus volume per unit surface area ($\mu m^3/\mu m^2$) and number of thrombi per unit surface area (thrombi/μm^2).

This meticulous approach has given invaluable information about mechanisms of thrombus formation, despite being limited to end-point measurements.

Pressure: This method measures the pressure difference between chamber entrance and exit which progressively develops when growing thrombotic masses gradually occlude the chamber[24,44]. Thus, thrombus formation is simultaneously recorded as it propagates on the thrombogenic surface in the chamber. However, the method does not differentiate between deposition of fibrin, platelets, leukocytes and erythrocytes. This elegant approach has so far been used only in the ex vivo model with the annular perfusion chamber[24,44].

Radiolabelled Platelets and Radiolabelled Fibrinogen

This method is used in the ex vivo models with the cylindrical[9] and the semi-cylindrical perfusion chambers[10]. Platelets, either from non-human primates or pigs, are labelled in vitro by ^{111}In and injected into the donor animal[9,10]. Thus, autologous radiolabelled platelets are used to measure platelet deposition on the thrombogenic surface. Also, ^{125}I-radiolabelled fibrinogen is used to quantify fibrinogen/fibrin deposition in combination with the ^{111}In-platelets in non-human primates[12]. These isotope methods are used as end-point measurements and as continuous measurements by means of a scintillation camera. However, the ^{111}In-method does not discriminate between platelets adherent to the surface and platelet aggregates comprising the bulk of the thrombi. The participation of leukocytes has not been assessed. The red cell content of thrombi is determined from measurements of total thrombus hemoglobin[48].

Immunological Determination of Fibrin Deposition

Quantification of fibrin deposition on artery subendothelium has also been performed by immunological methods following plasmin digestion of the thrombotic deposits[45]. The soluble fibrin fragment E of the plasmin digest is measured by radioimmunoassay, and taken as total amount of deposited fibrin and fibrinogen. The fibrinogen content of the plasmin digest is quantified by radioimmunoassay of fibrinopeptide A (FPA) following incubation of a part of the plasmin digest by thrombin. FPA is a peptide of 16 amino acids which is split off specifically from the α-aminoterminal fibrinogen chain by thrombin. The fibrin content of the plasmin digested thrombotic deposit is derived from the total amount of deposited fibrin and fibrinogen measured by fragment E minus fibrinogen measured by FPA.

KNOWLEDGE GAINED FROM EXPERIMENTS WITH EX VIVO THROMBOSIS MODELS

Much knowledge about mechanisms of thrombus formation has been gained from experiments with the ex vivo models. However, earlier in vitro perfusion experiments with the annular- and parallel-plate perfusion chambers[3,13,16,17,24] were of fundamental importance for the experimental design of the ex vivo experiments.

In particular, the blood shear rate at the thrombogenic surface is important for a variety of mechanisms leading to thrombus formation both in vitro and ex vivo. Generally, shear rate dependent mechanisms controlling platelet-surface adhesion and platelet-platelet interaction appear at lower shear rates in the in vitro models than in the ex vivo models[22,32,49]. Below, mechanisms of thrombus formation as derived from ex vivo perfusion experiments are discussed with respect to processes of platelet-surface adhesion, platelet-platelet interaction and coagulation.

Platelet-Surface Adhesion

One of the first events of thrombus formation in vivo and ex vivo is the adherence of platelets to the thrombogenic surface. Conclusive evidence for a role of von Willebrand factor (vWF) and the platelet membrane glycoprotein Ib (GP Ib) in platelet-surface adhesion has been demonstrated in ex vivo experiments[22,23,37,49,50]. The platelet-membrane glycoprotein complex IIb-IIIa (GP IIb-IIIa) is also involved in platelet adhesion, particularly at high arterial shear rate[22,51].

The effect of vWF on platelet adhesion to artery subendothelium and collagen is shear rate dependent[22,23,37,49,50]. Lack of vWF in plasma and platelets results in impaired platelet adhesion at a shear rate of about 1500 sec^{-1}; the adhesion defect is gradually enhanced with increasing shear rates.

GP Ib deficiency decreases platelet adhesion to artery subendothelium, also at low arterial shear rates where platelet adhesion is normal in patients lacking vWF[49]. Thus, deficiency of GP Ib seems to impair platelet adhesion over a broader range of shear than deficiency of vWF.

Deficiency of GP IIb-IIIa leads to impaired platelet-spreading on arterial subendothelium, particularly at arterial wall shear rates above 1300 sec^{-1}[22,51]. This adhesion defect is relatively mild, and much less pronounced than the defect seen in blood from patients suffering from deficiency of either GP Ib or vWF.

However, platelet adhesion is also affected by the growth rate of mural thrombi. Rapid incorporation of platelets into a growing thrombus at the upstream end of a thrombogenic surface depletes platelets from blood layers streaming adjacent to the surface, resulting in lower platelet concentration at the surface and decreased platelet adhesion at downstream areas[47]. This physical effect is part of the so-called "axial dependence phenomenon"[46,47], which also may include upstream depletion of clotting factors in the same blood layers.

Platelet-Platelet Interaction

Human and animal ex vivo perfusion experiments have identified four platelet components involved in platelet-platelet interaction: (A) GP IIb-IIIa[22,51-53], (B) vWF[23,50], (C) platelet storage granules[22], and (D) the prostaglandin metabolism[41,54].

GP IIb-IIIa is mandatory for normal platelet-platelet interaction irrespective of the shear rate. The interaction is apparently mediated by binding of vWF, fibrinogen and fibronectin to GP IIb-IIIa which bridge adjacent platelets[19,55]. The synthetic peptide RGDV or monoclonal antibodies directed against GP IIb-IIIa and which inhibit binding of these ligands to GP IIb-IIIa interrupt platelet-platelet interaction at arterial wall shear rates[52,53]. However, a patient suffering from afibrinogenemia, having about 1/50 of normal amounts of fibrinogen in plasma and platelets, has normal platelet-platelet interaction at arterial shear, at least during the first 5 min of thrombus formation[23,56,57]. This indicates that vWF and/or fibronectin may substitute for fibrinogen in the process of platelet-platelet binding.

vWF is necessary for both platelet adhesion and platelet-platelet interaction. These interactions are apparently mediated by different mechanisms of vWF, since impaired platelet-platelet interaction is observed at wall shear rates where vWF is not required for normal platelet-surface adhesion[23,50]. vWF mediation of platelet-platelet interaction occurs presumably by binding of vWF to GP IIb-IIIa[58].

Platelet storage pool deficiency, deficiency of dense granules and α-granules, also results in reduced platelet-platelet interaction[22]. Reduced amounts of released proaggregatory ADP from dense granules and of vWF, fibronectin, fibrinogen and thrombospondin from α-granules are apparently the cause of the reduced platelet-platelet interaction.

The prostaglandin metabolism is required for normal platelet-platelet interaction, but only at very high wall shear rates (>5200 sec^{-1}) which may be present at the apex of arterial stenoses. A single oral dose of 1.0 g of aspirin inhibits platelet cyclooxygenase irreversibly with more than 97%[59], and results in 42% reduction of the thrombus volume at 5200 sec^{-1}[41]. This inhibition is due to blocked production of the proaggregatory prostaglandin metabolite thromboxane A_2[60].

COAGULATION AND PLATELET FUNCTION IN THROMBUS FORMATION

The ex vivo perfusion models allow studies of thrombotic mechanisms at venous and arterial blood flow conditions in the absence of anti-coagulation. Much emphasis has been laid on the role of coagulation in ex vivo arterial thrombus formation at arterial blood flow conditions.

However, insertion of an arteriovenous shunt in an animal or puncture of a human antecubital vein may trigger blood clotting in the respective systems. Therefore, it is important to assess to which extent the devices and the procedures affect the clotting mechanisms. Such measurements have been performed, generally by measuring plasma levels of FPA. FPA plasma levels introduced by the arterio-venous shunt with the cylindrical chamber in baboons yield about 11 ng/ml[26]. The pre-chamber levels of the human models following 5 min blood drawing (10 ml/min) is about 3.7 ng/ml[14]. Post-chamber levels without the introduction of a thrombogenic surface in the respective chambers are about 37 ng/ml for the annular chamber[37] and about 20 ng/ml for the parallel-plate chamber. Plasma levels of FPA introduced by the arterio-venous shunt with the semi-cylindrical chamber have not been reported.

Apparently, the respective devices and/or the procedures of the human and baboon models generate some clotting activity, although not excessive. This activity is within the ranges of those reported in patients with disseminated intravascular coagulation (12-40 ng/ml) and with myocardial infarction (5-26 ng/ml)[61]. A plasma level of 3.5 ng/ml FPA is generally defined as the upper limit of healthy individuals.

Two approaches have been taken to investigate the role of coagulation in ex vivo induced thrombogenesis: (A) natural models; humans with a deficiency in a selected coagulation factor and (B) traditional and experimental inhibitors of various coagulation factors. Experimental inhibitors include monoclonal antibodies[35,40,52,62], synthetic compounds and bioengineered peptides[26,53,62,63], various fractions of heparin[24,27] and hirudin[64]. The human models have demonstrated that deficiency of F VII, F VIII and F IX, and inhibition of tissue factor activity in stimulated endothelium[40], in artery subendothelium[35], and in extracellular endothelial matrix[40,62] by a monoclonal anti-tissue factor antibody[65] inhibits deposition of fibrin on the respective surfaces at venous and arterial wall shear rates. Deficiency of F VIII and F IX reduces thrombus formation at a wall shear rate of 650 sec^{-1}[56], thus at flow conditions encountered in small coronary arteries[29]. However, deficiency of F VIII has no effect on thrombus formation on procoagulant artery subendothelium and on non-procoagulant collagen fibrils at wall shear rates above 2600 sec^{-1}, thus at flow conditions encountered at the apex of moderately stenosed arteries[23,50]. These observations do not necessarily imply that coagulation is without importance for thrombus formation at stenotic shear rates, since additional perfusion experiments with e.g. specific irreversible thrombin inhibitors and disturbed blood flow conditions such as at arterial stenoses in vivo are needed before any firm conclusion can be drawn.

Nonetheless, irreversible thrombin inhibition by D-phenylalanyl-L-prolyl-L-arginyl chloromethyl ketone at wall shear rates of 420-640 sec^{-1} strongly inhibits platelet thrombus formation[26]. Interruption of platelet-dependent thrombus formation is also observed with relatively high doses of heparin and low-molecular weight heparin at wall shear rates within the same range[24,27]. This effect is mediated by partial inhibition of F Xa and thrombin, by endothelial release of tissue factor pathway inhibitor[66,67] which inhibits the clotting activity of tissue factor/F VIIa and of F Xa[68-70], and apparently by affecting the platelet function through interference of the binding of vWF to platelet GP Ib[71]. Interruption of platelet-dependent thrombus formation at 450 sec^{-1} wall shear rate is also achieved by infusion of activated protein C in baboons which inhibits F Va and F VIIIa[72,73]. Thus, interruption of ex vivo arterial thrombus formation is not only achieved by blocking activated clotting factors, but can also be achieved by increasing the plasma level of naturally occurring coagulation inhibitors.

Much remains to be learned about the impact of coagulation on arterial thrombus formation, and in particular at high shear conditions and at blood flow conditions at arterial stenoses. Model systems allowing such studies have been developed and characterization of the thrombotic response at such blood flow conditions are currently underway.

SHEAR RATE DEPENDENT COLLAGEN-INDUCED THROMBOSIS: HUMAN EX VIVO MODEL WITH PARALLEL-PLATE PERFUSION CHAMBERS

Shear rate dependent blood-collagen interactions in the human model with the parallel-plate perfusion chambers are discussed in more detail. Particular emphasis is laid on interactions of platelets and fibrin with purified type III collagen fibrils, and on characterization of this ex vivo system in more detail than previously reported[14,47].

Collagen Characteristics

Purified type III collagen fibrils have in most instances been used as thrombogenic surface in this model[14,36]. The collagen is extracted from placenta by pepsin-digestion and subsequently purified by selective salt precipitation[74]. Fibrils are made by dialysis against 20 mM Na_2HPO_4[75]. The preparation is >95% pure as judged by densitometry scanning of SDS-PAGE gels. The fibrils induce optimal platelet aggregation in platelet-rich plasma in the aggregometer device at 3-4 μg/ml. Fibrils as well as monomers are non-procoagulant. The reproducibility of thrombus formation induced by this well defined chemical surface is good, and a density of 10 μg/cm^2 collagen fibrils gives optimal thrombus formation.

Type III collagen fibrils are present in healthy subendothelium of vessels, but more abundant in atherosclerotic plaques[76-78]. Rupture of such vascular lesions may expose type III collagen fibrils, also intermixed with type I collagen fibrils, to the blood stream. Direct demonstration of platelet adhesion to intimal collagen fibrils has been demonstrated at the level of transmission electron microscopy. Comparison of the thrombogeneity of collagen type III fibrils with intact artery subendothelium reveals small differences although much more fibrin is deposited on the latter surface[47]. This difference is attributed to subendothelial tissue factor which triggers clotting when complexed with plasma FVII(a)[79], in contrast to the collagen fibrils which are non-procoagulant[14].

Activation of Coagulation by the Human Ex Vivo Device

Generation of clotting activity by the human ex vivo model with the parallel-plate perfusion chambers was quantified by assaying plasma levels of FPA proximally and distally to the chamber. However, the FPA assay reflects only part of the thrombin generation in plasma, since this assay measures only the 16 amino acid fibrinogen peptide split off the α-aminoterminal chain by thrombin, but not the thrombin bound to platelets. Nevertheless, the FPA plasma level is a useful indicator of clotting activity in a variety of clinical situations[61].

Post- and pre-chamber FPA plasma levels were quantified in 2.3 ml blood samples collected into a 0.7 ml mixture of sodium citrate, heparin and aprotinin. Post-chamber blood samples were collected immediately distal to the chamber into a syringe mounted on a withdrawal pump (Harvard Apparatus, Infusion/Withdrawal pump, model 902A, South Natick, MA). The samples were drawn with the withdrawal pump switched on and the roller pump draining the blood from the vein switched off. Thus, blood was drained from the vein continuously at constant flow rate (10 ml/min) while the 2.3 ml post-chamber blood sample was collected. Further processing of blood samples for assaying of plasma FPA-levels was according to the manufacturer (Boehringer/Mannheim FRG).

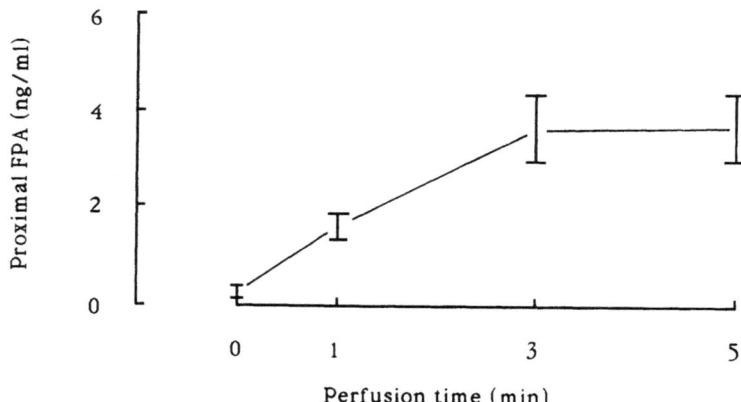

Figure 1. Plasma levels of FPA proximally to the parallel-plate perfusion chamber at 0, 1, 3, and 5 min blood drainage at 10 ml/min. The venepuncture was performed with a Butterfly infusion set (19 G, No 4590, Abbott Ireland Ltd, Rep. of Ireland) and the plastic tubing of the Butterfly infusion set was cut by a scissor to collect blood samples for FPA-measurements following blood perfusions at the indicated time points. Mean ± S.E., n=11.

Pre-chamber plasma levels of FPA were measured at 0, 1, 3 and 5 min blood drainage at 10 ml/min (Figure 1). The average levels of FPA increased from 0.3 ng/ml at 0 min to 3.7 ng/ml at 3 and 5 min. These values are slightly above the upper limit of the normal range (3.5 ng/ml), but are e.g. within the range reported for women on hormonal contraception[45]. Thus, activation of coagulation does occur proximally to the chamber, and is presumably due to the puncture of the vein. Exposure of tissue factor in the wall of the vein to plasma F VII(a) converts F X and F IX to F Xa and F IXa, respectively, leading to generation of FPA in plasma. The average wall shear rate at 10 ml/min in the needle and the tubing (Butterfly-19, No 4590, Abbot Ireland Ltd, Rep. of Ireland) draining blood from the vein to the chamber is calculated to be 3350 sec^{-1} according to formula III. Such a high wall shear rate limits the deposition of fibrin on vascular connective tissue to a minimum[14,24,37,47], indicating that fibrin monomers may accumulate in the blood stream and enter the parallel-plate perfusion chamber.

The average post-chamber FPA plasma level with non-collagen coated Thermanox™ cover slips positioned in the chamber is 20±3 ng/ml (mean ± S.E., n=4) at 4 1/2 min perfusion at a flow rate of 10 ml/min, which corresponded to a wall shear rate of 100 sec^{-1}. Less than 1% surface coverage with fibrin is measured on Thermanox™ under such experimental conditions[14]. Apparently, additional FPA is generated in the blood stream during passage through the chamber which increases from an average pre-chamber level of 3.7 ng/ml to an average post-chamber level of 20 ng/ml. This post-chamber level is within the range of the level measured in the extracorporeal shunt with the cylindrical perfusion chamber without insertion of a thrombogenic surface[26].

Fibrin Deposition and FPA Plasma Levels

The kinetics of fibrin deposition on non-procoagulant type III collagen fibrils were investigated by time-course studies at various wall shear rates. Plasma levels of FPA were measured both proximally and distally to the chamber in an attempt to see whether these levels were associated with the surface coverage with fibrin on the collagen fibrils.

Quantitation of the surface coverage with fibrin on the collagen fibrils[13,14] and of pre- and post-chamber plasma levels of FPA was performed as described above.

Significant deposition of fibrin (average of 17% surface coverage) at a wall shear rate of 100 sec^{-1} was not observed until the perfusion time was prolonged to 5 min (Figure 2B). Deposition of fibrin at 1 and 3 min covered less than 3% of the surface. The pre-chamber FPA values (Figure 2A) were not correlated to the fibrin deposition. The average post-chamber levels of FPA increased constantly with time, from 18 ng/ml at 1/2 min to 76 ng/ml at 4 1/2 min (Figure 2C), but were not correlated with the fibrin deposition on the collagen surface. Increasing the wall shear rate from 100 to 5200 sec^{-1} lowered gradually the already low fibrin deposition on the collagen fibrils, although only significant at 5200 sec^{-1} (Figure 3A). The post-chamber plasma levels of FPA decreased also gradually by increasing shear rate, and significantly at wall shear rates of 2600 and 5200 sec^{-1} (Figure 3B). However, the individual plasma levels of post-chamber FPA and surface coverage with fibrin at the respective shear rates were not significantly correlated.

The relatively low fibrin deposition on the non-procoagulant collagen fibrils, is apparently mediated by binding of activated clotting factors to plasma membranes of deposited platelets and leukocytes[80,81], which efficiently amplify the coagulation process[82] locally at the surface. Activated clotting factors are presumably originating at the site of the venipuncture and translocated to deposited platelets and leukocytes at the collagen surface. This view is supported by the morphological appearance of platelet thrombi and fibrin on the surface. Significant fibrin deposition at the wall

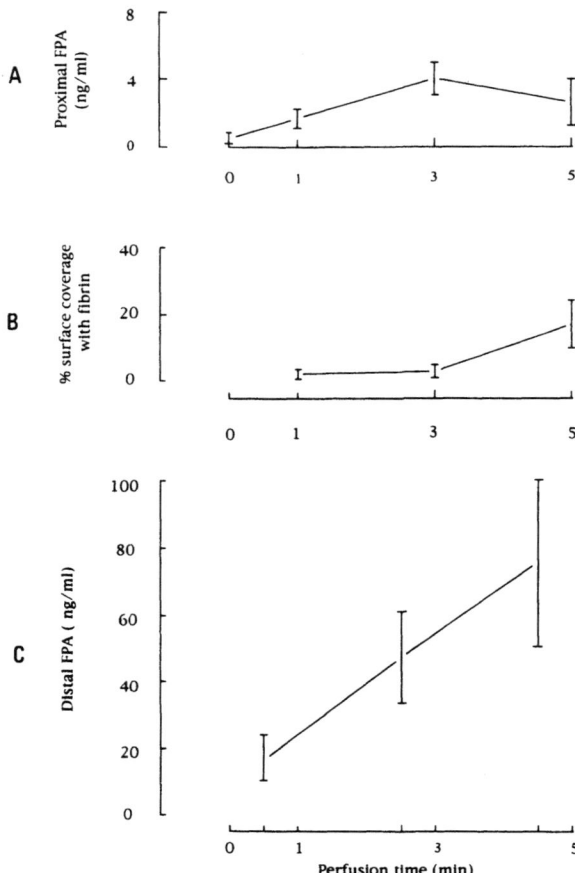

Figure 2. Time course of pre- and post-chamber FPA plasma levels and surface coverage with fibrin on collagen fibrils at a wall shear rate of 100 sec^{-1}. (A) pre-chamber FPA plasma levels at 0, 1, 3, and 5 min perfusion (mean ± S.E., n=4) and (B) surface coverage with fibrin 1, 3, and 5 min perfusions (mean ± S.E., n=4) are from the same perfusion experiments. (C) post-chamber FPA plasma levels at 1/2, 2 1/2, and 4 1/2 min perfusions (mean ± S.E., n=4) are from a different set of perfusion experiments.

shear rate of 100 sec^{-1} is first seen on collagen at 5 min when platelet thrombi start to form. Furthermore, morphological inspection of time course studies show that fibrin strands are frequently sprouting out from the platelet thrombi at the first 3 min and subsequently onto the collagen fibrils at 5 min. On the other hand, the gradual decrease of fibrin coverage on collagen by increasing wall shear rates may reflect lower amounts of thrombin generated at the surface as indicated by the corresponding lower post-chamber FPA plasma levels and/or enhanced removal of activated coagulation factors and fibrin monomers from the surface by the concomitantly increasing shear forces.

Lack of correlation between individual pre- and post-chamber FPA plasma levels and deposition of fibrin on the collagen may reflect the morphometric approach which only measures the surface coverage with fibrin, but not the varying thickness of the fibrin meshes on the surface. Immunological quantitation[45] of the fibrin mesh on the surface seems a more meaningful approach. Also, it appears that more detailed analysis of the coagulant events at the thrombogenic surface is needed before post-chamber FPA plasma levels may be taken as an indicator of coagulation at the reactive surface.

Platelet-Collagen Adhesion

The kinetics of platelet-collagen adhesion were studied by standard morphometry as described above[3,13,43] following 1, 3 and 5 min perfusions at wall shear rates of 100, 650, 1600, 2600, and 5200 sec^{-1} (Figure 4A).

Platelet adhesion increased gradually by increasing the wall shear rate from 100 to 1600sec^{-1}, but decreased when the wall shear rate was further increased from 1600 to 5200 sec^{-1} (Figure 4A). The rate of platelet adhesion at a wall shear rate of 100 sec^{-1} was constant during 5 min, while the rates at 650 and 2600 sec^{-1} were maximal at 1 min and decreased with about 50% at 3 and 5 min (Figure 5A).

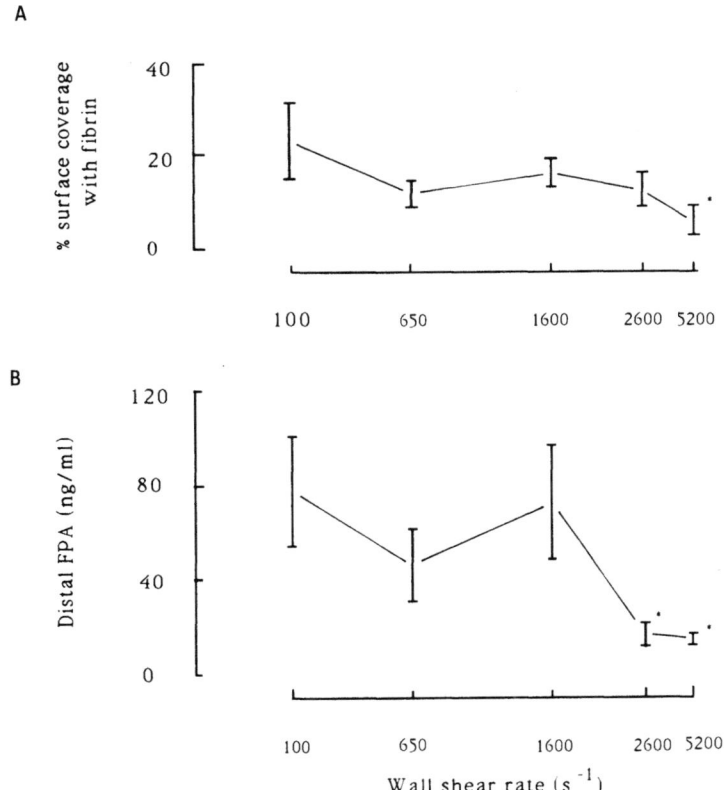

Figure 3. (A) surface coverage with fibrin on collagen and (B) post-chamber FPA plasma levels following 5 min perfusion at wall shear rates varying from 100 to 5200 sec^{-1}. Blood flow rates at wall shear rates of 100, 650, and 2600 were 10 ml/min, at 1600 sec^{-1} 6.3 ml/min and at 5200 sec^{-1} 20 ml/min. Surface coverage with fibrin is given as mean ± S.E., and plasma FPA levels as mean ± S.E., n=5-7. Student's t-test for unpaired data. *$p<0.05$ when compared to the respective values at 100 sec^{-1}

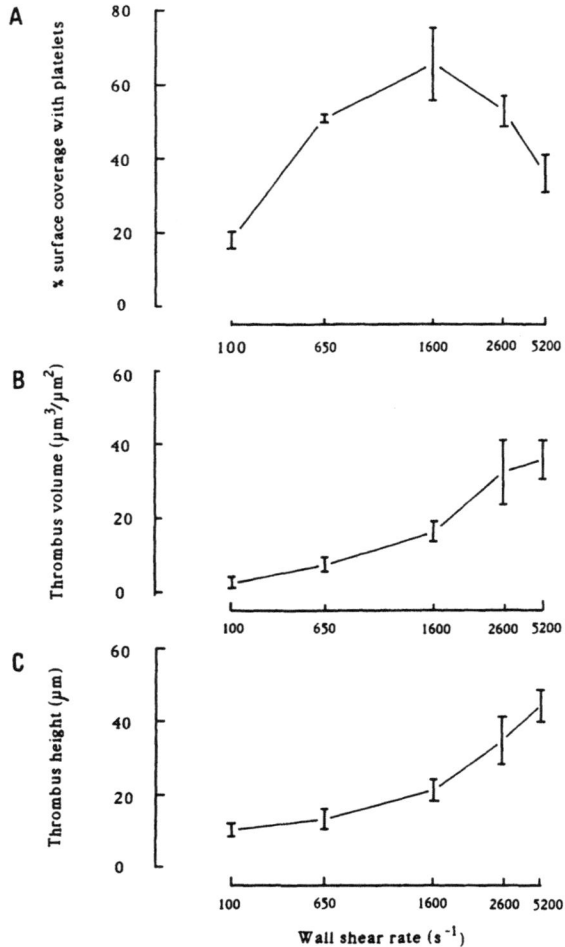

Figure 4. (A) surface coverage with platelets, (B) thrombus height, and (C) thrombus volume following 5 min perfusions at wall shear rates of 100, 650, 1600, 2600, and 5200 sec^{-1}. The blood flow rates are given in the legend to Figure 3. Mean ± S.E., n=4-13.

The kinetics of this platelet collagen adhesion are governed by physical and chemical factors[13,16,17,24,29]. The rate of platelet-surface adhesion is dependent on diffusion controlled arrival of platelets to the surface at wall shear rates below 600 sec^{-1}[83]. Above this shear condition, the adhesion appears gradually more reaction rate controlled as indicated by the requirements for vWF, GP Ib and GP IIb-IIIa[3,16,17,22,24,29,49,50]. Thus, the increase of platelet-collagen adhesion at 650 sec^{-1} relative to 100 sec^{-1} is apparently due to increased radial platelet diffusion. This enhanced diffusion is induced by higher concentration of erythrocytes in the axis of the blood flow at 650 sec^{-1} which increases platelet concentration locally at the collagen surface and the rate of platelet arrival at the collagen fibrils[84]. However, at the more reaction rate controlled range of wall shear rates, the levelling off of platelet-collagen adhesion at 1600 sec^{-1} and the drop of adhesion at 2600 and 5200 sec^{-1} are possibly due to high wall shear forces and rapidly growing thrombi which consume platelets from blood layers streaming adjacent to the collagen surface. High wall shear forces reduce the residence time of platelets at the collagen surface and may also remove platelets loosely attached to the collagen fibrils. Consumption of platelets by growing thrombi depletes the blood layers streaming adjacent to the collagen surface for platelets which

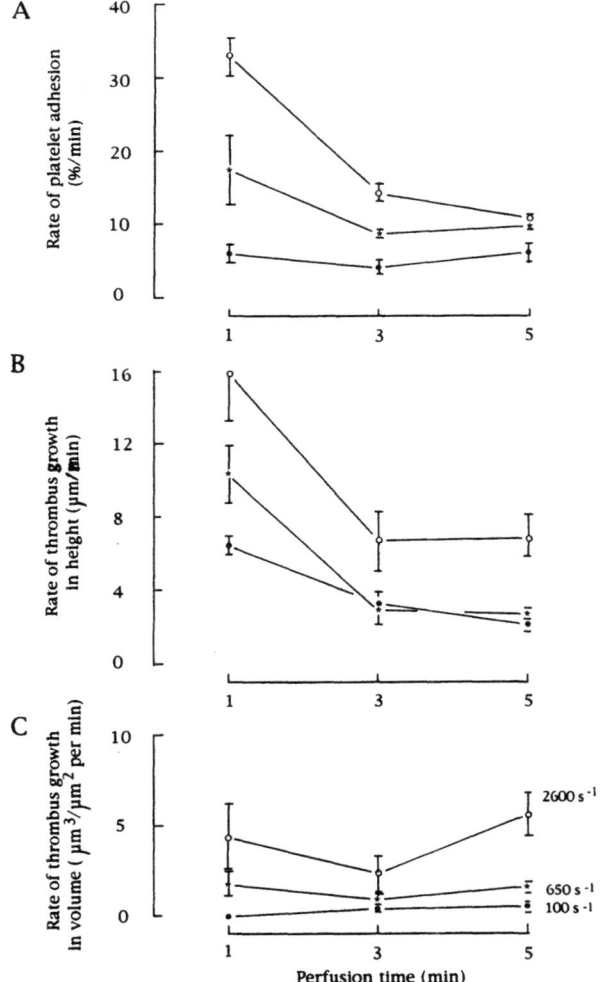

Figure 5. (A) rate of platelet adhesion and of (B) thrombus growth in height and (C) volume at 1, 3 and 5 min perfusions at wall shear rates of 100 (•), 650 (*), and 2600 sec^{-1} (o). Rates are expressed as average per min of the respective end-point measurements at 1, 3 and 5 min. Mean ± S.E., n=4-11.

results in lower platelet-collagen adhesion. Evidence for such a physical phenomenon was obtained in experiments showing that aspirin ingestion reduced the thrombus volume with 42% at 5200 sec^{-1} which resulted concomitantly in 69% enhancement of the platelet-collagen adhesion[41].

Analysis of platelet-collagen adhesion by time course studies revealed a constant rate of platelet adhesion during the first 5 min at 100 sec^{-1}. The diffusion controlled adhesion is relatively low, reaching 20% surface coverage at 5 min. This contrasts with the rate of platelet-collagen adhesion at 650 and 2600 sec^{-1} which drops with about 50% at 3 and 5 min. A possible explanation for this phenomenon may be related to a combination of platelet depletion and the gradually lower area of collagen fibrils available for platelets to adhere to, since 17 to 37% of the surface is already covered by platelets within the first min. However, local disturbances of the blood flow dynamics induced by the shape of the collagen-attached thrombi may have affected the adhesion rate, particularly in view of their large size at 3 and 5 min.

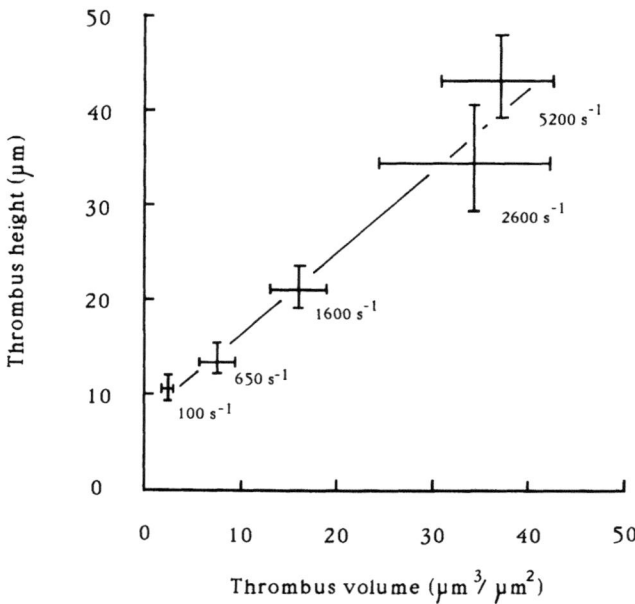

Figure 6. Correlation of thrombus height and thrombus volume from experiments summarized in Figure 4. Linear regression analysis, r=0.93 and p<0.01, n=4-13.

Thrombus Growth

The kinetics of thrombus growth were studied by computer-assisted morphometry as described above[42,43] following 1, 3 and 5 min perfusions at wall shear rates of 100, 650, 1600, 2600, and 5200 sec^{-1} (Figure 4B and C).

In contrast to the platelet-collagen adhesion, thrombus dimensions as volume and height increased with increasing shear throughout the 100 to 5200 sec^{-1} range of shear rates. The correlation between height and volume of these thrombi was highly significant, irrespective the wall shear rate (Figure 6). Thus, increasing thrombus height was reflected by increasing thrombus volume. However, the rate of thrombus growth in height was maximal at 1 min, but levelled off to about 50% at 3 and 5 min (Figure 5B). This contrasted with the rate of thrombus growth in volume, which appeared constant throughout the 5 min period (Figure 5C).

The continuous growth in height and thrombus volume when the wall shear rate was raised from 100 to 5200 sec^{-1} indicates that the mechanisms of shear dependent thrombus growth are different from those of the shear dependent platelet-collagen adhesion. However, the growth in height was less pronounced than the growth in volume, indicating isotropic growth initially in the direction perpendicular to the blood flow, turning anisotropic at later stages. This time-dependent change in growth pattern is substantiated by the growth rate in height and in volume, showing constant growth rate in volume, but about 50% reduced growth rate in height at 3 and 5 min. It is possible that high local shear stresses bend the thrombi in the direction of the blood flow, as previously observed in citrated blood[43] but apparently in a manner where the ratio of height and volume is constant. Whether this unexpected relationship is a result of the balance of thrombus stability and shear stresses at the respective shear conditions remains to be established.

Axial Dependence

The kinetics of platelet-collagen adhesion, number of thrombi per surface area, thrombus height and volume and fibrin deposition[3,13,42,43] were quantified at axial positions of 1 and 13 mm downstream for the blood flow inlet at the collagen coat (Table 2). The quantitations were performed following 1, 3, and 5 min perfusions at wall shear rates of 100, 650, and 2600 sec^{-1}.

Table 2. Axial dependence of blood-collagen interactions: % downstream average interactions relative to the respective upstream average interactions which were defined as 100%

Shear rate (sec^{-1})	Perfusion time (min)	Platelet adhesion[1]	No of thrombi[2]	Thrombus volume[3]	Thrombus height[4]	Fibrin deposition[5]
100	1	31 **	44 **	40 **	42 *	61
-	3	49 **	31 **	33 *	84	106
-	5	53 **	81	33 *	64	197
650	1	62 **	35 *	18 **	79	78
-	3	42 **	68	50 *	94	154
-	5	43 **	74	20 **	50	263
2600	1	63 *	76	31 *	65	98
-	3	64 *	86	39 *	67	91
-	5	66 *	103	43 *	60	332

[1] % surface coverage with platelets, upstream value defined as 100%
[2] No of thrombi/100 µm^2, upstream value defined as 100%
[3] µm^3/µm^2 thrombus volume, upstream value defined as 100%
[4] µm thrombus height, upstream value defined as 100%
[5] % surface coverage with fibrin, upstream value defined as 100%. Lack of significance of enhanced downstream fibrin deposition despite the high percentages at 5 min is due to considerable individual variation of the already low deposition along the axis. Student's t-test for paired data, *<0.05 and **p<0.01.

Platelet adhesion and thrombus volume were significantly decreased downstream at all times and shear conditions employed. The number of thrombi was generally decreased downstream, but only significantly at 1 and 3 min at 100 sec^{-1} and at 1 min at 650 sec^{-1}. The average thrombus height was also generally decreased downstream, but only significantly at 1 min at 100 sec^{-1}. The downstream fibrin deposition was not significantly changed from the relatively low upstream deposition, but was in general higher at the downstream position at 3 and 5 min.

This axial decrease in platelet adhesion and thrombus dimensions has been attributed to upstream depletion of platelets in the blood layers streaming adjacent to the surface by upstream growing thrombi[46,47]. Thus, lowered platelet concentration in these blood layers at the downstream surface results in reduced platelet adhesion and thrombus growth. It is interesting to note that this platelet depletion/axial dependency remains constant throughout the 5 min at all three wall shear rates, indicating that the strong thrombogenic properties of the collagen fibrils leads to a continuous incorporation of platelets into the growing thrombi.

The tendency of higher downstream fibrin deposition at 3 and 5 min is presumably due to translocation of activated clotting factors from upstream thrombotic deposits to platelets deposited on downstream locations. F Va binds to the membrane of activated platelets which makes binding sites available for F Xa, and together they convert prothrombin to thrombin, which converts fibrinogen to fibrin at the activated platelet surface[81]. However, translocation of fibrin monomers may also enhance downstream fibrin deposition.

These dynamic axial events of blood-collagen interactions are apparently triggered by upstream thrombus formation. The processes have been termed "axial dependence phenomena"[46,47], and may be explained in part by physical mechanisms.

CONCLUSIONS

Characterization of blood-collagen interactions at wall shear rates varying from 100 to 5200 sec^{-1} and maintained for 1, 3 and 5 min reveals interactions governed both by physical and chemical mechanisms. The wall shear rate plays an unambiguous role in these interactions, as well as platelet depletion for the axial dependence phenomena. However, the significance of factors as shear stresses, thrombus stability and translocation of thrombotic material are less understood, because no system is available to quantify these events. It should also be emphasized that the studies reported in this paper were carried out in laminar blood flow, but that disturbances of the blood flow patterns introduced by rapidly growing thrombi might have affected the blood-surface interactions as well.

FUTURE DIRECTIONS

Ex vivo models of thrombogenesis are attractive approaches for studying mechanisms of thrombosis and to evaluate traditional and experimental antithrombotics. More important and useful information might be expected from experiments with these devices in the future. However, it seems imperative to develop ex vivo devices with model systems of arterial stenoses, since the current knowledge of mechanisms of thrombogenesis at such lesions is poor. It also seems necessary to develop systems which quantify thrombus embolization. Nevertheless, development of ex vivo models has been a step in the good direction, since they have improved our understanding of thrombogenesis and helped improving our strategy in the battle against thrombotic disease.

ACKNOWLEDGEMENTS

This study was supported by the Norwegian Council on Cardiovascular Diseases,

The Anders Jahre Foundation and the Odd Fellow Foundation. Mrs. Bjørg Dalaaker and Torill Halvorsen are thanked for excellent secretarial support. Parts of the reported data were obtained during 1989/1990 at The Cardiovascular Research Unit of F. Hoffmann-La Roche Ltd., Basle, Switzerland, headed by Prof. Hans R. Baumgartner.

REFERENCES

1. Baumgartner HR: The role of blood flow in platelet adhesion, fibrin deposition and formation of mural thrombi, Microvasc. Res., 5:167, 1973.
2. Turitto VT, Baumgartner HR: Effect of physical factors on platelet adhesion to subendothelium, Thromb. Diath. Haemorrh. (Suppl), 60:17, 1973.
3. Baumgartner HR, Muggli R: Adhesion and aggregation: morphological demonstration and quantitation in vivo and in vitro, In: Platelets in Biology and Pathology (Ed., Gordon JL), Elsevier/North Holland Biomedical Press, Amsterdam, 1976.
4. Baumgartner HR: Effects of anticoagulation on the interaction of human platelets with subendothelium in flowing blood, Schweitz. Med. Wschr., 106:1367, 1976.
5. Muggli R, Baumgartner HR, Tschopp TB, Keller H: Automated microdensitometry and protein assays as a measure for platelet adhesion and aggregation on collagen coated slides under controlled flow conditions, J. Lab. Clin. Med., 95:195, 1979.
6. Sakariassen KS, Bolhuis PA, Sixma JJ: Human blood platelet adhesion to artery subendothelium is mediated by factor VIII-von Willebrand factor bound to the subendothelium, Nature, 279:636, 1979.
7. Sakariassen KS, Aarts PAMM, de Groot PhG, Houdijk WPM, Sixma JJ: A perfusion chamber developed to investigate platelet interaction in flowing blood with human vessel wall cells, their extracellular matrix, and purified components, J. Lab. Clin. Med., 102:522, 1983.
8. Frangos JA, Eskin SG, McIntire LV: Flow effects on prostacyclin production by cultured human endothelial cells, Science, 227:1477, 1985.
9. Hanson SR, Kotze HF, Savage B, Harker LA: Platelet interactions with Dacron vascular grafts. A model of acute thrombosis in baboons, Arteriosclerosis, 5:595, 1985.
10. Badimon L, Badimon JJ, Galvez A, Chesebro JH, Fuster V: Influence of artrial damage and wall shear rate on platelet deposition. Ex vivo study in a swine model, Arteriosclerosis, 6:312, 1986.
11. Badimon L, Turitto V, Rosemark JA, Badimon JJ, Fuster V: Characterization of a tubular flow chamber for studying platelet interaction with biologic and prosthetic materials: deposition of indium 111-labelled platelets on collagen, subendothelium, and expanded polytetrafluorethylene, J. Lab. Clin. Med., 110:706, 1987.
12. Cadroy Y, Horbett TA, Hanson SR: Discrimination between platelet-mediated and coagulation-mediated mechanisms in a model of complex thrombus formation in vivo, J. Lab. Clin. Med., 113:436, 1989.
13. Sakariassen KS, Muggli R, Baumgartner HR: Measurements of platelet interaction with components of the vessel wall in flowing blood, In: Methods in Enzymology, Academic Press Inc., New York, Vol. 169, 33, 1989.

14. Sakariassen KS, Joss R, Muggli R, Kuhn H, Tschopp TB, Sage H, Baumgartner HR: Collagen type III induced ex vivo thrombogenesis in humans: role of platelets and leukocytes in deposition of fibrin, Arteriosclerosis, 19:276, 1990.
15. Grabowski EF: Platelet aggregation in flowing blood at a site of injury to an endothelial cell monolayer: quantitation and real-time imaging with the TAB monoclonal antibody, Blood, 75:390, 1990.
16. Baumgartner HR, Muggli R, Tschopp TB, Turitto VT: Platelet adhesion, release reaction and aggregation in flowing blood: effects of surface properties and platelet function, Thromb. Haemostas., 35:124, 1976.
17. Sakariassen KS, Fressinaud E, Girma JP, Meyer D, Baumgartner HR: Role of platelet membrane glycoproteins and von Willebrand factor in adhesion of platelets to subendothelium and collagen, Ann. New York Acad. Sci., 516:52, 1987.
18. Meyer D, Fressinaud E, Sakariassen KS, Baumgartner HR, Girma JP: Role of von Willebrand factor in platelet adhesion, Ann. New York Acad. Sci., 509:118, 1987.
19. Fressinaud E, Baruch D, Girma JP, Sakariassen KS, Baumgartner HR, Meyer D: Von Willebrand factor-mediated platelet adhesion to collagen involves platelet membrane glycoprotein IIb/IIIa as well as glycoprotein Ib, J. Lab. Clin. Med., 112:58, 1988.
20. Tschopp TB, Weiss HJ, Baumgartner HR: Interaction of thrombostatic platelets with subendothelium: normal adhesion, absent aggregation, Experientia, 31:113, 1975.
21. Sakariassen KS, Nievelstein PFEM, Coller B, Sixma JJ, The role of platelet membrane glycoprotein Ib and IIb/IIIa in platelet adherence to human artery subendothelium, Br. J. Haematol., 63:681, 1986.
22. Weiss HJ, Turitto VT, Baumgartner HR: Platelet adhesion and thrombus formation on subendothelium in platelets deficient in glycoprotein IIb-IIIa and storage granules, Blood, 67:322, 1986.
23. Turitto VT, Weiss HJ, Baumgartner HR: Platelet interaction with rabbit subendothelium in von Willebrand's disease: altered thrombus formation distinct from defective platelet adhesion, J. Clin. Invest., 70:1730, 1984.
24. Baumgartner HR, Sakariassen KS: Factors controlling thrombus formation on arterial lesions, Ann. New York Acad. Sci., 454:162, 1985.
25. Weiss HJ, Baumgartner HR, Turitto VT: Regulation of platelet-fibrin thrombi on subendothelium, Ann. New York Acad. Sci., 516:380, 1987.
26. Hanson SR, Harker LA: Interruption of acute platelet-dependent thrombosis by the synthetic antithrombin D-phenylalanyl-L-prolyl-L-arginyl chloromethyl ketone, Proc. Natl. Acad. Sci. USA, 85:3184, 1988.
27. Cadroy Y, Harker LA, Hanson SE: Inhibition of platelet-dependent thrombosis by low molecular weight heparin (CY222): comparison with standard heparin, J. Lab. Clin. Med., 114:349, 1989.
28. Turitto VT, Weiss HJ, Baumgartner HR: Rheological factors influencing platelet interaction with vessel surfaces, J. Rheol., 23:735, 1979.
29. Turitto VT, Baumgartner HR: Platelet-surface interactions, In: Hemostasis and Thrombosis. Basic Principles and Clinical Practice (Eds., Colman RW, Hirsh J, Marder VJ, Salzman EW), J.B. Lippincott Company, Philadelphia, 1987.
30. Turitto VT: Mass transfer in annuli under conditions of laminar flow, Chem. Engn. Sci., 30;503, 1975.

31. Aarts PAMM, van den Broek JATM, Kuiken GDC, Sixma JJ, Heethaar RM: Velocity profiles in annular perfusion chamber measured by laser-doppler velocimetry, J. Biomech., 17:61, 1984.
32. Baumgartner HR: Effects of acetylsalicylic acid, sulfinpyrazone and dipyridamole on platelet adhesion and aggregation in flowing native and anticoagulated blood, Haemostasis, 8:340, 1979.
33. Baumgartner HR, Muggli R, Tschopp TB: Antiplatelet drugs and the interactions between platelets, fibrin and components of the vessel wall, In: Advances in Pharmacology and Therapeutics II (Eds., Yoshida H, Hagihara Y, Ebashi S), Pergamon Press, Oxford, 1982.
34. Sakariassen KS, Ottenhof-Rovers M, Sixma JJ: Factor VIII-von Willebrand factor requires calcium for facilitation of platelet adherence, Blood, 63:996, 1984.
35. Weiss HJ, Turitto VT, Baumgartner HR, Nemerson Y, Hoffmann T: Evidence for the presence of tissue-factor activity on subendothelium, Blood, 73:968, 1989.
36. Baumgartner HR: Eine neue Methode zur Erzeugung von Thromben durch gezielte Überdehnung der Gefässwand, Z. Gesamte Exp. Med., 137:227, 1963.
37. Weiss HJ, Turitto VT, Baumgartner HR: Role of shear rate and platelets in promoting fibrin formation on rabbit subendothelium. Studies utilizing patients with quantitative and qualitative platelet defects, J. Clin. Invest., 78:1072, 1986.
38. Roux SP, Sakariassen KS, Turitto VT, Baumgartner HR: Effect of aspirin and epinephrine on experimentally induced thrombogenesis in dogs: a parallelism between in vivo and ex vivo thrombosis models, Arteriosclerosis and Thrombosis, 11:1182, 1991.
39. Clozel M, Kuhn H, Baumgartner HR, Procoagulant activity of endotoxin-treated human endothelial cells exposed to native human flowing blood, Blood, 73:729, 1989.
40. Sakariassen KS, Aznar Salatti J, Anton P, Nemerson Y, Prydz H: Tissue factor/factor VIIa is the main initiator of coagulation of stimulated endothelium and extracellular matrix of stimulated endothelium, Medicom. Europe BV, Busum, in press.
41. Sakariassen KS, Cousot D, Hadvary P, Baumgartner HR: Aspirin reduces thrombus volume in human non-anticoagulated blood only at shear rates characteristic for stenosed arteries, Thromb. Haemostas., 65:813, 1991.
42. Baumgartner HR, Turitto VT, Weiss HJ: Effect of shear rate on platelet interaction with subendothelium in citrated and native blood. II. Relationships among platelet adhesion, thrombus dimensions, and fibrin formation, J. Lab. Clin. Med., 95:208, 1980.
43. Sakariassen KS, Kuhn H, Muggli R, Baumgartner HR: Growth and stability of thrombi in flowing citrated blood: assessment of platelet-surface interactions with computer-assisted morphometry, Thromb. Haemostas., 60:392, 1988.
44. Baumgartner HR, Kuhn H, Tschopp TB: Factors influencing the growth, stability and fate of arterial thrombi, In: Atherosclerosis IV (Eds., Schettler G, Gotto AM, Middelhoff G, Habenicht AJR, Jurutka KR), Springer Verlag, Berlin, 1983.
45. Inauen W, Baumgartner HR, Haeberli A, Straub PW: Excessive deposition of fibrin, platelets and platelet thrombi on vascular subendothelium during contraceptive drug treatment, Thromb. Haemostas., 57:306, 1987.
46. Sakariassen KS, Baumgartner HR: Axial dependence of platelet-collagen interactions in flowing blood: upstream thrombus growth impairs downstream platelet adhesion, Arteriosclerosis, 9:33, 1989.

47. Sakariassen KS, Weiss HJ, Baumgartner HR: Upstream thrombus growth impairs downstream thrombogenesis in non-anticoagulated blood: effect of procoagulant artery subendothelium and non-procoagulant collagen, Thromb. Haemostas., 65:596, 1991.
48. Cadroy Y, Hanson SR: Effects of red blood cell concentration on hemostasis and thrombus formation in a primate model, Blood, 75:2185, 1990.
49. Weiss HJ, Turitto VT, Baumgartner HR: Effect of shear rate on platelet interaction with subendothelium in citrated and native blood. I. Shear-dependent decrease of adhesion in von Willebrand's disease and the Bernard-Soulier syndrome, J. Lab. Clin. Med., 92:750, 1978.
50. Fressinaud E, Sakariassen KS, Rotschild C, Baumgartner HR, Meyer D: Strong impairment of thrombus formation in native blood from patients with von Willebrand disease, Blood, 74:32(A), 1989.
51. Weiss HJ, Turitto VT, Baumgartner HR: Further evidence that glycoprotein IIb-IIIa mediates platelet spreading on subendothelium, Thromb. Haemostas., 65:202, 1991.
52. Hanson SR, Pareti FI, Ruggeri ZM, Marzec UM, Kunicki TJ, Montgomery RR, Zimmerman TS, Harker LA: Effect of monoclonal antibodies against the platelet glycoprotein IIb/IIIa complex on thrombosis and hemostasis in the baboon, J. Clin. Invest., 81:149, 1988.
53. Cadroy Y, Houghten RA, Hanson SR: RGDV peptide selectively inhibits platelet-dependent thrombus formation in vivo. Studies using a baboon model, J. Clin. Invest., 84:939, 1989.
54. Weiss HJ, Turitto VT, Vicic WJ, Baumgartner HR: Effect of aspirin and dipyridamole on the interaction of human platelets with sub-endothelium: studies using citrated and native blood, Thromb. Haemostas., 45:136, 1981.
55. Plow EJ, Scrouji AH, Meyer D, Marguerie G, Ginsburg MH: Evidence that three adhesive proteins interact with a common recognition site on activated platelets, J. Biol. Chem., 259:5388, 1984.
56. Weiss HJ, Turitto VT, Vicic WJ, Baumgartner HR: Fibrin formation, fibrinopeptide A release, and platelet thrombus dimensions on subendothelium exposed to flowing native blood: greater in factor XII and XI than in factor VIII and IX deficiency, Blood, 63:1004, 1984.
57. Weiss HJ, Baumgartner HR, Turitto VT: Regulation of platelet-fibrin thrombi on subendothelium, Ann. New York Acad. Sci., 516:380, 1987.
58. Moake JL, Turner NA, Stathopoulos NA, Nolasco LH, Hellums JD: Involvement of large plasma von Willebrand factor (vWF) multimers and unusually large vWF forms derived from endothelial cells in shear stress-induced platelet aggregation, J. Clin. Invest., 78:1456, 1986.
59. Burch JW, Stanford N, Majerus PM: Inhibition of platelet prostaglandin synthetase by oral aspirin, J. Clin. Invest., 6:314, 1978.
60. Roth GJ, Majerus PW: The mechanism of the effect of aspirin on human platelets. I. Acetylation of a particulate fraction protein, J. Clin. Invest., 56:624, 1975.
61. Mombelli G, Monotti R, Haeberli A, Straub PW: Relationship between fibrinopeptide A and fibrinogen/fibrin fragment E in thromboembolism, DIC and various non-thromboembolic diseases, Thromb. Haemostas., 58:758, 1987.
62. Aznar Salatti J, Anton P, Nemerson Y, Prydz H, Sakariassen KS: The synthetic peptide arg-gly-asp-val inhibits binding of fibrin to the extracellular matrix in flowing non-anticoagulated blood, Thromb. Haemostas., 65:222, 1991.

63. Cadroy Y, Maraganore JM, Hanson SR, Harker LA: Selective inhibition by a synthetic hirudin peptide of fibrin-dependent thrombosis in baboons, Proc. Natl. Acad. Sci. USA, 88:1177, 1991.
64. Kelly AB, Marzec UM, Krupski W, Bass A, Cadroy Y, Hanson SR, Harker LA: Hirudin interruption of heparin-resistant arterial thrombus formation in baboons, Blood, 77:1006, 1991.
65. Carson SD, Ross SE, Bach R, Guha A: An inhibitory monoclonal antibody against human tissue factor, Blood, 70:490, 1987.
66. Sandset PM, Abildgaard U, Larsen ML: Heparin induces release of extrinsic pathway inhibitor (EPI), Thromb. Res., 50:803, 1988.
67. Lindahl AK, Abildgaard U, Staalesen R: The anticoagulant effect in heparinized blood and plasma resulting from interaction with extrinsic pathway inhibitor, Thromb. Res., in press.
68. Broze G Jr., Miletich JP: Characterization of the inhibition of tissue factor in serum, Blood, 69:150, 1987.
69. Rao LVM, Rapaport SI: Studies of a mechanism inhibiting the initiation of the extrinsic pathway of coagulation, Blood, 69:645, 1987.
70. Broze G Jr., Warren LA, Novotny WF, Higuchi DA, Gerard JJ, Miletich JP: The lipoprotein associated coagulation inhibitor inhibits the FVII-tissue factor complex also inhibits FXa: insights into its possible mechanism of action, Blood, 71:335, 1988.
71. Sobel M, McNeill PM, Carlson PL, Kermode JC, Adelman B, Conroy R, Margues D: Heparin inhibition of von Willebrand factor-dependent platelet function in vitro and in vivo, J. Clin. Invest., 87:1787, 1991.
72. Gruber A, Griffin JH, Harker LA, Hanson SR: Inhibition of platelet-dependent thrombus formation by human activated protein C in a primate model, Blood, 73:639, 1989.
73. Gruber A, Hanson SR, Kelly AB, Yan BS, Bang N, Griffin JH, Harker LA: Inhibition of thrombus formation by activated recombinant protein C in a primate model of arterial thrombosis, Circulation, 82:578, 1990.
74. Miller EF, Rhodes RK: Preparation and characterization of the different types of collagen, In: Methods in Enzymology, Academic Press Inc., New York, Vol. 82, 32, 1983.
75. Bruns RR, Gross J: High resolution analysis of the quarter stagger model of the collagen fibril, Biopolymers, 13:931, 1977.
76. McCullagh KA, Balion G: Collagen characterization and cell transformation in human atherosclerosis, Nature, 258:73, 1975.
77. Shekonin BV, Domogatsky SP, Idelson GA, Kopteliansky VE, Rukosuev VS: Relative distribution of fibronectin and type I, III, IV, V collagens in normal and atherosclerotic intima of human arteries, Atherosclerosis, 67:9, 1987.
78. Barnes MJ: Collagens in atherosclerosis, Coll. Rel. Res., 5:65, 1985.
79. Nemerson Y: Tissue factor and hemostasis, Blood, 71:1, 1988.
80. Miletich JP, Jackson CM, Majerus PW: Properties of the factor Xa binding site on human platelets, J. Biol. Chem., 253:6908, 1978.
81. Rosing J, Bevers EM, Confurius P, Hemker HC, van Dieijen G, Weiss HJ, Zwaal RFA: Impaired factor X and prothrombin activation associated with decreased phospholipid exposure in platelets from a patient with a bleeding disorder, Blood, 65:1557, 1985.
82. Tracy P: Regulation of thrombin generation at cell surfaces, Sem. Thromb. Haemostas., 14:227, 1988.

83. Turitto VT: Blood viscosity, mass transfer and thrombogenesis, In: Progress in Hemostasis and Thrombosis (Ed., Spaet TH), Grune and Stratton, New York, 1982.
84. Aarts PAMM, van den Broek SAT, Prins GW, Kuiken GDC, Sixma JJ, Heethaar RM: Blood platelets are concentrated near the wall and red blood cells in the center of flowing blood, Arteriosclerosis, 8:819, 1988.

CHANGES IN VASCULAR GEOMETRY IN ATHEROSCLEROTIC PLAQUE RUPTURE AND ITS RELATIONSHIP TO THROMBOSIS IN ACUTE VASCULAR EVENTS

Lina Badimon and Juan Jose Badimon

Unidad de Investigación Cardiovascular, Centro de Investigación y Desarrollo, Consejo Superior de Investigaciones Científicas, Barcelona, Spain; and Cardiovascular Biology Research, Cardiac Unit, Massachusetts General Hospital, Harvard Medical School, Boston, MA, USA

INTRODUCTION

Angiography in patients with unstable angina or myocardial infarction with subtotal coronary occlusion often reveals eccentric stenoses with irregular borders suggesting ruptured atherosclerotic plaques and thrombosis, as documented by angioscopy and at autopsy. We have studied these processes "ex vivo" in a perfusion chamber, "in vivo" in porcine and canine models, and in vitro in humans. Our results suggest that specific local risk factors at the time of plaque disruption influence the degree of thrombogenicity, and therefore the different clinical syndromes. These risk factors can be subdivided into two groups: local vessel wall related factors and systemic factors of local action[1]. There is clinical and experimental evidence to suggest that three systemic factors at the time of plaque rupture may enhance thrombogenicity: a) the levels of epinephrine (i.e., in stress and smoking), b) the level of serum cholesterol, and c) impaired fibrinolysis resulting from high serum lipoprotein (a). Among the local factors, besides the degree of vascular damage, there are rheological factors that significantly contribute to the outcome of thrombosis. We have experimentally demonstrated that the more severe the stenotic lesion after plaque rupture the higher the local shear rate with an enhanced platelet deposition and thrombus formation; and, that platelet deposition and thrombosis are particularly favored if the rupture includes the apex of the stenotic plaque because of its high shear rate. Plaque rupture produces a rough surface with variable geometry that stimulates platelet activation and occlusive thrombus which is enhanced depending on the degree of damage. The presence of a residual thrombus, after spontaneous or pharmacological reperfusion, induces geometrical changes with a further narrowing of the patent cross-sectional area open to flowing blood; additionally, the surface of the residual thrombus is very thrombogenic due to fibrin bound thrombin in the original fragmented thrombus.

We have investigated cellular interactions mimicking conditions subsequent to atherosclerotic plaque rupture in a controlled system of known variables taking into consideration the most relevant factor in cardiovascular physio-pathology, that is the fluid dynamic factors that modulate the cell-and fluid phase protein-interaction with the wall at the particular microenvironment of the ruptured plaque. Static systems, commonly used to elucidate biochemical interactions involved in atherosclerotic disease and its cell biology, disregard the physicochemical control of flow on such interactions in the cardiovascular system.

LOCAL VASCULAR FACTORS

Rheological factors

The role of fluid dynamics in the localization of atherosclerosis is the subject of a significant amount of research[2,3]. Atherosclerosis does not develop at random in the arterial vascular system but at certain locations, such as branching points and curvatures, where blood flow is disturbed and flow separations develop. In addition, blood flow is pulsatile and passes through vessels that are complex both in geometry and compliance. The geometric variability determines the local characteristics of blood flow, according to a tridimensional modification of fluid dynamics in the area. Progressive atherosclerosis, either by lipid infiltration or mural thrombotic complication, induces a local change in geometry, the formation of high grade stenosis, and a modification of the dynamics of blood cells-vessel interaction that may lead to reduction of blood flow or total occlusion. In this article we are dealing with the role of blood flow in the regulation of the complication of atherosclerotic plaque rupture, that is, thrombus growth and stabilization.

1) Flow-Shear Rate

To investigate the dynamics of platelet deposition and thrombus growth following vascular injury we have used a well characterized perfusion system with parallel streamlines that mimic shear rate conditions developing in different arteries and the microcirculation[4-9] (Figure 1).

Exposure of de-endothelialized vessel wall, mimicking mild vascular injury (Type II injury[10]), to perfusing blood induces platelet deposition to the exposed vessel[4]. Platelet deposition increased with both shear rate and exposure time, reaching a maximum within 5 to 10 minutes of exposure. However, at high shear rate, such as can develop in a stenosed artery, and long exposure time (over 10 minutes) the mural thrombus could be dislodged by the perfusing flowing blood, suggesting that the thrombus was labile. Exposure of native fibrillar collagen type I bundles (thus mimicking a deep type III injury[10]) to perfusing blood produced platelet deposits two orders of magnitude higher than that induced by subendothelium at the same rheologic conditions[6]. Even at high shear rate the deposited platelets were not dislodged by flow; they remained adherent to the surface even though the growing thrombus was protruding into the lumen. These results suggest that native collagen is much more reactive than subendothelium with respect to the growth and stability

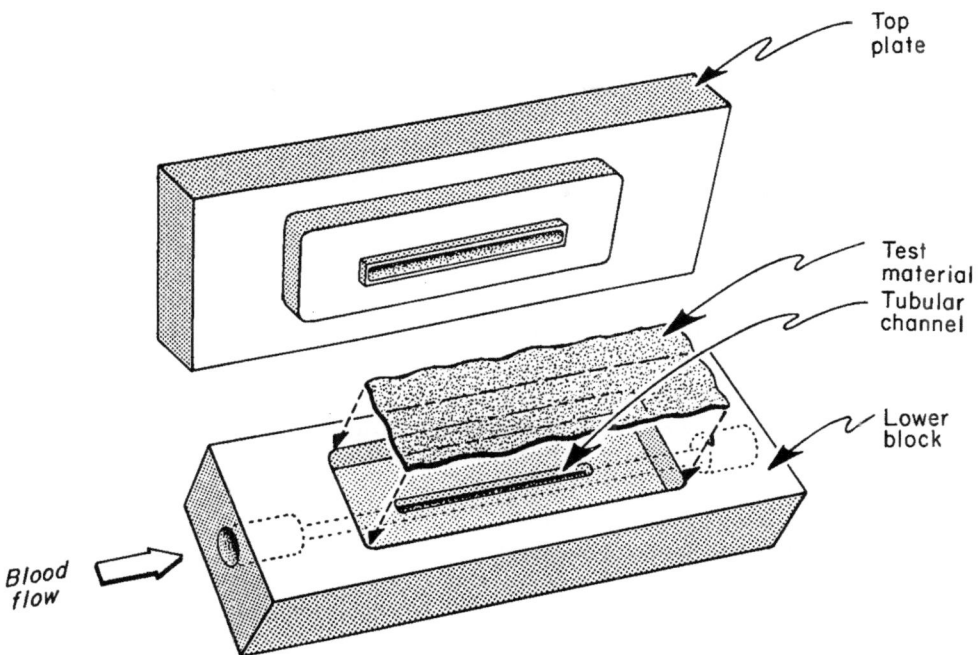

Figure 1. Diagram of the perfusion chamber developed to study cell and protein interaction with a substrate under controlled hemodynamic conditions. The perfusion chamber was designed to retain the cylindrical shape typical of the vasculature, to be flexible enough to accept a variety of biological and prosthetic materials, and to simulate a broad range of physiologic flow conditions.

of the thrombus, and moreover suggest that the flow mechanisms for depositing platelets on the two surfaces may be substantially different. It appears from our studies that the nature of the surface will determine the overall stability of the thrombus, whereas the local flow conditions will influence the rate of growth of the platelet thrombus. Similar experimental quantitative information regarding the role of the substrate in the final outcome of thrombus formation, growth and stabilization has now been acquired by continuous imaging and with various degrees of stenosis.

We have recently directly compared severely damaged vessel wall (Type III injury), mildly damaged vessel wall (Type II injury), and isolated (enzymatic digestion) fibrillar collagen type I to find that severely damaged vessel wall was the most potent inductor of thrombus build-up and that the inhibition of thrombin significantly reduced platelet accumulation[9]. In addition to collagen, other components of the deep layers of the vessel wall, such as tissue factor, exhibit a potent agonist effect for platelets.

Thrombogenesis in vivo occurs in the presence of both platelets and the entire spectrum of the coagulation zymogens and enzymes. Platelets accelerate clotting and act as membrane sites for the localization of proteins; conversely, thrombin generation dramatically influences platelet aggregation and secretion of granular contents. Intravascularly, untreated native blood circulates over damaged vascular surfaces, and ruptured atherosclerotic plaques, or even implanted prosthetic vascular materials with a complete battery of molecules ready for triggering the reparative or healing process. We have analyzed the importance of the availability of coagulation processes in thrombus growth and stabilization. On mildly damaged vessel wall, at high local shear rate (1690/s), the rate of platelet deposition from 1 to 5 minutes was much higher in native than in heparinized blood; in contrast at lower shear rates (212/s) no difference was detected between native and heparinized blood[11]. These findings suggested that under conditions of low thrombogenicity the presence of anticoagulation may be less important[9,11,12]. Similar findings were obtained when the same study was performed with the highly thrombogenic native collagen type I bundles; at low shear rate there was little difference but at high shear rate a significant difference in thrombus build-up was observed. These studies emphasize the importance of shear rate in thrombus growth. Exposure of plaque matrix may stimulate platelets both biochemically and by presentation of a topologically rough surface with the consequent induction of local flow disturbances and small intraplaque recirculation zones.

Overall, it is likely that mild injury to the vessel wall, with a limited thrombogenic stimulus, may induce a transient thrombotic mass. In association with endothelial-dependent vasoconstriction, may be the underlying cause of the onset of unstable angina. On the other hand, deep vessel injury secondary to severe plaque disruption or ulceration results in the exposure of deep vessel structures, that in association with the high shear conditions and endothelial-dependent vasoconstriction, may lead to persistent thrombotic occlusion and myocardial infarction. This oversimplified hypothesis tries to explain the gross thrombotic outcome of the acute events. Cellular interactions at the vessel (substrate interactions) and in the thrombotic mass (cell to

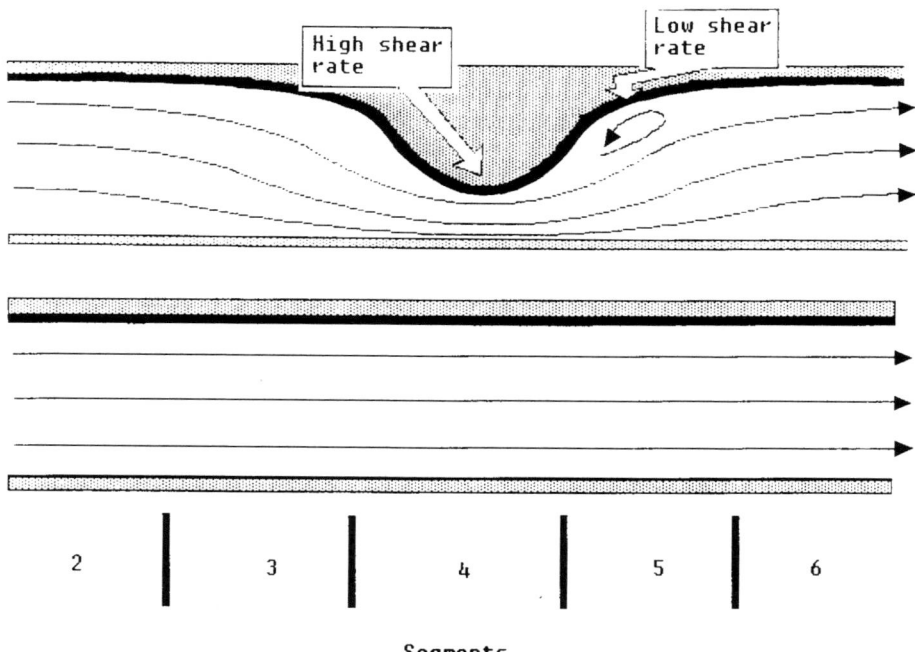

Figure 2. The parallel flow perfusion chamber (shown in figure 1) was modified to model different low and high grade stenosis. This diagram show the schematic flow lines in the chamber with the eccentric stenosis showing non-parallel streamlines (top) and in the regular chamber showing parallel streamlines (bottom).

cell interactions) are presently being studied and will probably help to better understand these very complex phenomena.

2) Geometry

Additionally, few studies have considered the influence of fluid dynamic factors on thrombosis in stenotic vessels where flow streamlines are not parallel. It is a general observation that extreme levels of shear stress exist in close proximity in the presence of advanced vascular lesions[13]. Such differing levels of shear may influence cell interaction mechanisms through inherently different pathways. Blood flowing through a vessel at a certain pressure in the axial direction is suddenly accelerated as it passes through the stenosis and then is immediately deaccelerated. This

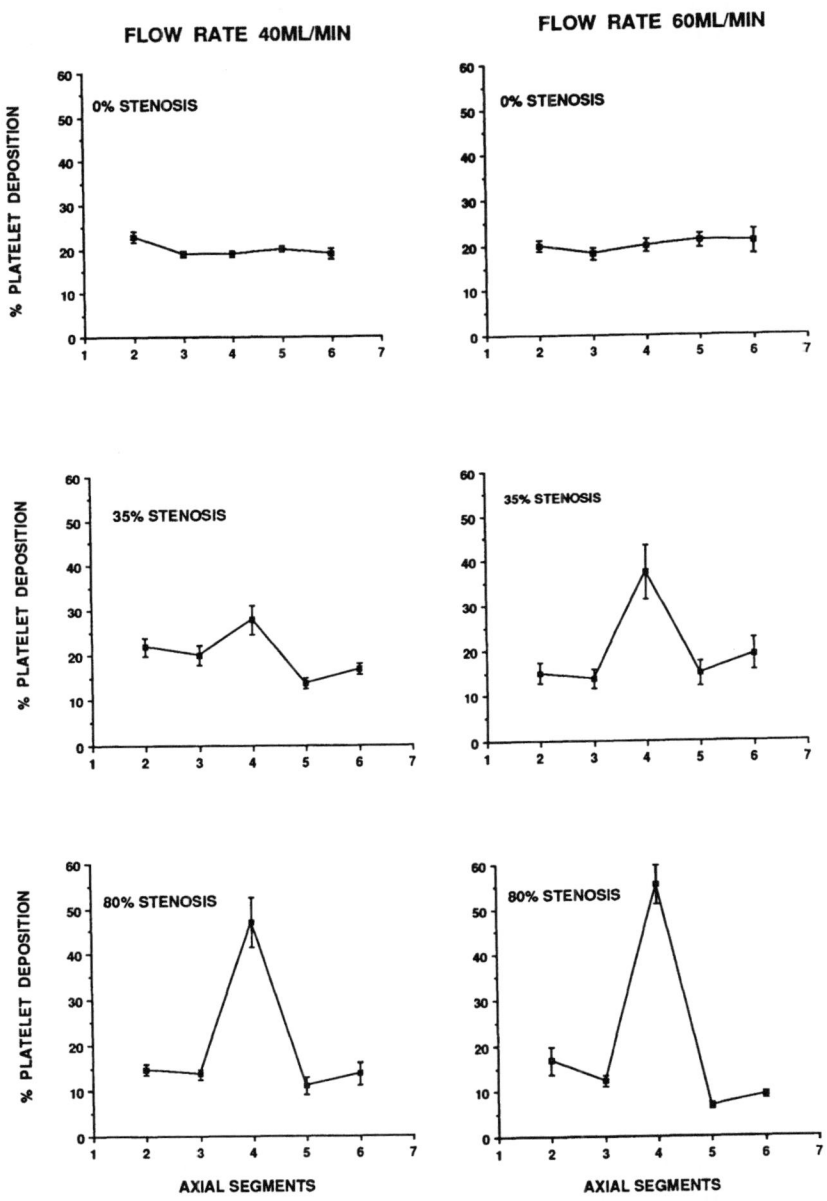

Figure 3. Axial distribution of platelet deposition along the axis of the blood flow. Segments 2-6 were separated according to the different local flow profiles. The study was done placing the stenotic perfusion chamber in an extracorporeal circulation in the porcine model. The substrate was severely damaged porcine vascular wall (Taken from Badimon L, Badimon JJ, J. Clin. Invest., 84:1134-1144, 1989).

deacceleration induces flow separations, recirculation zones (vortexes) with a sudden decrease in pressure downstream to the stenosis. The combination of low with high shear rate areas may induce activation of different mechanisms simultaneously and hence contribute to the acute coronary syndromes.

2a- Eccentric stenosis

In patients who underwent sequential coronary arteriography before and after the onset of unstable angina, progression of stenosis was commonly found in the second part of the study (after the onset)[14,15]. In contrast this progression was not seen in patients with sequential arteriograms whose symptoms remained stable. Arteriography has revealed that eccentric stenosis with scalloped or overhanging edges occur in the majority of patients with unstable symptoms, but not in patients with stable symptoms[16]. These complicated lesions were also seen using postmortem arteriography and histologic examination revealed plaque disruption[17].

We developed the first stenotic flow chamber to analyze the specific mechanisms of interaction in non-parallel flow systems mimicking stenotic vessels[18] (Figure 2). We have found that platelet deposition increases significantly with increasing stenosis in the presence of the same vascular injury, indicating a shear rate induced cell activation. In addition, analysis of the axial distribution of platelet deposition by end point analysis or by continuous imaging of the blood-vascular wall interaction and computer-assisted analysis indicated that the apex, and not the flow recirculation zone distal to the apex, is the segment of greatest platelet accumulation (Figure 3)[18,19]. These data suggest that the severity of the acute platelet response to plaque disruption will depend in part on the sudden changes in the degree of stenosis following rupture and on the occurrence of the rupture at the apex of the pre-existing plaque. Also, these results indicate the importance of flow, or better local shear conditions, as regulator of cell-wall and cell-cell interaction.

2b- Concentric stenosis

Sometimes stenosis may be concentric; therefore we have studied in in vivo models of carotid angioplasty the effect of a concentric high grade stenosis in blood borne cell-vessel wall interaction. We have used the canine and porcine model of angioplasty in the carotid arteries performed under fluoroscopy guidance. Shear rate conditions at the apex of the stenosis were increased up to 19-fold by externally superimposing a silicone ring (up to 75% stenosis). Angiographic filling defect or total occlusion was observed in 30% of the arteries with the stenosis as compared to 0% in the parallel flow non-stenotic vessels (Table 1 and 2). Maximal platelet deposition occurred at the apex of the stenosis (Table 3). In the presence of a thrombogenic stimulus, platelet deposition and thrombus formation will depend on the local shear rate conditions induced by the degree of residual stenosis. These findings may explain acute occlusion and re-thrombosis and may be relevant to the development of restenosis after angioplasty because the deposited platelets might release mitogenic factors and passively contribute to increase the stenotic narrowing. As found in the perfusion chamber model, the same arterial injury will induce a different outcome depending on the flow conditions. In low shear conditions, vascular damage will lead to mural thrombosis; when plaque damage or rupture occurs in geometrically narrowed areas, platelet deposition will be enhanced and thrombosis may progress to total occlusion[20,21].

Table 1. Angiographic filling defects and total occlusions in different degrees of high grade stenosis in the canine model.

PERCENT STENOSIS	>70%	<70%
THROMBUS PRESENT	6	1
NO THROMBUS PRESENT	2	13

Table 2. Angiographic filling defects and total occlusion in different degrees of high grade stenosis

RESIDUAL DIAMETERS	<1.5 mm	1.5-1.7 mm	>1.7mm
THROMBUS PRESENT	4	2	1
NO THROMBUS PRESENT	1	2	12

3) Flow - Geometry- Thrombosis

Exposed collagen and extracellular matrix from injured vessel wall and thrombin generated by the activation of the coagulation factors are powerful platelet activators. Platelets activated by these and other pathways, such as ADP, arachidonic acid, etc, expose their receptors for binding of specific ligands. After the initial contact of platelets with the activating surface-cell adhesion, the monolayer of adherent platelets initiate the interaction with overlying neighboring platelets that are brought by the circulation. The integrin glycoprotein IIb/IIIa (GPIIb/IIIa) receptor complex in the platelet membrane is exposed and binds to ligand adhesive glycoproteins to form bridges between the cells, supporting platelet-platelet interaction, aggregation and thrombus formation. Adhesive macromolecules including fibrinogen, von Willebrand factor (vWF), fibronectin, vitronectin, and laminin recognize the receptor and bind to it.

The receptor-mediated mechanisms relating to the platelet interaction in the thrombotic process around stenoses have not been directly studied, although studies in laminar parallel flow conditions have identified important receptor-function activities in thrombus formation. Platelet glycoprotein Ib (GPIb) is necessary for normal platelet adhesion to subendothelium at high shear rates, presumably through

Table 3. Platelet deposition ($\times 10^6/cm^2$) in the different segments into which the carotid arteries were divided after autopsy.

GROUPS	I	II	p
PROXIMAL	4.36±1.9	5.87±1.59	ns
SEGMENT 1	7.56±2.43	8.11±3.5	ns
SEGMENT 2	8.41±3.76	13.02±5.9	ns
SEGMENT 3	8.19±3.02	36.98±21.3	p<0.001
SEGMENT 4	8.49±5.4	13.06±7.22	ns
SEGMENT 5	8.20±3.61	13.71±14.04	ns
DISTAL	3.62±2.52	1.87±.99	ns

The angioplasty balloon was dilated in segments 1 to 5; the stenotic narrowing was located in segment 3 of group II.

its interaction with vWF. This adhesive molecule also binds to GPIIb/IIIa, and therefore has a double role in platelet adhesion (platelet interaction to the vessel wall) and in platelet aggregation (platelet to platelet interaction) and thrombus formation, mainly in areas under high shear rate. At high shear rate conditions, platelet GPIb and GPIIb/IIIa both appear to be involved in the events of adhesion, whereas GPIIb/IIIa may be involved predominantly in platelet-platelet interaction. These mechanisms have not been evaluated with respect to events in the vicinity of a geometrically created stenosis, but are currently under investigation by our group utilizing antibodies anti GPIIb-IIIa[22].

Residual thrombus

Spontaneous lysis of a thrombus appears to play a role not only in unstable angina[23], but also in acute myocardial infarction[24,25]. Both in spontaneous and pharmacological thrombolysis, the presence of a residual mural thrombus predisposes to recurrent thrombotic vessel occlusion[26-28]. We have experimentally identified two of the possible contributing factors. First, the residual mural thrombotic mass may protrude into the lumen, resulting in a small residual diameter and, as described before, an increased local shear rate, which facilitates platelet activation and deposition at the lesion area (rheological risk factor)[18,19]. Second, a residual thrombus is a highly thrombogenic surface (substrate reactivity risk factor)[29,30]. We have recently studied, by continuous monitoring, the kinetics of thrombus formation on stenotic severely damaged vessel wall and its possible mechanism of action in flowing blood conditions. Autologous platelets were labeled with ^{111}Indium, reinjected, and thrombus growth followed for 50 minutes with a gamma camera and imaging conditions selected to achieve the best spatial resolution. Images of the area of interest were acquired in a dynamic mode in cumulative periods of two minutes per frame and stored in an imaging computer. Blood from heparinized animals (average 1.5 IU/ml) produced significant thrombus accumulation when perfusing the thrombogenic substrate; the rate of thrombus growth was 2±0.5. Embolization or fragmentation of the evolving thrombus rendered a thromboactive surface that rapidly induced re-growth; the rate of growth was 8±1. The infusion of a specific thrombin inhibitor, r-hirudin, in the heparinized animal after 15 minutes of thrombus evolution, abolished thrombus growth. Axial dependence analysis evidenced the same pattern of events occurring at the apex of the stenosis and at the distal flow recirculation area. Therefore, we could show that heparin is not able to completely inhibit thrombus regrowth after fragmentation or embolization of a thrombus; that thrombin is locally available at the thrombus-blood interface to induce rapid re-growth. In situ-thrombin is not inhibited by systemic heparinization and, in the event of partial thrombus dislodgment, it is a very active local nidus for further platelet activation and acute platelet thrombus formation[31]. Thrombus-bound thrombin is not accessible to inhibition by heparin-antithrombin III complex but it is susceptible to inhibition by other specific and direct thrombin inhibitors[30-32].

Local vasoconstriction

Proximal and distal vasoconstriction to the angioplasty site was observed in an experimental carotid angioplasty model[33], and found to be directly related to the degree of platelet deposition. In addition, vasoconstriction was partially inhibited by pretreatment with aspirin, suggesting a platelet dependent vasospasm. Therefore, experimental data suggests that platelets could partially contribute to reduction of blood flow. Platelet dependent vasoconstriction and thrombin dependent vasoconstriction occur in the absence of endothelium suggesting a direct effect on the smooth muscle cells[34]. This could explain the angiographic observation of transient vasoconstriction whenever plaque disruption or fissuring and thrombosis occur in acute coronary syndromes[35]. In the model of angioplasty, we could observe that the inhibition of platelet function did not totally inhibit vasoconstriction, suggesting the existence of a secondary platelet-independent pathway to induce vasoconstriction in vivo. The importance of endothelium-dependent modulation of vascular tone has emerged in the last years since it was observed that endothelial cells are not only an inert barrier but important regulators of vascular homeostasis[36-39]. In normal conditions, the relaxing factors tend to predominate. An alteration in the endothelium may set the cells to generate increasingly proconstrictor mediators. Indeed, atherosclerotic arteries have an elevated vascular tone. The resulting increase in local shear rate conditions when there is no modification in distal vascular resistance, can contribute to local cell activation and thrombosis.

In the acute coronary syndromes there seems to exist a predisposition to platelet and thrombin dependent vasoconstriction at the site of plaque disruption and thrombosis. The presence of abnormal endothelium on or around the atherosclerotic plaque could contribute with an abnormal endothelium-dependent vasoconstrictor response to mediators that in normal vessel wall would induce vascular relaxation with a local and maybe temporal impairment of the rheological conditions.

SYSTEMIC RISK FACTORS

Local thrombosis on an atherosclerosis affected vessel, as we have discussed in the previous section, can lead to a focal hypercoagulable or thrombogenic situation that may favor progression or recurrence of the original atherothrombotic vascular disease. Increasing experimental and clinical evidence also suggest the existence of a primary hypercoagulable or thrombogenic state, which can favor thrombosis in vascular areas at risk. Systemic mediators of vasoconstriction and cell activation can converge locally at a risk area and cooperatively with rheological determinants induce or favour a thrombotic event.

An increased adrenergic activation is associated with acute cardiovascular syndromes[40]; the hypothesis is that a high circulating catecholamine concentration could stimulate the dynamics of arterial thrombus formation in areas at high local vessel wall risk[41]. Of no less importance is the increasing experimental evidence of the effects of hypercoagulability (fibrinogen, factor VII, poor fibrinolysis, etc) and of enhanced platelet reactivity at the site of vascular injury in hypercholesterolemia[42,43]. High plasma levels of homocysteine and of lipoprotein (a) are also beginning to be identified as powerful "thrombogenic risk factors." Lipoprotein (a) (Lp(a)) has been shown to be an important risk factor for ischemic heart disease, presumably for thrombotic occlusion particularly in familial hypercholesterolemia[44,45]. Lp(a) shows close structural homology to plasminogen, which could result in competitive inhibition of its fibrinolytic activity, thus predisposing to thrombotic complications[46].

SUMMARY

The pathophysiology of the acute coronary syndromes and of the progression of chronic coronary artery disease is complex, but in most cases it appears to be based on the rupture of underlying atherosclerotic plaques with superimposed thrombosis. The mechanisms of thrombus formation on atherosclerotic wall at the cellular and molecular levels are not fully understood, but it is clear that they involve local hemodynamic conditions, in addition to vascular and blood-borne factors. We have studied these processes and our results suggest that the combination of those factors in different cardiovascular situations will regulate cell-substrate and cell-cell interactions rendering different thrombotic responses and hence different clinical events.

An important regulatory factor is the local rheology at the site of rupture and the geometry of the residual plaque. Other contributing factors to the final outcome of thrombosis include the degree of damage and nature of the exposed material to the perfusing blood; the presence of residual thrombus or fragmentation of an evolving thrombus; and systemic factors, such as catecholamines, plasma lipids, etc. Platelet and endothelium dependent vasoconstrictive phenomena could contribute to an impairment of the rheological conditions that may favor thrombosis.

Thrombus formation, depending on the relative contribution of these factors, could be either occlusive or mural. Occlusive thrombus will lead to the presentation of occlusive clinical vascular events. Mural thrombus will contribute to the progression and growth of atherosclerosis by its organization into the wall. The study of the cellular and molecular interactions in the evolution of clinically relevant thrombosis is also helping in the development of future approaches for the prevention and management of athero-thrombotic events.

AKNOWLEDGEMENTS

The authors want to thank M. Cohen, R. Lassila, A. Merino and B. Stein for their collaboration in different parts of the studies mentioned in this article. This work was supported in part by NIH grants HL-39840, HL-38933 and HL-38393 and CICYT grant SAL91/0733.

REFERENCES

1. Badimon L, Badimon JJ, Cohen M, Chesebro JH, Fuster V: Vessel wall related risk factors in acute vascular events, Drugs, In press.
2. Nerem RM, Levesque MJ: Fluid dynamics as a factor in the localization of atherogenesis, Annals NY Academy Sciences, 416:709-717, 1983.
3. Asakura T, Karino T: Flow patterns and spatial distribution of atherosclerotic lesions in human coronary arteries, Circulation Research, 66:1045-1066, 1990.
4. Badimon L, Badimon JJ, Galvez A, Chesebro JH, Fuster V: Influence of arterial damage and wall shear rate on platelet deposition. Ex vivo study in a swine model, Arteriosclerosis, 6:312-320, 1986.
5. Badimon L, Turitto VT, Rosemark JA, Badimon JJ, Fuster V: Characterization of a tubular flow chamber for studying plalelet interaction with biological and prosthetic materials. Deposition of ^{111}Indium-labelled platelets on collagen, subendothelium, and Goretex (R), J. of Laboratory and Clinical Medicine, 110:706-718, 1987.
6. Badimon L, Badimon JJ, Turitto VT, Vallabhajosula S, Fuster V: Platelet thrombus formation on collagen type I. Influence of blood rheology, von Willebrand Factor and blood coagulation, Circulation, 78:1431-1442, 1988.
7. Badimon L, Badimon JJ, Galvez A, Turitto VT, Fuster V: Platelet interaction with vessel wall and collagen in pigs with homozygous von Willebrand's disease associated with abnormal collagen aggregation, Thrombosis and Haemostasia, 61:57-64, 1989.
8. Badimon L, Badimon JJ, Turitto VT, Rand J, Fuster V: Functional behavior of vessels from pigs with von Willebrand's disease, Arteriosclerosis, 9:184-188, 1989.
9. Badimon L, Badimon JJ, Lassila R, Heras M, Chesebro JH, Fuster V: Effects of thrombin inhibition in porcine platelet interaction with severely damaged vessel wall, mildly damaged vessel wall and isolated fibrillar collagen type I. Hirudin and r-hirudin vesus heparin in arterial thrombosis, Blood, 78:423-434, 1991.
10. Ip JH, Fuster V, Badimon L, Badimon JJ, Taubman MB, Chesebro JH: Syndromes of accelerated atherosclerosis: role of vascular injury and smooth muscle cell proliferation, J. of the American College Cardiology, 15:1667-1687, 1990.
11. Badimon L, Badimon JJ, Rand J, Turitto VT, Fuster V: Platelet deposition on von Willebrand factor-deficient vessels, The J. of Lab. and Clinical Medicine, 110:634-647, 1987.
12. Badimon L, Badimon JJ, Turitto VT, Fuster V: Role of von Willebrand factor in mediating platelet-vessel wall interaction at low shear rate: The importance of perfusion condition, Blood, 73:961-967, 1989.
13. Zarins CK, Giddens DP, Bharadvaj BK: Carotid bifurcation atherosclerosis. Quantitative correlation of plaque localization with flow velocity profiles and wall shear rate, Circulation Research, 53:502-514, 1983.
14. Moise A, Theroux P, Tarymans P, Descoings B, Lesperance J, Waters DD,

Pelletier GB, Bourassa MG: Unstable angina and progression of coronary atherosclerosis, N. Engl. J. Med., 309:685, 1983.

15. Ambrose JA, Winters SL, Arora RR, Eng A, Ricco A, Gorlin R, Fuster V: Angiographic evolution of coronary artery morphology in unstable angina, J. Am. Coll. Cardiol., 7:472, 1986.

16. Ambrose JA, Winters SL, Stern A, Fuster V: Angiographic morphology and the pathogenesis of unstable angina pectoris, J. of the Am. Coll. of Cardiology, 5:609-616, 1985.

17. Levin DC, Fallon JT: Significance of the angiographic morphology of localized coronary stenoses. Histopathological correlates, Circulation, 66:316, 1982.

18. Badimon L, Badimon JJ: Mechanisms of arterial thrombosis in non-parallel streamlines. Platelet thrombi grow on the apex of stenotic severely injured vessel wall, J. of Clinical Investigation, 84:1134-1144, 1989.

19. Lassila R, Badimon JJ, Vallabhajosulan S, Badimon L: Dynamic monitoring of platelet deposition on severely damaged vessel wall in flowing blood. Effects of different stenosis on thrombus growth, Arteriosclerosis, 10:306-315, 1990.

20. Merino A, Badimon L, Cohen M, Badimon JJ, Fuster V: Mechanisms of arterial thrombosis in non-parallel streamlines in vivo, Thrombosis and Haemostasis, 65:1399A, 1991.

21. Merino A, Cohen M, Badimon JJ, Fuster V, Badimon L: Synergistic action of severe wall injury and shear forces on thrombus formation in arterial stenosis: Definition of a shear rate threshold, J. Amer. Coll. Cardiol., In press, 1992.

22. Badimon L, Badimon JJ, Fuster V: Thrombogenesis and inhibition of platelet aggregation. Experimental aspects and future approaches, Zur. Kardiol., 79:133-145, 1990.

23. Fuster V, Chesebro JH: Mechanisms of unstable angina, New England J. of Med., 315:1023-1025, 1986.

24. Rentrop KP, Feit F, Blanke H, Sherman W, Thornton JC: Serial angiographic assesment of coronary artery obstruction and collateral flow in acute myocardial infarction, Circulation, 80:1166-1175, 1989.

25. Van de Werf F, Arnold AER, and the European Cooperative Study Group for Recombinant Tissue-Type Plasminogen Activator (r-tPA): Effect of intravenous tissue plasminogen activator on infarct sioze, left ventricular function and survival in patients with acute myocardial infarction, British Medical J., 297:374-379, 1988.

26. Fuster V, Stein B, Badimon L, Badimon JJ, Ambrose JA, Chesebro JH: Atherosclerotic plaque rupture and thrombosis. Evolving concepts, Circulation, 82:II47-II59, 1990.

27. Davies SW, Marchant B, Lyon JP, Timmis AD, Rothman MT, Layton CA, Balcon R: Coronary lesion morphology in acute myocardial infarction: Demostration of early remodeling after streptokinase treatment, J. of the Am. Coll. of Cardiology, 16:1079-1086, 1990.

28. Gulba DC, Barthels M, Westhoff-Bleck M, Jost S, Raffienbeul W, Daniel WG, Hecker H, Lichtlen PR: Increased thrombin levels during thrombolytic therapy in acute myocardial infarction. Relevance for the success ot therapy, Circulation, 83:937-944, 1991.

29. Badimon L, Lassila R, Badimon JJ, Vallabhajosula S, Chesebro JH, Fuster V: Residual thrombus is more thrombogenic than severely damaged vessel wall, Circulation, 78:119,474, 1988.

30. Badimon L, Badimon JJ, Lasilla R, Heras M, Chesebro JH, Fuster V: Thrombin

inhibition by hirudin decreases platelet thrombus growth on areas of severe vessel wall injury, J. of the Am. Coll. of Cardiology, 13:145A, 1989.

31. Badimon L, Badimon JJ, Lassila R, Merino A, Chesebro JH, Fuster V: Re-thrombosis on an evolving thrombus is mediated by thrombus-bound thrombin that is not inhibited by systemic heparin, Thrombosis and Haemostasis, 65:760, #321, 1991.

32. Weitz JI, Hudoba M, Massel D, Maragamore J, Hirsh J: Clot-bound thrombin is protected from inhibition by heparin-antithrombin III but is susceptible to inactivation by antithrombin III-independent inhibitors, J. of Clinical Investigation, 86:385-391, 1990.

33. Lam JYT, Chesebro JH, Steele PM, Badimon L, Fuster V: Is vasospasm related to platelet deposition? Relationship in a porcine preparation of arterial injury in vivo, Circulation, 75:243-248, 1987.

34. Cohen RA, Shepherd JT, Vanhoutte PM: Inhibitory role of endothelium in the response of isolated coronary arteries to platelets, Science, 221:273-274, 1983.

35. Hackett D, Davies G, Chierchia S: Intermittent coronary occlusion in acute myocardial infarction: Value of combined thrombolytic and vasodilatory therapy, New England J. of Medicine, 317:1055-1059, 1987.

36. Moncada S, Gryglewski R, Bunting S, Vane JR: An enzyme isolated from arteries transforms prostaglandin endoperoxides to an unstable substance that inhibits platelet aggregation, Nature, 263: 663-665, 1976.

37. Furchgott RF, Zawadzki JV: The obligatory role of endothelial cells in the relaxation of arterial smooth muscle by acetylcholine, Nature, 299:373-376, 1980.

38. Luscher TF: Endothelium-derived relaxing and contracting factors: potential role in coronary artery disease, European Heart J., 10:847-857, 1989.

39. Yanagisawa M, Kurihara H, Kimura S: A novel potent vasoconstrictor peptide produced by vascular endothelial cells, Nature, 332:411-415, 1988.

40. Muller JE, Toefler GH, Stone PH: Circadian variation and triggers of onset of acute cardiovascular disease, Circulation, 79:733-743, 1989.

41. Lassila R, Badimon JJ, Badimon L: Physiologically accesible epinephrine levels modulate platelet interaction to collagen in human blood at flow conditions typical of stenotic vessels, Submitted for publication.

42. Hunt BJ: The relation between abnormal hemostatic function and the progression of coronary disease, Current Opinion in Cardiology, 5:758-765, 1990.

43. Badimon JJ, Badimon L, Turitto VT, Fuster V: Platelet deposition at high shear rates is enhanced by high plasma cholesterol levels. In vivo study in the rabbit model, Arteriosclerosis, 11:395-402, 1991.

44. Dahlen GH, Guyton JR, Attar M, Farmer JA, Kantz JA, Gotto AM: Association of levels of lipoprotein (a), plasma lipids, and other lipoproteins with coronary artery disease documented by angiography, Circulation, 74:758-765, 1986.

45. Seed M, Hoppichler F, Reaveley D, McCarthy S, Thompson GR, Boerwinkle E, Utermann G: Relation of serum lipoprotein (a) concentration and apolipoprotein (a) phenotype to CHD pateients with familial hypercholesterolemia, New England J. of Medicine, 332:1494-1499, 1990.

46. Loscalzo J: Lipoprotein (a): A unique risk factor for atherothrombotic disease, Arteriosclerosis, 10:672-679, 1990.

CURRENT STATUS OF BIOMATERIALS: USE AND DEVELOPMENT

Paul Didisheim,

Head, Biomaterials Program
Devices and Technology Branch
Division of Heart and Vascular Diseases
National Heart, Lung, and Blood Institute
National Institutes of Health
Bethesda, MD 20892 U.S.A.

INTRODUCTION

The rationale for including a talk on biomaterials in a conference on cardiovascular engineering is that engineers and materials scientists have over the years designed devices to replace various organs or perform certain functions in the body, and that the properties of the biomaterials as well as the design of the device influence the events that occur when such devices come into contact with blood and tissues.

DEVICE UTILIZATION

In the U.S., close to 100,000 hip prostheses, a quarter of a million lens implants, and over 6 million artificial kidneys are used annually[1]. In the development of these devices, bulk and surface properties of the biomaterials including biocompatibility have had to be considered; problems of mild to severe nature may occur, although not with such frequency as to prevent them from being widely used and in general improving human health and quality of life. In the United States alone, more than 11 million persons have within their bodies implanted prosthetic devices[1].

Approximately 32,000 mechanical heart valves, and about 20,000 bioprosthetic valves, or valves of tissue origin, are implanted each year in the U.S. In addition, 144,000 pacemakers and 160,000 prosthetic vascular grafts are implanted annually, cardiopulmonary bypass is used about 260,000 times, and 31,000 intraaortic balloon pumps are used per year in the U.S. In contrast, only about 400 ventricular assist devices (VAD's) and a much smaller number of total artificial hearts (TAH'S) were implanted[2].

CLASSIFICATION OF DEVICES

Those devices which serve a primarily cardiovascular function may be classified[3] into two major categories: 1) external communicating and 2) implant devices. External communicating devices are subdivided into two categories: 1) those that serve as an indirect blood path and 2) those in contact with circulating blood.

Those that serve as an indirect blood path include cannulae, catheters, syringes, needles, and tubing used for administration of blood, fluids or medications or for collection of blood specimens for diagnostic or therapeutic purposes, and devices such as plastic bags for the storage of blood and blood products.

Those devices that come in contact with circulating blood include cardiopulmonary bypass equipment; extracorporeal membrane oxygenators (ECMO); hemodialysis equipment (artificial kidney); donor and therapeutic apheresis equipment (for removal of erythrocytes, leucocytes, or platelets, via a continuous-flow system with inflow and outflow connections to the person's circulation); devices such as affinity columns for absorption of specific substances from the blood, for example cholesterol, gamma globulin, macroglobulin, bilirubin, or barbiturates; interventional cardiovascular devices such as those for percutaneous transluminal balloon angioplasty of coronary (PTCA) or peripheral arteries, atherectomy devices, and percutaneous circulatory support systems including miniature pumps capable of aiding the work of the heart; and temporary pacemaker electrodes.

Implant devices include mechanical or tissue (bioprosthetic) heart valves; prosthetic or tissue vascular grafts; circulatory support devices such as intra-aortic balloon pumps, VAD's, and TAH's; inferior vena cava filters designed to trap embolizing thrombi; vascular stents intended to increase the diameter of a stenosed artery; arteriovenous shunts such as those used for access to hemodialysis equipment; conduits implanted to correct anatomic or functional abnormalities of the circulation in infants and children with congenital cardiac defects; internal drug or nutrient delivery catheters such as those for total parenteral alimentation; and permanently implanted pacemaker electrodes.

With this vast array of cardiovascular devices, most of them functioning well with few side-effects, it may be surprising that the nature of biocompatibility of biomaterials is so poorly understood. Unfortunately, very little science has been associated with biomaterials research and development until recently; for example, the first artificial kidney contained a membrane available for food packaging; the first experimental artificial arteries were made of available commercial plastic tubing; and even more recently prosthetic vascular grafts have been made of materials such as nylon, dacron, and teflon, including expanded polytetrafluoroethylene (EPTFE) commercially available under various names including Gore-Tex; these materials are more commonly used for the fabrication of shirts, ropes, tents and ski jackets. Only in the last few years have efforts begun to design materials with functions specific to the intended use of the device.

CLASSIFICATION OF BLOOD-MATERIAL INTERACTIONS

Over the years it has become apparent that a multitude of events occur both in

the blood and on or within biomaterials when these two come in contact. The nature and severity of these interactions are determined by the physical and chemical properties of the material, the time of contact with blood, the precise flow conditions, and other factors. They may be considered in two categories:
1) Those interactions which mainly affect the device and may have an undesirable effect on device function and consequently on the subject;
2) Those interactions which mainly affect the blood and tissues and have a potentially undesirable effect on the subject.

A classification of blood-material interactions is presented in Table 1[3].

Table 1   A classification of blood-material interactions.

1. Interactions which mainly affect the device:
a) Adsorption of plasma proteins, lipids, calcium or other substances from the blood onto the surface of the device, or absorption into the device
b) Adhesion of platelets, leucocytes, or erythrocytes onto the surface of the device, or absorption of their components into the device
c) Formation of a pseudointima or tissue capsule on the surface of the device; this pseudointima is composed of fibrin, collagen, other extracellular matrix proteins, macrophages, and endothelial cells
d) Alterations in mechanical or other properties of the device or of the materials contained in the device

2. Interactions which mainly affect the blood and tissues:
a) Activation of platelets, leucocytes or other cells, or activation of the coagulation, fibrinolytic, complement, or other pathways, including any component of the immune system, resulting in immunosuppression, immunopotentiation, or immunomodulation
b) Formation of thrombi on the device surface
c) Embolization of thrombotic or other material from the device's luminal surface to another site within the circulation
d) Injury to circulating blood cells resulting in anaemia, haemolysis, leucopenia, thrombocytopenia, or altered function of blood cells
e) Injury to cells and tissues adjacent to the device
f) Intimal hyperplasia on or adjacent to the device, resulting in reduced flow or affecting other functions of the device
g) Adhesion and colonization of bacteria or other microorganisms on or near the device

PREDICTIVE TESTS OF BIOCOMPATIBILITY

Biocompatibility and related aspects of the interactions of biomaterials and devices with blood and other tissues have recently become topics for the development of standards by the International Organization for Standardization (ISO) based in Geneva. ISO's Technical Committee on "Biological Testing of Medical and Dental

Materials and Devices" is composed of 12 working groups; one of these is entitled "Selection of tests for interactions with blood." An international standard on this topic has been under preparation since 1990, with input from many experts from 20 countries[3]; final international approval of this standard is expected in 1992. In the process of developing this standard, the working group debated the issue of what tests should be recommended to evaluate interactions of materials and devices with blood. After numerous meetings, the group concluded that the state of knowledge in the field had not yet advanced sufficiently to allow a prediction based on results of any test or group of tests, of the likelihood that a device in contact with blood would be tolerated by the body and would not cause clinically significant adverse effects, in other words, that the device would be biocompatible. Nevertheless, the committee recommended that certain tests should be conducted, and that in order to optimize sensitivity, the categories of tests should be distributed across the spectrum of mechanisms affected which were classified as: 1) thrombosis, 2) platelets and platelet functions, 3) blood coagulation and fibrinolytic mechanisms, 4) hematologic effects (hemolysis, effects on erythrocytes or leucocytes), and 5) complement and other components of the immunologic mechanism.

DEFINITIONS OF BIOCOMPATIBILITY

The following definition was proposed at a Consensus Development Conference held in 1986 under the auspices of the European Society for Biomaterials[4]: "Biocompatibility is the ability of a material to perform with an appropriate host response in a specific application." It was further recommended that since biocompatibility has to be defined in terms of the conditions of use, that is the specific application, the term "biocompatible material" is inappropriate and should not be used. The reason for this recommendation is based on observations that the same biomaterial may appear to be biocompatible or bio-incompatible depending upon the specific experimental conditions employed, which include the design of the device of which the biomaterial is composed, its particular location and application in the body, and the nature of the blood flow field across its surface.

Slightly different from the above is the definition of biocompatibility proposed here: biocompatibility is the ability of a device to perform with a clinically tolerable response in a specific application. This definition emphasizes that in order for a device to be biocompatible, the response it elicits should be compatible with life, health, and normal function of the device and organism.

BIOMATERIALS RESEARCH SUPPORTED BY NHLBI

The following is a description of the research projects dealing with cardiovascular biomaterials and related topics that are currently being supported by the National Heart, Lung, and Blood Institute (NHLBI). These activities are administered within the Division of Heart and Vascular Diseases of NHLBI, in the Devices and Technology Branch of that Division. Currently, approximately 70 grants are being supported in this area. The titles of a few of these are listed below. They include the following individual investigator-initiated research grants:

- Substrate Properties and Human Endothelial Growth
- Platelet Activation and Polymer Surface Thrombogenicity
- Biomedical Polymers - Cell Activation and Interleukin 1
- Calcification in the Cardiovascular System
- Immune Response Changes with Blood Pump Use
- Thrombosis on Biomaterials: Role of Vascular Components
- Cellular Mechanisms of Vascular Graft Failure
- In Vitro Evaluation of Surface-Induced Thrombogenesis
- Analysis of Blood Trauma from Microporous Oxygenators
- A Prototype Gene Therapy Delivery System

Some grants have particular relevance to the main theme of this conference, as the following titles indicate:

- Direct Observations of Interfacial Processes
- Augmented Protein Transport in Sheared Suspensions
- Wall Shear Stress in the Cardiovascular System
- Biomechanics of Blood Cells, Vessels, and Microcirculation
- Effect of Flow and Pressure on Cultured Endothelial Cells

In addition to the individual investigator-initiated research grants which comprise by far the largest fraction of the grant support in this area, two other grant mechanisms are worthy of mention. The first is Requests for Applications (RFA's); the second is Small Business Innovations Research Grants (SBIR's). The RFA is a mechanism whereby advisory committees to NIH, together with NIH staff, identify areas underrepresented among currently funded projects and considered to be of special significance and promise. As funds permit, a certain amount of funds is set aside from the general pool and is designated for a particular program. The total budget of all RFA's represents less than 5% of the Institute's budget. In the area of biomaterials and related topics, three such RFA's have been developed and released in the past 3 years. The first was awarded in 1988 and is entitled Vascular Healing: Cell and Rheologic Factors. Eight awards were made and their topics include:

- Growth Factor Mediation of Healing of Vascular Grafts
- Neutrophil Attack on Vascular Graft Endothelial Cells
- Rheologic and Geometric Factors in Vascular Homeostasis
- Biomechanical Factors in Anastomotic Intimal Hyperplasia
- Elastomeric Polypeptide Vascular Materials

Awards for a second RFA were announced in 1991; its title is Mechanisms of Damage Caused by Cardiopulmonary Bypass. The program was developed jointly with NHLBI's Division of Blood Diseases and Resources. Seven awards were made and the project titles include:

- Prevention of Organ Injury During Cardiopulmonary Bypass
- Control of Blood Activation by Synthetic Surfaces
- Brain Function and Protection During Cardiopulmonary Bypass
- Platelet-Leucocyte Pathobiology in Cardiopulmonary Bypass
- Neutrophil Adherence Reactions in Cardiopulmonary Bypass

A third RFA in the biomaterials area is entitled Cardiovascular Device-Centered Infections. Awards were also announced in 1991. This RFA was developed jointly with the National Science Foundation, the first time such a joint venture between NIH and NSF has been undertaken. The topics of the 3 awards made are:

- Role of Polymer Structure in Bacterial and Neutrophil Adhesion
- Role of Matrix Proteins in Bacterial Adhesion to Surfaces
- Structure of Bacterial and Plasma Protein Ligands in Staphylococcal Adhesion

The SBIR program has attracted a number of grant applications in the biomaterials area. This program is available only to small businesses, and is mandated by the Congress of the United States to dedicate a fixed percentage, 1.25%, of the total budget allocated for extramural research grants by each institute at NIH; extramural grants are those made to institutions outside NIH as opposed to grants for research to be conducted within NIH (intramural). Even though this is a small program, it has enabled a number of clinically useful devices and techniques to be developed. Titles of currently funded grants in the biomaterials area include:

- Immobilized Chelator for Extracorporeal Lead Removal
- Repopulation of Xenograft Heart Valves with Fibroblasts
- Novel Vascular Prosthetic Device
- Insulation of Implant Devices by Plasma Polymer Coatings
- Improved Pacemakers by Surface Metallized Polyurethane

FUTURE RESEARCH DIRECTIONS

Extensive research on vascular physiology and cell and molecular biology in the past 20 years has demonstrated that a blood vessel is not just a passive tube that transports blood from one place to the other but is a complex organ endowed with many functions which are essential for the maintenance and regulation of blood flow. These discoveries have led to the development of a hypothesis concerning the ideal vascular graft, which is as follows: for a synthetic vascular graft to function as a normal blood vessel, it must contain some or all of the activities and properties which make natural blood vessels function normally. These include prostacyclin (PGI_2), endothelium-derived relaxing factor (EDRF), tissue plasminogen activator (tPA), heparin and other glycosaminoglycans, thrombomodulin, ADPase, compliance, and undoubtedly other as yet unknown factors. Approaches to vascular graft development have changed markedly in recent years as a result of the new vascular biology. Whereas early grafts were made of rigid, smooth, non-porous surfaces, now they are flexible, non-kinking, textured, and porous. The ideal future graft may be lined with autologous endothelial cells[5]; or it may have compounds bound to or released from its surface which stimulate adhesion of endothelial cells or inhibit smooth muscle cell proliferation[6]. Alternatively, it may be made of bioabsorbable materials which gradually disappear as host cells grow into the graft and replace it[7]; or it may be synthesized from elastomeric polypeptides capable of controlling cell growth[8]. Another futuristic idea being investigated currently is based on new genetic engineering techniques. Both local and systemic applications of this approach can be envisaged: locally, implanted valves, vascular grafts, or stents could be lined prior to implant with autologous endothelial cells which have been previously genetically engineered to produce increased concentrations of a desirable substance such as tPA which can aid in preventing thrombosis, or growth factor inhibitors which suppress smooth muscle cell proliferation. Feasibility of this approach has already been reported[9]. Systemically, a hybrid organ or "organoid" composed of a large filamentous network of synthetic fibers could be covered with endothelial or other cells that have been genetically engineered prior to implant to express LDL receptor or to release insulin; the application of such a pseudoorgan would be in the treatment of hypercholesterolemia or diabetes. Preliminary studies suggest the feasibility of this approach as well[10]. Numerous technical hurdles remain, but available genetic engineering technology combined with advances in materials science make it likely that some of these seemingly futuristic ideas will become a reality.

It appears likely that for long-term implants of the future, biocompatibility can be best assured by developing biomaterials and devices which are lined with autologous tissue or which promote adhesion and colonization of autologous cells and matrix components in such a way that the foreign material is not in direct contact with the host and therefore is recognized by the host as self rather than as a foreign material.

REFERENCES

1. Moss AJ, Hamburger S, Moore RM, Jeng LL, Howie LJ: Use of selected medical device implants in the United States, Advance Data, 191, 1988.
2. Didisheim P, Watson JT: Application of materials in medicine and dentistry: cardiovascular, In: Biomaterials Science: An Introductory Text (Eds., Ratner BD, Hoffman AS, Lemons JE, Schoen FJ), Academic Press, 1992, in press.
3. Biologic Testing of Medical Devices - Part 4. Selection of tests for interactions with blood. International Standard ISO10993-4, 1992, in press.
4. Williams DF: Consensus and definitions in biomaterials, In: Implant Materials in Biofunction, (Eds., DePutter C, deLange GL, DeGroot K, Lee AJC), Advances in Biomaterials, 8:11-71, Elsevier, Amsterdam, 1988.
5. Herring MB: The use of endothelial seeding of prosthetic arterial bypass grafts, Surg. Ann., 23(pt 2):157-171, 1991.
6. Hubbell JA, Massia SP: Endothelial cell-selective materials for tissue engineering in the vascular graft via a new receptor, Bio/Technology, 9:568-572, 1991.
7. Greisler HP, Endean ED, Klosak JJ, Ellinger J, Dennis JW, Butte K, Kim DU: Polyglactin 910/polydioxanone bicomponent totally resorbable vascular prostheses, J. Vasc. Surg., 7:697-705, 1988.
8. Long MN, King VJ, Prasad KU, Freeman BA, Urry DW: Elastin repeat peptides as chemoattractants for bovine aortic endothelial cells, J. Cell. Physiol., 140:512-518, 1988.
9. Dichek DA, Neville RF, Zwiebel JA, Freeman SM, Leon MB, Anderson WF: Seeding of intravascular stents with genetically engineered endothelial cells, Circulation, 80:1347-1353, 1989.
10. Thompson JA, Haudenschild CC, Anderson KD, DiPietro JM, Anderson WF, Maciag C: Heparin-binding growth factor 1 induces the formation of organoid neovascular structures in-vivo, Proc. Natl. Acad. Sci. USA, 86:7928-7932, 1989.

CARDIAC VALVE REPLACEMENT WITH MECHANICAL PROSTHESES: CURRENT STATUS AND TRENDS

F. Javier Teijeira and Adel A. Mikhail

*Cardiovascular & Thoracic Surgery Division, University of Sherbrooke, Sherbrooke, Quebec, JIH 5N4, Canada
+MDR&R Consultants, 2332 West 111th Street, Bloomington, Minnesota, 55431, USA

In this chapter, an attempt is made to review the current state of the art regarding the design aspects of mechanical valves and their correlation with clinical performance. Literature is reviewed and analyzed for pressure gradients and thromboembolic rates of monoleaflet and bileaflet valve designs. Two ten-year studies are focused on as examples of long-term results. Finally, current trends in valve design and materials are discussed.

STATE OF THE ART REVIEW

The past thirty years have witnessed exciting flurries of activity in the field of cardiac valve design resulting in remarkable advances, which, along with innovative surgical and cardiac preservative techniques, helped save hundreds of thousands of heart valve patients worldwide. The ball and cage, pioneered by Hufnagel in 1952, is considered the parent design. It became commercially available as the Starr-Edwards valve (Figure 1) in 1961. This design was criticized mainly because of its high profile, elevated pressure gradients, ball variance, and relatively high thromboembolic complications. The Starr-Edwards valve went through several reiterations (Figure 2). These included metal balls/bare metal cages criticized for being a noisy mechanism, silastic balls criticized for size variance, cracking and sticking due to lipid absorption, and cloth-covered cages to dampen noise criticized for cloth wear and fraying which resulted in embolism. These settled into the current design models aortic 1260 and mitral 6120 employing properly fabricated silastic balls and bare metal cages.

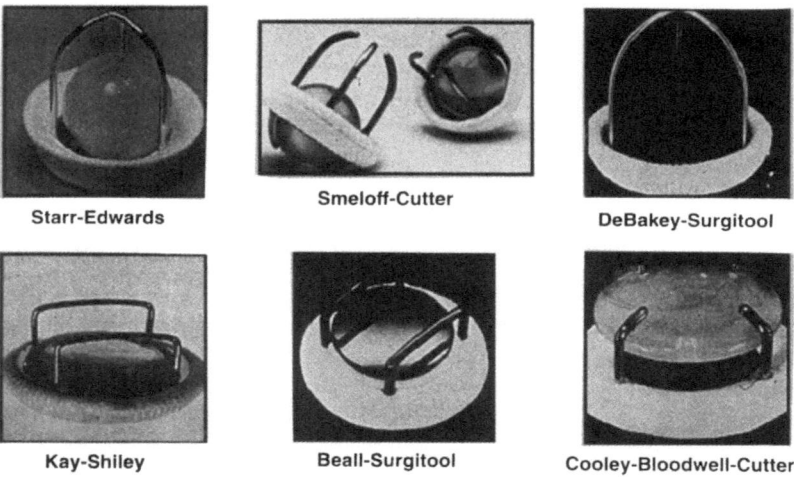

Figure 1. Lateral Flow Valves

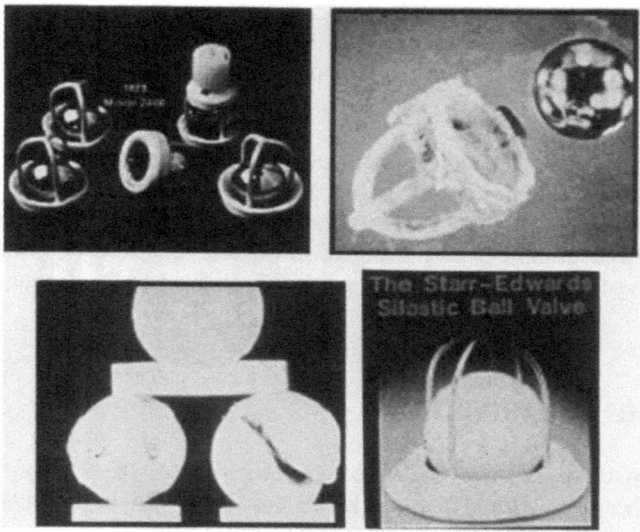

Figure 2. Starr-Edwards Valves

Other early valve designs also employed the idea of a vertically-floating occluder (ball or disc) in order to alleviate some of the above-mentioned objectionable features. For example, the Smeloff-Cutter ball and cage valve was designed to reduce valve profile and allow some leakage by dropping the ball's resting position to within the housing ring. The DeBakey-Surgitool valve employed a pyrolytic carbon-coated ball with the objective of minimizing thrombogenicity. The floating-disc designs like Kay-Shiley, Beall-Surgitool and Cooley-Bloodwell-Cutter (Figure 1) were introduced mainly to reduce valve profile and alleviate ball variance. A common denominator in these designs employing vertically-floating occluders is the fact that blood flow has to change to a lateral direction and turn around the occluder. This lateral flow, or non-central flow, was implicated in the high pressure gradients, turbulence, and cell damage associated with these designs. In addition, some of these valves suffered structural failure and high wear.

Central Flow Designs, Pivoting Disc Valves

The problems associated with lateral flow designs led to the birth of central flow designs in the form of pivoting discs and hinged bileaflet configurations. The first generation of pivoting (or tilting) disc valves appeared in the late 1960's and early 1970's. These valves included the Lillehei-Kaster (LK) and the Delrin-disc Bjork-Shiley (BSDD) prostheses. The latter was soon modified to incorporate a pyrolytic carbon spherical disc (BSSD) (Figure 3). The LK, BSDD, and BSSD employ a pivoting disc design with no fixed hinges. The disc rotates freely within the housing mechanism and pivots on two side struts (LK) or two U-shaped struts extending into the orifice (BSDD and BSSD). The disc opens with a varying combination of angular and lateral (outward) motion, and is retained within the housing by two prongs (LK), or by the outflow strut extending into a circular depression on the disc's outflow surface (BSDD and BSSD). The LK housing was manufactured of titanium while BSDD and BSSD used Haynes 25 alloy (Stellite) for housing material. These two free-rotating (hingeless) pivoting disc configurations are considered to be the parent designs for pivoting disc valves which have in the past few years acquired the nomenclature of "monoleaflet" valves.

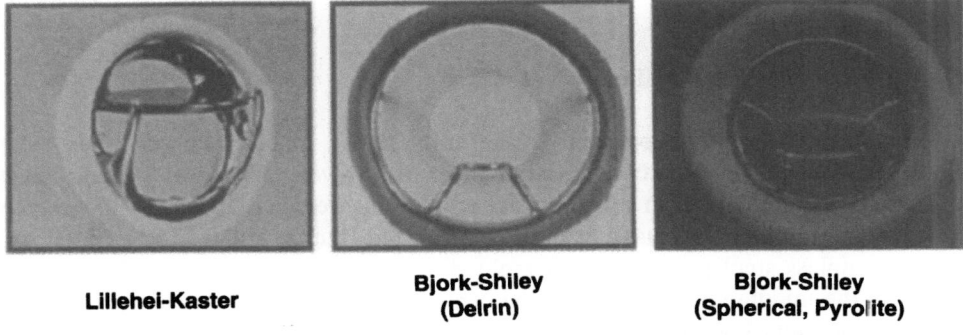

Lillehei-Kaster   Bjork-Shiley (Delrin)   Bjork-Shiley (Spherical, Pyrolite)

Figure 3. Central Flow Valves. Pivoting Disc Designs, First Generation

A second generation of pivoting disc valves evolved between the mid 1970's and mid 1980's as designers and manufacturers tried to alleviate the criticisms leveled against the first generation, mainly, high pressure gradients (particularly in smaller sizes) and high thromboembolic complication rates. This second generation (Figure 4) includes the Bjork-Shiley with a convex/concave disc (BSCC) and its successor, the Bjork-Shiley Monostrut (BSMS), the Medtronic Hall (MH), the Omniscience (OS)[1] and its pyrolytic carbon analog, the Omnicarbon (OC). Of these valves, only the MH and OS are currently marketed in the United States. The BSCC was withdrawn from

---

[1]The Omniscience valve underwent some design modifications prior to approval by the FDA in 1985. The current design has been available since 1982.

the market due to structural failure. The BSMS and OC have not been granted approval by the Food and Drug Administration (FDA) for marketing in the United States.

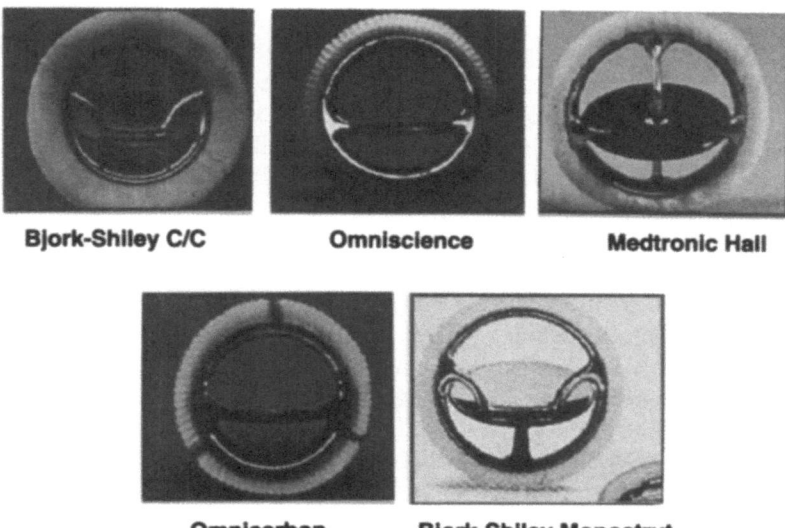

Figure 4. Central Flow Valves. Pivoting Disc Valves, Second Generation

Central Flow Designs, Bileaflet Valves

Bileaflet designs also appeared in the early 1960's. Their evolution has been almost parallel to the evolution of the pivoting disc valves. The Gott-Daggett valve (Figure 5) was implanted in several hundred patients between 1963 and 1966. The valve includes two leaflets made of silastic-vulcanized Dacron attached to a center bar by two stainless steel pins held by a molded superstrut. Since the two leaflets are joined in the middle and hinged over the central strut, the flow characteristics of this valve are not very favorable, resulting in the creation of stasis and turbulent areas

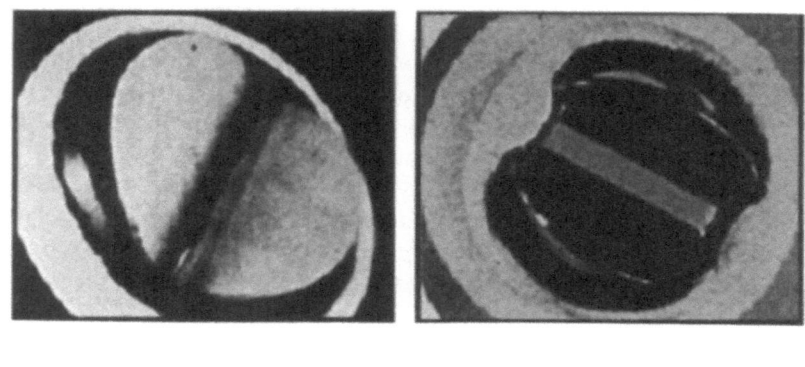

Figure 5. Central Flow Valves. Bileaflet Valves, First Generation

downstream. The valve suffered structural problems and high embolic rates which resulted in its withdrawal from the market shortly after 1966. The Kalke-Lillehei valve more resembles the current bileaflet designs in which two semi-circular leaflets are hinged at their tips by pin-like structures into the housing. On opening, the valve creates three flow channels and exhibits good hydrodynamic characteristics. Neither of these two valves enjoyed much success.

A second generation of bileaflet valves evolved in the mid to late 1970's (Figure 6). The St. Jude Medical valve (SJM) became available in 1977. This valve is manufactured from pyrolytic carbon-coated graphite material and is composed of two leaflets that are hinged at their tips, similar to the Kalke-Lillehei valve with its butterfly-shaped grooves in the housing. The valve produces three flow channels of approximately equal surface area. The hydrodynamics and clinical results of the SJM valve have proven to be much more favorable than the first generation bileaflet valves. Two more bileaflet prostheses were introduced into the marketplace in the early and mid 1980's, the Duromedics (DRM) and the CarboMedics valves. Both valves are manufactured of pyrolytic carbon and incorporate Biolite suture rings. The design of these three valves varies in certain parameters, mainly the opening angle, leaflet curvature, pivot-axis location, and pivot mechanism. The DRM valve was approved and marketed in the United States in 1986 but was removed from the marketplace shortly afterwards for design review because of structural failures. The CarboMedics valve has not been granted approval by the FDA for marketing in the United States.

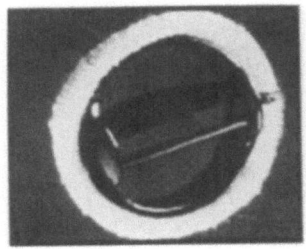 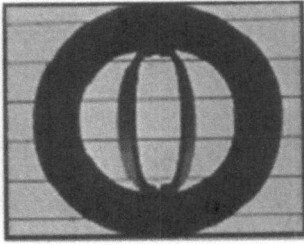 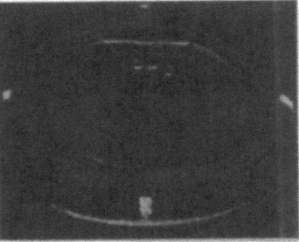

St. Jude Medical   **Duromedics**   **CarboMedics**

Figure 6. Central Flow Valves. Bileaflet Valves, Second Generation

Design Parameters and Their Hydrodynamic Influence

One or more of the following design parameters (Figure 7) were studied and optimized in the evolution course of second generation valves, mainly to improve flow characteristics.

Flow Channel Diameter: This involves maximizing the ratio of orifice diameter to tissue annulus diameter (Do/TAD). This ratio is generally in the range of 0.73-0.83, being lowest in the smallest size. For example, a size 19mm TAD valve would have an orifice diameter of approximately 14mm (Do/TAD=0.74). This lower ratio contributes significantly to the higher pressure gradients (ΔP) observed in smaller sizes. Maximization of this ratio necessitates designing a housing and suture ring thinner than 2.5mm, as given in the above example. An incrementally larger orifice radius will be significantly beneficial in smaller aortic valves. In larger mitral sizes, this contribution is less significant (Poisseille Law) and its benefit should be weighed against the risk of higher interference potential between the moving disc and ventricular structures. As the thickness of the housing/suture ring is made smaller, the clearance between the disc path and ventricular tissue (Clt) becomes smaller and interference potential increases. This is particularly risky when the mitral valve is implanted with the disc path towards the posterior wall of the ventricle.

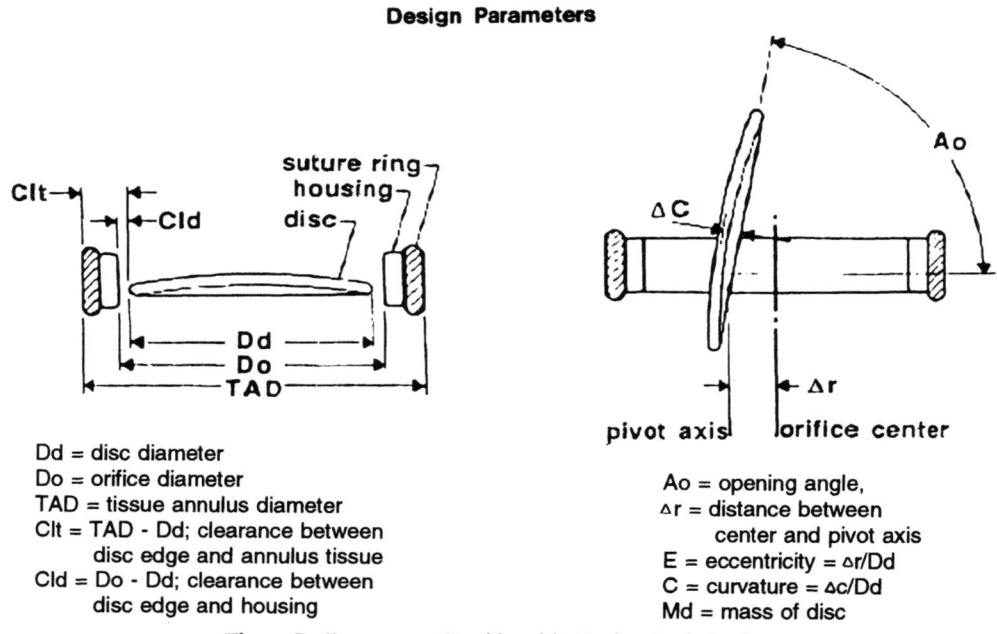

Figure 7. Parameters Considered in Design Optimization

Opening Angle and Disc Curvature: Optimizing the opening angle (Ao) is trickier and has to be considered along with other design parameters. A basic premise is that the magnitude of Ao is inversely proportional to forward flow energy loss and directly proportional to closing phase energy loss. A schematic relationship is given in Figure 8 which depicts an optimum range around 60°-70°. In reality, the opening angle in the second generation pivoting disc valves ranges from 60° (BSCC) to 80° (OS). The BSCC valve's angle (60°) and that of its short-lived successor (BSCC 70°) were optimized through experimental measurements of Ao vs ΔP (forward flow energy loss) and total regurgitation (closing plus leakage) at various flow rates[1].

Researchers have linked the optimization of disc curvature to its opening angle. It is suggested that curvature be of such extent as to place the leading edge of the disc as parallel as possible to the inflow axis. This arrangement minimizes flow separation and turbulence downstream. Earlier experiments using dyes or particles and more recent techniques using Laser Doppler Anemometry (LDA) are employed to map fluid velocity profiles in the vicinity of open valves during forward flow. These techniques are utilized to optimize disc curvature and opening angle.

Eccentricity Obviously, eccentricity as defined in Figure 7, $E = \Delta r/Dd$, is an important factor since it determines the disc pivoting axis (moment arm). The greater

E is, the lower is the force required to open or close the disc. However, this advantage is counterbalanced by the creation of a smaller minor orifice which, if smaller than a certain optimum size, will have poor flow characteristics, i.e., lower velocity and higher turbulence. Optimization of pivot axis eccentricity is usually

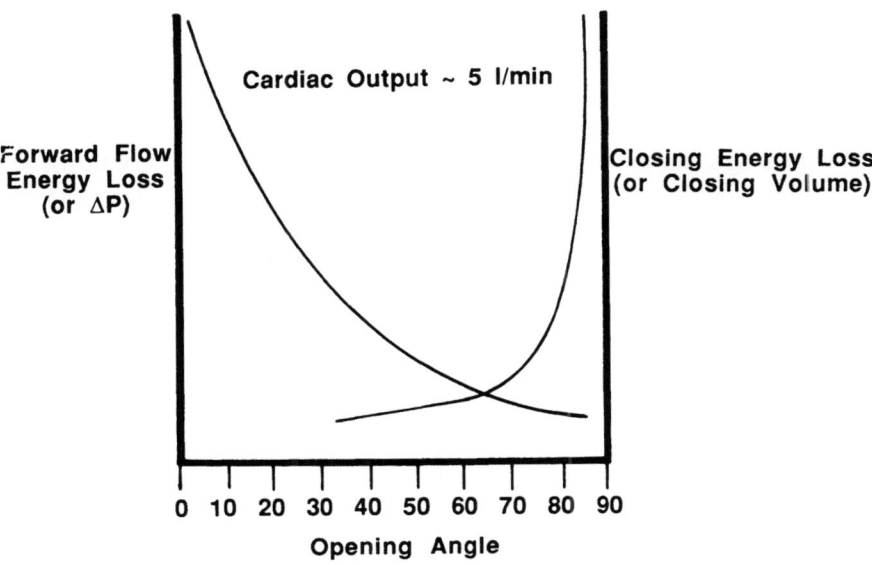

Figure 8. Schematic Relationship of Opening Angle to Forward Flow and Closing Energy Losses

performed by mapping flow velocity profiles in valve flow channels using flow visualization and/or LDA techniques[2,3].

Table 1 summarizes measured values of opening angle, curvature, and eccentricity for some second generation valves. As shown, eccentricity (E) varies between 0.17-0.25 in these valves.

Table 1. Comparative Geometric Parameters

Valve Type	SJM	MH	BSCC	OS
Opening angle (maximum)	86°	70°-75°	59°	80°
Curvature	0.00	0.00	0.12	0.05
Eccentricity	0.23	0.19	0.25	0.17

SJM = St. Jude Medical, MH = Medtronic Hall, BSCC = Bjork-Shiley Convex/Concave, OS = Omniscience.

Disc Clearance and Closing Angle: The disc clearance is defined as the gap between the disc edge and the housing's inside wall (Cld) (Figure 7). The size of this gap is a determinant of the leakage flow during the closed phase of the valve cycle. In order to minimize the leakage energy loss, designers try to make this disc clearance (or gap) as small as possible. This gap width is usually kept between 50-100 micrometers. However, as the gap width gets smaller, the risk of intrinsic jamming (sticking) becomes greater. Obviously, this sticking or jamming between the disc and housing is a function of the variation in the roundness of the disc edge and of the housing's inside circumference, as well as the variation in roughness/smoothness of these surfaces. The latter are essentially micro variations in roundness.

The importance of the closing angle is neglected in some designs. In principle, by the time the disc edge reaches the rim of the housing backward flow causing disc closure ends. From this point, additional backward flow constitutes leakage, which is dependent on the gap width as mentioned before. The further motion of the disc from the housing's rim to its midpoint (0° closing angle) does not benefit the closing cycle and is essentially useless. Hence, the heart energy expended during this phase and its reverse (from 0° to housing's rim) is wasted (Figure 9).

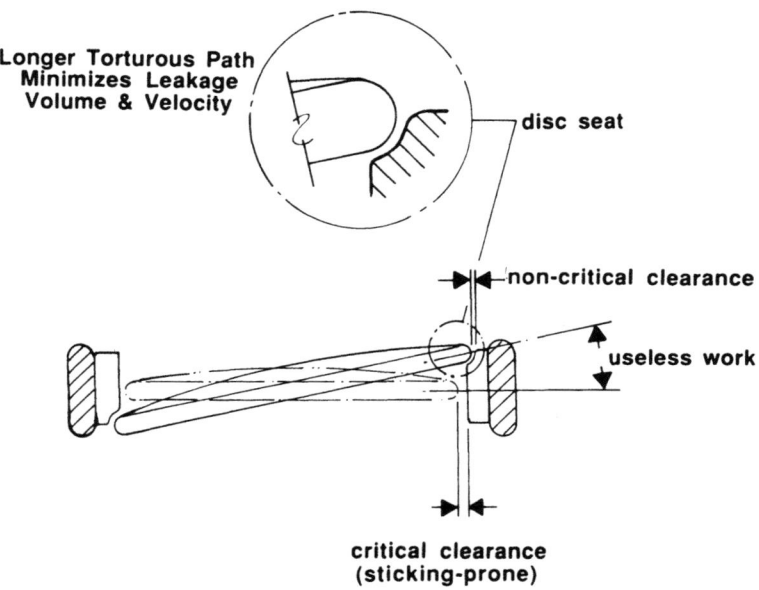

Figure 9. Closing Angle Considerations

In addition, the disc seat feature which is present in some valve designs (OS, OC, and DRM) provides a tortuous path for the backflow, which reduces its velocity and quantity. This disc seat, therefore, acts as a control for back leakage and minimizes jet velocities through these gaps[4].

Jamming or sticking, sometimes caused by extraneous material (for example suture tails, protein or tissue buildup), is much more likely to occur in the case of disc closure at 0°. Closing at the rim of the housing minimizes this jamming potential. Jamming in the closed position can have very serious consequences including death[5,6]. Overcoming this jamming often requires relatively high pressures, possibly as high as 340 mmHg[7].

In Vitro Hydrodynamics, Forward Flow Resistance

Optimization of design parameters had measurable impact on the hydrodynamics of mechanical valve prostheses, generally resulting in lower energy losses. During each opening/closing cycle of a mechanical valve, energy is lost as three distinct components during forward flow (opening stage), backward flow (closing stage), and leakage (closed stage). Only the first component is quantitated in the clinical setting as pressure gradients either by catheterization, or more recently, by Doppler echocardiography measurements. For this reason, we will concentrate in this section on the forward flow energy loss component as it relates to various valve designs (the other two components are discussed later under "Current Trends").

The literature includes numerous good publications which report on forward energy losses. Three reports are chosen for discussion in this section as examples. A report from Japan[8] is of interest since it includes the parent ball and cage valve (Starr-Edwards), some first generation pivoting disc valves, and some second generation prostheses (Figure 10). A distinctly lower pressure gradient (ΔP) is noted for pivoting disc valves compared to the Starr-Edwards, which underscores the lower flow resistance of central flow designs. Also of interest in this figure are the second generation valves, namely the SJM, OS, MH, and BSCC demonstrating flow resistance (ΔP) that is somewhat lower than their first generation counterparts LK and BSSD.

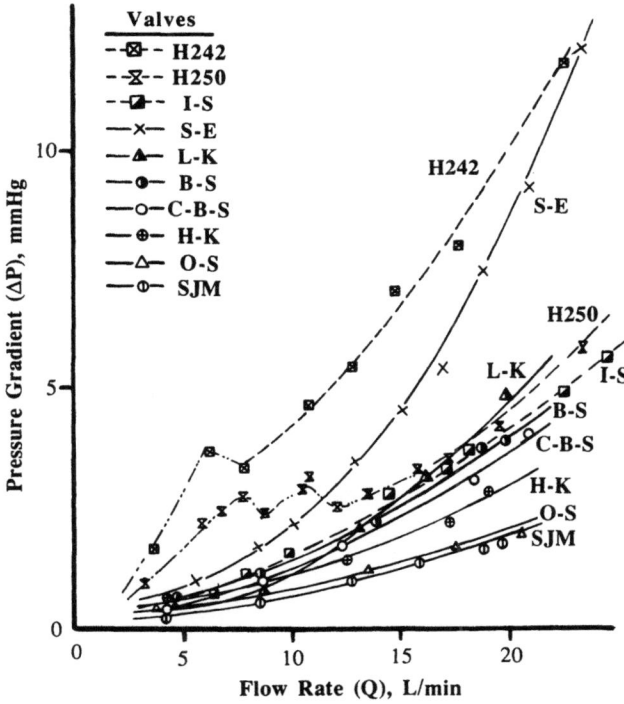

Figure 10. Forward Flow Energy Loss, In Vitro, Steady Flow[8]. H242 & H250=Hancock Porcine, I-S=Ionescu-Shiley, S-E=Starr-Edwards, L-K=Lillehei-Kaster, B-S=Bjork-Shiley Spherical Disc [BSSD in text], C-B-S=Bjork-Shiley Convex/Concave [BSCC in text], H-K=Medtronic Hall [MH in text], O-S=Omniscience, SJM=St. Jude Medical.

Another extensive study was published[9] in which several first and second generation valves were hydrodynamically tested. More recently, Knott et al.[4] reported forward flow energy losses at four flow rates in an aortic channel for ten mechanical valves (Figure 11). These valves include SJM and DRM of second generation bileaflet design and MH, OS, OC, and BSMS of the second generation pivoting disc valves. Except for the SJM which exhibits somewhat lower energy losses, all of these valves appear to show very comparable pressure gradients.

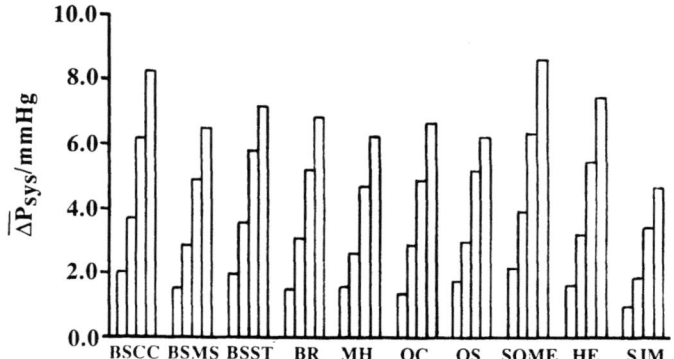

Figure 11. Forward Flow Energy Loss, In Vitro, Pulsatile Flow[4]. ΔP at Cardiac Outputs of 3.0, 4.5, 6.5, and 8.0 l/min. BSCC = Bjork-Shiley Convex/Concave, BSMS = Bjork-Shiley Monostrut, BSST = Bjork-Shiley Spherical Disc [BSSD in text], BR = Bicer, MH = Medtronic Hall, OC = Omnicarbon, OS = Omniscience, SOME = Sorin, HE = Duromedics [DRM in text], SJM = St. Jude Medical.

From these results and a mammoth amount of published data, one observes a dramatic leap in design, accompanied by a sizable drop in pressure gradients, from the lateral flow to central flow designs. However, improvements are much less dramatic going from the first to the second generation central flow valves.

Clinical Performance of Central Flow Valves, Second Generation

The objective of this section is to review the clinical performance of the second generation monoleaflet and bileaflet valves which have clinical histories long enough to provide meaningful data. The road to such review is quite bumpy. The distinct lack of uniformity in definitions and complication reporting in the clinical literature renders such a task very difficult and conclusions somewhat risky.

The two clinical parameters which are believed to be relatively more related to valve design are pressure gradients and thromboembolic rates. The design relationship of the former parameter is obvious and can be, at least in theory, substantiated by in vitro measurements. The latter, the thromboembolic rate, is more subtly and indirectly related to design. Areas of poor blood flow or stagnation are more prone to thrombus formation. Similarly, areas with high turbulence or shear stresses may induce thrombosis through blood cellular damage. Consequently, pressure gradients and thromboembolic rates are the two parameters chosen for this comparative review.

In addition, analysis of the accumulated data in this section allows us to examine two popular perceptions: a) the bileaflet design is superior with regard to forward flow resistance (ΔP) and thrombogenicity, and b) pyrolytic carbon is less thrombogenic than titanium and Stellite. The second perception is inevitably related to the first in that the examined bileaflet valve is also the all-pyrolytic carbon design.

Clinical Pressure Gradients, ΔP

The risk associated with cardiac catheterization has justifiably discouraged researchers from carrying out meaningful comparative ΔP studies. The advent of noninvasive Doppler echocardiography techniques rendered this objective more attainable and, as a result, a few studies have appeared in recent years.

Doppler pressure gradients of normally functioning SJM (bileaflet) and Bjork-Shiley, BS[2] (monoleaflet) valves, were measured in 27 patients[10]. Mean gradients in these aortic valve replacement (AVR) patients range between 7-21 mmHg for the BS valve (sizes 31-19 mm TAD), and 13-23 mmHg for the SJM valve (sizes 23-19 mm TAD). Comparative data by valve size (Table 2) indicate no apparent difference in the mean ΔP values reported for these two valves. The author reports no statistically significant difference between the overall average values of these patient groups. A similar study[11] demonstrates comparable peak and mean ΔP values for BS and SJM valve patients. This study involves 46 AVR and 52 mitral valve replacement (MVR) patients (Table 3). A third report[12] studied 65 patients having these same two valves (Table 4). Extremely close average values of peak gradients are reported by these authors for AVR (29 mmHg for BS vs 30 mmHg for SJM) and for MVR patients (10 mmHg for BS and SJM).

Table 2. Echo Doppler ΔPmean Comparison of SJM vs. BS, AVR[10]

Valve Size (mm)	SJM Avg.±SD	(n)	BS Avg.±SD	(n)
19	23	(1)	21±7	(3)
21	16±5	(5)	16	(1)
23	13±3	(5)	14±4	(4)
25			14±7	(4)
27			9±4	(2)
29/31			7±6	(2)

SJM = St. Jude Medical, BS = Bjork-Shiley, AVR = Aortic Valve Replacement, ΔP in mmHg.

Gibbs et al.[13] reported detailed data by valve size for 53 AVR and 60 MVR patients implanted with two monoleaflets, the BS and MH valves, and a newer bileaflet, the DRM (Table 5). The authors conclude that there are no statistically significant differences between the measured ΔP values in their three patient groups.

Doppler hemodynamic measurements in carefully selected patients with SJM and MH valves were more recently reported[14]. Patients with comparable age, valve size, and left ventricular ejection fraction (LVEF) were selected. Their Doppler ΔP results

---

[2]The authors did not specify the type(s) of Bjork-Shiley valve involved.

Table 3. Echo Doppler ΔP Comparison SJM vs. BS[11]

AVR		SJM n=38	BS n=8
ΔP peak	avg. ± SD	22±12	27±9
	range	4-61	13-36
ΔP mean	avg. ± SD	12±7	14±6
	range	2-32	6-23
MVR		SJM n=44	BS n=8
ΔP peak	avg.± SD	11±4	10±3
	range	4-24	5-16
ΔP mean	avg. ± SD	5±2	5±2
	range	1-14	2-7

SJM = St. Jude Medical, BS = Bjork-Shiley, AVR = Aortic Valve Replacement, MVR = Mitral Valve Replacement, ΔP in mmHg.

Table 4. Echo Doppler ΔPpeak Comparison SJM vs. BS[12]

AVR	n	ΔP peak
SJM	15	30±14
BS	20	29±15
MVR	n	ΔP peak
SJM	17	10±4
BS	13	10±4

SJM = St. Jude Medical, BS = Bjork-Shiley, AVR = Aortic Valve Replacement, MVR = Mitral Valve Replacement, ΔP in mmHg.

demonstrated no statistically significant differences in either AVR or MVR patients (Tables 6 & 7).

A similar study[15] involved AVR and MVR patients implanted with two monoleaflet valves, OS (19 patients) and MH (19 patients). Patients were carefully selected to have comparable valve size and LVEF. The ΔP mean and effective orifice area were determined for AVR and MVR patients. The authors conclude that there are no significant differences between these measured parameters for the OS and MH valve patients.

These various studies report pressure gradient comparisons involving monoleaflet (BS, MH, and OS) and bileaflet valves (SJM and DRM). Generally, no significant differences are reported between these valves in either the aortic or mitral positions. This leads us to conclude that under clinical conditions, no difference can be detected between monoleaflet and bileaflet valves in ΔP Doppler measurements. Given that some measurable differences exist under in vitro conditions, one has to conclude that in vitro differences are too small to be detected under clinical conditions, i.e., these differences are likely buried in the baseline noise of clinical measurements.

Table 5. Echo Doppler ΔPpeak Comparison DRM vs. BS vs. MH[13]

Valve Size (mm)		ΔP peak Avg (n) [Range]		
		DRM	BS	MH
AVR	19	52 (1)	56 (1)	
	21		40 (2) [31-49]	
	23	28 (6) [23-42]	25 (5) [19-36]	22 (5) [3-41]
	25	22 (9) [15-36]	16 (4) [14-21]	17 (5) [13-25]
	27	14 (6) [4-29]	10 (2) [9-10]	18 (2) [15-20]
	29	18 (1)		10 (1)
	31		19 (1)	
MVR	23		12 (1)	
	25		12 (4) [10-14]	
	27	10 (4) [4-16]	10 (6) [5-19]	8 (1)
	29	8 (4) [4-10]	7 (9) [4-13]	10 (5) [8-12]
	31	8 (8) [3-10]	8 (11) [4-13]	8 (7) [6-11]

DRM = Duromedics, BS = Bjork-Shiley, MH = Medtronic Hall, AVR = Aortic Valve Replacement, MVR = Mitral Valve Replacement, ΔP in mmHg.

Table 6. Echo Doppler ΔPmean Comparison SJM vs. MH, AVR (n=42)[14]

	SJM	MH	Difference
Age (years)	50±18	55±14	NS
Mean Valve Size (mm)	24±3	24±3	NS
LVEF (%)	55±11	54±7	NS
ΔP mean-rest	11±4	9±4	NS
ΔP mean-exercise	18±7	15±6	NS

SJM = St. Jude Medical, MH = Medtronic Hall, AVR = Aortic Valve Replacement, LVEF = Left Ventricular Ejection Fraction, NS = Not Statistically Significant, ΔP in mmHg.

Thromboembolic Complication Rates

Thromboembolic complications are believed to be related to blood flow characteristics through cardiac valves. Areas of stagnation, eddies, or high shear stresses may induce blood clotting mechanisms resulting in thrombus formation on the valve surface or in its vicinity. Such thrombus may dislodge resulting in embolic episodes in one organ or another. Peripheral and central nervous system (CNS) emboli are known to be associated with prosthetic valve replacement. CNS

Table 7. Echo Doppler ΔPmean Comparison SJM vs. MH, MVR (n=32)[14]

	SJM	MH	Difference
Age (years)	57±15	52±9	NS
Mean Valve Size (mm)	29±3	29±3	NS
LVEF (%)	53±9	53±6	NS
ΔP mean-rest	2.5±1.4	3±1.1	NS
ΔP mean-exercise	5.1±3.5	7.0±2.9	NS

SJM = St. Jude Medical, MH = Medtronic Hall, MVR = Mitral Valve Replacement, NS = Not Statistically Significant, ΔP in mmHg.

thromboembolism may be transient (few hours) with certain symptoms, e.g., blurred vision, slurred speech, or heaviness/numbness of an arm or a leg. More serious CNS thromboembolism (e.g. a stroke) may completely clear with prompt and proper treatment or may leave some deficit.

A thrombus on the valve may also stay adhered to the surface and continue to attract more fibrin and blood cells until it reaches such a size as to cause valve thrombosis. This would then manifest itself as severe valve stenosis or insufficiency or a combination of both.

Conventionally, according to the recent "Guidelines for Reporting Morbidity and Mortality, after Cardiac Valvular Operations"[16], all these complications, i.e., transient ischemic episodes, serious ischemic episodes (strokes), and valve thrombosis fall under the term thromboembolism. For clarity in this text, TE denotes thromboembolism whether transient or serious, VT denotes valve thrombosis, ΣTE denotes overall thromboembolism (according to guidelines), i.e., ΣTE = TE + VT.

Patients with mechanical prostheses are treated continuously with anticoagulant (coumarin-type) therapy to reduce the incidence of thromboembolism by reducing blood coagulability. This treatment results in an increase of blood coagulation time as measured by the prothrombin time (PT time). Historically, the anticoagulation therapy target was to increase PT time to 1.5-2.5X normal coagulation time. More recently, the trend is to keep this range closer to 1.5-2X to reduce the incidence of hemorrhagic complications.

Anticoagulation therapy presents a dilemma to the physician because of the tight therapeutic index of coumarin. Slightly higher doses can cause serious anticoagulation-related hemorrhagic (ACH) complications. This therapy, therefore, has to be individually tailored and closely monitored to minimize this trade-off of one problem, ΣTE, for another, ACH. This intricate relationship between the two complications mandates looking at both of them when clinical results are analyzed. Quite often, a paper reporting relatively low ΣTE rates also reports relatively high ACH rates and vice-versa. Therefore, the sum of ΣTE + ACH is chosen as another comparative clinical parameter and is denoted in this text as ΣTH.

On reviewing the clinical literature for valve complication rates, one soon realizes that prospective comparative clinical studies based on proper protocols with randomized patient selection and close follow-up are extremely rare. In the absence of such ideal investigations, one has to rely on individual valve studies. However, comparison between individual studies is rather risky because of variations in the demographics of patient cohorts, patients' preexisting disease, surgical technique, time period of the study, anticoagulation therapy level, patient compliance, and follow-up protocols. For instance, anticoagulation interruption or poor patient compliance, atrial fibrillation[17-19], history of thromboembolism, and rheumatic disease[20] have been correlated to higher incidence of thromboembolism. In MVR the orientation of monoleaflet valves in the annulus is an important factor. Orientation of the disc path (large orifice) toward the ventricular posterior wall is implicated in higher TE rates via interference/poor flow mechanism[21,22]. As a result, most researchers and manufacturers recommend an anterior orientation (large orifice toward aortic outflow tract) for MVR patients to minimize interference potential. These factors can vary significantly between one study and another. Consequently, comparing results of individual studies lacks scientific approach and is inconclusive at best.

The next best approach is to combine as many clinical studies as possible for each prosthesis and to use the cumulative complication rates for comparison purposes. In comparing these results one has to assume, still with some risk, that these variations will even out with the large number of patient-years of follow-up for each valve. A similar approach (without comparison of cumulative data) has been taken before[23,24] and provided valuable information.

To illustrate the difficulties encountered in comparing individual studies, a long-term (10 years) clinical experience with a monoleaflet design in one institution (OS,[25]) is compared with an apparently similar 10-year experience with a bileaflet valve (SJM,[26]) in another hospital. A summary of patient demographics and cohort characteristics for the two patient groups (Table 8) indicates some similarities, e.g., the implant period and male/female ratio. However, the mean age (57 vs. 63), which appears to be different, could not be statistically tested because of the lack of raw data in the SJM report. This is not unusual in this literature, because numerous examples can be found of patient parameters that cannot be statistically analyzed when individual studies are compared.

Table 8. 10-year Study Comparison, Monoleaflet vs. Bileaflet, Demographics and Follow-up

	Monoleaflet OS, Teijeira[25]	Bileaflet SJM, Czer et al[26]
Implant Period	1980-1989	1978-1988
Number of Patients	225	616
% of Males	44	45
Mean age (yrs)	57±12	63±15
%AVR-MVR-DVR	42-51-7	47-41-12
Total Follow-up (pt-yrs)	1030	2031
Mean Follow-up (yrs)	5.0±2.1	3.3
% Follow-up	100	95

OS=Omniscience, SJM=St. Jude Medical, AVR=Aortic Valve Replacement, MVR=Mitral Valve Replacement, DVR=AVR+MVR.

Pertinent complication rates (Table 9) indicate generally comparable results for the two valves. In particular, TE rates show no significant difference. On the surface, however, the OS exhibits significantly lower ACH, ΣTH, and late mortality. Nevertheless, this could be a typical example of hasty conclusions since the SJM patient group is older, a fact which may explain the higher propensity to hemorrhagic complications and the higher late mortality.

Table 9. 10-Year-Study Comparison of Monoleaflet vs. Bileaflet, Complication Rates

Rate %/pt-yr	Monoleaflet OS, Teijeira[25]	Bileaflet SJM, Czer et al[26]	Statistical Comparison
TE (%/pt-yr)	1.4	2.0	NS
VT (%/pt-yr)	0.1	0.3	NS
ΣTE (%/pt-yr)	1.5	2.3	NS
ACH	1.0	2.6	S
ΣTH	2.4	4.8	S
Late mortality	2.7	6.4	S

OS = Omniscience, SJM = St. Jude Medical, TE = Thromboembolism, VT = Valve Thrombosis, ΣTE = TE + VT, ACH = Anticoagulant-related hemorrhage, ΣTH = ΣTE + ACH, NS = Not significant ($p > 0.05$, Cox F-Test), S = Significant

Analyzing cumulative rates of several studies involving thousands of patient-years minimizes possible patient population differences and the probability of erroneous conclusions. The literature was reviewed with the objective of including as many studies as possible for each prosthesis. Using the National Library of Medicine's (Bethesda, Maryland) database, a search of the literature was performed for second generation central flow valves. Articles that provided linearized complication rates (%/pt-yr) for overall patients and with a minimum of 3.5 years of follow-up were selected. Only articles written in English with reasonably clear complication definitions were considered. If linearized rates were not available, they were calculated from data given in the article whenever possible. Publications reporting experience with only the early OS design (pre-FDA approval model, discontinued since 1982) are excluded from this analysis[27-29].

Rates of TE, VT, and ACH complications for overall patients (AVR+MVR+AVR/MVR) obtained from 25 individual studies are summarized (Table 10). The abundance of values between parentheses, i.e., the calculated values, underscores the lack of uniformity in reporting these complications. Linearized rates (%/pt-yr) range between 0.7-3.6 for TE, 0-1.2 for VT, and 0.6-5.7 for ACH complications. Cumulative complication rates for each of the four prostheses are illustrated in Figure 12.

Statistical comparison of cumulative complication rates ΣTE and ΣTH for monoleaflet and bileaflet valves are given in Tables 11 and 12. Considering the rates of ΣTE in Table 11, they are found to range between 2.2-3.1%/pt-yr. The rate for the bileaflet SJM is not significantly different from those of the monoleaflet OS or BSMS valves, but significantly different from that of the MH. The significantly higher ΣTE

Table 10. Central Flow Prostheses, Second Generation: Complication Linearized Rates (%/Pt-Yr), Overall Patients

Valve	Reference	Study (yrs)	Pt-Yrs	TE	VT	ACH	ΣTH
MH	30	4.5	2676	3.3	1.2	0.7	(5.1)
MH	31	7.5	1225	2.1	0	1.2	(3.3)
MH	32	7.5	2640	(3.1)	0	(0.6)	(3.8)
MH	33	6	566	2.3	(0.2)	0*	(2.5)
MH	34	8	(312)	(1.0)	(0)	(1.0)	(2.0)
MH	35	5	(947)	(1.8)	(0.3)	(0.3)	(2.4)
MH	36	4.5	237	1.7	(0.8)	NG	NG
OS	37	6	408	1.2	0.2	3.7	5.1
OS	38	5	262	(3.1)	(0.8)	3.4	7.2
OS	39	6	1076	(2.4)	0.3	1.6	(4.3)
OS	40	5	650	(1.1)	(1.2)	(0.6)	(2.9)
OS	25	10	1030	1.4	0.1	1.0	2.4
BSMS	41	3.5	497	3.2	0	2.2	(5.4)
BSMS	42	5	541	1.5	0	0.1	(1.7)
BSMS	43	4	(1268)	(2.4)	(0.2)	(1.8)	(4.4)
SJM	44	9	2562	1.8	0.2	3.2	(5.2)
SJM	45	5.5	1972	2.1	(0.3)	1.3	(3.7)
SJM	46	7	1897	1.7	(0.2)	1.3	(3.2)
SJM	26	10	2031	2.0	0.3	2.6	(4.8)
SJM	47	5	(642)	(3.6)	(0.2)	(0.9)	(4.7)
SJM	48	5	2111	2.5	0.5	0.5*	(3.5)
SJM	49	8	(770)	1.0	(0)	0.9	(1.9)
SJM	50	5	1238	2.6	(0)	5.7	(8.3)
SJM	51	10	(650)	(2.8)	(0.8)	(0.8)	(4.4)
SJM	52	4	(300)	0.7	0.3	1.0	(2.0)

Pt-Yrs = Patient-Years, TE = Thromboembolism, VT = Valve Thrombosis, ACH = Anticoagulation-Related Hemorrhage, ΣTH = TE + VT + ACH. MH = Medtronic Hall, OS = Omniscience, BSMS = Bjork-Shiley Monostrut, SJM = St. Jude Medical. Numbers between ( ) are calculated from data given in the report. NG = Not given and unable to calculate. *Report includes fatal bleeding only.

of the MH can be traced to the inclusion of the South African report by Antunes et al[30], which contributed 2676 pt-yrs and the highest ΣTE rates (4.5%/pt-yr). The patient population in this paper was noted to have unusually poor anticoagulation therapy compliance. When MH reports are considered excluding that of Antunes, the rate drops to 2.5%/pt-yr and the difference is rendered insignificant.

The range of the more meaningful ΣTH (TE + VT + ACH) is 3.8-4.2 %/pt-yr. Statistical analysis reveals no significant difference between the rates for the bileaflet SJM and any of the other monoleaflet valves. Similarly, when the cumulative data for

AVR patients and MVR patients are analyzed separately in the same fashion, no statistically significant differences could be detected between bileaflet and monoleaflet valve groups.[3]

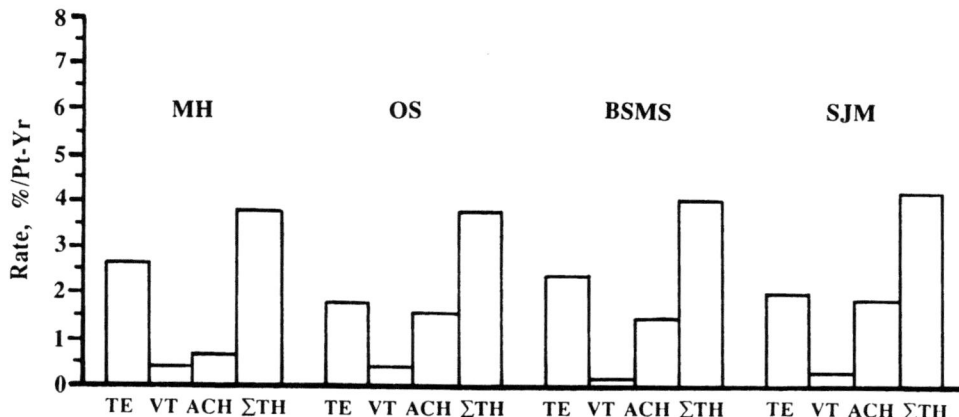

Figure 12. Second Generation Valves. Literature Cumulative Linearized Rates, Overall Patients, Thromboembolism and Anticoagulant-Related Hemorrhage. MH=Medtronic Hall, OS=Omniscience, BSMS=Bjork-Shiley Monostrut, SJM=St. Jude Medical, TE=Thromboembolism, VT=Valve Thrombosis, ACH=Anticoagulant-related hemorrhage, ΣTH=TE+VT+ACH.

Table 11. Cumulative Complication Rates, ΣTE, and Statistical Comparison of Bileaflet vs. Monoleaflet Valves.

Bileaflet	Monoleaflet	
	MH	3.1 (270/8603) S
	MH**	2.5 (151/5927) NS
SJM 2.3* (323/14,173)	OS	2.2 (75/3426) NS
	BSMS	2.5 (57/2306) NS

ΣTE=Thromboembolism+Valve thrombosis, SJM=St. Jude Medical, MH=Medtronic Hall, OS=Omniscience, BSMS=Bjork-Shiley Monostrut, NS=Not significant ($p > 0.05$, Cox F-Test), S=Significant, *Linearized rate in %/pt-yr followed by (no. of events/total pt-yrs), **MH papers excluding Antunes, 1988 report.

Table 12. Cumulative Complication Rates, ΣTH, and Statistical Comparison of Bileaflet vs. Monoleaflet Valves.

Bileaflet		Monoleaflet
	MH	
		3.8 (320/8366)
		NS
SJM	OS	
4.2*		3.8 (130/3426)
(596/14,173)		NS
	BSMS	
		4.0 (92/2306)
		NS

ΣTH = Thromboembolism + Valve thrombosis + Anticoagulant-related hemorrhage, SJM = St. Jude Medical, MH = Medtronic Hall, OS = Omniscience, BSMS = Bjork-Shiley Monostrut, NS = Not significant (p > 0.05, Cox F-Test), *Linearized rate in %/pt-yr followed by (no. of events/total pt-yrs).

These striking findings lead us to conclude that the differences in design between the bileaflet and monoleaflet valves do not impact thromboembolic rates to a significantly measurable level. Thrombotic formations are more commonly observed on parts of the housing rather than on the moving component (the disc or the leaflet). This is expected due to the generally more intricate design composition of the housing (pivots, shields, retaining structures, etc.) which results in poor flow areas. Taking into consideration that the housings of the valves in Table 10 are made of titanium, Stellite, and pyrolytic carbon, it can also be concluded that these materials possess very close thrombogenicity.

Interestingly, similar conclusions are arrived at in two rare comparative studies that randomized patients between a monoleaflet and a bileaflet prosthesis. Vogt et al[53] compared the performance of the BSSD and SJM valves in 178 patients (778 pt-yrs). The authors report no differences in terms of survival, postoperative improvements, or incidence of valve-related complications. In another study[54] 182 patients (559 pt-yrs) were prospectively randomized between SJM and MH valves. Thromboembolic rates (%/pt-yr) are reported to be 4.3 and 3.3 for SJM and MH valve patients, respectively. The authors conclude "In summary, the results obtained from this study appear to indicate that, in the medium term, the performances of the St. Jude and Medtronic-Hall prostheses are identical in this compliant population group..." "The choice of one or the other is, then, a matter of preference or may be linked to other factors such as ease and regular availability and price."

Similar conclusions involving a first-generation pivoting disc valve are worth noting.

---

[3]Details of this analysis are the subject of a future publication.

Olesen et al[55] reported long-term (9-17 yrs, 2301 pt-yrs) results for 262 AVR patients implanted with LK valves. This study shows ΣTE and ΣTH rates (%/pt-yr) to be 1.6 and 2.3, respectively. The authors, comparing the literature data of other more contemporary valves to their results state, "Furthermore, the incidence of valve-related complications hardly differs from the finding in our series with the Lillehei-Kaster prosthesis." ..."Therefore, we conclude that even if further improvements in the design of mechanical prosthesis are likely, the possibility that such innovations will more than marginally improve long-term results seems small."

CURRENT TRENDS

The past three decades have witnessed enormous exciting progress in open heart surgery and in the design of cardiac valve prostheses. Most of the evolutionary advances in valve design occurred, roughly, in the first half of this period. In the last decade (1980's), we have witnessed only minor modifications and design fine-tuning of cardiac prostheses rather than new breakthroughs in safety or effectiveness. The central flow monoleaflet and bileaflet mechanical valve designs, conceived in the late 1960's and early 1970's, still dominate the field today. The current prostheses are manufactured from the same materials employed in that period (Stellite, titanium, pyrolytic carbon). These materials require anticoagulation therapy. Warfarin (a coumarin derivative) is still almost universally prescribed for the lifetime of mechanical valve recipients in spite of its very serious side effects and potential drug/drug interactions. No safer substitute has been discovered yet.

The race is still on for two major objectives: a) attaining a mechanical prosthesis which requires no anticoagulation and b) a tissue prosthesis with durability as high as mechanical prostheses. The progress towards a nonthrombogenic mechanical valve is hardly measurable. Current research is more involved in revealing the intricacies of valve designs and materials used at present. In the following discussions we will highlight some obviously active research areas.

Mechanical Valve Durability

Mechanical valves constitute the majority of implants worldwide. They are selected mainly for their durability advantage compared to tissue valves. This makes the durability issue of paramount importance in mechanical valves.

One of the profound lessons learned in the 1980's is that welded components must be completely avoided at the design stage. The high incidence of structural failures of the BSCC (Figure 13) led to its withdrawal from the marketplace and to numerous liability suits in the USA. Failure of the welded struts, through fatigue mechanism, resulted in disc escape with frequent catastrophic sudden death[56].

Pyrolytic carbon, PC, (a ceramic material) has been the material of choice for valve components for more than twenty years. Until recently, the common perception was that ceramics do not fail through fatigue mechanism. Ritchie et al[57] reports that cyclic loading experiments on composite samples of graphite coated with PC revealed true cyclic fatigue behavior leading to structural failure. The author describes his methodology of creating subcritical cracks and of measuring crack propagation velocity. The author reports "...sub-critical crack velocities under cyclic loading were found to be many orders of magnitude faster than those measured under equivalent monotonic loads and to occur at typically 45% lower stress-intensity level...' Based on these findings, he concludes that "...cyclic fatigue in pyrolytic carbon-coated graphite is reasoned to be a vital consideration in the design and life-prediction procedures of prosthetic devices manufactured from this matter." Intraoperative or early postoperative structural failures of PC valve components are invariably believed to

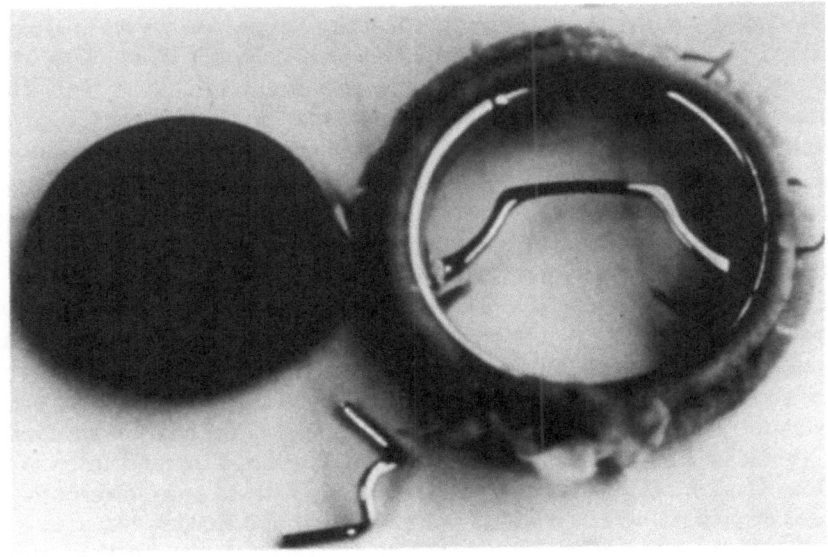

Figure 13. Outlet Strut Fracture in a Bjork-Shiley Convex/Concave Valve

result from mishandling during the surgical procedure. Sharp instruments could cause a deep scratch leading to fracture shortly afterwards, once the valve starts to function. Understandably, this plausible explanation is embraced by manufacturers but reluctantly agreed to by surgeons. For late postoperative structural failures, occurring months or years after implant, this mishandling/deep scratch explanation becomes more credible if a cyclic fatigue mechanism is accepted for this ceramic material. A subcritical deep scratch, material defect, or inconspicuous microfracture is then perceived to propagate slowly under physiological cyclic loading until it reaches the "critical" dimension which causes fracture.

Cavitation erosion is also thought to play a role in inducing subcritical cracks and in assisting crack propagation. This phenomenon has been noted to occur in accelerated wear tests and cavitation-like damage was noted in explants. Kafesjian et al[58], using water jets, reported high initial cavitation resistance in PC components followed by accelerated weight loss with time and the formation of microcracking. He also noted, in ultrasonic horn experiments, that the degree of cavitation varies considerably even over the surface of a single sample.

Cavitation erosion was reported[59] on MH mitral valve discs (PC component) used in a ventricular assist device under physiological conditions. Gross et al[60] reported that in a constantly closed SJM valve, under steady leakage flow and physiological conditions, flow velocity is insufficient to cause venturi-induced cavitation. Such velocity, however, is expected to vary with the width of the gap between the closed leaflet and housing. It is not clear in this report whether the whole range of gap widths was tested.

The issue of PC components' durability and their failure mechanisms has come into focus in the past few years because of reported late postoperative structural failures. The exact incidence of these structural failures cannot be known with certainty since one cannot safely assume that all such incidences are reported in the literature or through the "Medical Device Reporting" (MDR) system of the FDA. Certainly, not all death cases are of known cause and not all are autopsied.

Generally, the rate of late postoperative structural failure of PC components seems to be acceptably low in most cases. Only one case of disc fracture is reported in the literature for the LK valve[61]. It is estimated that 50,000 LK valves have been implanted since 1971. The MH valve (100,000 estimated implants), has 4 disc fractures reported[62,63]. In spite of these reported low incidences, a certain manufacturing lot of the MH valve had to be recalled. Eleven cases of structural failures are reported for the SJM[64-71], apparently a very low incidence in view of the 300,000 valves implanted so far. The DRM valve, with a shorter history, has a relatively higher incidence (25) of structural failure[72-79]. The DRM valve was withdrawn from the marketplace by the manufacturer for further studies. Apparently, these failures precipitated much tougher demands by the FDA regarding PC component testing and quality assurance.

In view of the new studies of PC materials, e.g. in the cyclic fatigue area, the earlier, almost unconditional, faith in PC components should be substituted with more meticulous design of PC components, particularly at stress-concentration points. Extreme care should be taken in the manufacturing and quality assurance of these components and surgeons should be further alerted to the possibility that a deep scratch during handling may produce a microcrack that may propagate under normal physiological conditions and lead to structural failure via a fatigue mechanism.

Alternative Materials

As mentioned before, there have been no breakthroughs or major advances in mechanical valve materials since the early 1970's. Stellite, titanium, and pyrolytic carbon are still the materials of choice in spite of the need for life-long anticoagulation therapy.

Angelini et al[80] describes "glassy carbon" material, belonging to the same family of turbostatic carbons, as PC, which can be obtained by controlled pyrolysis of certain polymeric resins. This methodology's major advantage is the ability to manufacture the prosthesis by a molding process. The material is noncorrodible, biocompatible, and can potentially give smoother surfaces. Certain shortcomings are observed, however; these include the need to control shrinkage and the difficulty in avoiding microbubble formation within the material matrix which can lead to microfractures. This "glassy carbon," in general, has slightly inferior mechanical properties compared to the currently used PC material.

Metal surface modification by coating with microspheres to render it microporous was recently reported[81-83]. Microporosity of 40 micrometers or less is reported to provide an optimal surface for the formation of a thin translucent neointima which gets its nutritives through a diffusion process. This neointima is reported not to thicken, thus avoiding potential interference with disc motion. The author recommends limiting the porous surface to low-flow areas adjacent to the suture ring leaving the high-flow areas with a polished surface.

Imachi et al[84] reported high blood compatibility for the "jellyfish" valve which has a thin polymer membrane, made of Cardiothane fixed at its center on a valve seat. He incorporated these valves in artificial heart blood pumps for 1-125 days without the use of anticoagulant or antiplatelet drugs and reported no thrombus formation on these valves or around the valve seats.

The Carmeda Company (Sweden) has developed a technique for bonding heparin covalently to certain solid surfaces. The attached molecules retain their biological activity for a period of time that allows limited time use, e.g. in extracorporeal circulation, cardiac support systems, catheters, etc. While these surfaces may be of some short-term value in an implantable device like a heart valve, long-term benefits are doubtful.

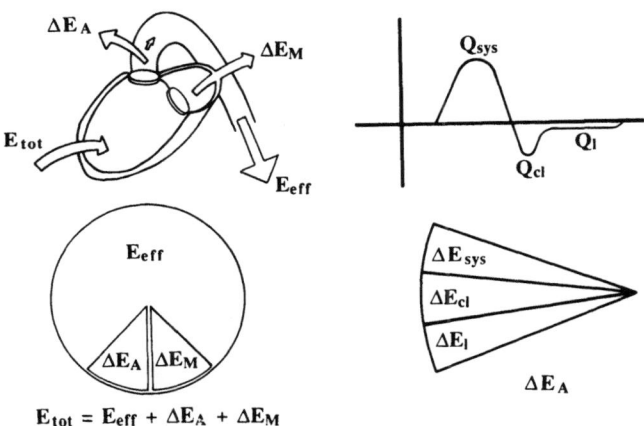

Figure 14. Cardiac Energy Balance and Flow Stages Through Valves

Regurgitation Energy Losses: Revisiting The Forgotten

During valve closing and the closed segments of the pulse cycle, blood flows in the opposite direction, resulting in loss of cardiac energy. These energy losses, which are measured with a good degree of accuracy in laboratory experiments, cannot be quantitated in the clinical setting yet. The lack of such ability imparted some degree of indifference among clinicians and that, in turn, caused valve designers to relax their objectives regarding these energy losses.

Some of the energy generated with every left ventricular contraction is absorbed (lost) by each of the two valves (aortic and mitral) to effect their opening/closing cycle. The net energy is released to the circulation to induce blood flow (Figure 14).

Such energy losses, ΔEa (aortic) and ΔEm (mitral) are negligible in the case of healthy native valves. Mechanical valve prostheses, by virtue of their design, are inescapably stenotic and regurgitant (insufficient). In vitro energy losses indicate that as much as 10% of the ventricular energy can be lost by each mechanical prosthesis, i.e., a double valve replacement patient (AVR/MVR) may lose as much as 20% of left ventricular (LV) energy for these valves to function. The effective net energy to the circulation, Eeff, in this case is only 80% of LV energy.

For each prosthetic valve, energy is lost in three distinct stages. For example, in the aortic position, these three energy losses are: a) resistance of forward flow during systole, E_{sys}, b) reverse flow during closing, E_{cl}, and c) leakage flow during closed phase, E_l.

These three components of energy losses were carefully studied in ten mechanical valves[4].

$$\Delta E_a = E_{sys} + E_{cl} + E_l \qquad (1)$$

The total energy loss ranged from 6-10% of effective energy for the studied valves (Figure 15). Their regurgitant energy loss ($E_{cl}+E_l$) which accounts on the average for 60% (range 45-75%) of total energy loss, is the component that cannot be quantitated in patients yet. Consequently, as mentioned before, regurgitant energy loss is ignored by most clinicians and downplayed by manufacturers.

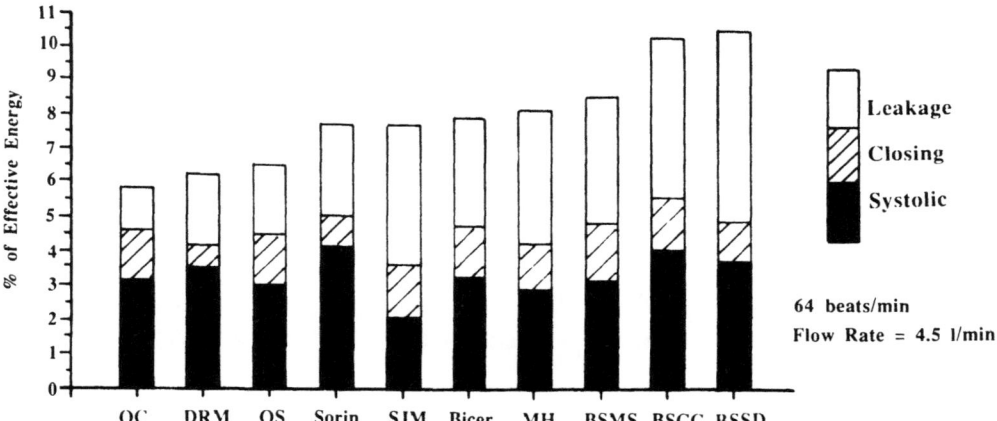

Figure 15. Systolic Energy Loss (Pressure Gradients) vs. Total Energy Loss[4]. OC=Omnicarbon, DRM=Duromedics, OS=Omniscience, SJM=St. Jude Medical, MH=Medtronic Hall, BSMS=Bjork-Shiley Monostrut, BSCC=Bjork-Shiley Convex/Concave, BSSD=Bjork-Shiley Spherical Disc

The leakage energy loss, E_{cl}, is often presented by manufacturers as a "built-in" feature to "backwash" or "self-wash" valve components, and that this "backwash" is necessary to reduce thromboembolism. This seemingly attractive concept is widely accepted but has no scientific support. Literature review reveals no studies that prove "backwash" lowers the incidence of thromboembolism. It is no more than a nice "hand-waving" explanation that is propagated as a good excuse for high leakage energy losses.

Conversely, scientific evidence reveals the serious nature of leakage energy losses. Knott et al[4] states, "Leakage flow and the associated energy loss must be seen as the most serious source of cell damage and thrombus generation. Leakage gaps...may lead

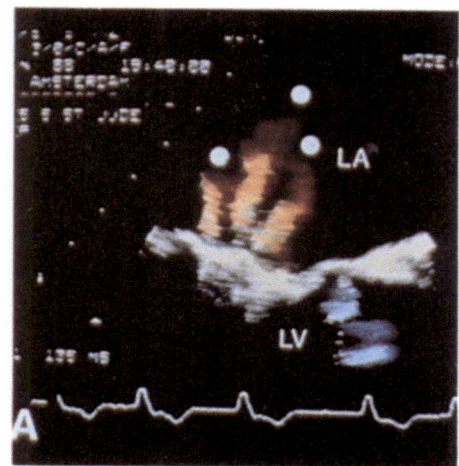

Figure 17. Clinical Detection of Leakage Jets. St. Jude Medical valve and its echocardiographic image.[85]

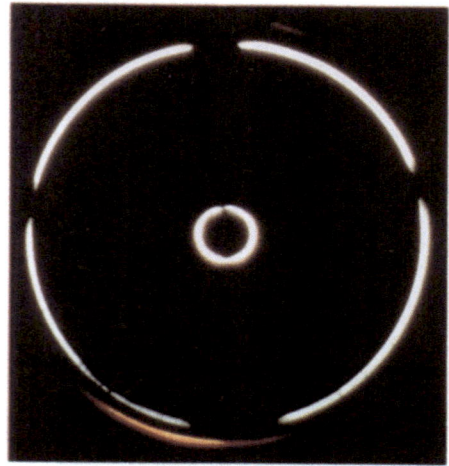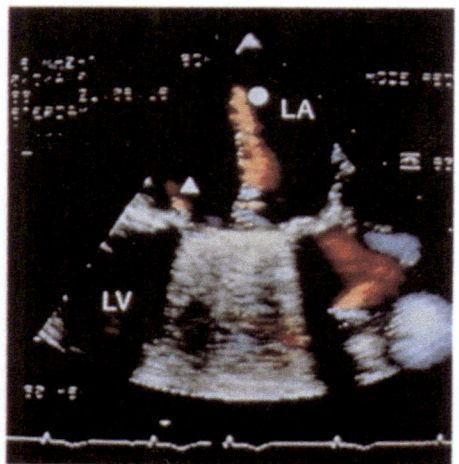

Figure 18. Clinical Detection of Leakage Jets. Medtronic Hall valve and its echocardiographic image.[85]

to increased hemolysis because of high shear stresses within the gap flow and within the turbulent mixing region of the backflow jet. The leakage jet velocities are about 3 to 5 times higher than the peak velocities during systole..." Leakage energy losses exhibit about five-fold variation among the ten tested valves, ranging approximately 1.2%-5.5% of effective energy (Figure 16). The lower leakage in some designs is attributed, at least in part, to the presence of a disc seat whose surface mates with the disc edge contours to create a more resistant leakage path (Figure 9).

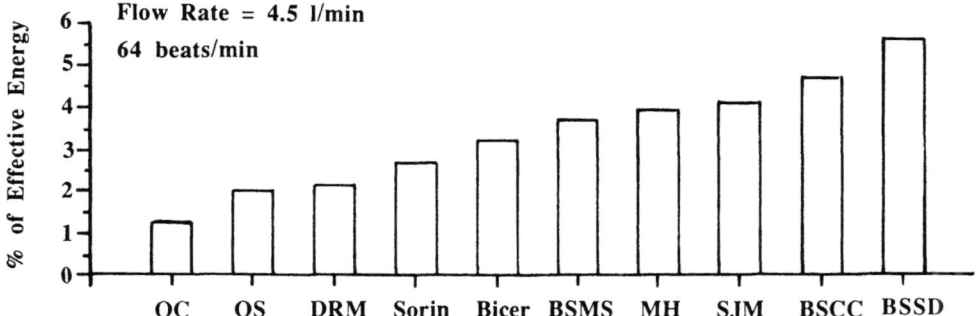

Figure 16. Leakage Energy Loss[4]. OC=Omnicarbon, OS=Omniscience, DRM=Duromedics, BSMS=Bjork-Shiley Monostrut, MH=Medtronic Hall, SJM=St. Jude Medical, BSCC=Bjork-Shiley Convex/Concave, BSSD=Bjork-Shiley Spherical Disc

During the past few years, clinicians were able to observe and study leakage jets with the aid of color Doppler echocardiography techniques. Using the transesophageal flow imaging approach in patients with mitral valve prostheses, systolic intravalvular leakage jets were observed in all SJM and MH patients[85]. The SJM, for example, exhibits 3 holosystolic leakage jets which correspond to the one central and two peripheral gaps (Figure 17). The MH valve shows, as expected, a strong central leakage jet (Figure 18). No such leakage jets were observed with the Starr-Edwards valve since, in the closed position, there are no gaps between the silastic ball and the housing rim.

One important clinical manifestation of high shear stress in leakage jets is red blood cell (RBC) damage (hemolysis) which is measured as elevated lactate dehydrogenase enzymes (LDH), reticulocyte counts, bilirubin, and depressed levels of haptoglobin. Hemolytic anemia may result in severe cases. The great majority of mechanical valve patients exhibit elevated LDH levels. These levels frequently reach 2-6 times the upper normal value. Most of these cases are considered "subclinical", which is not a well-defined term, and are not taken seriously unless signs of anemia are also noted. There are no studies which delineate the long-term ill effects of chronic "subclinical" hemolysis or grade its seriousness below the anemia level. Published studies sporadically report LDH values without much comment. However, a few literature reports describe cases of hemolytic anemia sometimes necessitating reoperation in patients with SJM valves[48,52,86-88] and with MH valves[89]. It is noteworthy that both valves are reported to have high in vitro leakage[4,9].

The velocity and shear stress in leakage jets are expected to be a function of disc clearance, i.e., the width of the gap between the leaflet(s) and the housing (see Cld, Figure 7). This relationship needs to be studied thoroughly in order to shed some light on the varying degrees of subclinical hemolysis and hemolytic anemia observed among patients with the same valve design. In addition, it is necessary to assess the long-term effects of elevated LDH levels and low-grade (borderline) hemolytic anemia.

Regurgitation ($E_{cl}+E_l$) varies within approximately 2.8%-6.5% ($\Delta=3.7\%$) of effective energy. The effect of saving as much as 3.7% of cardiac energy is not known. There is, therefore, a greater need to study the long-term effects of such mild regurgitation exhibited by all mechanical valves to varying degrees. In addition to the gross parameter of survival, progressive trends in more sensitive parameters, e.g. heart size, exercise tolerance, LV function, and quality of life, need to be measured over several years for patients with various valves.

It is hoped with further advancements in Doppler technology that valve closing and leakage flows could be quantitated in the clinical setting. When more clinical attention is given to these energy losses, manufacturers and design engineers will be stimulated to further optimize valve designs.

Discussing current trends in heart valves cannot be complete without pondering a difficult question. Under the current extremely tough regulatory requirements many believe this area is overregulated. It is also quite obvious, as far as medical liability is concerned, that the USA is the most litigious society. In this atmosphere, is it possible for innovations in heart valve design to thrive in the USA?

REFERENCES

1. Bjork VO: The optimal opening angle of the Bjork-Shiley tilting disc valve prosthesis, Scand. J. Thor. Cardiovasc. Surg., 15:223, 1981.
2. Yoganathan AP, Sung H-W, Woo Y-R, et al.: In vitro velocity and turbulence measurements in the vicinity of three new mechanical aortic heart valve prostheses: Bjork-Shiley Monostrut, Omni-Carbon, and Duromedics, J. Thorac. Cardiovasc. Surg., 95:929, 1988.
3. Giersiepen M, Krause U, Knott E, et al.: Velocity and shear stress distribution downstream of mechanical heart valves in pulsatile flow, The International J. of Artificial Organs, 12:261, 1989.
4. Knott E, Reul H, Knoch M, et al.: In vitro comparison of aortic heart valve prostheses, J. Thorac. Cardiovasc. Surg., 96:952, 1988.
5. Pai GP, Ellison RG, Rubin JW, et al.: Disc immobilization of Bjork-Shiley and Medtronic Hall valves during and immediately after valve replacement, Ann. Thorac. Surg., 44:73, 1987.
6. Effler DB: Valve sticking: Complication of Medtronic-Hall prosthesis, Thai. J. Surg., 8:41, 1987.
7. van Son JAM, Steinseifer U, Reul H, et al.: Jamming of prosthetic heart valves by suture trapping: Experimental findings, Thorac. Cardiovasc. Surgeon, 37:288, 1989.
8. Fujita T, Kawazoe K, Umezu M, et al.: Valve characteristics and its clinical application, especially on the biological valve, Japanese Annals of Thoracic Surgery, 1:30, 1981.
9. Kohler J: A comparison of pressure- and volume-loss of technical and biological heart valve prostheses, Z. Kardiol., 75 (Suppl. 2):272, 1986.
10. Kisanuki A, Tei C, Arikawa K, et al.: Continuous wave Doppler echocardiographic assessment of prosthetic aortic valves, J. of Cardiography, 16:121, 1986.

11. Panidis IP, Ross J, Mintz GS: Normal and abnormal prosthetic valve function as assessed by Doppler echocardiography, J. Am. Coll. Cardiol., 8:317, 1986.
12. Bhatia S, Moten M, Werner M: Frequency of unusually high transvalvular Doppler velocities in patients with normal prosthetic valves, J. Am. Coll. Cardiol., 9(Suppl A):238A, 1987.
13. Gibbs JL, Wharton GA, Williams GJ: Doppler ultrasound of normally functioning mechanical mitral and aortic valve prostheses, Int. J. Cardiol., 18:391, 1988.
14. Tatineni S, Barner HB, Pearson AC, et al.: Rest and exercise evaluation of St. Jude Medical and Medtronic Hall prostheses: Influence of primary lesion, valvular type, valvular size, and left ventricular function, Circulation, 80(Suppl I):I-16, 1989.
15. Plehn J: Dartmouth-Hitchcock Medical Center, Hanover, New Hampshire, Private Communication, 1990.
16. Edmunds LH Jr., Clark RE, Cohn LH, et al.: Guidelines for reporting morbidity and mortality after cardiac valvular operations, Ann. Thorac. Surg., 46:257, 1988.
17. Sherman DG, Goldman L, Whiting RB, et al.: Risk of thromboembolism in patients with atrial fibrillation, Arch. Neurol., 41:708, 1984.
18. Caplan LR, D'Cruz I, Hier DB, et al: Atrial size, atrial fibrillation, and stroke, Ann. Neurol., 19:158, 1986.
19. Petersen P, Godtfredsen J: Embolic complications in paroxysmal atrial fibrillation, Stroke, 17:622, 1986.
20. Steele PP, Rainwater J: Favorable effect of sulfinpyrazone on thromboembolism in patients with rheumatic heart disease, Circulation, 62:462, 1980.
21. DeWall RA, Ellis RL: Implantation Techniques: A primary consideration in valve surgery, Ann. Thorac. Surg., 48:S59, 1989.
22. Mikhail AA, Ellis R, Johnson S: Eighteen-year evolution from the Lillehei-Kaster valve to the Omni design, Ann. Thorac. Surg., 48:S61, 1989.
23. Edmunds LH Jr.: Thromboembolic complications of current cardiac valvular prostheses, Ann. Thorac. Surg., 34:96, 1982.
24. Edmunds LH Jr.: Thrombotic and bleeding complications of prosthetic heart valves, Ann. Thorac. Surg., 44:430, 1987.
25. Teijeira FJ: 10-years clinical experience with the Omniscience mechanical prosthesis, Proceedings of the Fifth Annual Meeting of the Mediterranean Association of Cardiology and Cardiac Surgery, September 23-27, Alexandria, Egypt, 369, 1991.
26. Czer LSC, Chaux A, Matloff JM, et al: Ten-year experience with the St. Jude Medical valve for primary valve replacement, J. Thorac. Cardiovasc. Surg., 100:44, 1990.
27. Fananapazir L, Clarke DB, Dark JF, et al.: Results of valve replacement with the Omniscience prosthesis, J. Thorac. Cardiovasc. Surg., 86:621, 1983.
28. Mestres CA, Igual A, Murtra M: Clinical performance of the Omniscience prosthetic heart valve with Dacron sewing ring, Thorac. Cardiovasc. Surgeon, 33:296.
29. Cortina JM, Martinell J, Artiz V, et al: Comparative clinical results with Omniscience (STM1), Medtronic-Hall, and Bjork-Shiley convexo-concave (70 degrees) prostheses in mitral valve replacement, J. Thorac. Cardiovasc. Surg., 91:174, 1986.
30. Antunes MJ, Wessels A, Sadowski RG, et al: Medtronic Hall valve replacement in a third-world population group: A review of the performance of 1000 prostheses, J. Thorac. Cardiovasc. Surg., 95:980, 1988.

31. Beaudet RL, Poirier NL, Doyle D, et al: The Medtronic-Hall cardiac valve: 7-1/2 years' clinical experience, Ann. Thorac. Surg., 42:644, 1986.
32. Butchart EG, Lewis PA, Grunkemeier GL, et al: Low risk of thrombosis and serious embolic events despite low-intensity anticoagulation: Experience with 1,004 Medtronic Hall valves, Circulation, 78(Suppl I):I-66, 1988.
33. Keenan RJ, Armitage JM, Trento A, et al.: Clinical experience with the Medtronic-Hall valve prosthesis, Ann. Thorac. Surg., 50:748, 1990.
34. Lopez JF, Bharadwaj B, Lal S: Experience with Hall-Kaster valves, Vascular. Surgery, 24:16, 1990.
35. Nitter-Hauge S, Semb B, Abdelnoor M, et al.: A 5 year experience with the Medtronic-Hall disc valve prosthesis, Circulation., 68(Suppl II):II-169, 1983.
36. Starek PJK, Murray GF, Keagy BA, et al.: Clinical experience with the Hall pivoting disk valve, Thorac. Cardiovasc. Surgeon, 31/II:66, 1983.
37. Akalin H, Corapcioglu ET, Ozyurda U, et al: Clinical evaluation of the Omniscience cardiac valve prosthesis: Follow-up of up to 6 years, J. Thorac. Cardiovasc. Surg., 103:259, 1992.
38. Carrier M, Martineau J-P, Bonan R, et al: Clinical and hemodynamic assessment of the Omniscience prosthetic heart valve, J. Thorac. Cardiovasc. Surg., 93:300, 1987.
39. Damle A, Coles J, Teijeira J, et al: A six-year study of the Omniscience valve in four Canadian centers, Ann. Thorac. Surg., 43:513, 1987.
40. DeWall R, Pelletier LC, Panebianco A, et al: Five-year clinical experience with the Omniscience cardiac valve, Ann. Thorac. Surg., 38:3, 1984.
41. Aris A, Padro JM, Camara ML, et al: Clinical and hemodynamic results of cardiac valve replacement with the Monostrut Bjork-Shiley prosthesis, J. Thorac. Cardiovasc. Surg., 95:423, 1988.
42. Nakano S, Kawashima Y, Matsuda H, et al.: A five-year appraisal and hemodynamic evaluation of the Bjork-Shiley Monostrut valve, J. Thorac. Cardiovasc. Surg., 101:881, 1991.
43. Thulin LI, Bain WH, Huysmans HH, et al.: Heart valve replacement with the Bjork-Shiley Monostrut valve: Early results of a multicenter clinical investigation, Ann. Thorac. Surg., 45:164, 1988.
44. Arom KV, Nicoloff DM, Kersten TE, et al: St. Jude Medical prosthesis: Valve-related deaths and complications, Ann. Thorac. Surg., 43:591, 1987.
45. Baudet EM, Oca CC, Roques XF, et al: A 5-1/2 year experience with the St. Jude Medical cardiac valve prosthesis: Early and late results of 737 valve replacements in 671 patients, J. Thorac. Cardiovasc. Surg., 90:137, 1985.
46. Burckhardt D, Striebel D, Vogt S, et al: Heart valve replacement with St. Jude Medical valve prosthesis: Long-term experience in 743 patients in Switzerland, Circulation, 78(Suppl I):I-18, 1988.
47. DiSesa VJ, Collins JJ, Cohn LH: Hematological complications with the St. Jude valve and reduced-dose Coumadin, Ann. Thorac. Surg., 48:280, 1989.
48. Kinsley RH, Antunes MJ, Colsen PR: St. Jude Medical valve replacement: An evaluation of valve performance, J. Thorac. Cardiovasc. Surg., 92:349, 1986.
49. Mattila SP, Mattila EJ, Harjula AL: Long-term follow-up of St. Jude Medical valve, Scand. J. Thor. Cardiovasc. Surg., 24:121, 1990.
50. Myers ML, Lawrie GM, Crawford ES, et al.: The St. Jude valve prosthesis: Analysis of the clinical results in 815 implants and the need for systemic anticoagulation, J. Am. Coll. Cardiol., 13:57, 1989.

51. Nair CK, Mohiuddin SM, Hilleman DE, et al.: Ten-year results with the St. Jude Medical prosthesis, Am. J. Cardiol., 65:217, 1990.
52. Panidis IP, Ren J-F, Kotler MN, et al.: Clinical and echocardiographic evaluation of the St. Jude cardiac valve prosthesis: Follow-up of 126 patients, J. Am. Coll. Cardiol., 4:454, 1984.
53. Vogt S, Hoffmann A, Roth J, et al.: Heart valve replacement with the Bjork-Shiley and St. Jude Medical prostheses: A randomized comparison in 178 patients, Eur. Heart J., 11:583, 1990.
54. Antunes MJ: Clinical performance of St. Jude and Medtronic-Hall prostheses: A randomized comparative study, Ann. Thorac. Surg., 50:743, 1990.
55. Olesen K, Rygg I, Wennevold A, et al.: Aortic valve replacement with the Lillehei-Kaster prosthesis in 262 patients: An assessment after 9-17 years, Eur. Heart J., 12:680, 1991.
56. Lindblom D, Bjork VO, Semb BKH: Mechanical failure of the Bjork-Shiley valve: Incidence, clinical presentation, and management, J. Thorac. Cardiovasc. Surg., 92:894, 1986.
57. Ritchie RO, Dauskardt RH, Yu W, et al.: Cyclic fatigue-crack propagation, stress-corrosion, and fracture toughness behavior in pyrolytic carbon-coated graphite for prosthetic heart valve applications, J. of Biomedical Materials Research, 24:189, 1990.
58. Kafesjian R, Chahine G, Frederick G, et al.: Characterization of the cavitation potential of pyrolytic carbon, Proceedings of the International Symposium on Surgery for Heart Valve Disease, June 12-16, London, England, 38, 1989.
59. Lamson TC, Stinebring DR, Deutsch S, et al.: Real-time in vitro observation of cavitation in a prosthetic heat valve, ASAIO Transactions, 37:M351, 1991.
60. Gross JM, Guo GX, Hwang NHC: Venturi pressure cannot cause cavitation in mechanical heart valve prostheses, ASAIO Transactions, 37:M357, 1991.
61. Pilichowski P, Gaudin Ph, Brichon P-Y, et al.: Fracture and embolization of a Lillehei-Kaster mitral valve prosthesis disc: One case successfully operated, Thorac. Cardiovasc. Surgeon, 35:385, 1987.
62. DDL (Devices & Diagnostics Letter): Washington Business Information, Inc., Arlington, Virginia, 1, July 1, 1988.
63. MH MDR (Medtronic Hall, Medical Device Report), (FDA M219190), 1991.
64. Hasse J: Escaped leaflet in a St. Jude Medical mitral prosthesis, In: Advances in Cardiac Valves, Clinical Perpectives, Yorke Medical Books, 1983.
65. Hjelms E: Escape of a leaflet from a St. Jude Medical prosthesis in the mitral position, Thorac. Cardiovasc. Surgeon, 31:310, 1983.
66. Odell JA, Durandt J, Shama DM, et al.: Spontaneous embolization of a St. Jude prosthetic mitral valve leaflet, Ann. Thorac. Surg., 39:569, 1985.
67. Orsinelli DA, Becker RC, Cuenoud HF, et al.: Mechanical failure of a St. Jude Medical prosthesis, Am. J. Cardiol., 67:906, 1991.
68. SJM MDR (St. Jude Medical, Medical Device Reports), (FDA M116225), 1985.
69. SJM MDR (St. Jude Medical, Medical Device Reports), (FDA M125685), 1986.
70. SJM MDR (St. Jude Medical, Medical Device Reports), (FDA M148399), 1987.
71. SJM MDR (St. Jude Medical, Medical Device Reports), (FDA M182613, M184089, M189945, M202719, M205267), 1990.
72. Alvarez J, Deal CW: Leaflet escape from a Duromedics valve, J. Thorac. Cardiovasc. Surg., 99:372, 1990.
73. Dimitri WR, Williams BT: Fracture of the Duromedics mitral valve housing with leaflet escape, J. Cardiovasc. Surg., 31:41, 1990.

74. Klepetko W, Moritz A, Mlczoch J, et al.: Leaflet fracture in Edwards-Duromedics bileaflet valves, J. Thorac. Cardiovasc. Surg., 97:90, 1989.
75. DDL (Devices & Diagnostics Letter): Washington Business Information, Inc., Arlington, Virginia, 3, June 3, 1988.
76. DRM MDR (Duromedics, Medical Device Reports): (FDA M150116, M152554, M152555), 1988.
77. DRM MDR (Duromedics, Medical Device Reports): (FDA M175128, M177663, M178926), 1989.
78. DRM MDR (Duromedics, Medical Device Reports): (FDA M186679, M209449, M209571, M217126), 1990.
79. DRM MDR (Duromedics, Medical Device Reports): (FDA M223051, M223464, M223758, M232445, M246900, M248799), 1991.
80. Angelini G, Price C, Jenkins G: Glassy carbon: A new material for the manufacture of mechanical heart valve prostheses, Proceedings of the International Symposium on Surgery for Heart Valve Disease, June 12-16, London, England, 40, 1989.
81. Bjork VO: Development of an artificial heart valve, Ann. Thorac. Surg., 50:151, 1990.
82. Bjork VO, Sternlieb JJ, Kaminsky DB: Optimal microporous surface for endothelialization of metal heart valves in the blood stream, Scand. J. Thor. Cardiovasc. Surg., 24:97, 1990.
83. Bjork VO, Sternlieb JJ, Kaminsky DB: Modified porous metal-surfaced Bjork-Shiley Monostrut heart valve, Scand. J. Thor. Cardiovasc. Surg., 24:101, 1990.
84. Imachi K, Mabuchi K, Chinzei T: Blood compatibility of the jellyfish valve without anticoagulant, ASAIO Transactions, 37:M220, 1991.
85. van den Brink RBA, Visser CA, Basart DCG, et al.: Comparison of transthoracic and transesophageal color Doppler flow imaging in patients with mechanical prostheses in the mitral valve position, Am. J. Cardiol., 63:1471, 1989.
86. Jones EJ: St. Jude Medical prosthesis, J. Thorac. Cardiovasc. Surg., 81:642, 1981.
87. von der Emde J, Kockerling F, Rein J, et al.: Measures of prevention and technical problems during reoperations in cardiac surgery, Thorac. Cardiovasc. Surgeon, 34:5, 1986.
88. Taggart DP, Spyt TJ, Wheatley DJ, et al.: Severe haemolysis with the St. Jude Medical prosthesis, Eur. J. Cardio-thorac. Surg., 2:137, 1988.
89. Kinsley RH, Colsen PR, Antunes MJ: Medtronic-Hall valve replacement in a third world population group, Thorac. Cardiovasc. Surgeon, 31/II:69, 1983.

ABBREVIATIONS USED IN THIS CHAPTER

ACH:	Anticoagulation-related hemorrhage	OC:	Omnicarbon valve
AVR:	Aortic valve replacement	MH:	Medtronic Hall valve
BSCC:	Bjork-Shiley, Convex/Concave disc valve	OS:	Omniscience valve
BSDD:	Bjork-Shiley, Delrin disc valve	ΔP:	Pressure gradient
BSMS:	Bjork-Shiley, Monostrut valve	PC:	Pyrolytic carbon
BSSD:	Bjork-Shiley, Spherical disc valve	RBC:	Red blood cells
DRM:	Duromedics valve	SJM:	St. Jude Medical valve
FDA:	Food and Drug Administration, USA	TAD:	Tissue annulus diameter
LDA:	Laser Doppler Anemometry	TE:	Thromboembolism, transient and serious
LDH:	Lactate dehydrogenase enzyme		

ΣTE:	Overall thromboembolism including valve thrombosis (TE+VT)	LK:	Lillehei-Kaster valve
		VT:	Valve thrombosis
LVEF:	Left ventricular ejection fraction	ΣTH:	Overall thromboembolism + hemorrhagic complications (ΣTE+ACH)
MDR:	Medical Device Reporting system, FDA		
MVR:	Mitral valve replacement		

FLOW THROUGH MECHANICAL HEART VALVES AND THROMBOSIS: VISUALIZATION BY WASHING TEST

Bernard Brami

Centre d'Etudes Coeur et Fluides
22, rue Boulainvilliers
75016 Paris

FOREWORD

Since the emergence of the first Hufnagel valve about thirty years ago, valve prostheses have continuously, but slowly, been improved; however, such devices still are associated with a number of problems. The choice between mechanical and tissue valves is still controversial and it is clear that the ideal prosthetic valve does not as yet exist[1-3]. With respect to mechanical valves there have been a number of stages in their development. The initial stage was characterized by the use of many improper valve designs, which led to disastrous consequences until the development in 1960 of the Starr-Edwards valve. This device proved safer than many earlier ones and consequently it became adopted by the surgical community. During this period, long term usage of these devices revealed several defects inherent to this type of prosthesis, specifically related to mechanical and thromboembolic failures. In 1973 pyrolitic carbon was developed for valve components. This material limited excessive valve wear and also reduced the level of thromboembolic events[4].

In 1978 the St. Jude valve appeared, which incorporated several new concepts, one of which being the bileaflet system which rotated around a fixed axis. The remarkable mechanical behaviour of this valve led to its widespread usage in the medical field, and it became the standard for valve performance in the 1980's. It also probably led psychologically to a bad tolerance of any case of fracture or dysfunction of other designs. This aspect applies even to tissue valves which are generally superior with respect to thromboembolic complications, but where acceptance is limited by their structural failure characteristics, even though such failures are slowly progressive and less quickly fatal than mechanical failures.

The market of mechanical valves is now undergoing changes, but thromboembolic complications are still the major difficulty[5,6]. Such accidents are often difficult to quantify, but they seem more or less equivalent regardless of the type of mechanical valve used. In general a decreasing rate of thromboembolism is observed as one proceeds from ball-caged valves, monodisc, and then bileaflets, the latter being the less thrombogenic with an 0.8% to 5% rate of events per year/patient, depending on a variety of factors such as surgical site, patient conditions etc. Thromboembolic accidents are presently accepted by the medical community as inevitable. They remain the major problem linked to mechanical valves and clearly represent a challenge for the future.

INTRODUCTION

The understanding of thrombosis associated with mechanical valves prostheses thromboses advanced significantly around 1970, when it was realized that this pathology was related to abnormal flow disturbances. Until then, analysis of flow was primarily done in vitro, by means of two techniques: (1) a general visualization of the outflow downstream to the prostheses, which showed differences between the types of outflow (uniform or turbulent) and (2) measurements of velocity profiles by mean of Laser Doppler Anemometry (LDA), at various distances downstream of the prostheses[7-10]. The latter method permitted a precise determination of areas of stagnation near the prosthesis, reduced velocity upstream and a calculation of shear stress near the valve components which could have been damaging to blood cells. But neither of these two methods, considered "classical," allows an evaluation of the direct contact flow-prosthesis nor an assessment of the importance of friction forces on the total surface of the valve, both of which are dependent on flow speed coming into contact with surfaces of the prostheses.

Such factors are important because we know that downstream to turbulence a stabilization of the flow by means of reattachment may occur at a distance from the obstacle[11]. Thus, a "classical" observation of the flow may fail to appreciate flow separation zones on the device. It is why we have developed the concept of exploring the flow coming into contact with all the elements of a prosthesis whatever may be the complexity of its architecture.

THE WASHING TEST

The goal of this test is to directly visualize on the mechanical valves surfaces the effect of friction of the flow which crosses them during their opening time. For this we use a well known technique in aeronautic research which consists in coating a selected profile with a coloring agent, preferably white. After careful cleaning, valves are completely covered by a spray or by a silky thin brush. Each valve thus prepared is introduced in a left artificial ventricle and therefore will function during a certain time in a pulsatile pattern (70 bts/mn), with standard hydrodynamic conditions, in aortic and mitral position. One must mention that the coloring agent does not restrict the mobility of the valves, and it does not suffer from any direct effect of the blood analog fluid (aqueous glycerin solution).

During the functional cycle, one observes a progressive washing of some parts of the prostheses, and after some time a stabilization of the washing effect. Then, the prostheses are pulled out from the simulator and we can observe that on each valve some zones are perfectly washed, and some others are poorly washed.

This technique is developed taking in account the following observations:

- The nature of the coloring governs the washing effect of the fluid and the time necessary to obtain a significant result. The goal is to obtain a good compromise between a total and rapid dissolving of the coloring or an incomplete washing.

- The objective is also to obtain a significant result in a relatively short time in order to facilitate repetition of manipulations. The coloring agent has been previously developed in order to give a significant result within three minutes, after which the washing effect is stabilized and doesn't significantly progress.

- At present this technique doesn't allow verification of the depth of coating. But after several manipulations and despite the imperfect circumstances of this technique, the washing effect is finally the same for every valve, with only small changes in the time needed for the result, but always within the time limit test.

- This method explores only the mechanical components of the valve and is not applied to the sewing cuff, for which the responsibility in some thromboembolic accidents is known.

RESULTS

Most of current mechanical valves are thus tested, but we shall stress our study on two clinically well known, the Starr-Edwards (1620 aortic and 6120 mitral) and the Björk-Shiley convexo-concave 60° valve[12,13].

Starr-Edwards valve (aortic position): a perfect washing of the entry orifice and of all the surface of the sphere is noted. On the other hand at the level of the cage struts, one notes a perfectly washed intermediate zone and traces of coloring immediately under the sewing cuff and at the level of the apex (Fig. 1).

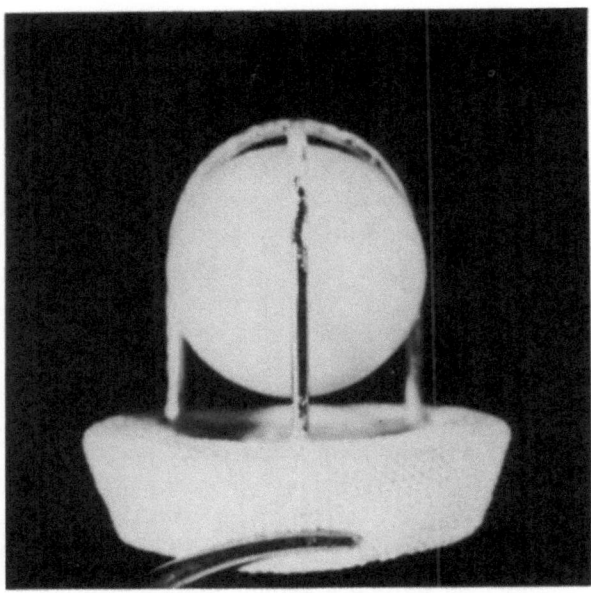

Figure 1. Starr-Edwards valve after washing test (aortic position).

Starr-Edwards valve (mitral position): results are approximately the same as in aortic position, nevertheless with badly washed zones slightly more spread and less distinct transitions with the parts of the cage which are well washed.

The Björk-Shiley CC 60°: in aortic position a perfect washing is noted on the upstream side of the valve. By the same token, on the downstream side there exists a badly washing effect in the central hollow and its periphery, and at the level of the struts (Fig. 2). In mitral position, the upstream side is well washed, but the whole downstream side exhibits poor washing.

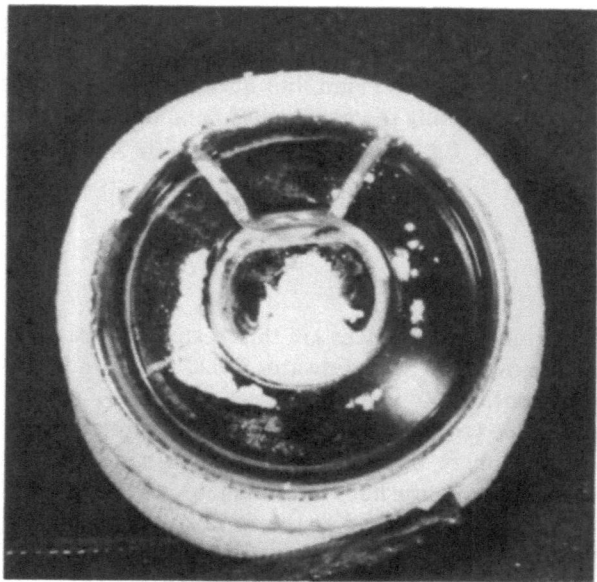

Figure 2. Björk-Shiley valve after washing test (aortic position).

ANALYSIS OF THE RESULTS

If the difference in washing between the upstream and downstream sides of each type of valve may be noteworthy, we note also some differences in washing within the downstream side itself. Regarding the mechanics of fluids, well washed zones correspond to the maintenance of enough speed of the flow near the surface (persistence of the boundary layer) to bring about a washing effect. Badly washed zones are due to a stagnation zone corresponding to a severe fall of the velocity (flow separation effect).

But the main interest of this technique is to show on these two well known valves the correlation which may exist between the badly washed zones shown by the test and the preferred zones of initial thrombus formation.

If thrombus localizations on the sewing cuff are excluded, Starr-Edwards valves thromboses are localized primarily at the origin of the cage struts and the apex (Fig.3)[14]. The hydrodynamics of the flow explains the poor washing effect on the apex, as it is well known that beyond a certain value of Reynolds number, on the downstream of any sphere introduced in a flow, a swirling zone is formed with return flows and a tendency to stasis[15]. Velocity curves by LDA which study the flow through Starr-Edwards valve have confirmed this phenomenon[16]. The ball of the valve, which is constantly in rotation on itself is submitted to a permanent washing. On the other hand, the apex of the cage is more or less blinded by the sphere when its extreme position is achieved, and so it is in a stasis zone where there is a tendency to thrombose. As for the poor washing of the initial struts of the cage, it seems to be due to their slightly off-center position with respect to the direction of the jet canalized by the inlet ring.

Figure 3. Clinical example of Starr-Edwards valve clotted.

The Björk-Shiley valve downstream face is well known to be the preferred site of thrombus formation, with occasional localizations in the central hollow or on the struts (Fig.4)[17,18]. All of these localizations are observed in the washing test. The velocity curves studying this valve have confirmed already that the area close to the downstream side was stagnant with its classical aspect of "camel back" of the curve[8,19]. But the washing test gives a direct view of the stasis zone obviously caused by the occluder, even when it is maximally opened.

DISCUSSION AND COMMENTS

Since there is a good correlation between the stasis zones identified by the washing test and the clinical sites where thromboses can be initiated, some remarks can therefore be expressed:
- The results obtained are purely visual and not quantified as yet. We intend to use this method, after development and quantification to compare with clinical aspects, and in vitro studies.
- All the mechanical valves, included those not mentioned in this work but studied according this protocol, exhibit poorly washed sections. This may be related

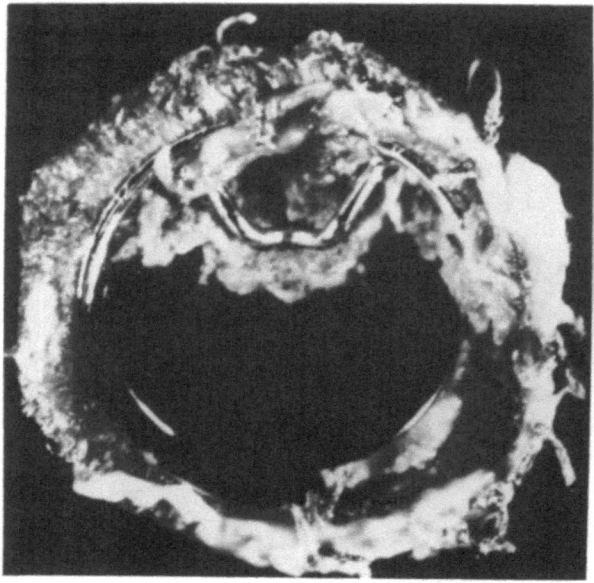

Figure 4. Clinical example of Björk-Shiley valve clotted.

to the fact that all mechanical valves are at risk for thrombosis. On the other hand, it doesn't seem possible, taking in account the badly washed surface or its visual aspect, and the clinical rate of thrombosis accidents. On the contrary, one can simply note that the significant overevaluation of the poor washing effect which is observed in mitral position compared to aortic position, corresponds to a well known clinical reality.

- Finally, we should indicate that only one element is accounted for in thrombus formation with this test; however we know that the process is multifaceted and related to a number of clinical considerations (left ventricular function, rhythm disorders, coagulation disorders, etc.). Such factors need to be considered but it is clear local transport is one important mechanism which is linked to the design of mechanical prostheses.

CONCLUSIONS

In so far as this test is referred to as a novel technique, it may be destined to improvements and to more advanced correlations with transvalvular flow studies and clinical funds. Therefore, in the light of the work, we can draw some basic conclusions:

1) All the mechanical prostheses present poorly washed zones on their downstream side, localized on similar sites, both in mitral or aortic position, but more extended in mitral position. These zones are due to a flow separation effect.

2) Correlation between this test and downstream flow analysis is good in most valves, the benefit of this test being to show precise zones where friction is insufficient, and which are not always shown by the so called "classical" methods, particularly at the orifices of valves.

3) The poorly washed zones identified by the test seem to correspond to thrombogenic sites clinically observed, when thromboses are at an early stage. This

may give this test a predictive analytic value to a thrombogenic risk, but without any statistically apparent value. No deduction can be made from the importance of the poorly washing surface of one valve in comparison with another one. On the contrary, the difference between aortic and mitral position on the same valve corresponds to a realistic clinical view.

4) As with all other methods exploring flow dynamics through prostheses, this test confirms the influence of the prostheses design in the formation of thromboses.

REFERENCES

1. Acar J: Le choix entre les différents types de valves, L'information Cardiologique, 9:307-308, 1985.
2. Grunkemeier GL, Rahimtoola SH: Artificial heart valves, Ann. Rev. Med., 41:251-263, 1990.
3. Symposium: 25 ans de protheses valvulaires cardiaques, L'information Cardiologique, Vol. 12, 1988.
4. Bokros J, Haubold A, Shim H: Carbon in medical devices, In: Biocompatibility of Clinical Implant Materials (Ed., Williams DR), CRC Press, Boca Raton, 1981.
5. Edmunds LH, Jr: Thrombotic and bleeding complications of prosthetic heart valves, Ann. Thorac. Surg., 44:430-445, 1987.
6. Farah E, Enriquez-Sarano M, Vahanian A, Houlegatte JP, Boubaker A, Roger V, Acar J: Thromboembolic and haemorrhagic risk in mechanical and biological aortic prostheses, Eur. Heart J. Suppl. D, 43-47, 1984.
7. Wright JTM: A flow visualization study of prosthetic and mitral heart valves in a model of the aorta and the left heart, Engeneering in Medicine, 6(2):31, 1977.
8. Bruss KH, Reul H, Van Gilse J, Knott E: Pressure and velocity fields at four mechanical heart valve prostheses. Björk-Shiley Standard Björk-Shiley Concave-Convex, Hall-Kaster and St. Jude Medical, Life Support Syst., 1:3-15, 1983.
9. Farahifar D, Cassot F, Bodard H, Pelissier R: Velocity profiles in the wake of two prosthetic heart valves using a new cardiovascular simulator, J. Biomech., 18:789-802, 1985.
10. Yoganathan AP, Corcoran WH, Harrison EC: In vitro velocity measurements in the near vicinity of aortic valve prostheses, J. Biomech., 12:135-152, 1979.
11. Rebuffet P: Aerodynamique Experimentale, Tome 2, Dunod, Paris, 1969.
12. Grunkemeier GL, Starr A: Twenty five year experience with Starr-Edwards heart valves: follow-up methods and results, Can. J. Cardiol., 4:381-385, 1988.
13. Lindblom D: Long-term clinical results after aortic valve replacement with the Björk-Shiley prosthesis, J. Thorac. Cardiovasc. Surg., 95:568-577, 1988.
14. Lelguen C, Fernandez F, Gerbaux A, Neveux E, Bickert P, Maurice P, Louvet J, Farah E, Acar J: Thromboses tardives sur protheses mitrales, Arch. Mal. Coeur., 72(7):730-738, 1979.
15. Modi VK, Akutsu T: Flow visualization studies with Starr-Edwards heart valve prosthesis. Flow visualization II, Hemisphere Publishing Corporation, Washington, 593-597, 1980.
16. Yoganathan AP, Reamer HH, Corcoran WH, Harrison EC, Schulman TA, Parnassus W: The Starr-Edwards aortic ball-valve: flow characteristics, thrombus formation and tissue overgrowth, Artificials Organs, 5(1):6-17, 1981.
17. Sabbagh AH: Clotted Björk-Shiley mitral valve prostheses: early detection and surgical management, Cardiovasc. Cent. Bull., Apr-Jun, 21(4):101-107, 1983.

18. Wright JO, Hiratzka LF, Brandt B, Doty DB: Thrombosis of the Björk-Shiley prosthesis: illustrative cases and review of the litterature, J. Thorac. Cardiovasc. Surg., Jul, 84(1):139-144, 1982.
19. Yoganathan AP, Corcoran WH, Harrison EC, Carl JR: The Björk-Shiley aortic prosthesis flow chracteristics, thrombus formation and tissue overgrowth, Circulation, 58(1):70, 1978.

DESIGN, DEVELOPMENT AND TESTING OF BLOOD PUMPS

H. Reul

Helmholtz-Institute for Biomedical Engineering at the RWTH Aachen
Pauwelsstr. 30, D-5100 Aachen, Germany

INTRODUCTION

The term "blood pump" is very comprehensive. Practically, it describes all blood pumping devices from the roller pump to the artificial heart.
On one hand there are a number of physiological requirements, concerning pump-characteristics and the limits of blood damage, which are valid for all kinds of blood pumps. On the other hand, deviations from the optimal conditions are sometimes unavoidable for different reasons, or may even be preferred when other possible complications, which may be independent of the pump itself, are taken into account. In those applications where alternatives are possible, it is to be carefully evaluated whether the expenditure, either technical or surgical, is to be of reasonable benefit to the patient. Of course, this decision is only left to the physician and cannot be discussed here.
It is, however, the responsibility of the blood pump designer, to take into account all known rheological, material and physiological data and to integrate them into his construction. Before these requirements are discussed in detail, the range of applications of blood pumps is first presented.

BLOOD PUMPS - OVERVIEW

The following four tables present a systematic overview on short term and intermediate term applications of blood pumps, including physiological demands, advantages and disadvantages as well as future perspectives. (Tables 1 to 4)

PULSATILE OR NON-PULSATILE FLOW

The discussion on the necessity of pulsatile flow for circulatory support is very controversial in literature. On one hand there are a number of results indicating that long-term non-pulsatile pumping has several detrimental physiological effects. On the other hand, there are strong arguments that these effects are only transient for a period of about three weeks[1].

Table 1. Classification of blood pumps

Characteristics	BLOODPUMPS I				
Appl. time	Short-term < 12 h		Intermediate term 12 h - 4 weeks		
Type	Roller pump	Centrifugal pump	Pneumatic displacement pump	El. mechanical displacement pump	Centrifugal pump
Drive-system	Electromechanical	Electromechanical	Pneumatic	Electromechanical	Electromechanical
Pump-body	Displacement type	Centrifugal-Type	Membrane type	Piston type	Centrifugal-type
Energy-transmission	Roller, Tube	Rotor, Spiral-Housing	Membrane, Compressed Air	Membrane, Pusher plate	Rotor, Spiral-Housing
Materials	PVC, Tygon, PUR	Steel, Titanium, Carbon	Housing: PUR, Silicone, AL; Membrane: PUR		Steel, Carbon
Applications	HLM, Dialysis, ECMO		LVAD, RVAD, BVAD, ECMO, after HLM, after Infarction		

Table 2. Physiological demands

Characteristics	BLOODPUMPS II				
Appl. time	Short-term < 12 h		Intermediate term 12 h - 4 weeks		
Type	Roller-pump	Centrifugal-pump	Pneumatic displacement-pump	El. mech. displacement-pump	Centrifugal-pump
Demands Physiological	HLM: Q ≤ 8 l/min p ≤ 400 mmHg; Dialysis: 0.15 l/min ≤ Q ≤ 0.25 l/min p ≤ 200 mmHg; ECMO: 2 l/min ≤ Q ≤ 4 l/min p ≤ 200 mmHg		LVAD: 2 l/min ≤ Q ≤ 6 l/min p ≤ 150 mmHg, f ≤ 150/min Filling pressure ≤ 15 mmHg; RVAD: 2 l/min ≤ Q ≤ 6 l/min p ≤ 40 mmHg, f ≤ 150/min Filling pressure ≤ 15 mmHg		
	Compatibility with blood and tissue (non-toxic, non-corrosive, no thromboembolic complications, no pannus-formation), sufficient pressure and flow				
Anatomical	None, extracorporeal		Extracorporeal, paracorporeal, intrathoracic, intraabdominal, no compression of tissue and large vessels		
Pumping-mode	Non-pulsatile, pulsatile		Pulsatile	Pulsatile	Non-pulsatile, pulsatile
Flow-regulation	Rotational speed occlusion	Rotational speed	Stroke-volume, pump-rate		Rotational speed
Control	None, ECG-triggering	None, ECG-triggering	Stroke-volume, pump-rate or both		Rotational speed
Technical	Efficiency, power, fluid-mechanics of blood compartments (stagnation, turbulence, high velocity gradients), heat dissipation, fatigue-life, reliability, vibration, noise, size, weight, handling, serviceability				

Table 3. Advantages of blood pump types

Characteristics	BLOODPUMPS III				
Appl. time	Short-term < 12 h		Intermediate term 12 h < 4 weeks		
Type	Roller-pump	Centrifugal-pump	Pneumatic displacement-pump	El. mech. displacement-pump	Centrifugal-pump
Valves	None	None	Check-valves, mechanical, biological		None
Advantages	Classical, well-known, easy to operate and to sterilize; Disposable tubing, inexpensive and reliable, no valves	Small size, small filling volume; Allows pumping time longer than 12 h, no flexible parts, no valves	Best approximation of physiological pressure and flow waveforms; Good control by signals from: Drive-line	Position-sensor implantability; transcut. energy transmission	Small size, small filling volume, no valves, no flexible parts, implantability

Table 4. Problems and perspectives of blood pump types

Characteristics	BLOODPUMPS IV				
Application-time	Short-term < 12 h		Intermediate term 12 h - 4 weeks		
Type	Roller-pump	Centrifugal-pump	Pneumatic displacement-pump	El. mech. displacement-pump	Centrifugal-pump
Problems and Disadvantages	High hemolysis, max. application time about 6 h, high dp/dt necessary for pulsatility	Thrombi at shaft-seal caused by high velocity gradients and local heat, less hemolysis than roller-pump, low efficiency, heat dissipation	Extracorporeal, bulky drive-system, big percutaneous drivelines, anatomical fit for small patients problematic. Thrombus formation at valves may cause brain infarction and strokes	Extracorp. drive-system, percut. leads, battery-op. only short-term, anatomical fit problematic because of implantable energy-converter and compliance chamber	Thrombi at shaft-seal by local heat generation, application only as RVAD because of non-pulsatile characteristics
Perspectives	Reduce hemolysis, provide pulsatile pumping mode	Drive without shaft-seal or by incapsulation of seal, increase pump efficiency, reduce stases, provide pulsatile mode	Smaller drive-systems use pistons instead of magnetic valves for better control of dp/dt. Development of improved valves in terms of hemodynamics and durability	Development of improved batteries (zinc-air?), development of fully implantable systems	Battery development, no shaft-seal, improvement of efficiency, provide pulsatile mode

Table 5 indicates data collected from literature, which describe the detrimental aspects of non-pulsatile flow.

Kidney function

Senning et al.[2] reported that renal function is significantly depressed during non-pulsatile perfusion despite a high flow rate. Finsterbusch[3] noted changes in renal arteriograms, such as narrowing, straightening and loss of the normal configuration during steady perfusion. Kohlstaedt et al.[4] showed that the release of renin is triggered by the decrease of pulse pressure amplitude rather than by the reduction of mean arterial pressure. Agishi et al.[5] reported similar results and related the reduced perfusion rate, during non-pulsatile pumping, to the viscoelastic wall properties. The data of Many et al.[6] indicate a significant decrease in urine volume, in sodium excretion, in osmolar clearance and an increase in urine osmolality during bilateral renal depulsation.

Table 5. Physiological effects on non-pulsatile flow

PHYSIOLOGICAL EFFECTS OF NON-PULSATILE FLOW	
Kidney	Significantly decreased function, deformation of renal arteries, reduced renin-secretion, urine-volume and sodium-secretion
Capillary circulation	Slowing down of erythrocyte-motion, increased aggregation tendency of erythrocytes
Cell-metabolism	Strong decrease of lymphatic flow by elimination of pulsatile component
O_2-consumption	Strong decrease in O_2-consumption caused by elimination of convective diffusion between cells and surrounding fluid
Brain-function	Ischemic damage at nerve-cells and cell-swelling already after 2 hours
Peripheral resistance	Increase of mean pressure in systemic and pulmonary circulation, increase of peripheral resistance, decrease of pH, increase of lactate-level
Baro-receptors	p and dp/dt are important for physiological function, missing pressure pulse causes increase of peripheral resistance

Capillary circulation

Ogata et al.[7] microscopically observed a marked slowing of red cell movement in the capillaries after twenty minutes of steady perfusion. An intravascular aggregation of erythrocytes was also noted by Long et al.[8]. Petrow[9] developed a theoretical model of the microcirculation and found that the blood pressure amplitude of the pulsatile blood supply evokes elastic interactions in the tissue, in which the noncompressible tissue fluid functions as a mediator of strength. The particular morphological structure around the tissue capillary gives rise to the suggestion that a tissue pump is working which supports the blood flow, as well as the exchange of substances at the capillary wall. The effectiveness of this pump depends upon the elastic tissue properties.

Cellular metabolism

Lymph flow has the specific task of removing such substances which cannot enter the capillaries from the interstitial fluid. Parsons and MacMaster[10] showed in animal experiments that there was almost no lymph flow during non-pulsatile pumping. Wilkens et al.[11] concluded that changes of interstitial pressure caused by the surrounding pulsatile fluctuations of arteries would be of sufficient magnitude to support the flow of the interstitial fluid and lymph.

Oxygen consumption

Shepard and Kirklin[12] conducted comparative animal studies on oxygen consumption during pulsatile and non-pulsatile cardiopulmonary bypass. They found an increase, during pulsatile flow, of about 25% and related it to the "jiggling" of the tissues, which may be causing changes in the boundary layer of interstitial fluid around cell membranes and thus enhancing diffusion. Simultaneously the peripheral resistance decreased about 20%.

Brain damage

Sanderson et al.[13] and Wright et al.[14] examined brains of dogs subjected to total cardiac pulsatile and non-pulsatile bypass for signs of ischaemic nerve-cell damage. They found that, after two and three hours non-pulsatile perfusion, cell swelling occurred in the Purkinje cells; further ischaemic cell changes were found mainly in the cerebral cortex and again in the Purkinje cells.

Peripheral resistance

Mandelbaum and Burns[15] found a signifcant increase in mean pressure in both the systemic and pulmonary circulation during non-pulsatile pumping. Systemic pressure increased 25% on an average and pulmonary pressure 27% on an average. Trinkle et al.[16] reported an increase in peripheral resistance as well and found some other statistically significant advantages of pulsatile flow: higher flow rate, higher arterial pH, less hemolysis, lower transfusion volume, lower arterial lactate. The results of Soroff[17] indicate that, when the aortic arch and the carotid arteries are depulsated, the aortic mean pressure increases by 11%.

Baroreceptor control

The baroreceptors are distributed along the walls of the large systemic arteries, the most important and sensitive receptors lie on both sides of the wall of the carotid arteries above the carotid bifurcation and in the aortic arch.

Guyton[18] explains the function of the baroreceptors as follows: "The baroreceptor impulses inhibit the vasomotor center of the medulla and excite the vagal center. The net effects are (1) vasodilation throughout the peripheral circulatory system and (2) decreased cardiac rate and decreased strength of contraction. Therefore, excitation of the baroreceptors by pressure in the arteries causes the arterial pressure to decrease. Conversely, low pressure has opposite effects, causing the pressure to rise back towards normal." The reaction of the baroreceptors, i.e. the number of delivered impulses, is influenced by two different factors, first by the absolute magnitude of arterial pressure and second by the rate of pressure change. When the baroreceptors are excited by pulsatile pressure changes, bursts of impulses occur during a rapid rise of pressure. Thus, the discharge rate of the receptors is increased not only by the mean arterial pressure but also by the rate of rise of pressure[18-20]. This effect informs the vasometer center not only of the actual mean arterial pressure but also of instantaneous pressure changes.

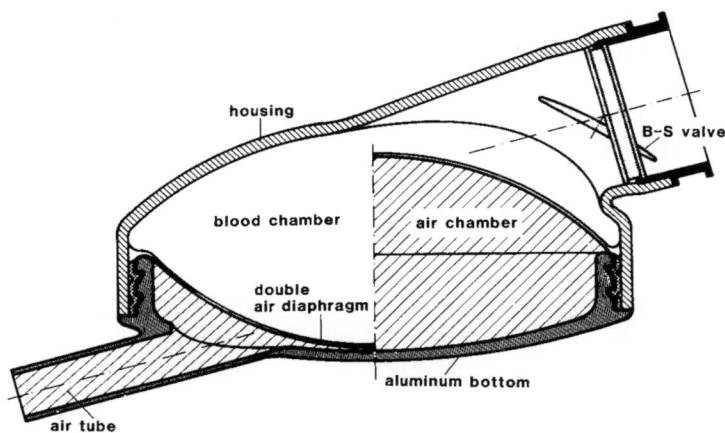

Figure 1. Berlin pneumatic VAD

BLOOD PUMP TYPES

Pneumatic displacement pumps

The importance of pneumatically driven diaphragm- or sac-type blood pumps is indicated by the increasing number of successful clinical applications. Over the years a large number of these pump types has been developed world-wide. The following Figures 1 to 6 show a selection of various configurations.

Until now the best results are obtained when these blood pumps are used as a bridge to transplant. The success rate for ventricular assist devices (VAD) and total heart replacement (TAH) are not significantly different[21].

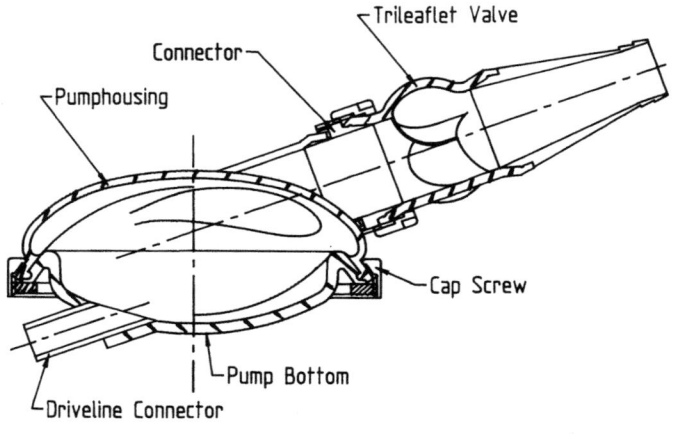

Figure 2. Helmholtz-Institute, Aachen, Pneumatic VAD

A further important application for VAD is the temporary use as assist-device during recovery of the reversibly damaged natural heart. In many cases both the left and the right ventricle need to be supported by separate pumps. Normally, the pumps are applied over a period of 12 h to two weeks. In special cases an extension up to several months may be necessary. For VADs, an average application time of four weeks can be assumed.

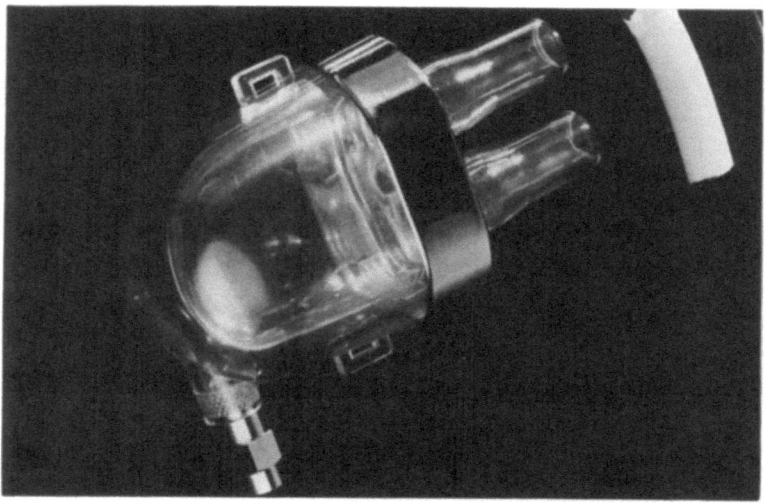

Figure 3. Nippon-Zeon pneumatic VAD

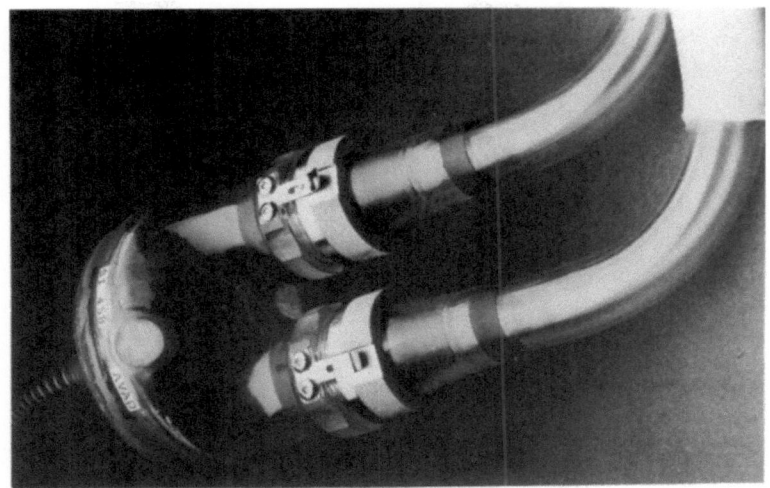

Figure 4. Symbion pneumatic VAD

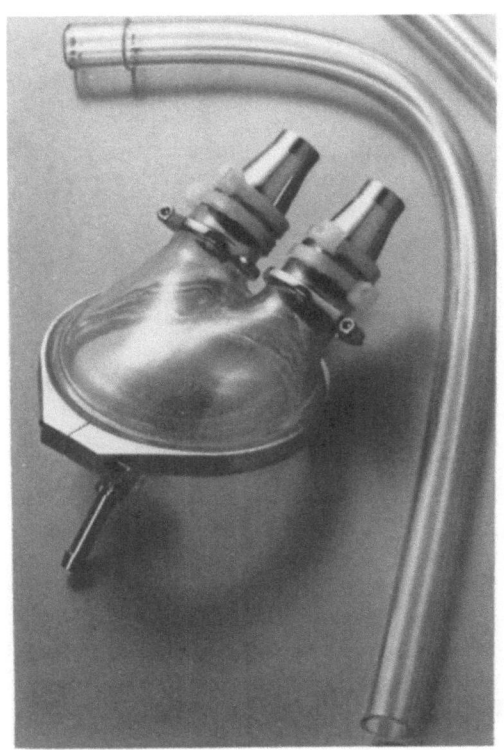

Figure 5. Toyobo pneumatic VAD

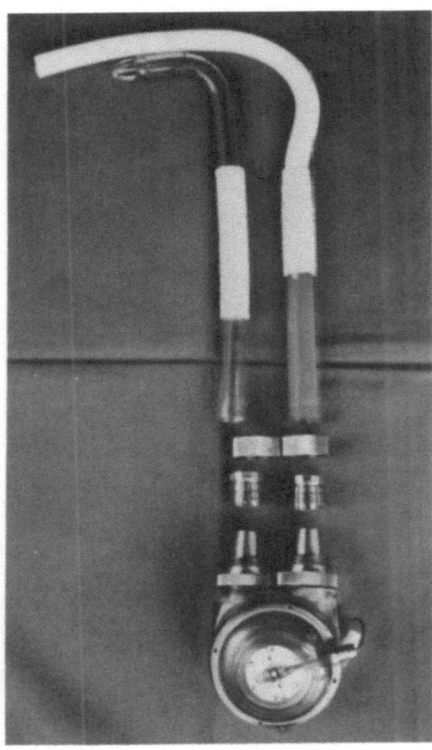

Figure 6. Thoratec pneumatic VAD

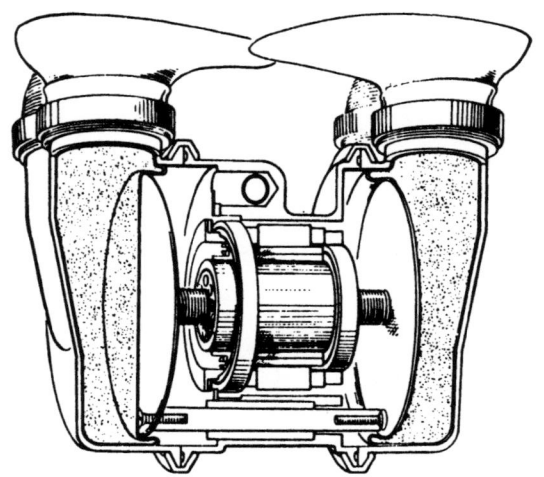

Figure 7. Penn-State roller-screw artificial heart

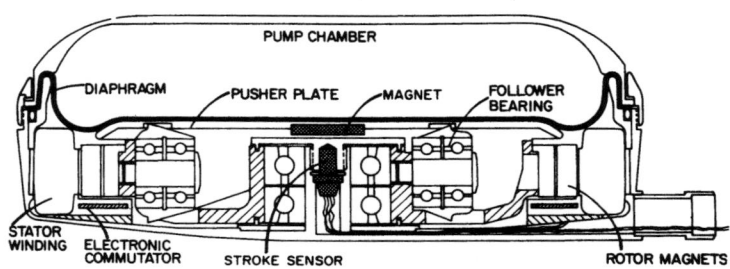

Figure 8. Thermo-Cardiosystems pusher plate pump

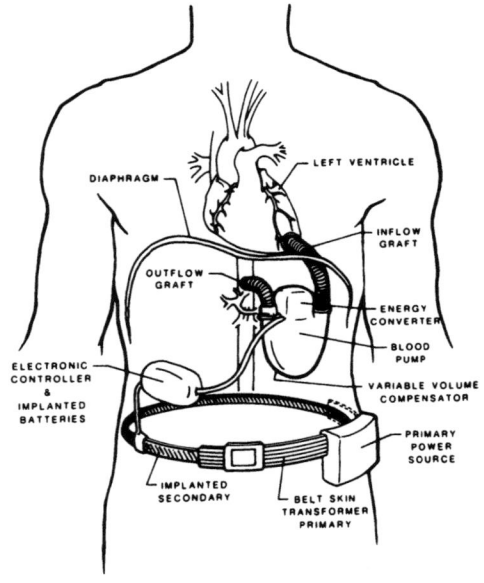

Figure 9. Totally implantable Novacor pusher plate LVAD

Figure 10. SARNS-3 M Centrifugal blood pump

Pusher plate pumps

From the large variety of pusher plate pumps (electromechanical, electrohydraulic, electromagnetic, thermoelectric) only a small selection can be presented here. They are represented by the three types shown in Figs. 7, 8, and 9. For further information, the interested reader may be referred to the excellent volume Assisted Circulation[22].

Figure 11. Biomedicus centrifugal blood pump

245

Rotational pumps

Currently two types of centrifugal pumps are in clinical use: The SARNS-3M (Fig. 10) and Biomedicus pump (Fig. 11). Their application ranges from HLM to ECMO and ventricular assist. Advantages of centrifugal pumps are: less blood damage than roller pumps, low filling volume, weight and size, no spallation products. Disadvantages are: potential clotting at the sealing area due to heat generation and poor washout.

A new rotational pump version is the axial Hemopump (Fig. 12), which is a miniaturized high speed pump for intra-aortic left ventricular assist applications.

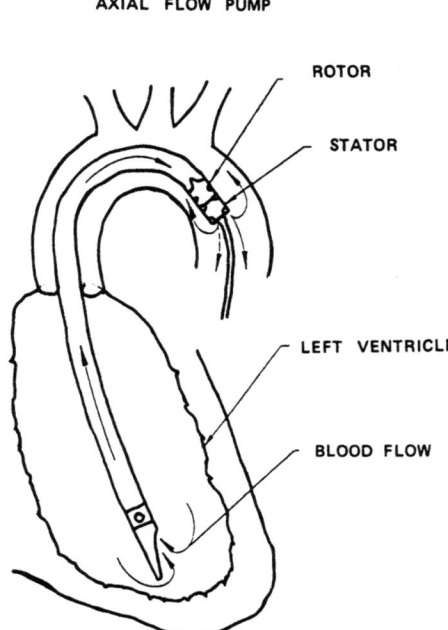

Figure 12. High speed axial Hemopump

NEW TECHNIQUES FOR BLOOD PUMP DEVELOPMENT

CAD-design

Over the last years, a number of engineering techniques have been developed, like computer system design and validation by finite element analysis.

Design and fabrication of technical parts by means of CAD (Computer Aided Design) is already an established engineering tool. Improved design results can be obtained by conveying parameter studies, optimizing algorithms and more transparent displays (for example 3-D views) before actual manufacturing of the respective parts. First, a 2-D geometry is generated. The parts with rotational-symmetry are generated by rotating the earlier generated 2-D images. The parts with plane surfaces are generated by translation, rotation and reflection. The generated 3-D geometry can directly be transferred to CNC-manufacturing (Computer Numeric Controlled). After definition of surface-tolerances and tool-parameters, the toolpathes of the milling-tool can be simulated and displayed on the screen.

Figure 13 shows the 3-D sectional drawing of a completed blood pump with all components. Figure 14 shows the CAD-design of the in- and outlet valves integrated into a sinus-shaped housing.

FEM-analysis

The thin movable membranes of the pump and the valves belong to the highest stressed parts during blood pump operation. For dimensioning the membranes, several counter-current effects have to be taken into account: thick membranes are stressed by high bending stresses (buckling folds) during motion. The same effect also applies to the valves.

In addition, the hemodynamic properties deteriorate with increasing membrane (leaflet) thickness in terms of increasing opening pressures and pressure losses. If the membranes are to be optimized with respect to geometry, thickness distribution and material properties, the stress strain distributions within the membranes during the whole pump cycle have to be calculated as accurately as possible. Since the calcification of PUR-surfaces is enhanced by mechanical stresses, the stress distribution of the membranes should be known. The particularly complicated geometries and the large deformations of the membranes suggest the application of Finite-Element-Methods.

Figure 15 shows, as an example, deformations together with the associated strains of a pump diaphragm (HIA pump) during operation.

The highest, and also critical, values occur directly in the constraint in the top position (end systolic) of the diaphragm under internal pressure. The modified Riks process permits bulging processes to be generated between intermediate layers corresponding to actual bulging observed by photographs during normal pump operation. The deformations of the diaphragm between the end positions was simulated taking into account the contact with the toroidal pump base rim. The material stresses in the bulges are twice as high as in a previously computed ideal roll folding movement.

Manufacturing

Figure 16 shows the existing pump family with stroke volumes of 10, 20, 50, 70 and 90 ml, developed at the HIA.

The pump bodies mainly comprise components thermoformed from commercially available, semi-finished materials. The rigid bottom of the pumps is made of polyvinylchloride (PVC), the flexible pump parts (like pump housing and valve housing and the diaphragm on the air side) are manufactured from thermoplastic polyurethanes (TPU).

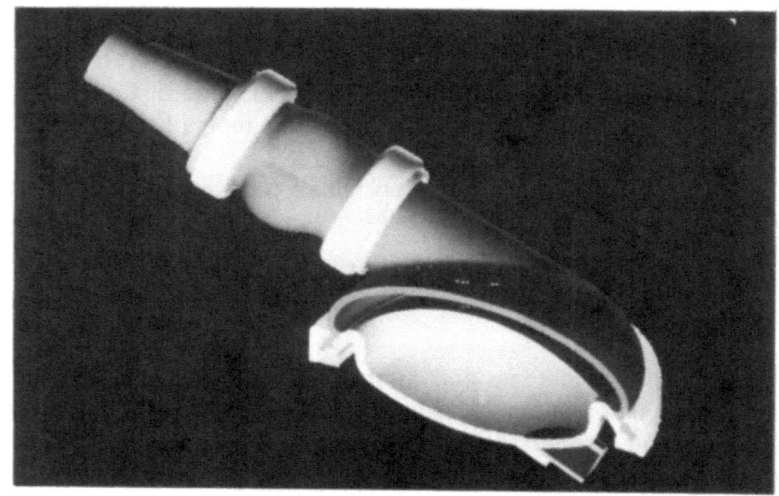

Figure 13. 3D-CAD study of HIA pneumatic displacement pump with trileaflet valves

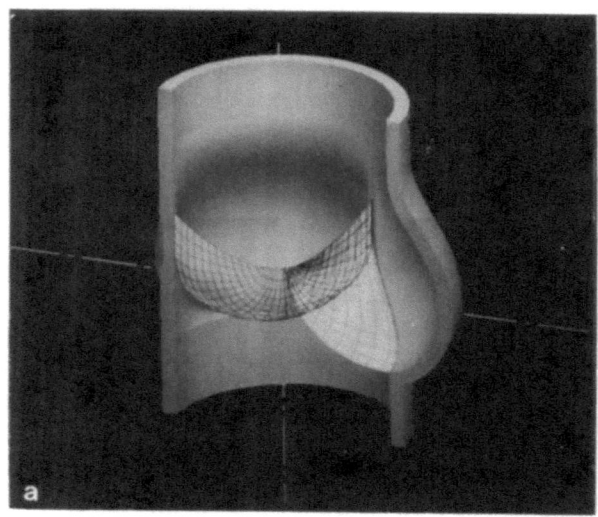 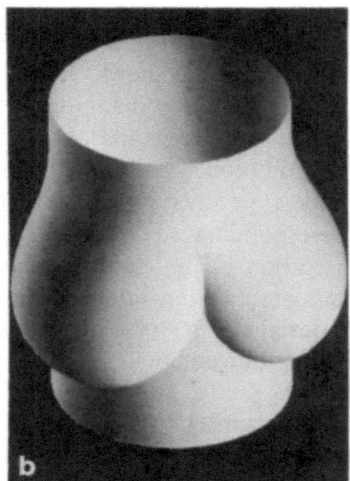

Figure 14. 3D-CAD study of the valve with sinus shaped housing and valve leaflets in a cut-away (a) and an exterior (b) view

FEM-Simulation of the Pump Diaphragm Loading

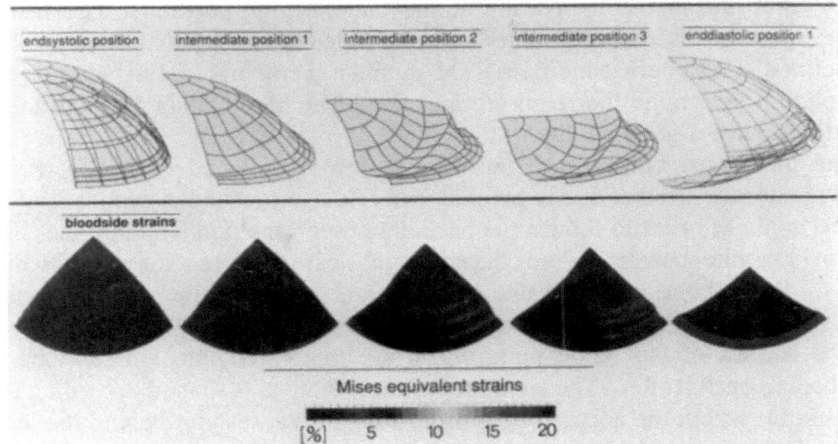

Figure 15. FEM loading simulation of the pump diaphragm in end and intermediate positions (Mises comparative strains)

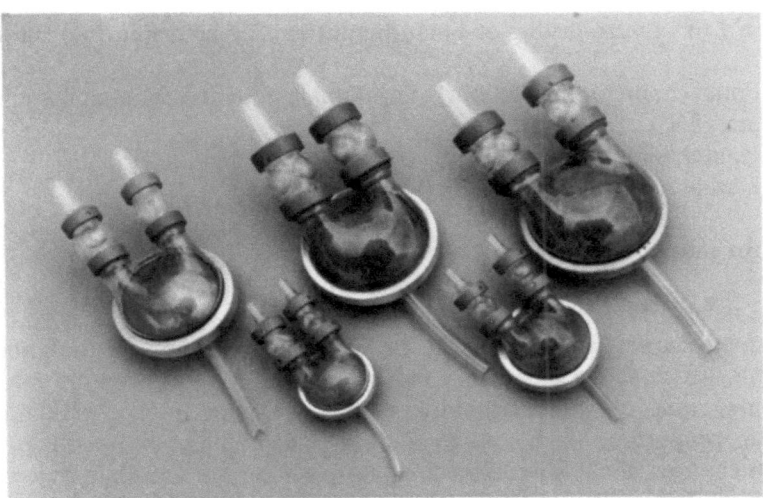

Figure 16. HIA pneumatic pump family with 10, 20, 50, 70 and 90 ml stroke volumes

All components on the blood side are coated with a biocompatible polyurethane (Akzo, Obernburg, Germany). The inside coating of the housing and the blood coating membranes are simultaneously manufactured by dipcoating, providing a seamless diaphragm-housing-junction.

For this purpose, the thermoformed housing is fixed onto a mold, the geometry of which conforms to the membrane at the end-diastolic position. The resulting hollow element is rinsed several times with the biocompatible polyurethane solution under defined atmospheric conditions. The number of coatings and the concentration of the solution determine the required thickness. The blood contacting surfaces are air-dried and have a good quality.

The membrane boundary at the d-h-junction is designed in such a way that extreme membrane stresses due to bending and folding are avoided by means of large curvature radii. In order to reduce the friction between the pump membranes during operation, graphite powder is introduced as lubricant between membranes during assembly. Pump housings and valve conduits are connected by specially designed screw-on connectors.

The leaflets of the blood pump valve are integrated into a PUR-tube with sinuses behind each leaflet. The geometry of these sinuses is shaped according to the natural sinuses within the aortic root in order to improve washout behind the leaflets by vortex formation and also to support the dynamics of valve closure. Currently, economic conditions significantly limit the widespread use of ventricular assist devices. The combination of refined engineering techniques such as CAD and FEM, low manufacturing costs, high quality and performance may eventually lead to extended clinical applications.

Further details are given in Knierbein et al.[23].

IN VITRO-TESTING OF BLOOD PUMPS

The field of in vitro-testing of blood pumps can be classified into three main categories:
1) Pump characterization - operating domain, function curves, pressures, flow etc.
2) Hemolysis testing
3) Durability testing

Mock loop for pulsatile flow testing

Before newly developed blood pumps can be tested in vivo extensive in vitro-tests for system characterization have to be performed. Important tools for these measurements are so-called circulatory mock loops or pulse duplicators, which are used for simulation of the circulatory system. During pump evaluation the hemodynamic properties of the pump and the adaptability of its drive unit to physiological changes of the simulated circulatory system are the main objectives. A circulatory mock loop with reproducibly adjusted test parameters also facilitates the comparison of different pump designs under reproducible conditions.

Simulating the varying properties of the natural circulation involves a number of problems. The physiological load against which the heart normally pumps is governed by the mass inertia of blood, by the elastic properties of the vascular system, and by friction forces which are all individually distributed throughout this system.

These correspond to masses, resistances and compliances which are not lumped but distributed parameters.

A mathematical model can be formulated by a continuous description of pressures and flows in different physiological conditions of the vascular system and in consideration of the pumping work of the heart. Many of the boundary conditions such as the viscoelastic vascular properties or the active contraction of muscle wall fibres cannot be described in a linear fashion. Therefore, a physical circulatory mock loop can only be a simplified approximation adapted to the problem investigated.

Over the last 30 years several physical models have been developed to simulate the complex dynamic behavior of the vascular system in a simplified way[24-31]. In the field of blood pump testing the so-called windkessel models have proved as useful. Their design comprises in-series elements which represent compliance and friction in lumped form and, if taken into consideration, also the mass-inertia of vessels and blood.

All previous concepts[24,27-31] used similar windkessel elements for simulation of arterial elastic properties. Resistance and inertia are simulated by means of flow resistors and tubes of the appropriate diameters and lengths. Some mock loops have a compact design[25]. In others, the various elements are coupled in series by tube connections[27-29].

For testing blood pumps the mock loop represents a hydraulic load which should be as close as possible to the dynamic properties of the physiological load. This is a prerequisite to achieve flows and pressures at the intersection between pump and mock loop. The pump dynamics as well as the dynamic behavior of the physical simulator have to be adapted to each other.

The requirements for such a pulsatile mock loop can basically be formulated as follows:
- At the intersection between pump and mock loop a physiological load has to be simulated, which is characterized by the input and output impedance of the system;
- Resistances and volumes must be variable in a range which for different pump sizes and different circulatory conditions approximates physiological conditions;
- Measurements, adjustment or control of the system parameters must be sufficiently accurate;
- Handling of the system must be simple and reliable in order to facilitate reproducibility of test parameters;
- The simulation of biventricular assist and/or TAH should be possible;
- It should be transportable and have a clearly arranged and compact design.

The complicated functions of the vascular system under various physiological influences need not necessarily be transferred to the model in all details. The same applies to the complex phenomena of pulsewave transmission and reflexion, which can be largely overlooked when the conditions in the coupling plane between pump and loop are matched to generate physiological pressure and flow waveforms. A lumped parameter approach can largely satisfy these requirements.

The system may be finally reduced to a three-element model (Figure 17):
- One windkessel with variable air volume for simulating arterial compliance,
- One adjustable resistor for simulating peripheral resistance,
- One open tube system for simulating the venous system and atrium.

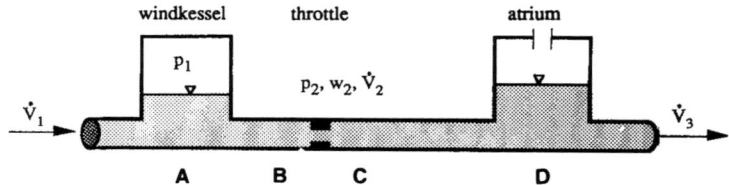

Figure 17. Schematic diagram of the mock loop

This model can be numerically treated and the according design parameters in terms of compliances and resistances can be obtained[32]. The input into the numerical model is the periodic volume flow, which is modelled according to known physiological flow as function of time. The output is a pressure-time function which represents aortic or pulmonary pressure.

Figure 18 shows the mock loop, based on the foregoing model analysis. This version was designed with a view to simple function and handling. It has the following advantages:
- Each loop consists basically of three plexiglas cylinders of identical diameter and length; one represents the compliance chamber, one the atrium and the third serves as a fluid reservoir.
- Each mock loop can be used for either LVAD or RVAD pumps, since they are identical.
- The glueing areas have been largely reduced, simplified or eliminated.
- Sealing has been simplified by introduction of defined seals and sealing areas.
- The compliance range has been increased, now providing air volumes from 0.5 to 5.5 liters, covering the whole range of potential values.
- Simplified manufacturing because of identical cylinders and cover plate sealing design.

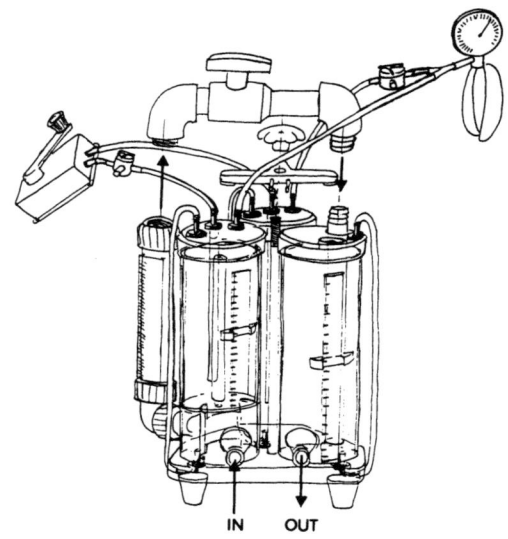

Figure 18. Principle sketch of the mock loop

During systole the fluid ejected by the pump enters the windkessel chamber. For adjusting the different compliances, the air volume within the windkessel can be adjusted without changing the fluid level. Fluid is simply pumped from the third reservoir into the compliance chamber by an auxiliary pump. This facilitates simple adjustment of different compliances. The fluid then passes from the compliance chamber through a rotameter and an adjustable throttle into the second chamber which represents the atrium. The fluid level here dictates the pump inlet pressure. It can also be adjusted directly over a wide range.

For biventricular applications two loops can be coupled and the atria can be interchanged by connecting the two pumps so that the left atrium now represents the right and vice versa. For simulation of the higher left ventricular flow a special bypass line can be connected with the opposite atrium and can be separately controlled by an additional throttle valve.

Practical experience has shown that using this mock loop the hydrodynamic properties of the attached pump and the adaptability of its drive unit to load changes can be reproducibly tested. The loop parameters like compliance, resistance, and atrial pressure can be varied to simulate pathological circulatory situations. Thus, the mock loops are very suitable tools for in vitro training of cardiac surgeons and technicians before in vivo work. They can learn which parameters of the drive unit must be changed if the natural circulation has to be stabilized in critical situations.

Mock loop for steady flow testing

The requirements for steady flow testing which applies to all kinds of rotational pumps are much simpler. A sample of a typical mock loop is given in Figure 19.

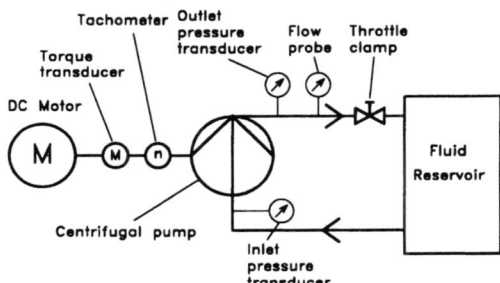

MOCK LOOP FOR CENTRIFUGAL PUMPS

Figure 19. Mock loop for steady flow testing (rotational pumps)

For rotational pumps the following hydrodynamic measurements have to be performed:
- Determination of the throttle curves (i.e. pump output as function of backpressure and rotational speed),
- Determination of efficiency.

Therefore, the mock loop itself can be reduced to a fluid reservoir, connecting in- and outlet tubing and a simple throttle clamp for adjusting the backpressure. Pressures are measured at pump inlet and outlet and flow at an appropriate location within the circuit. From these parameters the hydraulic pump work can be calculated. On the drive- unit side torque and rotational speed are measured for the determination of the efficiency of the pump head.

Hemolysis testing

The quality of a blood pump depends on the extent of damage to blood that it induces. This blood damage includes its affect on blood components and their general functions, for example coagulation or transport of oxygen. For the characterization of hemolysis recently a new Index of Hemolysis (IH") was suggested on the International Workshop on Rotary Blood Pumps at Obertauern (Austria), 1988 by H. Engelhardt (Helmholtz-Institute, Aachen) and R. M. Müller (2nd Dept. Surgery and Boltzmann Institute of Cardiac Surgery, Vienna, Austria). This new Index of Hemolysis includes the total hemoglobin in blood and represents a nondimensional index.

$$IH'' = \frac{V \cdot (1 - HK) \cdot PHb \cdot 10^6}{F \cdot t \cdot HB}$$

V: Blood volume in the mock loop [lit]
Hk: Hematocrit [-]
PHb: Increase of free plasma hemoglobin [mg/100 ml Serum]
F: Flow [lit/min]
t: Pumping time [min]
Hb: Total amount of hemoglobin [g/100 ml blood]

A potential set-up for hemolysis testing is shown in Figure 20.

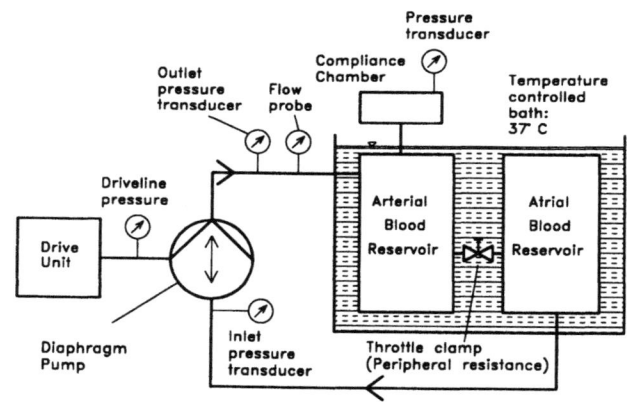

Figure 20. Mock loop for hemolysis testing of pulsatile pumps (schematically)

Blood flows from the atrial reservoir through a tube to the blood pump which is provided with a pressure transducer in the drive line. Another pressure transducer is coupled to the compliance chamber to measure the pressure in the arterial reservoir. Blood then is propelled through another tubing to the arterial reservoir. The outlet tubing is furnished with an inductive flow probe (Zepeda Instruments, Seattle, USA). The arterial reservoir is coupled to a compliance chamber of 1000 ml in order to simulate the aortic elasticity. Between arterial and atrial reservoir a clamped short piece of tube serves as a peripheral resistance. For the investigations, a flow of 5 l/min against a pressure of 120/80 mmHg is commonly used. This setting simulates the maximum requirement for a blood pump during left heart assistance. The pumps are always driven in "full-empty" mode.

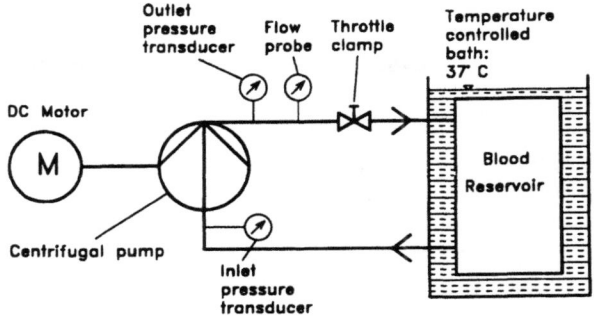

HEMOLYSIS MOCK LOOP FOR CENTRIFUGAL PUMPS

Figure 21. Mock loop for hemolysis testing of steady flow pumps (schematically)

A second mock loop for hemolysis testing of steady flow pumps is depicted in Figure 21.

The set-up consists of a blood bag with inlet and outlet connectors; it serves as a blood reservoir. A tube leads from the reservoir to the pump inlet. The pump is provided with two pressure transducers at the inlet and outlet. Another tube connects the pump outlet with the reservoir. This outlet tube is supplied with a tube clamp which throttles the pump to deliver a flow of 4 l/min against an afterload of 180 mmHg. This adjustment simulates the pressure-flow-relations during cardiopulmonary bypass. Further details on hemolysis testing are given in Westphal et al.[33,34].

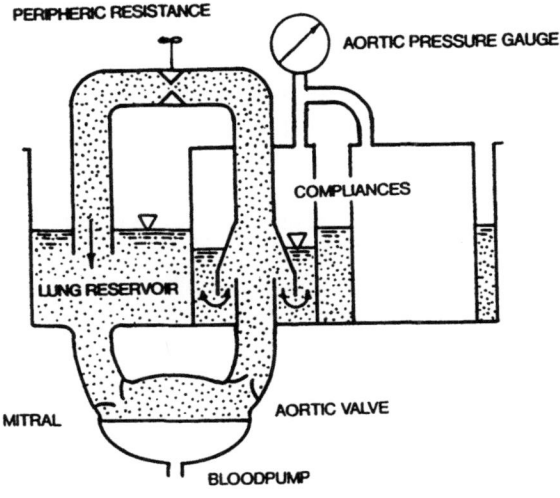

Figure 22. Schematic of fatigue test mock loop for pulsatile pumps

Durability testing

According to the FDA guidelines[35] for in vitro-testing of VAD and TAH, durability testing serves for the definition of failure criteria. These criteria should not be limited to component failure but should extend to circumstances when the system is unable to meet specific clinical requirements.

The devices must meet the following objectives:
- short term: at least four devices under specified test conditions with 0 failure of a period of at least twice the intended clinical application.

The guidelines also require the simulation of physiological pressures and flow and suggest that testing to be done in real time.

A suitable mock loop is shown in Figure 22.

The blood pump delivers fluid into an adjustable compliance chamber, which serves for the simulation of the physiological input impedance of the loop. Mean aortic backpressure is measured within the compliance chamber by means of a manometer. A throttle serves as peripheral resistance and an open fluid reservoir provides the appropriate pump inlet pressure.

SUMMARY

First, an overview on short term and intermediate term applications of blood pumps, including physiological demands, advantages and disadvantages as well as future perspectives is given.

The second part covers the controversial aspects of pulsatile versus non pulsatile pumping.

In the fourth part a selection of various blood pump designs is presented and part five covers new techniques for blood pump development including CAD-design, FEM-analysis and manufacturing aspects. Finally, part six deals with in vitro-testing, including hydrodynamic pump characterization, hemolysis- and fatigue-testing.

REFERENCES

1. Nosé Y: The need for a nonpulsatile pumping system, Int. J. of Artificial Organs, 12(2):113-115, 1988.
2. Senning A, Andres J, Bornstein P, Norberg B, Andersen NM: Renal function during extracorporeal circulation at high and low flow rates: experimental studies in dogs, Ann. Surg., 151:63, 1960.
3. Finsterbusch W: Renal arteriography during extracorporeal circulation in dogs, with preliminary report upon effects of low molecular weight Dextran, J. Thorac. Cardiovasc. Surg., 41:252, 1961.
4. Kohlstaedt KG, Page IM: Liberation of renin by perfusion of kidneys following reduction of pulse pressure, J. Exp. Med., 71:201, 1940.
5. Agishi T, Peirce EC, Kent DB: A Comparison of pulsatile and nonpulsatile pumping for ex vivo renal perfusion, J. Surg. Res. 9:623, 1969.
6. Many M, Soroff HS, Birtwell WC, Giron F, Wise H, Deterling RA: The physiologic role of pulsatile and nonpulsatile blood flow. II. Effects on renal function, Arch. Surg., 95:762, 1967.

7. Ogata T, Ida Y, Nonoyama A, Takeda J, Sasaki H: A comparative study on the effectiveness of pulsatile and nonpulsatile blood flow in extracorporeal circulation, Arch. Jpn. Chir., 29:59, 1969.
8. Long D (Discussion of Sanger PW et al.): Vasomotor regulation during extracorporeal circulation and open-heart surgery, J. Thorac. Cardiovasc. Surg., 1:355, 1960.
9. Petrow JM: Theorie der Mikrozirkulation. Teil 3: Die Rolle der Blutdruckamplitude bei der Mikrozirkulation im Gewebe, Zeitschrift für die gesamte Innere Medizin, 45(21):633-638, 1990.
10. Parsons RJ, MacMasters PD: Effects of pulse upon formation and flow of lymph, J. Exper. Med., 68:353, 1938.
11. Wilkens H, Regelson W, Hoffmeister FS: The physiologic importance of pulsatile blood flow, N. Engl. J. Med., 267:443, 1962.
12. Shepard RB, Kirklin JW: Relation of pulsatile flow to oxygen consumption and other variables during cardiopulmonary bypass, J. Thorac. Cardiovasc. Surg., 58:694, 1969.
13. Sanderson JM, Wright G, Sims FW: Brain damage in dogs immediately following pulsatile and nonpulsatile blood flow in extracorporeal circulation, Thorax., 27:275, 1972.
14. Wright G, Sanderson JM: Brain damage and mortality in dogs following pulsatile and nonpulsatile blood flow in extracorporeal circulation, Thorax., 27:738, 1972.
15. Mandelbaum J, Burns WH: Pulsatile and nonpulsatile blood flow, JAMA 1918, 121, 1965.
16. Trinkle JK, Helton NE, Bryant LR, Griffen WO: Pulsatile cardiopulmonary bypass: Clinical evaluation, Surgery, 68:1074, 1970.
17. Soroff HS, Many M, Birtwell WC, Giron F, Deterling RA: Hemodynamic effects of pulsatile and non pulsatile blood flow, Arch. Surg., 98:321, 1969.
18. Guyton AC: Textbook of Medical Physiology, W.B. Sanders, Co., Philadelphia, London, Toronto, 1971.
19. Rushmer RF: Cardiovascular dynamics, Saunder Philadelphia, 1970.
20. Myers GH, Parsonnet V: Blood pressure regulation (Chapt. 3), In: Engineering in the heart and blood vessels (Eds., Myers, Parsonnet), Wiley Interscience Publ., New York, 1969.
21. Rokitansky A, Wolner E: Total artificial heart and assist devices as a bridge to transplant, Intern. J. of Artificial Organs, 2:77-84, 1989.
22. Unger F: Assisted Circulation I, II, III, Springer Verlag Berlin, Heidelberg, New York 1979, 1984, 1989.
23. Knierbein B, Rosarius N, Reul H, Rau G: New methods for the development of pneumatic displacement pumps for cardiac assist, Intern. J. of Artificial Organs, 13(11):751-759, 1990.
24. Arabia M, Akutsu T: A new test circulatory system for research in cardiovascular engineering, Annals of Biomedical Engineering, Pergamon Press Ltd., 12, 29-48, 1984.
25. Donavan FM: Design of a hydraulic analog of the circulatory system for evaluation of artificial hearts, Medical Devices and Artif. Org., 3(4):439-49, 1975.
26. Kolff WJ: Mock circulation to test pumps designed for permanent replacement of damaged hearts, Cleveland Clin. Quart., 26:223-227, 1959.

27. Reul H, Runge JA: A hydraulic analog of the systemic and pulmonary circulation for testing artificial hearts, preliminary report, Proceedings ESAO I, 159-162, 1974.
28. Reul H, Runge JA: A hydraulic analog of the systemic and pulmonary circulation for testing artificial hearts, progress report, Proceedings ESAO II, 120-127, 1975.
29. Rosenberg G, Phillips WM, Landis DL, Pierce WS: Design and evaluation of the Pennsylvania State University mock circulatory system, ASAIO J., 4:41-49, 1981.
30. Swanson WM, Clark RE: A simple cardiovascular system simulator: design and performance, J. of Bioengineering, 1:135-145, 1977.
31. Wallner F: Development of a circulation mock up, Langenbecks Arch. Chir. 335 by Springer Verlag, 65, 1974.
32. Knierbein B, Reul H, Eilers R, Lange M, Kaufmann R, Rau G: Compact mock loops of the systemic and pulmonary circulation for blood pump testing, Intern. J. of Artificial Organs, 15(1):40-48, 1992.
33. Westphal D, Reul H, Rau G: Development and in vitro test results of the Helmholtz Centrifugal Pump, Proceedings of the Int. Workshop on Rotary Blood Pumps, Baden/Vienna Austria, Sept. 1991.
34. Westphal D, Rasche A, Reul H, Rau G: Blood damage in blood pumps: theoretical background and experimental results, Proceedings of the BME Workshop on in vitro testing of blood pumps, Aachen, April 1991 (in press).
35. FDA: Guidance for the preparation of applications to the FDA for ventricular assist devices and total artificial hearts, Silver Spring, Maryland, Dec. 3, 1987.

LASER MEASUREMENTS IN CARDIOVASCULAR FLOW DYNAMICS RESEARCH

Shi-Kang Wang and Ned H. C. Hwang

Cardiovascular Engineering Laboratory
Department of Biomedical Engineering
Memphis State University
Memphis, Tenn 38152 USA

Flow dynamic measurement in cardiovascular systems usually refers to the measurement of the time dependent pressure, blood flow velocity, and the volumetric flow rates. Generally, flow dynamic measurements in cardiovascular systems are rather difficult. This is not only because that the flow is basically unsteady, but also the physical frame of reference is always moving. During the past decades, various types of flow measurement instruments have been designed and developed. Among which, laser Doppler anemometer (LDA) has been one of the most favorite tool in laboratory studies.

LDA has definite advantages over most other types of flow velocity measuring devices. These include: that it does not require the physical presence of any probe (or transducer) in the flow field, that it has very high frequency response to the changes in flow velocities, that it requires no calibration as LDA measurement is practically unaffected by the temperature, density, or other physical properties of the medium. However, LDA has its physical limitations, particularly in application to the cardiovascular flow dynamic studies. In order to overcome these difficulties, several new laser measurement techniques have been developed in this laboratory in recent years. These techniques include:

(1) The optic-electro-hybrid feedback LDA (OEHF-LDA) utilizes the controlled frequency shift to achieve a significantly higher signal/noise ratio as compare to the conventional LDA system. Thus allow effective application of LDA in regions near solid, opaque boundaries. Application of OEHF-LDA to measure phasic regurgitation flow at distance few hundred micrometers from the moving surface of heart valve leaflet is demonstrated.

(2) A laser sweeping device (LSD) is designed to monitor the valve occluder movement. LSD is particularly design to measure the closing velocity of the occluder

within the final few microseconds before closure. It measures the time-displacement history with microseconds precision. The potential of cavitation in mechanical heart valves was evaluated by the computed squeeze flow field based on the valvular geometry and the measured occluder closing velocity.

(3) The two-dimension OEHF-LDA system can functionally replace the conventional two-component LDA system. It allows simultaneous measurement of the two components of the velocity vector in an unsteady flow field without the prohibitive cost of the conventional commercial two-component LDA.

(4) A laser-optic system (LOS) is developed to monitor the vascular wall motions without physically touching the wall. It measures the outside diameter and the diameter changes as functions of time for free-standing vascular conduits. Restricting the vessel motion to any fixed reference frame is not required during measurement.

INTRODUCTION

Over the previous decades, the results of flow dynamics study of the mammalian cardiovascular system undertaken by various investigators have been extensive. These studies were mostly performed in engineering laboratories. They have made significant contributions to our understanding of the cardiovascular systems, its normal functions and its progress in diseases. From an engineer's view, the cardiovascular system provides a field rich with flow dynamic problems. Engineering researchers have been attracted to the field from various fronts. While substantial advances have been made to formulate new methods and techniques in discovering new causes and cures of certain cardiovascular diseases, other research activities remain confused and frustrating experiences.

Flow dynamic measurements in cardiovascular systems are generally carried out to determine the time dependent pressure, blood flow velocities and the volumetric flow rates. Accurate measurement of velocity in the vascular flow field is particularly difficult. This is not only because the flow is basically unsteady, but also the physical frame of reference is always moving. The coupling of the fluid and solid boundaries and their movements formed the central feature in cardiovascular dynamics research.

In biological and physical sciences, prediction or interpretation of prototype phenomenon may also be made by observations of models. Cardiovascular models can generally be classified into three categories: the mathematical or computer models, the physical models, and the animal models. A combination of two or more of the above has also been used.

Various types of flow measurement instruments have been designed and developed for both laboratory and clinical applications. Among these, laser Doppler anemometer (LDA) has become one of the most favorite tool for velocity measurement in laboratory model studies in recent years. This chapter discusses several recently developed LDA techniques and their applications to cardiovascular studies. Other new laser devices and their applications are also presented.

MEASUREMENT OF REGURGITANT FLOW IN PROSTHETIC HEART VALVES (PHV)

Retrograde flow through PHV can usually be divided into two categories: (a)

The volume of fluid required to "wash" the valve occluder to the close position is related to the design and the dynamic behavior of the occluder. It is commonly known as the "dynamic closure volume," and (b) The PHV leakage after valve closure. This is directly related to the manufactural tolerance in fitting the various valve parts together.

Leakage Flow Characteristics Under Steady Flow Conditions

Leakage flow after valve closure may occur through the gaps between the leaflet and the valve body, and that between the leaflets. Regurgitant jet from the small gap space dissipates rapidly into the surrounding fluid which is generally at rest (Figure 1).

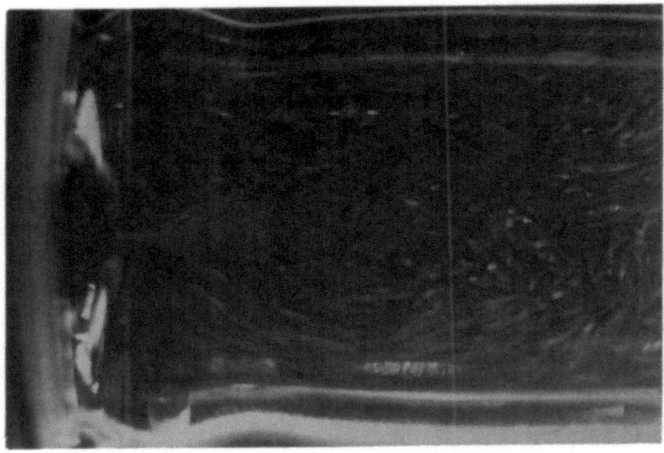

Figure 1. Regurgitant jet from the small gap space between the leaflets dissipates rapidly into the surrounding fluid.

Two 29mm bileaflet type mechanical PHVs were used in the steady leakage flow study. Leaflets in each of the testing valves were glued to the valve body and major radii sealed to provide a fix gap between the two leaflets (B-Datum) to allow steady leakage through the gap under constant pressure heads. Based on typical PHV assembly tolerance, the gap widths were made at 0.002" and 0.008", respectively.

The test valves were installed in the steady flow chamber with the B-Datum oriented vertically. Measurements were made at constant transvalvular water pressure heads (ΔP) equivalent to 20, 100, 150 and 200 mmHg. Flow visualizations and LDA mappings were made to determine the configuration of the leakage jet. Using a conventional LDA system (TSI Model 9184), velocity profile surveys were made at seven cross-sections at distances of 0.5, 1.0, 5.0, 10.0, 20.0, 50.0 and 100.0mm from the downstream surface of the B-Datum gap. Maximum jet velocities detected under all experimental conditions were less than 4.5 m/sec (Figure 2). The maximum velocity did not vary significantly between the two gap widths tested.

In this type of application, the conventional LDA system may be limited by its signal/noise (S/N) ratio, particularly when measurements were made very close to a reflective solid boundary.

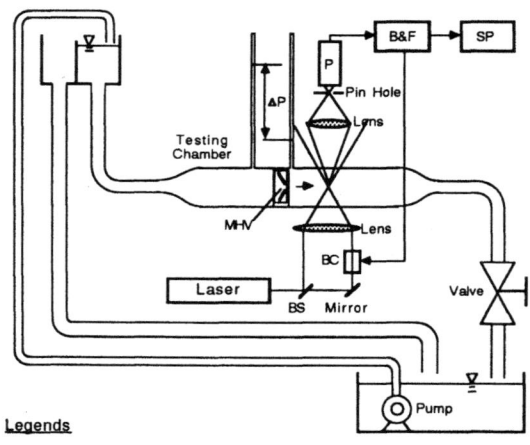

Legends

BC --- Bragg Cell
B&F --- Bragg Cell Driver & Frequency Mixer
BS --- Beam Splitter
ΔP --- Transvalvular Pressure
P --- Photodetector
SP --- Signal Processor

a. Steady Flow Chamber

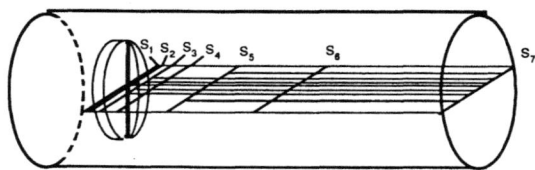

b. Seven LDA Servey Sections (S1 to S7)

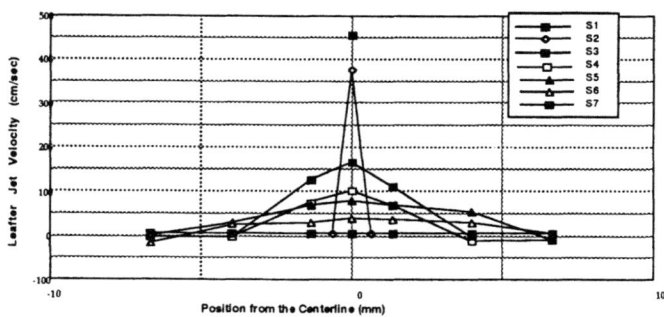

c. Velocity Profile of 0.002" Gap aat 100 mmHg

Figure 2. Steady flow chamber (a) with measuring sections (b) and jet velocity profiles (c).

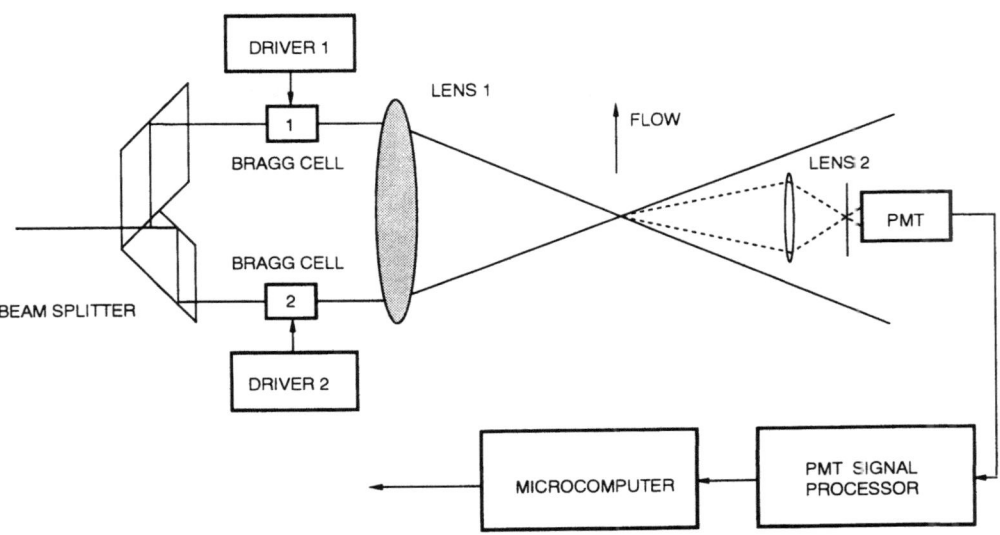

Figure 3. Conventional laser Doppler system with fixed frequency shift.

Limitations in a Conventional LDA System

Normally, an LDA system may be considered as made up of two independent parts, the optics and the signal processor. Figure 3 shows a typical laser Doppler anemometer system with dual Bragg cells in operation. The two Bragg cells serve to introduce a fixed frequency shift difference $\Delta f = f_1 - f_2$ which serves to discriminate the directions of flow. In this arrangement, the scattering volume is illuminated simultaneously by two equal intensity focused beams which intersect at an angle θ. While the tracer particles suspended in the fluid pass through each interferometric plane in the focal volume, the scattering of light, with both Doppler shift f_D and frequency shift Δf occurs. The frequency of the signal received and converted by the photodetector is given by

$$f = |\Delta f + f_D| \tag{1}$$

where

$$f_D = \frac{2\sin(\theta/2)}{\lambda} u \tag{2}$$

λ is the laser wavelength and u is the flow velocity.

Both Δf and f_D are frequency differences. They can take either positive or negative values. But the signal frequency f cannot be negative, as denoted by the absolute sign in Equation (1). This results in one of the major limitations in conventional LDA as explained below.

A particle moving opposite to the fringe movement yields a higher frequency and a particle moving in the same direction as the fringe movement yields a lower frequency (see Figure 4). The range of the measurable flow velocity in the same direction as the fringe movement is limited by the value of the frequency shift. The measurable flow velocity opposite to the fringe motion is also limited by the dynamic range of the signal processor and photodetector.

From Figure 4 we can also see that the range of directional ambiguity is reduced but not totally eliminated, and that the photodetector responds to the frequency increase with the increase of flow velocity. With increasing the flow velocity the bandwidth of the Doppler signal processor response increases correspondingly. In a commercial product, the conventional LDA processor must be designed to operate over a wide range of frequency to accommodate a maximum range of flow velocity application.

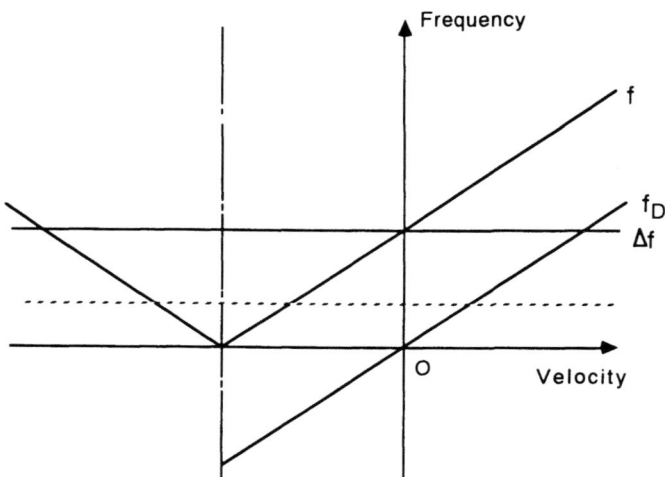

Figure 4. Frequency versus velocity diagram illustrating directional discrimination in a conventional LDA system with fixed frequency shift.

LDA with Controllable Acousto-Optic Frequency Shift

In a dual-cell LDA operation, if the acousto-optic frequency shift Δf can be controlled automatically, the frequency of the signal received from the photodetector f may be kept at a certain fixed value f_0. Under this condition, however, both sides of equation (1) will remain unchanged, in spite of the changing Doppler shift f_D. An acousto-optic frequency shift with controllable frequencies (CF-AOFS) as shown in Figure 5 can thus be realized by driving the Bragg cells with the voltage controlled drivers (VCD).

Variation of the Bragg frequency affects the first-order diffracted beam in two ways. On one hand, the efficiency of diffraction decreases, normally by a few per cent, as the Bragg frequency changes a few MHz from its center frequency f_{BC}. The variation of the Bragg frequency is limited by the S/N ratio, which may be too low due to the poor visibility in the probe volume. This problem exists also in conventional dual cell LDA operation, in which the frequency in one of the Bragg cells must be changed manually so that the measured range can be selected (e.g. in TSI Model 9184 the center Bragg cell frequency is 40 MHz and the maximum variation 5 MHz or 12.5% of f_{BC}).

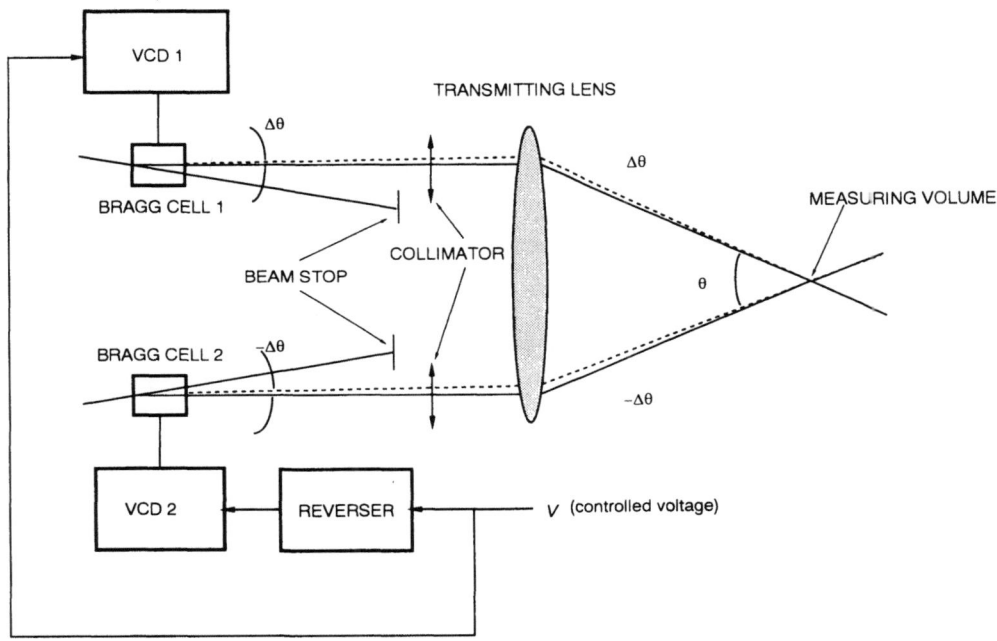

Figure 5. Controllable acousto-optic frequency shift system.

On the other hand, the beam direction will change if the Bragg frequency f_B changes. The relationship between the angular variation of the first beam $\Delta \alpha$ and the variation of Bragg frequency Δf_B may be approximated by the following equation[1]:

$$\Delta \alpha = (\lambda/V_s)\Delta f_B \qquad (3)$$

where V_s is the sonic speed in the crystal of the Bragg cell. At room temperature V_s is approximately 3600 m s^{-1}.

In order to avoid significant power loss in the laser beam, the variation Δf_B should be kept within a few MHz. For example, at 5 MHz the blue beam argon-ion laser has wavelength 488nm and the red beam He-Ne laser has wavelength 632.8 nm, their angular variations are 0.7 and 0.9 mrad, respectively. These values are very close to the divergence angle of the laser beam used in most LDA systems. A collimator can thus be used at a given position along the beams to keep the first-order beam motions parallel to each other (see Figure 5). Parallel beams should then converge on a certain point by means of a transmitting lens. The sample volume therefore no longer changes its position, as the beam cross angle gets an additional variation $\Delta\theta$, which is produced simultaneously by the opposite variation of the other Bragg frequency. This can be achieved simply by positioning the Bragg cells opposite to each other as shown in Figure 5. The same effect can also be achieved by rotating the two Bragg cells at 90° around their own beam axis.

Utilizing the CF-AOFS[2] technique, we combined the conventionally separated optical and electronic units into a closed loop with hybrid feedback signals in a new LDA system which we termed "optic-electro-hybrid feedback LDA" (OEHF-LDA[3]). In short, the key technique used in OEHF-LDA is a CF-AOFS in which the frequency shift is controlled by the scattering Doppler signals. These signals track the fluctuations of the flow velocity to keep the frequency of scattered light at a nearly constant value.

The Conception of an OEHF-LDA

The optic-electro-hybrid feedback LDA (OEHF-LDA) is designed using the CF-AOFS technique and the hybrid signal feedback containing both optic and electronic signals. Figure 6 shows the optical components and the electronic units organized in a closed loop by means of the hybrid feedback signals. Two Bragg cells are used to introduce each of the changeable frequency shifts into a laser beam with the values of frequency f_1 and f_2 respectively. A set of moving interferometric planes forms in the focal measuring volume. As the tracer particles pass through the measuring volume the photodetector picks up the scattering light frequencies which correspond to the relative velocity between the moving interferometric planes and scattering particles. The scattering radiations resulting from the Doppler frequency shift f_D and the changeable frequency difference between two laser beams f_1-f_2 are superimposed on each other with a nearly constant frequency f_0. This frequency is then emitted onto the photodetector.

When f_D varies with the fluctuation of flow velocity, the balance in the closed loop is disturbed. The output signal from the photodetector, including the information on velocity variation, is amplified by an intermediate frequency stage (IF) selected at frequency f_0 with narrow bandwidth. The amplified signal is then applied to a frequency discriminator in which a circuit with highly sensitive frequency response is set to discriminate the variation in the Doppler frequency f_D. The output from the discriminator is again amplified and used to control the voltage controlled driver (VCD) in both channels to change the driving frequency of the Bragg cells. A new frequency difference between the two laser beams will thus be formed to reduce the frequency deviation δf, which is produced from the change of Doppler frequency and supplied to the photodetector. This process is continued until the new balance of the closed loop is established. When the loop is in tracking, the feedback signal ensures that the output of the discriminator is close to zero and the feedback loop is locked at the intermediate frequency f_0. The variations of the frequency difference of the VCD, i.e. $\Delta f = f_1 - f_2$, depend only on the frequency signal f_D. Hence, the resulting scattering frequency f from the measuring volume is nearly a constant value f_0, no matter how the flow velocity changes.

Block Diagram of OEHF-LDA

Figure 7 is a signal flow chart which also illustrates the connection between the different parts in the OEHF-LDA system. The Doppler signal f_D is linearly proportional to the flow velocity. It can be considered as a 'disturbance source' which excites the controlled object (i.e. the fringes motion in the measurement volume, seen in Figure 7). The frequency of the scattering light changes from f to $f + \delta f$ compared with the desired frequency f_0 in the frequency discriminator. The 'error' voltage e resulting from the comparison of the two frequencies from the frequency discriminator is fed to a DC amplifier. The amplified 'error' voltage V is then used to drive the VCD to change the frequency shift difference Δf. The change of the frequency shift

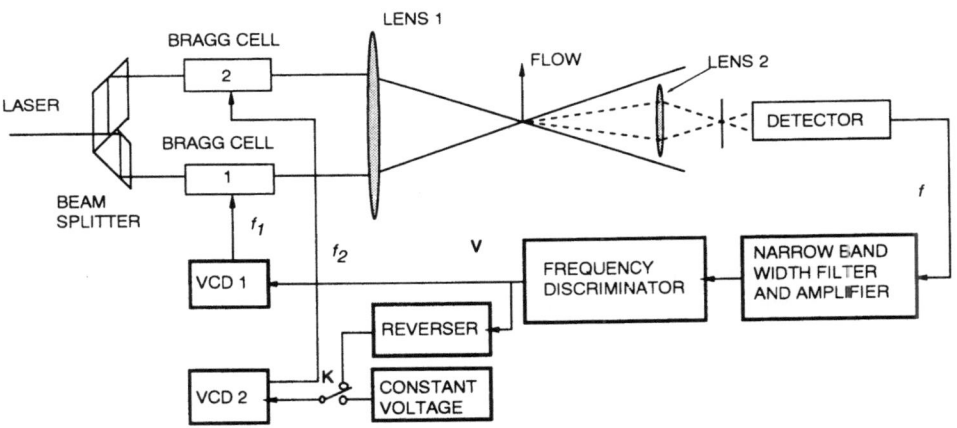

Figure 6. OEHF-LDA combining the optical and electronic units.

difference $\delta(\Delta f)$ counterbalances the effect from the 'disturbance' value δf_D to keep the frequency of the scattering light at a constant value.

In OEHF-LDA the quantitative relationships among the three frequencies (i.e. the Doppler frequency f_D, the optical changeable frequency shift Δf, and the signal frequency emitted from the measuring volume f) are identical to that of the conventional dual cell LDA operation mentioned before. There is no difference in optics, so that Equation (1) can also be used in OEHF-LDA. The only difference in OEHF-LDA is that Δf follows f_D, and that f is kept at a fixed value f_0, which is selected so the absolute sign in the equation can be omitted. The relationship between frequency and velocity are shown diagramatically in Figure 8.

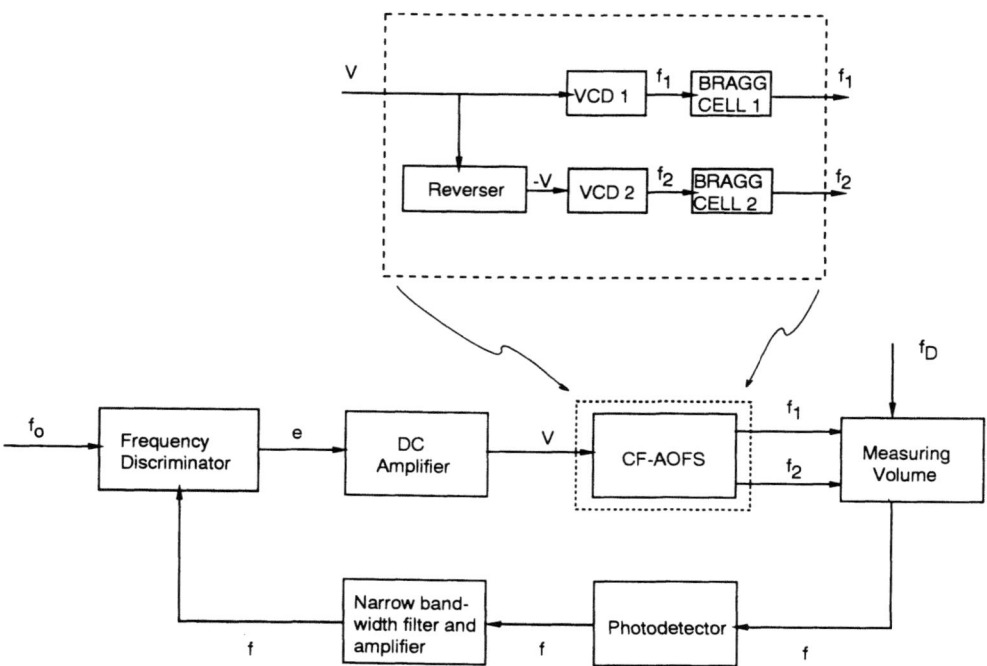

Figure 7. Signal flow chart of OEHF-LDA system.

As a result the directional ambiguity of flow velocity which is an inherent problem in the conventional laser Doppler system does not appear in the OEHF-LDA system. The most important difference between the OEHF-LDA and the conventional LDA is that the frequency of scattering radiation received by the photodetector is a constant value for the OEHF-LDA system, while the Doppler frequency f_D varies with the flow velocity. In the conventional LDA, however, the frequency of the received scattering signal in the photodetector always varies with the measured flow velocity.

Dynamic Range of OEHF-LDA

The dynamic range of OEHF-LDA generally depends on the center Bragg frequency f_{BC} in which the Bragg cell can get the best efficiency of the diffraction. It is reasonable to control the variation of the Bragg frequency Δf_B within 10% of f_{BC}. The signal frequency $f \sim f_0$, compared with f_{BC} is so small that it can be neglected in the dynamic range analysis. If both the Bragg cells in the optical system are on the same center frequency, the dynamic range of OEHF-LDA, Δf_{BR}, can be calculated as follows:

$$\Delta f_{max} - f_{max\ 1} - f_{min\ 2} - (1 + 10\%)f_{BC} - (1 - 10\%)f_{BC}$$

$$\Delta f_{min} - f_{min\ 1} - f_{max\ 2} - (1 - 10\%)f_{BC} - (1 + 10\%)f_{BC}$$

$$\Delta f_{BR} - \Delta f_{max} - \Delta f_{min}$$

$$- 20\%f_{BC} - (-20\%)f_{BC} - 40\%f_{BC}.$$

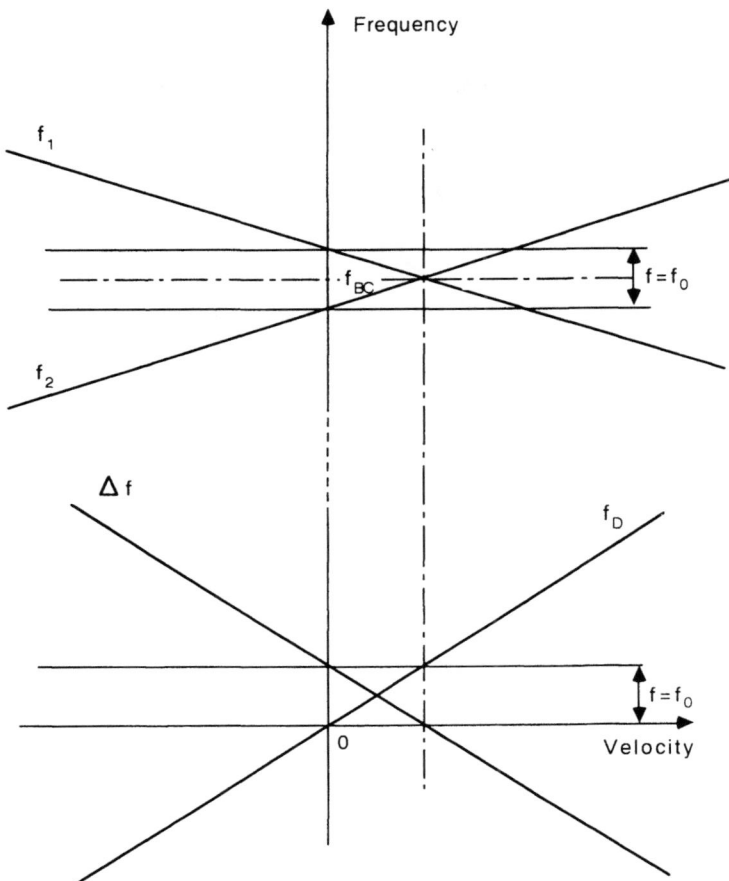

Figure 8. Frequency versus velocity diagram (OEHF-LDA).

For example, if the center Bragg frequency f_{BC} is 40 MHz, the dynamic range Δf_{BR} is 16 MHz.

The difference of the dynamic range of velocity measurement between the conventional LDA and the OEHF-LDA can be illustrated by comparing the two diagrams of frequency against velocity in Figure 9.

Figure 9(a) shows the conventional LDA with fixed frequency shift. There are two measurable regions of velocity separated by a 'dead' region, which is inherent from the bandwidth of the signal processor. For the OEHF-LDA Figure 9(b) shows that there is no 'dead' region along the velocity (v) axis. The range bounded by the absolute limit values of the positive frequency and negative frequency in Figure 9(b) is the dynamic range of velocity for the OEHF-LDA system. This represents the summation of the 'dead' region and two measurable regions of Figure 9(a).

Characteristics of OEHF-LDA

The characteristics of OEHF-LDA can be summarized as follows:

(i) Instead of a wide frequency bandwidth filter, a narrow one has been used in the signal processor of the OEHF-LDA system. Thus, the signal-to-noise ratio is significantly improved over that of the conventional LDA under the same frequency dynamic range.

(ii) While locking the hybrid feedback loop, the scattering signal is almost a

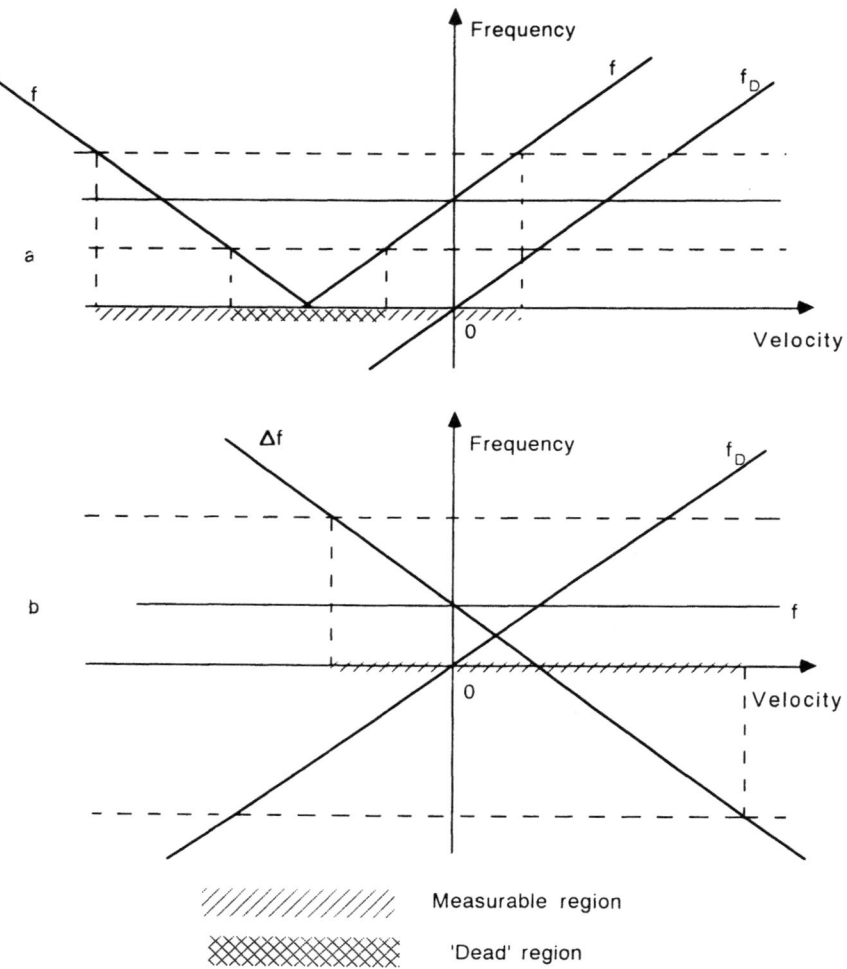

Figure 9. Comparison of (a) LDA with fixed frequency shift (b) OEHF-LDA (frequency vs velocity plots).

constant value determined by selecting the intermediate frequency f_0. Because f_0 is usually 5% of the full scale range and much less than the maximum Doppler frequency value. The photodetector and electronic components with the lower frequency response may be used for the Doppler signal processing in measuring the high flow velocity.

(iii) The OEHF-LDA system is similar to a frequency tracker system used for the conventional Doppler signal processing. The 'drop out' of signal still exists. However, the construction of the OEHF-LDA system is much simpler in design than the conventional LDA system. The optical signals are mixed in the measuring volume before the radiation transfer to the photocurrent, i.e. the measuring volume is used as an 'optical mixer' to create an optical, local oscillating signal with a controlled frequency. This controllable optical radiation always tracks the Doppler signal of frequency f_D.

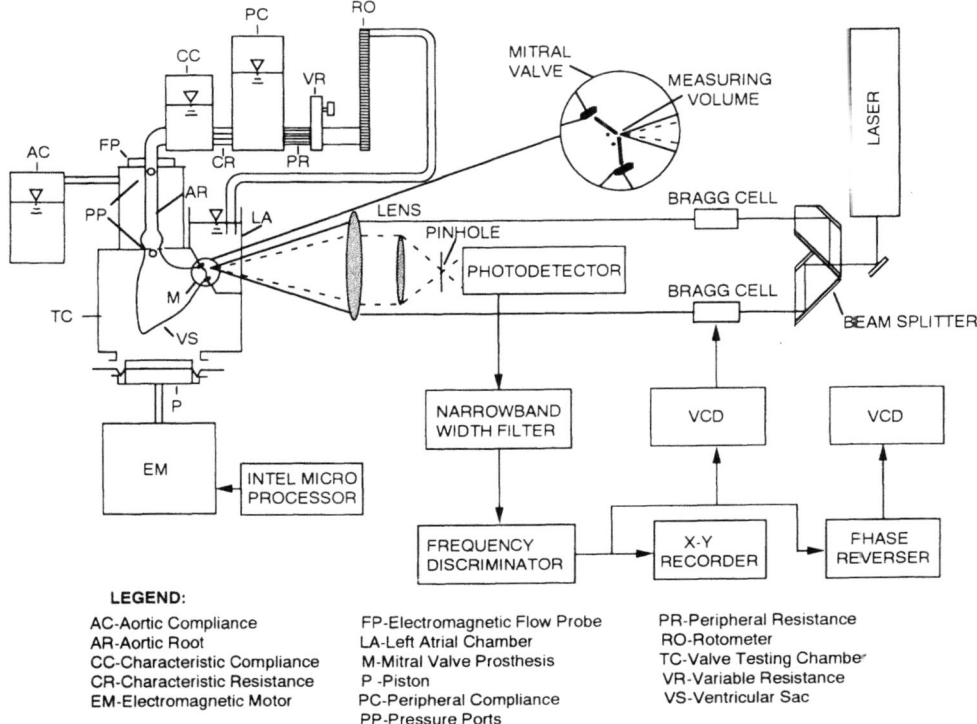

Figure 10. Heart valve testing loop and OEHF-LDA system for the mitral valve regurgitation velocity measurement.

Dynamic Regurgitant Flow Under Pulsatile Flow Conditions

The application of the OEHF-LDA system in unsteady flow velocity measurements is demonstrated in dynamic PHV regurgitant flow measurement using the backward scattering mode.

Detailed studies of PHV flow dynamics have been carried out by several research groups in vitro by installing the testing valve in a mock circulatory flow loop similar to that shown in Figure 10. The mock circulatory systems are usually designed to reproduce the physiological pressure and flow waveforms, and the geometries of the heart and its great vessels. In most of these studies, the pulsatile flow patterns and the velocity distributions in the vicinity of PHV were mapped by using conventional LDA systems[4-7].

Recently, several investigators have demonstrated that the flow structures in certain close regions of PHV such as the vicinities of the valve occluder, the valve seat, the hinges of the valve leaflets etc, are of particular importance. And that they may be related to valvular thrombosis and hemolysis.

Application of the conventional LDA to these close regions has been severely limited due to the following facts: (a) forward scattering LDA has not been possible since the scattered laser light cannot penetrate the opaque leaflet or valve body as it opens and closes periodically, and (b) application of back scattering LDA measurement has been limited due to poor signal-to-noise ratio in the conventional LDA systems, particularly when the measuring volume is placed within a short (micrometers) distance from the valve surface.

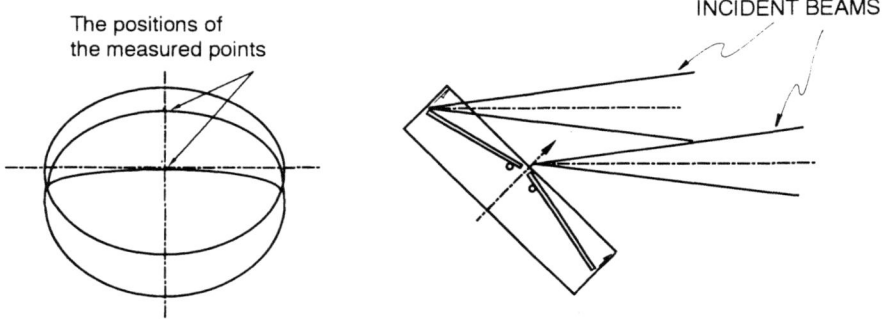

Figure 11. LDA measurement positions on the inflow side of the mitral valve.

The OEHF-LDA experiment presented below involved a 29 mm bileaflets PHV. The testing valve was installed in the atrioventricular (mitral) position of the valve testing chamber (see figure 10, M). Experiments were performed at the heart rate of 70 beats/min and a cardiac output of 5 liters/min. The afterload units in the mock circulatory system were adjusted to provide a physiological aortic pressure waveform at 120/80 mm Hg (1 mm Hg ≈ 133 Pa). Figure 11 shows schematically the LDA measurement positions.

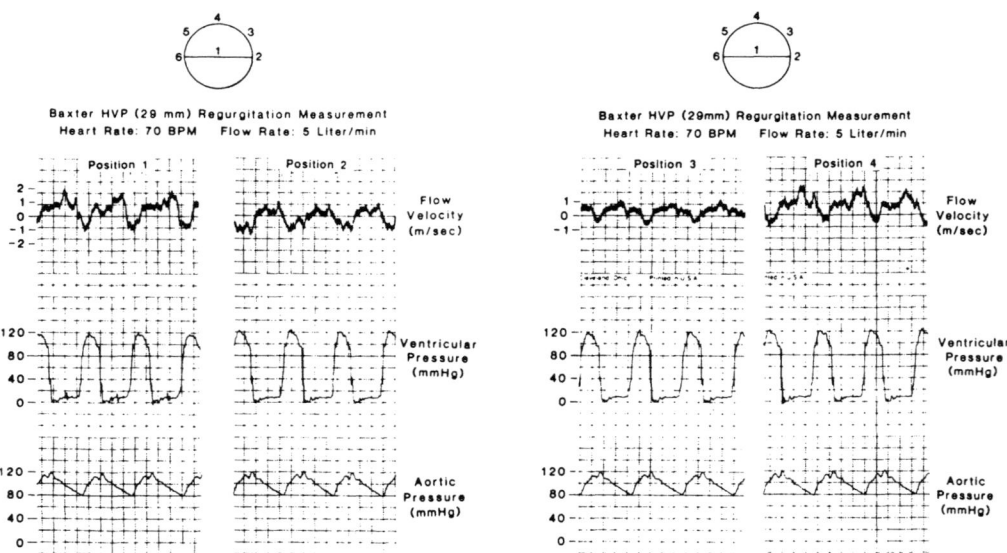

Figure 12. Typical OEHF-LDA velocity measurement.

A 30 mW argon-ion laser beam was used as a light source and 2μm silicone carbide particles were seeded into the fluid as scattering particles. The laser beams transmit through a 70 mm beam splitter, two Bragg cells and a 280 mm transmitting lens and enter the flow field from the inflow (atrial) side of the atrioventricular valve. The scattered light is collected by the same transmitting lens and is then focused onto the photodector by the other smaller lens. The ellipsoidal sampling volume is 0.15 mm in diameter and 1.2 mm in length. The fringe spacing is 1.97 μm. The center frequency of the VCDs is 60 MHz, and the frequency of the signal received from the photodetector is fixed at 500 kHz. The optical components were arranged so that the photodetector was situated on the same side as the incident laser beams (back scattering mode). The LDA measure volume is located approximately 500 μm from the gap surface of the two leaflets in the closed position, and/or 1 mm from the corner of the valve body and one leaflet in the closed position (see Figure 10). The angle between the velocity component measured and main flow direction is 45°.

The typical results are shown in Figure 12. Each measurement position is labelled at the top of the figure. In the time-velocity trace, the positive values indicate forward flow from the left atrium to the left ventricle, while the negative values indicate the regurgitant back flow from the ventricle back into the atrium.

LASER MEASUREMENT OF PHV LEAFLET VELOCITIES

Background and Significance

Surface erosion of Pyrolite® carbon mechanical heart valve prostheses (MHV) in patients was practically unnoticed until a small number of Edwards-Duromedics MHV (Baxter Edwards CVS Division, Irvine, CA 92714) clinical explants appeared with surface damage in 1987[8-10]. Electron micrographs of the surface indicated damage morphology similar to that of cavitation erosion[11]. Cavitation erosion of Pyrolitic® carbon MHV has been reported to occur under in vitro accelerated testing conditions[12] and in cardiac assist devices[13,14]. The potential of MHV cavitation is believed to be related to dynamic and flow dynamic behavior of the valve.

Venturi effect due to the regurgitant leakage jet, liquid separation from the leaflet surface due to water hammer pulse at closing, the squeeze flow field between the closing leaflet(s) and the valve body, and the propagation of vibration stress wave inside the valve component have each been considered as a contributing factor to MHV cavitation. It has been postulate that the MHV leaflet terminal closing velocity, which is linked with the valve sound, is a key to all the above mentioned factors.

Cavitation occurs when the local pressure in a flowing fluid falls below the vapor pressure of the fluid, creating small vapor bubbles in the fluid. When the bubbles flow into a region of higher pressure, they collapse violently. If this violent collapse (implosion) takes place near a material surface, it may cause localized surface pitting which may eventually lead to surface erosion.

Conceptually, the following causes may be considered as the contributing factors to MHV cavitation:

- The regurgitant leakage flow between the closed occluder and the valve body, or between the two occluder leaflets (in the case of the bileaflet MHV), may form jets of relatively high velocity through the narrow gap spaces. Local pressure will fall in proportion to the square of the jet velocity according to the Bernoulli Principle. This physical phenomenon is commonly known as the "Venturi effect."

- The moving fluid column tends to separate from the in-flow-side surface of the occluder as it suddenly closes behind the column. A negative pressure is generated with magnitude proportional to the instantaneous velocity change and the fluid mass involved. This effect is known as the "water hammer," a transient pressure phenomenon, commonly observed in the sudden closure of a valve in a water pipeline.

- Pressure may also decrease in the regions of accelerated flow.

- During the final phase of the occluder closing, the fluid mass in the gap space between the closing occluder and the valve body is squeezed in motion by the approaching boundaries. The instantaneous flow patterns in the narrow gap space could be rather complex, as can be analogous to the fluid between a ship and the pier during the final minutes of docking. The complex flow patterns (termed "squeeze flow") are directly related to the detailed geometry of the approaching boundaries (i.e., the occluder and the seating lip on the valve body), and the velocity of the closing occluder. It is speculated that isolated regions of high velocities may exist in or adjacent to the gap field to cause regional pressure decline.

The above considerations suggest that the risk of PVH cavitation may be higher at valve closing than opening, that high ventricular dp/dt may promote MHV cavitation and that larger valves may have more cavitation potential than smaller ones.

Laser Assessment of Leaflet Closing Velocity

Most of the current leaflet type mechanical PHVs are designed to operate with a rotational motion combined with a slightly translational motion. The speed at which the leaflet slams against valve body at closing was thought to be related to the "water hammer" and other phenomena which might contribute to MHV cavitation. To quantify the occurrence of these phenomena, a laser sweeping method (LS) was developed in this laboratory to measure the precise leaflet motion immediately before its closure.

The basic principle of the laser sweeping method for measuring leaflet motion is shown in Figure 13[(15)]. The moving leaflet may be used as a rotating mirror within the defined range of a small excursion. When the incident laser beam is aimed at a fix small spot on the leaflet surface with a small angle of rotation, the direction of the reflected beam is changed due to the changing incident angle. A convex lens was placed where the distance between the lens and the reflection point on the leaflet was equal to the focal length of the lens. The angular motion of the reflected beam was thus converted into a parallel motion through the lens.

The relationship between the rotating angle of the leaflet, α, the focal length, F, and the traveling distance of the reflected beam behind the lens may be expressed as:

$$L = F \sin 2\alpha \quad (4)$$

For a small angle of rotation ($\alpha \leq 3°$ in the present study), Equation (4) may be simplified to the following:

$$L = 2F\alpha \quad (5)$$

A linear grid used as a light fringe was placed behind the lens. As the moving beam swept across the grid, a series of flashes were generated behind the grid plane. Thus, the continuous leaflet motion in space domain was converted to a series of flashing signals collected by a photodetector in time domain. The reflected beam travels from one fringe to the next as the leaflet rotates through a certain angular displacement, $\Delta\alpha$, which, according to Equation (5), can be expressed as follows:

$$\Delta\alpha = \frac{\Delta L}{2F} \tag{6}$$

Where ΔL is the distance between two adjacent fringes on the linear grid.

The linear displacement of a certain point, P, on the leaflet, Δs, can thus be calculated as follows:

$$\Delta s = R\left(\frac{\Delta L}{2F}\right) \tag{7}$$

Where R is the distance measured from P to the axis of leaflet rotation.

If the time interval between the two adjacent peaks of the photodetector signal trace is Δt, the velocity of the point P on the leaflet (V_p) and the angular velocity (V_a) of the leaflet can be determined by the following relationships, respectively:

$$V_p = \frac{\Delta s}{\Delta t}, \tag{8}$$

$$V_a = \frac{\Delta\alpha}{\Delta t} \tag{9}$$

The testing MHV was installed in the mitral position of the pulsatile mock circulatory testing loop. In order to meet certain optic requirements, a transparent atrial chamber was constructed to allow free passage of both the incident and reflected laser beams. An Edwards-Duromedic 29 mm mitral MHV (ED) was used in the present study, with a St. Jude Medical 29 mm mitral valve (SJ) as the study control. The schematics of the experimental apparatus is shown in Figure 13.

A 5 mw Argon-ion laser beam was used as the light source (wavelength λ = 488 nm). The laser beam enters the flow field from the atrial side of the testing valve via the submerged mirror (M). Lens L_1 was placed in front of the atrial chamber to collect the reflected laser beam in rotation, and converted the rotational motion into parallel motion. The beam reflected through lens L_1 first converged and then diverged to form an optical waist at the focal plane behind the lens (Figure 14). The diameter of the waist may be determined by the following[16]:

$$d_3^{-2} \approx \frac{4F\lambda}{\pi D_e^{-2}} \tag{10}$$

Where the focal length F = 280 mm, and D_e^{-2} is the e^{-2} diameter of the unfocused illuminating beam (D_e^{-2} = 1.5mm). From Equation (10) the minimum waist diameter of the laser beam (115 µm) was calculated. A linear optic grid with a 1.1 mm fringe

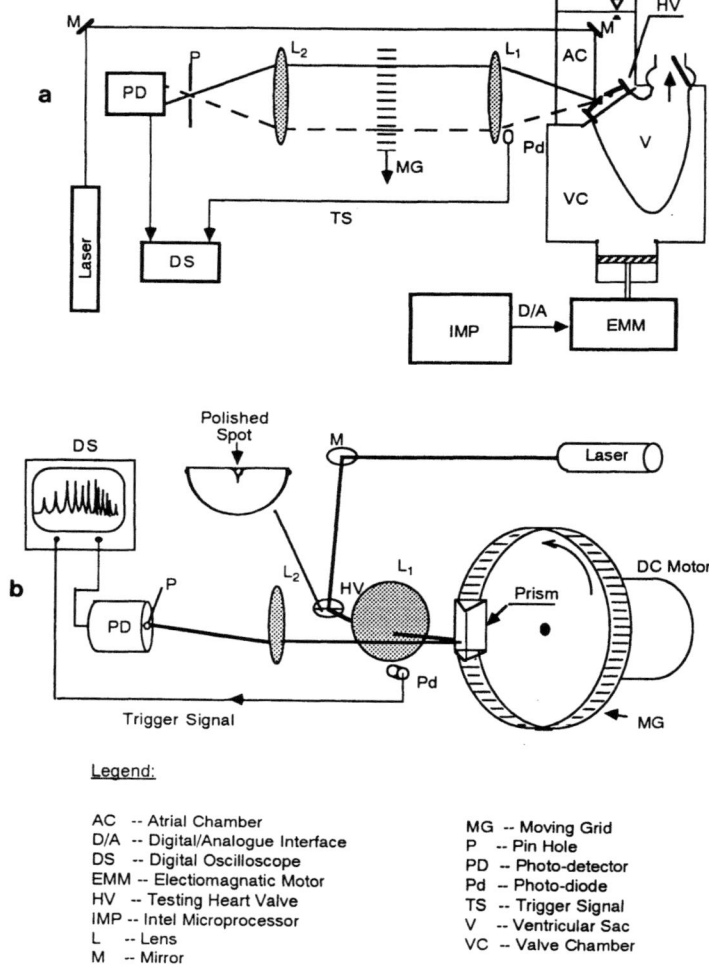

Figure 13. Schematics of (a) the laser sweeping (LS) system (b) the optic arrangements.

interval was placed at the laser beam waist perpendicular to the beam. From Equation (6) the corresponding angular displacement of the leaflet rotation was 0.11° between each fringe interval. Behind the grid, a second lens, L_2, with the same focal distance, was installed at one focal distance away from the grid. Through lens L_2, the laser beam was again collimated, and the flashing laser signal was focused to activate the photodetector surface, which was located at one focal distance downstream from the grid through a pinhole. The moving beam modulated by the grid was then transferred to a stationary flash spot on the photodetector.

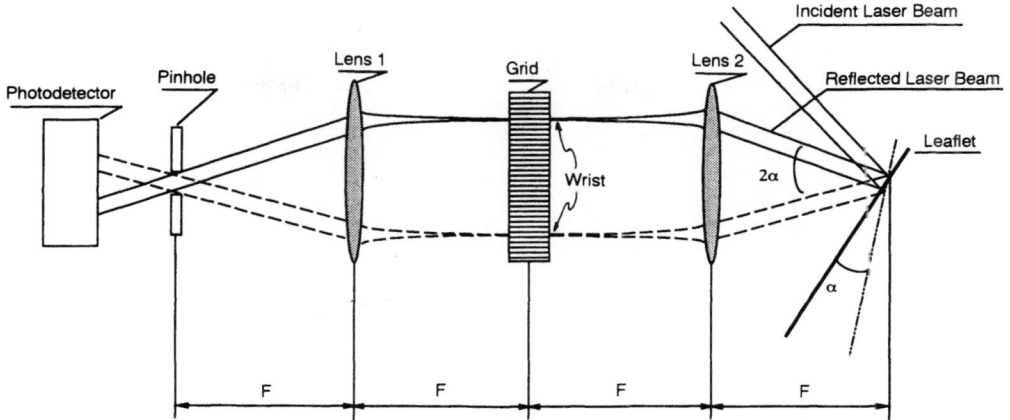

Figure 14. Optical system for leaflet velocity measurement.

A Tektronix 2230 Digital Memory Oscilloscope (Tektronix, Inc., Beaverton, OR) was used to monitor the photodetector output signal, and the high frequency oscilloscope registered and stored the instantaneous signals displayed on the screen. Time intervals between any pair of adjacent signal peaks were measured by means of a movable cursor. A 35 mm camera and a VHS video camera were used to photograph and record the signals on the oscilloscope screen for hard copy analysis.

In the experiment, the testing valve opened and closed periodically. An Intel 8085 Microprocessor (Intel Corporation, Santa Clara, CA) was utilized to produce the cardiac cycle drive function. Because the valve closure period is rather short compared with the whole cardiac cycle, it was difficult to acquire and store the respective flash signal by means of a manual trigger control; thus, a small circuit was designed to synchronize the recording of the leaflet motion with the ventricular pressure waveforms. The synchronizing signal was produced by using the Intel 8080 Microprocessor through an I/O interface.

Figure 15 shows a typical signal trace displayed on the oscilloscope screen. The horizontal scale between any two peaks represents the time required for the reflected laser beam to travel one corresponding optic fringe.

Substituting R with the major radius of the valves, Equation (7) gives linear displacement of the leaflet apex during this period. The displacement has a 20 μm resolution corresponding to each 1.1 mm laser beam displacement on the grid. Figure 16 exhibits the closing velocities of ED and SJ leaflets within the final 3° rotation before closure. The maximum closing velocity for these two valves at different heart rates are summarized in Table 1.

Simultaneous Measurement of Bileaflet Motions

Asynchronous motions of bileaflet PHV occluders has been observed both clinically and in testing loops[17]. The dynamic and fluid dynamic origins of leaflet flutters has recently been investigated[18,19].

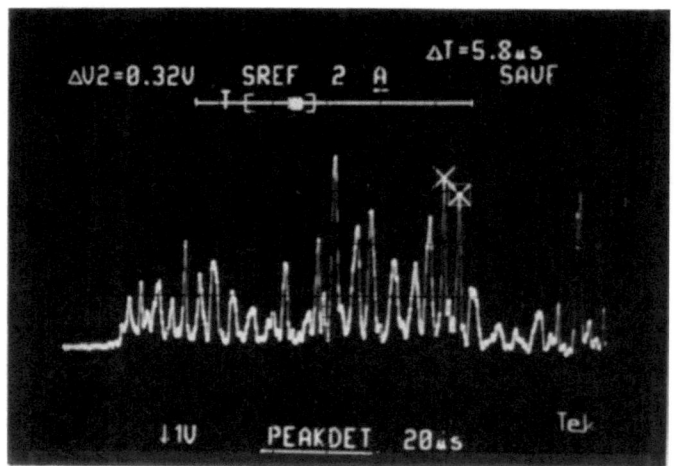

Figure 15. Typical oscilloscope display of leaflet closing signals.

The optic system proposed for simultaneous measurement of the two leaflets in a bileaflet PHV is shown schematically in Figure 17. Based on the same principle of the single leaflet measurement as discussed in previous section, a beam splitter (BS) will be added to the original set-up to split the incident laser beam into two equal strength beams in different directions, each aimed at the polished spot on the respective leaflet surface. The reflected beams from the surfaces will pass through the same rotating linear grid (G) and be redirected by two optical prisms (P_1 and P_2) each to a downstream lens (L_2 and L_3) and photo-detector (PD) respectively as shown.

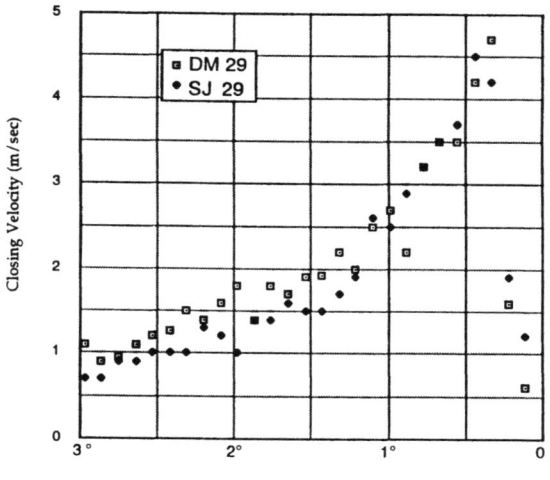

Figure 16. Terminal closing velocities for SJ29 and ED29 mitral valves.

Table 1. Maximum Closing Velocities of DM29 and SJ29.

Heart Rate (beats/min)	dp/dt (mmHg/sec)	Flow Rate (L/min)	Maximum Closing Velocity (m/sec)
Baxter-Duromedic 29 mm mitral MHV			
70	1,800	5.0	0.97±0.04
90	3,000	6.0	1.69±0.06
120	5,600	7.5	4.8 ±0.5
St. Jude Medical 29 mm mitral MHV			
70	1,800	5.0	0.9 ±0.04
90	3,000	6.0	1.64±0.05
120	5,600	7.5	4.5 ±0.4

Computation of the Squeeze Flow Field

The fluid mass in the gap space between valve body and the leaflet or in the gap between the two leaflets (B-Datum) were forced into motion when the boundaries were closing in on each other. The velocity (and pressure) distribution in the squeeze flow field is a time function which can be determined by using the geometry and the motion of the boundaries as the input. With the leaflet motion precisely determined and the geometry of the MHV given, the velocity distributions in the gap flow field were computed using the potential flow theory, which is justified due to the low Reynolds number expected[20].

A typical example of squeeze flow field computation performed on the ED 29 mm mitral MHV is shown in Figure 18. The two consecutive computations were made based on the relative positions (gap width: 36 μm and 17 μm) and the corresponding closing velocities (3.6 m/sec and 1.4 m/sec) respectively measured. Flow field velocity on the order of 16 m/sec can be spotted in Figure 18a. The spacing between the streamlines on the leaflet surface represents the velocity of the leaflet (3.6 m/sec). Narrower width between streamlines elsewhere represents proportionally the higher velocity.

The distribution of the ejected fluid from each end of the squeezed space may be approximated by the levels of the transient pressures measured at two flush pressure ports adjacent to the line of contact at closure.

NEW LDA METHODS IN VASCULAR DYNAMICS RESEARCH

Vascular stenoses (VS) are responsible for most of the vascular failures, both in natural and in substitute vascular grafts. Factors related to the development of VS include hemodynamic stresses, mismatch of the mechanical properties, low velocity or high frequency flow fluctuations. Graft to host vessel diameter difference and turbulent flow at the anastomotic region may also contribute to the uneven distributions of shear stress on the endothelia and subendothelial myocytes, which may also incite stenosis.

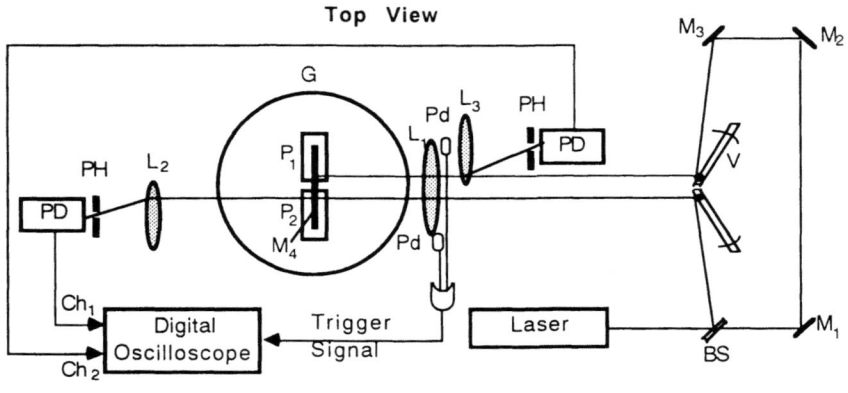

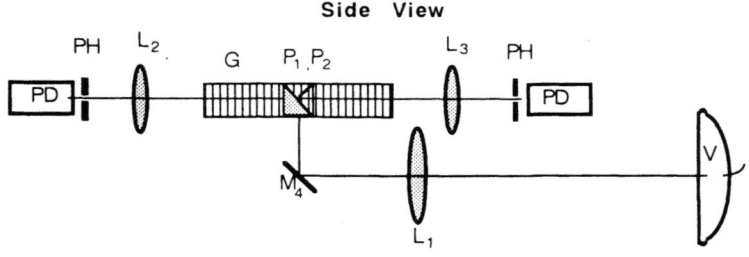

Figure 17. Optic system for simultaneous measurement of bileaflet motions.

The Arteriovenous Hemodialysis Angioaccess Loop Graft (AVLG) Model

To better understand the role of hemodynamics in vascular lesion development, we adapted an end-to-side arteriovenous hemodialysis angioaccess loop graft (AVLG) as the hemorrheologic model. AVLG is a surgically implanted conduit that facilitates hemodialysis in patients with chronic renal failure. The surgically implanted AVLG is anastomosed by one end to a host artery and by the other end to a vein to create a fistula conduit which brings high pressure arterial blood flow into an otherwise slow flowing vein. Abnormal hemodynamic characteristics are created lead to changes in both pressure and flow in the host vein to which the graft is anastomosed.

AVLG system provides an ideal model with a well-defined flow field in which the distributions of velocity, wall shear stress and structure of turbulence could be investigated in details. Based on the data of an earlier animal study, we designed an in vitro bench-top flow model to study the detailed flow dynamics of the AVLG system.

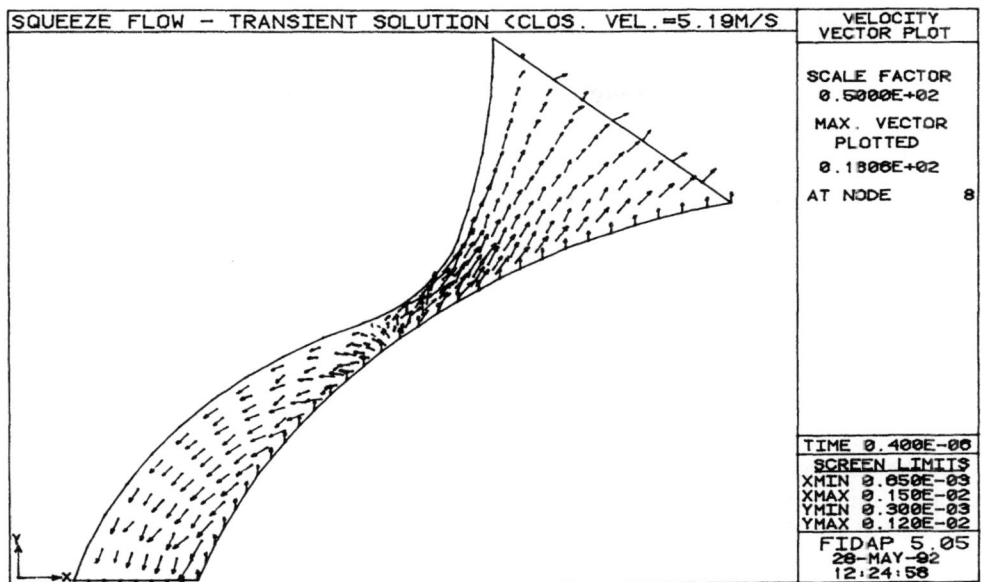

Figure 18. Computational solution for squeeze flow between valve leaflet and housing at final closing.

Most of the in vitro vascular dynamic models reported in the literature used rigid conduits, either using one-to-one or enlarged scales[21-26]. However, the wall motions may significantly alter the flow field inside blood vessels[27-29].

Recently, several studies using elastic transparent flow models have been reported[30-34]. We have developed an elastic transparent flow model to mimic the venous anastomosis of AVLG based on room temperature vulcanizing (RTV) silicone rubber casts obtained from femoral-to-femoral AVLG implanted in dogs (Noon and Hwang, 1987). Of particular interest were the hemodynamic characteristics of three AVLG flow models with different area reductions (AR), namely light (l), moderate (m) and significant (s), in the venous anastomotic regions. The consequences of pulsatility and wall motions upon flow features in the three flow models were compared under identical input flow conditions produced in an in vitro flow loop.

The Animal Model and the Flow Model Studies

Expanded polytetrafluoroethylene (ePTFE) loop grafts were implanted bilaterally between the femoral artery and the femoral vein in mongrel dogs (of average weight 30kg) to mimic the AVLG system created between the brachial artery and the cephalic vein in patients. The standard AVLG configuration used in this study was a 6 mm diameter, 25 cm long loop graft with both ends bevelled to 45° (anastomotic angle). The dogs were reoperated upon at different times for hemodynamic and pathophysiological studies. These detailed surgical procedures and in vivo studies have been reported elsewhere[35,36].

To acquire the luminal geometry of the AVLG system in situ, RTV silicone rubber was injected under physiological pressure to fill the AVLG system up to a distance of approximately 5 cm measured from both the proximal and distal ends of the anastomosis. The cast was obtained after a 24 h curing period in situ.

The selected RTV casts were used to fabricate elastic transparent Silastic flow models. On average, five Silastic models were made from each RTV cast. Due to the model fabrication tolerance, one was selected as the flow model, based on the dynamic compliance. It is believed that the matched compliance should help to duplicate the in vivo flow pattern inside the in vitro flow model. The flow model fabrication techniques have been presented elsewhere[37](Shu et al, 1987).

The in vitro AVLG system was constructed by connecting the arterial anastomosis model and the venous anastomosis model with a 6 mm PTFE GorTex loop graft (Gore Inc, Flagstaff, AZ, USA).

The experimental flow loop consisted of a constant head storage tank which provided inlet fluid to the model; a pulsatile wave duplicator; and a preload system which consisted of an adjustable linear resistor and an adjustable compliance unit to tune the arterial inlet flow waveforms. The box containing the flow model was installed immediately downstream of the compliance unit. An afterload system was provided at the distal artery and the distal vein to tune the arterial and venous flow waveforms.

Before carrying out the detailed LDA flow field survey, flow visualization was made to obtain general information about the flow field. Pulsatile hydrogen bubbles were generated at the upstream end of the proximal artery using a 6V DC square wave. The pulses of the hydrogen bubbles were synchronized with the pulse frequency of the AVLG system. Four platinum wires (0.08 mm) were stretched across the diameter of the model inlet as the cathode. The cathode wires were arranged in parallel with 0.5 mm spacings between.

A 15mW He-Ne laser with a wavelength = 632.8 nm (visible red) was used as the light source. The laser beam was enlarged by a microscope lens (x20). A 50μm pinhole located at the front side of the lens was used to filter the laser light. The filtered laser beam was then projected on a cylindrical lens. A light plane generated by the cylindrical lens was used to illuminate the flow field.

To facilitate laser Doppler anemometer (LDA) measurements, the flow model was mounted on a motorized two-dimensional transverse table which had an accuracy of 16 μm per dial division in both X and Y directions. The LDA used in the experiment was a TSI (Thermal Systems, Inc., St. Paul, MN 55164) Model 1090 tracker system with Bragg cells. Using an achromat focusing lens of 120 mm focal length, the elliptical measuring volume was calculated to have major and minor axes of 0.4 mm and 0.75 mm, respectively. The elliptical measuring volume consists of 64 fringes, spaced 1.38 μm apart.

Velocity surveys were made at seven different sections in the venous and arterial host vessels in the flow models. The sections were spaced 7 mm (approximately one diameter) apart.

At each cross-section surveyed, the starting position at the wall was manually adjusted to the upper side of the arterial wall and at the lower side of the venous wall. The "at the wall' locations were determined without flow in the loop. Near the wall, the first three sampling points along the diameter were spaced 100 μm from each other, and the fourth sampling point was 400 μm from the third. For all the remaining points along the same diameter until near the opposite wall 800 μm spacings were maintained. At this time, the same measuring procedure was repeated from the opposite wall. Certain overlap of the positions surveyed was necessary to check the consistency of the measurements.

To assess the value of wall shear rate, the distensible host vessel was instrumented with ultrasonic dimension gauges to measure changes in the segment diameter[35]. Two crystals were glued to the external surface on the opposite sides of the flow model and aligned along the diameter. The velocity signals, pressure signals and the ultrasonic dimension gauge signals were simultaneously recorded on a magnetic analogue tape using a FM record (Hewlett-Packard Model 3968A, band-width 0-10 kHz) at a tape speed of 19.05 mm s^{-1}. The signals were recorded for a minimum of 2 min at each sampling point and later played back for digitization using a PDP 11/23 digital computer (16 bits per word) at a sampling rate of 2500 points s^{-1}. According to the Nyquist criterion for signal processing, this would guarantee a frequency response of 1000 Hz in the flow measurement. The digitized data were processed on a Micro Vax 3.4 computer. Wall movement as a function of time was calculated from the variation of the recorded ultrasonic signal.

Since the voltages of the velocity signals were increased by a frequency shifter to distinguish positive from negative flow, the velocity data needed to be corrected before proceeding with data analysis. These velocity signals resulting from the function of a frequency shifter were corrected by a constant value recorded during the data acquisition period. The constant value varied with the location of the data acquisition. Assuming quasi-steady conditions, the velocity obtained at each point could be expressed as:

$$U_j(t) = \frac{1}{N}\sum_{n=1}^{N} u_j(t+nT) - \frac{1}{N}\sum_{n=1}^{N} [u_j(t+nT) - C_j] \quad (11)$$

$$j = 1, 2, \ldots, k \quad (12)$$

where j indicates the surveyed point at each cross-section, k is the number of measured points at each section, C_j is the constant frequency shift value of the jth point surveyed, T is the period of a cycle and N is the number of ensembling cycles. The pressure waveform at the proximal artery was used as a time trigger to identify the beginning of a cycle. The digitized velocity signals were 'ensemble averaged' for 30 cycles of continuous pulses at each measuring point. The wall shear stress distribution was calculated from the slope of the measured velocity profiles 'at the wall.' Since the wall motion at each cross-section surveyed was recorded simultaneously with the velocity profiles, the distance $dr_i(t)$ between the vessel wall and the nearest point to the wall is a function of time. Thus we have:

$$\tau_i(t) = \frac{\mu}{N}\sum_{n=1}^{N}\left[\frac{du(t+nT)}{dr_i(t)}\right]_{y=R} \quad (13)$$

where i = 0, 1, ..., 6, and μ is the viscosity of the testing fluid.

Typical LDA velocity profiles and wall shear stress distribution are presented in Chapter 385.

The Scanning LDA System

Blood flow velocity in arteries is a function of both space and time due to the pulsatile nature of the arterial system. The velocity of any one point in the general flow field can be measured by means of a conventional LDA. However, to obtain the velocity profiles, a precision transversal frame is usually required to carry either the whole optics or the flow model from one point to the next. In such case, the velocity at any particular phase of pulsatile flow field can only be obtained after the ensumbled average of significant numbers of cycles at each point is made. The dynamic velocity profile obtained from the different cycles is meaningful only if the quasisteady flow conditions could be established[39]. For the measurements with a conventional LDA, a relative mechanical frame between the LDA optics and the flow field must be provided to obtain space information of the flow field.

Mechanical scanning LDA techniques have been reported[40,41]. Each of which involves certain moving parts in the optical system. There is always the concern in mechanical vibrations. In the meantime, the mechanical scanning frequency is rather limited as compared with that of the pulsatile flow. Thus, both the quantity and quality of the measured data are restricted.

A primary acousto-optic scanning LDA (AOS-LDA) system[42,43] was developed in this laboratory for the measurement of pulsatile flow in tubes. The displacement of scanning laser measuring point was controlled by an electrical signal. The "instantaneous" velocity profiles were directly obtained from an intelligent flow analyser (IFA 550 Model, Thermal System Inc., St. Paul, MN 55164). There is no moving parts in the AOS-LDA system. Theoretically, there is no limitation of scanning frequency due to the large range of central working frequency of the Bragg cells (BC typically 40-80MHz) and the very small time lag in BC (less than 1 microsecond). Practically, the scan rate is limited by the flux of the scattering particles suspended in the fluid.

BC is an acousto-optic element which causes the laser beam to deflect when a frequency shift is applied upon its transducer by means of an electric driver. The angular displacement α may be approximated by the following equation[1]:

$$\alpha - (\lambda/V_s)f_B \qquad (14)$$

where λ is the laser wavelength, V_s is the speed of sound in the crystal of the BC, and f_B is the frequency shift. In a LDA system, both λ and V_s are constant. The value of α may be changed by variation of the Bragg frequency shift f_B.

Figure 19 shows the three different types of AOS-LDA transmitting systems with different characteristics as listed in Table 2. We chose Y-direction scanning model to measure the pulsatile velocity profiles in tube flows.

Figure 20 shows a practical AOS-LDA system developed for tube flow velocity profile measurements. The triangular waveform from a function generator passes through one of the voltage controlled drivers (VCD) and then formed a modulated frequency driving signal f_1 to drive the BC_1 (scanner$_1$). The frequency tracker loop included a frequency mixer, a frequency discriminator and another VCD to provide a second driving signal with a frequency f_2 for the BC_2. The frequency difference between f_1 and f_2 always has a constant value f_0, no matter how the f_1 changed under scanning. The two scanners, acting simultaneously and, changed the angles of the incident beams inversely respective to each other. The measuring volume moves, due to the variation of beam angles, to scan in the direction along the diameter of the field repeatedly.

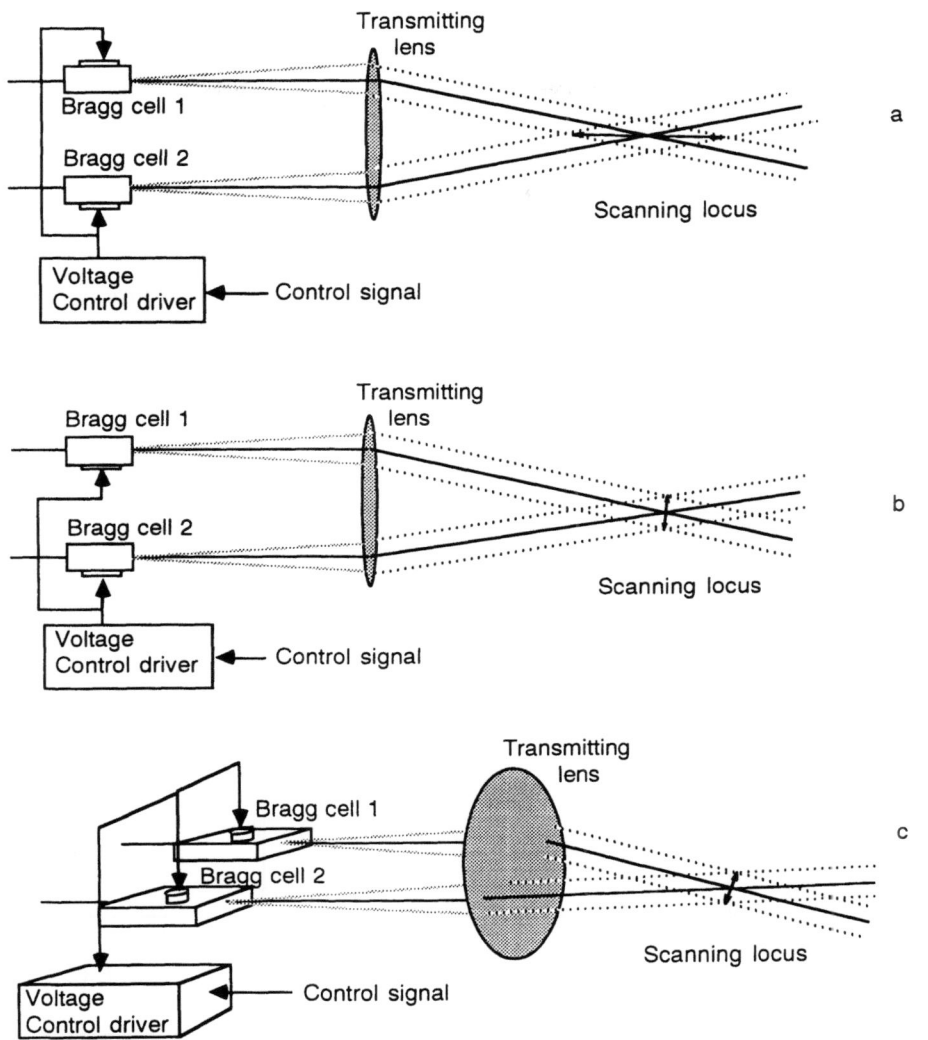

Figure 19. Acousto-optic LDA scanning systems (a) y-direction scanning model, (b) x-direction scanning model, (c) z-direction scanning model.

The optical wedges served to collimate the two beams parallel to one another at the central frequency of the Bragg cells. Compared with the single signal controlled scanning modes as shown in Figure 19, the synchronized tracking model can provide a fixed optical frequency shift f_0 to distinguish the flow direction. The bandwidth of the Doppler signal can be adjusted when the flow velocity varied in a large range[16], so that the low velocity near the boundary can be measured together with the high velocities in the midstream of the conduit. Instead of a pin hole in front of the photodetector of a conventional LDA system, a narrow slit was used as an optical aperture in the AOS-LDA system.

Table 2. Characteristics of three AOS-LDA models.

scanning model	x-direction	y-direction	z-direction
scanning locus	straight line	straight line	straight line
fixed frequency shift	available	available	available
measuring volume	constant	depends on scan distance	constant
forward and back scattering real image	good	depends on depth of field	good
distortion of image in circular conduits	no	no	yes
measure type	c, e, f, g	a, b, d, f, h	b, d, f, g,

a. instantaneous velocity profile of fluid flow in a rectangular cross section
b. instantaneous velocity profile of fluid flow in a circular cross section
c. instantaneous velocity profile in the main flow direction
d. shear rate perpendicular to the main flow direction
e. velocity gradient in the main flow direction
f. flow structure and scale
g. material surface motion
h. diameter of transparent conduit

A preliminary flow experiment using AOS-LDA is demonstrated by carrying out velocity profile measurement in a glass tube of 0.25 inches internal diameter and 8 inches in length. The flow model was placed inside a transparent box filled with the same testing fluid (distilled water). The fluid flow rate was adjustable within a range of 30-500 ml/min. An Innova 70 argon-ion laser (Coherent Laser Products Division, Palo Alto, CA 94303) was used as a light source and 6 μm polystyrene latex particles were seeded into the fluid as the light scattering particles. Between the laser and the optics mentioned above, a single transmitting fiberoptic cable (TSI Model 9265) was placed. The output laser power from the cable was about 30 mW. The signal processor in the system was an intelligent flow analyser (Model IFA 550, TSI). The scanning laser probe passed alternately through the tubing along its diameter within the 10mm scan distance. The results of the experiments were recorded on a scope with hard copies made by a printer. The traces shown in Figure 21 are the typical instantaneous velocity profiles and the triangle scanning signal which is proportional to the displacement in the scanning range.

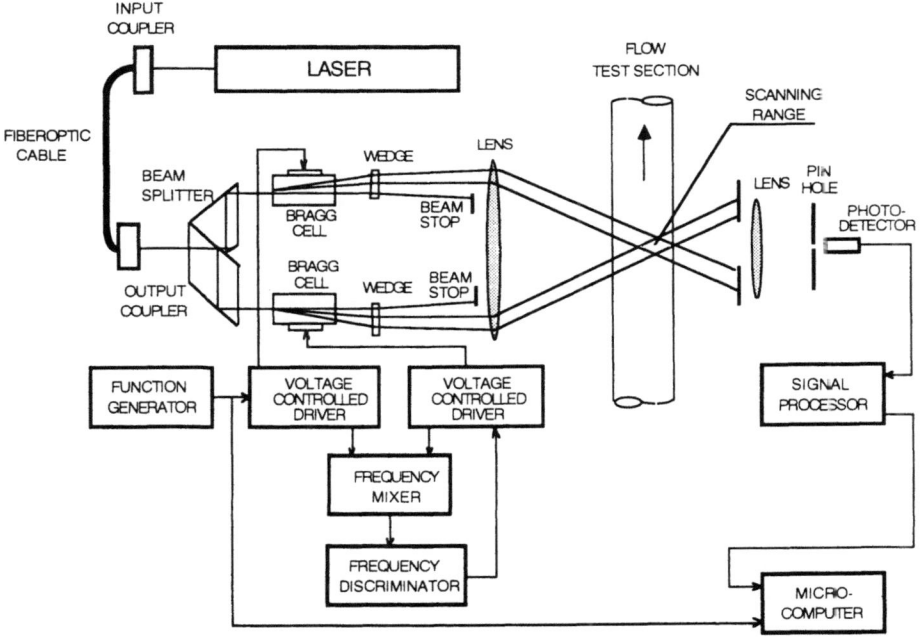

Figure 20. Schematics of AOS-LDA system for the tube flow measurement.

Utilizing AOS-LDA, we were able to measure the instantaneous velocity profiles in a pulsatile tube flow. The diameter of the conduit could also be measured simultaneously. The pulsatile flow in a transparent tube with moving wall may also be measured with the AOS-LDA technology. The current AOS-LDA system still needs more development to improve the signal-to-noise ratio so that the dimension of the flow channel (particularly for the inside diameter of the conduit with flexible walls) could be defined more precisely.

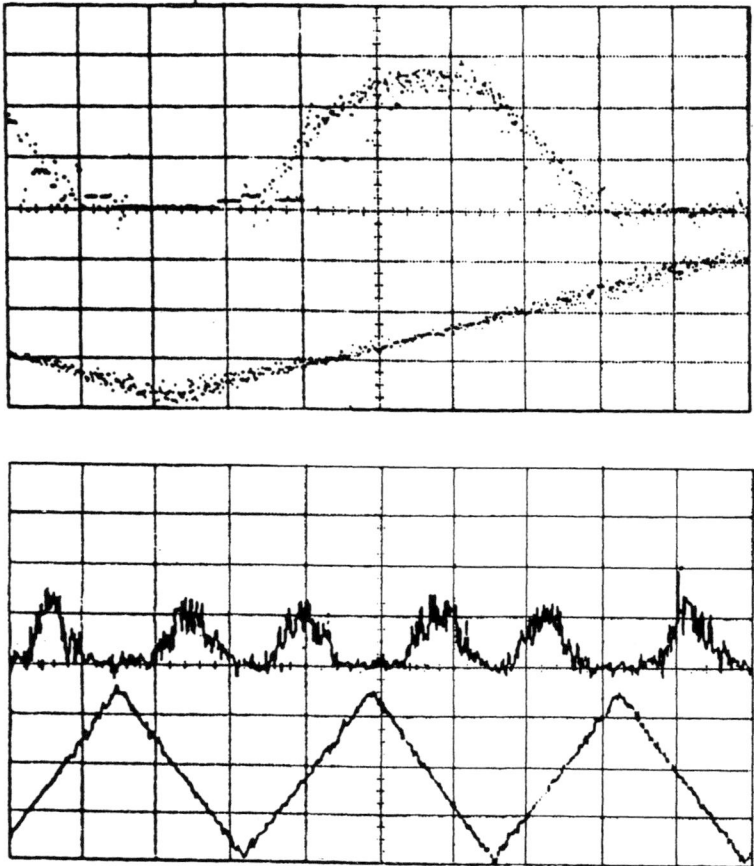

Figure 21. Typical AOS-LDA velocity profiles.

The Two-Dimensional OEHF-LDA

Accuracy in measurement of shear stress and Reynolds shear stress in the flow field has been a major concern in vascular dynamics research. In a pulsatile cardiovascular flow field, the fluid shear stress and the Reynolds shear stress (RSS) are both variables of space and time. Applying a conventional LDA system, the measurement of RSS can only be accomplished by using the two component laser doppler anemometer.

A two component OEHF-LDA system (see Section 1-3) was designed[44], in which the signals from the two components are discriminated by means of different intermediate frequencies of the two closed loops containing both the optical and the electric signals. The signals of the two velocity components are picked up by a single detector and separated by the two narrow-band filters. The number of optical elements and electronics are thus reduced significantly in comparison with the conventional two-component LDA system, which employs either a two-color argon laser, or the technique of polarization discriminations.

The configuration of the two-component OEHF-LDA system is shown in Figure 22. It consists of three basic parts: (a) laser transmitting optics and detector, (b) signal processor, and (c) microprocessor control and data acquisition unit.

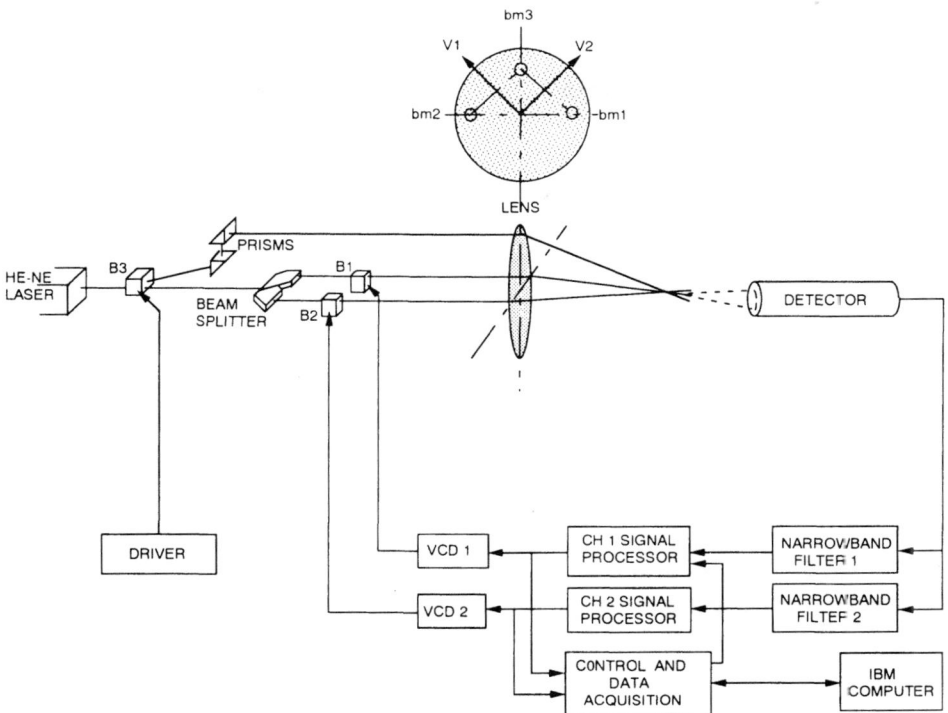

Figure 22. Schematics of the two-component OEHF-LDA system.

A two-component LDA can be set up by using a variety of optical arrangements[45]. In the two-component OEHF-LDA system, a three-beam scheme is adopted. Figure 22 shows that the beam from a 25mW He-Ne laser first passes through a Bragg cell (B3) which is driven by an oscillator with a constant frequency of 60MHz. The first order diffracted is chosen and refracted into the transmitting optics by a pair of prisms. The undiffracted part of the incident beam is divided into two parallel beams with equal intensity by a beam splitter. Another two Bragg cells, B1 and B2, driven by VCD1 and VCD2 respectively, are inserted to introduce changeable differential shifts. The three shifted beams (BM1, BM2 and BM3, in Figure 23) are focused on a common point by the transmitting lens, to form the measuring volume. The position of BM1, BM2, and BM3, on the transmitting lens is shown at the top of Figure 22. BM1 and BM3 are used to measure the velocity component V_1 which is in the plane perpendicular to the optical axis of the system; BM2 and BM3 measure the velocity component V_2 which is perpendicular to V_1.

An ordinary photo-diode, combined with a pre-amplifier is used as the photo-detector of the two-component OEHF-LDA system. In most cases, the photo-diode can work satisfactorily. Only when the scattering light is extremely weak, such as in the case of backscattering mode, a photomultiplier tube (PMT) is necessary.

The configuration of the two-component OEHF-LDA consists essentially two one-component OEHF-LDA systems, each has its own intermediate frequency. Each closed-loop satisfies Equation (1) individually.
where

$$f_{10} - f_1 - f_3 + f_{D1} \quad (15)$$

$$f_{20} - f_2 - f_3 + f_{D2} \quad (16)$$

$$f_{D1} - \frac{2\sin(\theta/2)}{\lambda} V_1 \quad (17)$$

$$f_{D2} - \frac{2\sin(\theta/2)}{\lambda} V_2 \quad (18)$$

f_{10}, f_{20} are the intermediate frequencies of the two closed-loops, respectively.
f_1, f_2 are the changeable frequency shifts produced by B1 and B2, respectively, f_3 is the fixed shift produced by B3, and θ is the crossing angle between BM1 and BM3, or BM2 and BM3.

Because the three incident beams all have the same polarization; the combination of BM1 and BM2 forms also a set of interferometric plane. It produces a scattering light signal when a particle passes through the measuring volume with a velocity vector, V. The frequency of this scattering light signal, f' is determined by

$$f' - f_2 - f_1 + f_{D3} \quad (19)$$

and

$$f_{D3} - \frac{2\sin(\varphi/2)}{\lambda} V_3 \quad (20)$$

where
φ is the cross angle between BM1 and BM2, V_3 is the component of V, on the direction joining BM1 and BM2 (see Figure 23)
Subtracting Equation (15) from (16), we obtain:

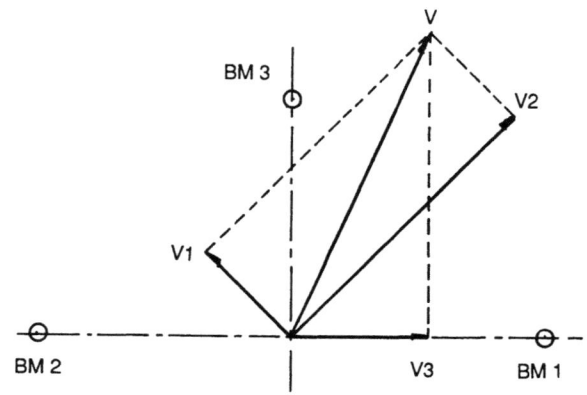

Figure 23. Relationship between beam-configuration and velocity components.

$$f_{20} - f_{10} - f_2 - f_1 + (f_{D2} - f_{D1}) \tag{21}$$

or

$$f_{20} - f_{10} - f_2 - f_1 + \frac{2\sin(\theta/2)}{\lambda}(V_2 - V_1) \tag{22}$$

From Figure 23 and the geometry pattern of three incident beams, the following relationships can be demonstrated

$$V_3 - \frac{1}{\sqrt{2}}(V_2 - V_1) \tag{23}$$

$$\sin(\varphi/2) - \sqrt{2}\sin(\theta/2) \tag{24}$$

Thus, Equation (19) can be rewritten as

$$f' - f_2 - f_1 + \frac{2\sin(\varphi/2)}{\lambda}V_3 - f_2 - f_1 + \frac{2\sin(\theta/2)}{\lambda}(V_2 - V_1) \tag{25}$$

Comparing with Equation (21), we have

$$f' - f_{20} - f_{10} \tag{26}$$

This indicates that the spectrum of the signal detected by photo-detector not only contains two intermediate frequencies, f_{10} and f_{20}, but also includes their difference, f_{20}-f_{10}. The spectrum distribution is shown in Figure 24. The two signals with center frequencies of f_{10} and f_{20}, respectively, should be separated from the detected signals before they are processed. Because f_{10} and f_{20} are kept at nearly constant during the measurement, signal separation can be realized easily by using two narrow-band filters with different central frequencies. In the design of the electronics, we selected f_{10} and f_{20} to equal 150 kHz and 500 kHz, respectively. The separated signal spectra are shown in Figure 24 and 24.

The separated signals are fed to a two-channel signal processor, each channel contains a frequency discriminator which has the same function as that used in the one-component OEHF system (see Section 1-3-1). The two-channel output signal voltages of the processor are fed to VCD1 and VCD2 respectively, to control the frequency shifts, f_1 and f_2, forming two closed-loops with hybrid feedback. In order to avoid the interference of the signal drop-out, a protection circuit is built in each channel of the signal processor.

At the beginning of the measurement, it is necessary to search the "tracking point" of each closed-loop in order to enter the loop in a locked-in state. Because there are two closed-loops working simultaneously, the recognition of a correct tracking point for each loop becomes complex, sometimes may even be difficult. To solve the problem, a microprocessor is adopted to search tracking points and control the signal processor in the two-component LDA system.

To operate, we first set the VCD2 at the maximum frequency. The micro-

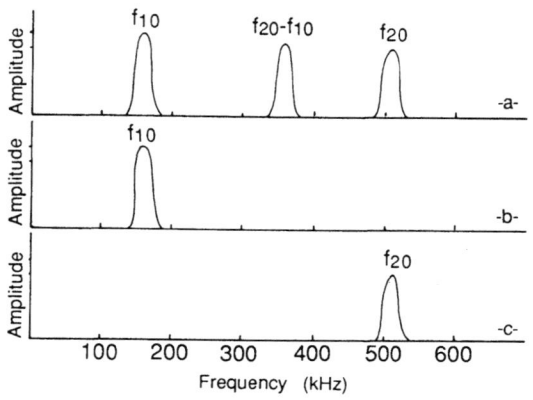

Figure 24. Photodetected signal spectrum of the two-component OEHF-LDA.

processor gives VCD1 a scanning voltage through a D/A converter. In the meantime, the microprocessor supervises whether or not the scattering light signal is in resonance with the filter 1. In the scanning period, the resonance will occur twice, the correct tracking point is recognized by the microprocessor. The digit corresponding to the tracking point is kept in the memory. Then VCD1 is set at the maximum frequency, the same operation is applied to VCD2 and the tracking point of another loop is obtained. Afterwards, the microprocessor applies the voltages corresponding to tracking points, VCD1 and VCD2, respectively, and makes these two loops close in sequence. When the microprocessor ensures that the two closed-loops are locked in, it informs the operator and begins to wait for the command to start the data acquisition unit. The whole operation procedure can be completed in a rather short period for quick response.

The data acquisition unit contains four high speed (12bit A/D) converters, which can be programmed either to operate independently, or to form in groups to convert one or two channel analog signal(s). During measurements, blocks of 32 kbytes RAM are utilized to store raw data. The data transfer from A/D converters to the memory is made using the direct memory access (DMA). A general purpose interface bus (GPIB) is provided in the data acquisition unit to send the measurement data to an IBM computer to obtain the mean velocity, turbulence intensity and power spectrum of each component. The Reynolds stress can also be computed.

LASER OPTICAL SYSTEM TO MONITOR VASCULAR WALL MOTIONS

Model study of arterial blood dynamics frequently involves flow models with flexible walls. The motion of the fluid inside the vascular conduit is coupled with the wall motions monitoring wall motions is especially important in the study of vascular graft dynamics.

Several techniques of measuring the vessel wall motions have been reported previously. These include the cantilever device developed by Patel et al[46], the laser Doppler velocity technique[32], the ultrasonic methods[47], and the many optic devices[48]

(Vio, 1980). While each of these previously reported techniques has its unique applications and advantages in measurement, they all practically require that the measured object (vessel) be placed at a relatively fixed distance or position with respect to the transducer(s). This is usually accomplished by artificially tying down the two ends of the object vessel to a certain fixed reference frame. By doing so, one may likely to impose restrictions on the natural wall motions, and thus create artifacts to the motion of the fluid flow inside the vessel. It is thus desirable to design an optical system which can provide precise measurement of free standing vessel wall motions.

Optic wall motion monitoring system (OMS) developed[49] in this laboratory involves basically a 30 milli Watt He-Ne laser and an optic arrangement as shown in Figure 25. The red laser beam first passes through a pin hole (P) pierced in a rotating disk to reach a cylindrical lens (CL), which converts the incident laser beam into a thin sheet of laser light. A parallel light sheet is formed after the laser light passes through an aperture (A), followed by the downstream spherical lens (L_1). The object vessel may be placed at any convenient location within the domain of the parallel light to cast a shadow region behind it. After the parallel light sheet passes through lenses L_2 and L_3, a well defined shadow of the object with clear parallel boundaries is formed. This shadow is then projected on the linear photodiode array (LPA)[50]. The length of three different segments, AB, BC, and CD can be monitored simultaneously from the LPA. The length of segment BC corresponds to the vessel diameter. As the diameter of the vessel changes, the length of segment BC takes different values. Monitoring these values at time intervals of Δt will provide information about the change in vessel diameter as it expands or contracts with time. If the circular cross section of the vessel can be assumed to be uniform in structure, and that the intravascular pressure is kept higher than that of the extravascular, the above diameter measurement can be interpreted directly into vessel wall motions.

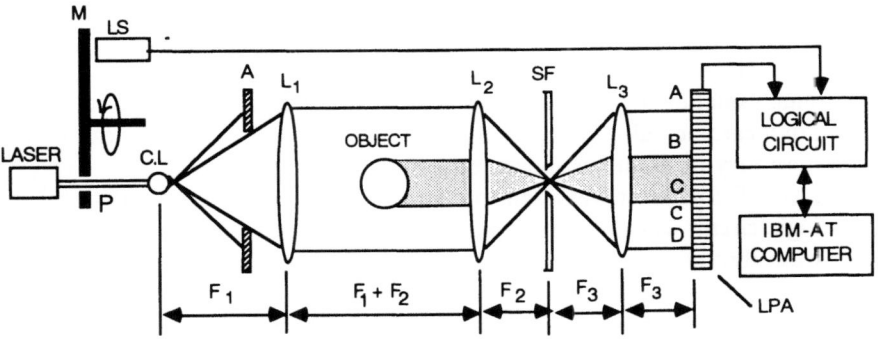

Figure 25. Laser optical system for monitoring vessel wall motions.

In the above optical system, the use of the aperture (A), the lenses L_2 and L_3, along with the special filter (SF) allowed great improvement in the quality of the parallel beam as well as the signal-to-noise ratio. The high resolution achieved through the use of laser can not possibly be matched had an ordinary white light been used as a light source. Since the signal output from the LPA is proportional to the number and not the position of the diodes activated, it is evident therefore, that the measurement can not be affected by the position of the object as long as it remains exposed to the incident laser light sheet.

Laser beam modulation by means of the rotating disk is crucial in eliminating the effect of vessel movement over the diameter measurement. When an unmodulated laser sheet is used to measure the fixed diameter of a moving vessel within a time interval Δt, its projected shadow AB will shift to a new position A'B' (Figure 26). The segment AA' will no longer be in the dark region and consequently, its diodes are now been activated. In the meantime, the segment BB' of the LPA had already been exposed to the laser before the vessel changed position. As a result, only segment A'B remains in the dark region during the entire Δt would reflect the length of A'B instead of the vessel diameter represented by AB. The optical modulator has proven to be very effective in eliminating this source of error.

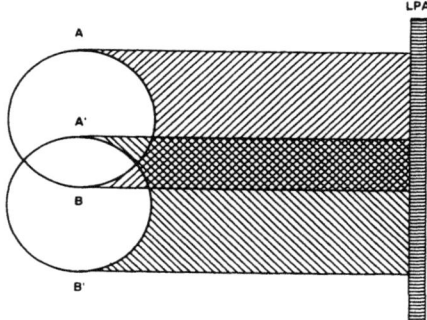

Figure 26. Diagram illustrating the influence of vessel imovements on diameter measurement.

The optical modular (M) consists of a disk of radius R rotating at a constant angular frequency. A pin hole of small radius r is pierced along the radius R and alined with the laser beam. The Laser beam is allowed to pass through the pinhole every time the disk decrees a complete revolution. This makes the measurement time interval Δt to be the inverse of the rotation frequency of the disk. In the present experiment, the laser beam is emitted for only a short duration of $(r/\pi R * \Delta t)$ second. Since r is much smaller than R, the exposition time is extremely short compared with the period of a cardiac pulse cycle. Consequently, the vessel movement can no longer effect the accuracy of diameter measurement of the moving vessel.

An IBM AT computer was used to control and process the output signal of the LDA. A logic circuit was also employed to synchronize the data acquisition with the laser emission through a light activated sensor (LS) mounted along with the optical modulator.

Preliminary calibration of the system was made by recording the diameter measurement of a rigid cylinder moving up and down across the laser sheet. Measurements have been taken with and without the optical modulator been incorporated with the measuring apparatus. Results are shown in Table 3.

Table 3. Diameter measurements of rigid cylinder taken with and without the optical modulator.

Condition	Speed 0	19mm/s	27mm/s
without modulator	2.498mm	2.149mm	2.024mm
with modulator	2.498mm	2.498mm	2.498mm

A uniform silicon rubber tube with diameter of 6 mm and wall thickness of 0.5 mm was also tested. ΔT in this experiment was chosen to be 0.01sec and the ratio of $r/\pi R$ is 0.04.

Pulsatile pressure waves at 60 beats per minute were provided to propel the testing fluid through the conduit. In Figure 27 the pressure wave in the tube is shown on the upper parts. The lower curve in represents the diameter change of the tube. It is obvious that in the case of the silicon rubber tube, the diameter change was more closely related to the pressure wave than was the case with the graft. This suggests that higher resolution photo diode array should be used for more rigid object vessels. A sudden rise in the diameter size has happened during the pressure fall at similar instances.

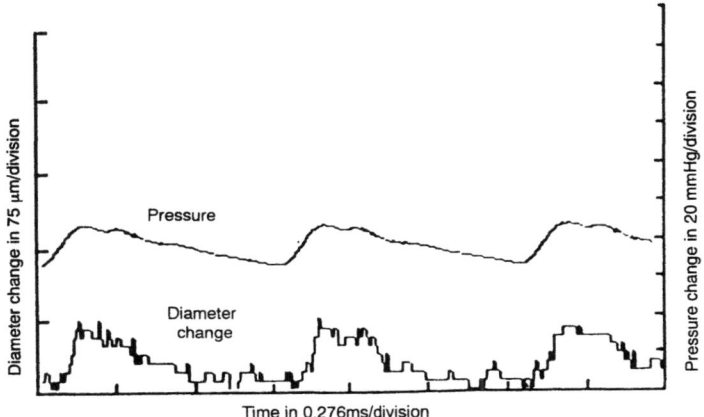

Figure 27. Typical signals of vessel diameter change vs pressure waves.

ACKNOWLEDGEMENTS

This work is partially supported by PHS/NIH Grant RR05511. The Chinese Natural Science Foundation provide support for SKW to visit Memphis State University where part of this is performed.

REFERENCES

1. Gordon EI: A review of acousto-optical deflection and modulation devices, Proc. IEEE, 54:1391-401, 1966.
2. Wang SK, Guo XM, Shi SX: Optic-electro-modulation feedback laser Doppler anemometer, Chinese patent No. 85-108397, 1985.
3. Guo XM, Wang SK, Liu CW, Hwang NHC: A new LDA system utilizing the optic-electro-hybrid feedback technique, J. of Physics E., Meas. Sci. Technol., 1:265-271, 1990.
4. Hwang NHC, Lu PC, Sallam AM: Measurements of turbulence in aortic valve prostheses, In: Prosthetic Heart Valves (Ed., Yoganathan AP), AAMI, CIT Press, Pasadena, California, 91-120, 1979.
5. Bruss KH, Reul H, Van Gilse J: Pressure drop and velocity fields at four mechanical heart valve prostheses, Life Support Systems, 1:3-22, 1983.

6. Woo YR, Yoganathan AP: In vitro pulsatile flow velocity and turbulent shear stress measurements in the vicinity of mechanical aortic heart valve prostheses, Life Support Systems, 3:283-312, 1985.
7. Schwarz AC, Tiederman WG, Phillips WM: Influence of cardiac flow rate on turbulent shear stress from a prosthetic heart valve, J. Biomech. Eng. ASME, 111:123-8, 1988.
8. Klepetko W, Moritz A, Khunl-Brady G, Schreiner W, Schlick W, Mlczoch J, Kronik G, Wolner E: Implantation of the Duromedic bileaflet cardiac prostheses in 400 patients, Ann. Thoraxic Surgery, 44:308-309, Sept. 1987
9. Klepetko W, Mortiz A: Leaflet fracture in Edwards-Duromedic leaflet valves, J. Thorac. and CV Surg., 97:90-94, 1989.
10. Quijano RC: Edwards-Duromedic dysfunctional analysis, Proceedings of Cardiostim: 6th International Congress, Monte Carlo, Monaco, 1988.
11. Kafesjian R, Wieting DW, Ely J, Chalhine G, Frederick G, Watson R: Characterization of the cavitation potential of Pyrolitic carbon, Proceedings of International Symposium Heart Valve Diseases, London, UK, June 12-16, 1989.
12. Bokros JC, LaGrange LD, Schoen FJ: Control of structure of carbon for use in bioengineering, In: Chemistry and Physics of Carbon (Ed., Walker PL), Dekker, New York, 103-171, 1972.
13. Tokuno T, Dube CM, Walker WF: Cavitation near moving prosthetic surfaces, Artif. Organs (Supp. II), 166-168, 1978.
14. Lamson CL, Stinebring DR, Deutsch S, Tarbell JM: Real-time in vitro observation of cavitation in a prosthetic heart valve, Trans. ASAIO, (in press, 1991).
15. Guo X, Xu CC, Hwang NHC: The closing velocity of Baxter-Duromedic heart valve prosthesis, Trans. Am. Soc. Artif. Intern. Organs, 36:529-532, 1990.
16. Adrian RJ: Laser velocimetry, In: Fluid Mechanics Measurement (Ed., Goldstein RJ), Hemisphere Publishing Corporation, Berlin, 155-244, 1983.
17. Feldman HJ et al.: Noninvasive in vivo and in vitro study of the St. Jude Medical mitral valve prosthesis, Am. J. Cardiol., 49:1101-09, 1982.
18. Prahbu A, Hwang NHC: Dynamic analysis of flutter in disc type mechanical heart valve prostheses, J. Biomech., 21(7):585-90, 1988.
19. Reif TH, Schulte TJ, Hwang NHC: Estimation of the rotational undamped natural frequency of bileaflet cardiac valve prostheses, J. Biomech. Engr., ASME, 112:327-32, 1990.
20. Hele-Shaw HJS: Investigation of the nature of the surface resistance of water and of streamline motion under certain experimental conditions, Trans. Inst. Naval Architects, 40, 1898.
21. Ku DN, Giddens DP: Laser Doppler anemometer measurements of pulsatile flow in a model carotid bifurcation, J. Biomech., 20:407-21, 1987.
22. Khodadadi JM, Valchos NS, Liepsch D, Moravec S: LDA measurements and numerical prediction of pulsatile laminar flow in a plane 90-degree bifurcation, J. Biomech. Eng., 110:129-36, 1988.
23. Sallam AM, Hwang NHC: Human red blood cell hemolysis in a turbulent shear flow - Contribution of Reynolds shear stresses, J. Biorheol., 21:783-97, 1985.
24. Lutz RJ, Hsu L, Menawat A, Zrubek J, Edwards K: Comparison of steady and pulsatile flow in a double branching arterial model, J. Biomech., 16:753-66, 1983.
25. Einav S, Avidor J, Vidne B: Haemodynamics of coronary artery-saphenous vein bypass, J. Biomed. Eng., 7:305-9, 1985.

26. Abdallah SA, Hwang NHC: Arterial stenosis murmurs: An analysis of pressure and flow fields, J. Acoust. Soc. Am., 83:318-344, 1988.
27. Goldsmith HL, Karino T: Mechanically induced thromboemboli, In: Quantitative Cardiovascular Studies (Eds., Hwang NHC, Gross DR, Patel DJ), University Park Press, Baltimore, Maryland, 1978.
28. Pei H, Xi BS, Hwang NHC: Wall shear stress distribution in a model human aortic arch: Assessment by an electrochemical technique, J. Biomech., 18:645-56, 1985.
29. Karino T, Motomiya M: Flow patterns in the human carotid artery bifurcation, Stroke, 15:50, 1984.
30. Liepsch D, Moravec S: Pulsatile flow on non-Newtonian fluids in distensible models of human artery, J. Biorheol., 21:571-86, 1984.
31. Liepsch D: Flow in tubes and arteries, a comparison, J. Biorheol., 23:395-433, 1986.
32. Deters OJ, Bargeron CB, Mark FF, Friedman MH: Measurement of wall motion and wall shear stress in a compliant arterial cast, J. Biomech. Eng., 108:355-358, 1986.
33. Shu MCS, Noon GP, Hwang NHC: Phasic flow patterns in a hemodialysis venous anastomosis, J. Biorheol., 24:711-722, 1987.
34. Shu MCS, Noon GP, Hwang NHC: Flow profiles and wall shear stress distribution at a hemodialysis venous anastomosis: preliminary study, J. Biorheol., 24:723-35, 1987.
35. Zamora JL, Gao ZR, Weilbaecher G, Navarro L, Ives CL, Hita C, Noon GP: Hemodynamic and morphologic feature of arteriovenous angioaccess loop grafts, Proc. ASAIO, Atlanta, Georgia, USA, May, 1985.
36. Noon GP, Hwang NHC: Hemorheologic contribution to thrombosis, Devices and Technology Branch Contractors Meeting, NHLBI-NIH, December, 1987.
37. Shu MCS: Hemodynamics study of angioaccess venous anastomoses, Ph.D. Thesis, University of Houston, Houston, Texas, 1988.
38. Gentile BJ, Gross DR, Chuong CTJ, Hwang NHC: Segmental volume distensibility of the canine thoracic aorta in vivo, Cardiovasc. Res., 22:385-9, 1988.
39. Wang LC, Guo GX, Tu R, Hwang NHC: Graft compliance and anastomotic flow patterns, Trans. Am. Soc. Artif. Intern. Organs, XXXVI:1-5, 1990.
40. Gartrell LR, Rhodes DB: A scanning Laser-velocimeter technique for measuring two-dimensional wake-vortex velocity distribution, NASA Technical paper, NASA: TP-1661, 43, 1980.
41. Hino M, Nadaoka K, Kobayashi T, Hironaga K, Muramoto T: Flow structure measurement by beam scan type LDA, Fluid Dyn. Res., 1:177-190, 1987.
42. Wang SK et al: Acousto-optical scanning laser Doppler anemometry, Chinese patent No. 88-2165836, 1988.
43. Guo GX, Li W, Hwang NHC: Measurement of tube flow velocity profiles utilizing acousto-optic scanning LDA, ASME Winter Annual Meeting, Dallas, Texas, USA, Nov. 25-30, 1990.
44. Li EB, Wang SK: Two-component LDA system with optic-electro-hybrid feedback, Proc. Intern. Conf. on Fluid Dynamics Measurement and Its Applications, Beijing, China, Oct. 25-27, 1989.
45. Oldengarm J: Two-dimensional laser Doppler velocimetry, Proceedings of the LDA-Symposium, Copenhagen, 1975.

46. Patel DJ, Vaishnav RN: Mechanical properties of arteries, In: Cardiovascular Flow Dynamics and Measurements (Eds., Hwang NHC, Normann NA), University Park Press, Baltimore, Maryland, 1987.
47. Betram CD: Ultra sonic transit-time system for arterial diameter measurement, Med. Biol. Eng. Comput., 15:589-499, 1977.
48. Begel DH: The static elastic properties of the arterial wall, J. Physol., 156:445-457, 1961.
49. Zhou JS, Wahab SA, Hwang NHC: Monitoring vascular wall motions with a laser optical system, ASME Winter Annual Meeting, Dallas, Texas, USA, Nov. 25-30, 1990.
50. Image Sensing Products Manual: EG&G Reticon Company, Salem, Massachusetts, 01970, 1989.

FLOW MODELS STUDIES OF HEART VALVES

Klaus Affeld, Klaus Schichl, and Andreas Ziemann

Biofluidmechanik Labor, UKRV
Freie Universität Berlin
1 Berlin 19, Germany

INTRODUCTION

In order to simulate the blood flow through an artificial heart valve, a device is required that generates the appropriate flow and pressure conditions, i.e., a mock circulation. Despite the many years of research on artificial heart valves, a consensus among researchers about an optimal flow model has not yet been attained. As the many references and also the guidelines of the International Standard Organization[1] show, there are no well defined specifications on the testing of artificial heart valves. The reason for the absence of agreement on this matter is possibly that the problems to be investigated regarding the flow model are inordinately complex. In this paper we will suggest two models that we believe reduce this complexity, thus facilitating the determination of the appropriate flow. Accordingly, we divide the flow models into two groups:
 - flow models to measure the bulk hydrodynamic qualities of the valve; and
 - flow models to research the valve's thromboembolic qualities.

The bulk hydrodynamic properties determine how well a valve will function if implanted in the body. They describe the acute function. However, no information can be gathered from these parameters regarding the function of the valve over an extended period of time. We cannot determine whether the valve will damage the blood, or whether a thrombus will be generated. To answer these questions more details of the flow are required. Since the flow through a valve has nearly all the complexity a hydrodynamicist can ask for, different methods obviously have to be applied as compared to the methods used to investigate the bulk hydrodynamic qualities. Consequently instead of trying to answer all questions with one flow model we suggest employing two different flow models, one for each set of problems.

The one for measurement of the bulk hydrodynamic properties should be better defined than the models generally employed; the one for the research into the thromboembolic qualities should permit the investigation of the detailed blood flow and render more refined information about the time and the space of the flow.

METHODS AND RESULTS

FLOW MODEL FOR BULK HYDRODYNAMIC QUALITIES

Bulk hydrodynamic measurement parameters are defined as follows: systolic pressure difference is the pressure drop across the valve during systolic flow when using an aortic valve. This difference in pressure is mainly caused by a flow contraction due to the protrusion of elements of the valve into the cross section of the duct. The closing volume is that which flows backwards when the valve closes. The leakage volume is the time integrated flow found in most valves during diastole as a result of incomplete closing of the occluder. The closing time of the valve is the time span between the initiating and the completion of the closing. From these parameters the hydraulic energy loss can be calculated. Energy is lost during the systole, closing, and during the diastole. The energy loss is the product of instantaneous flow and pressure difference.

In general a flow model for artificial heart valves - a mock circulation - imitates the natural heart, the natural aorta and the natural peripheral resistance. Modeling the heart, the artificial ventricle possesses an inlet and an outlet valve, one of them being the valve to be investigated. A pneumatic pressure pulse or a hydraulic volume pulse usually drives the ventricle. An elastic vessel models the aorta and a resistance generates a pressure difference. Figure 1, left, shows such a system. From this design a number of difficulties arise. Firstly it is not possible to obtain a precise volume pulse through the valve to be investigated. If the driving system is a pneumatic driver, the elasticity of the air does not permit one to generate a flow curve which is independent of the valvular resistance and a well defined flow curve cannot be obtained. Each valve to be tested generates its individual flow curve according to its flow properties. However, for comparison one wants to provide the same flow curve for all valves and then compare the pressure difference on the basis of the same flow. The effect of the compressibility of the air can be avoided if a liquid is used for the generation of the flow pulse. A mechanical device such as a cam or an electronic control[2,3] has been used to generate a well defined volume displacement in the artificial ventricle. However, this does not guarantee that the flow through the valve to be tested is the same as the displacement of the ventricle or the piston. Let us assume that the outlet valve is the one to be tested, then the inlet valve during its closure directs a part of the flow back into the reservoir and with this, the valve changes the flow curve through the outlet valve. How fast the inlet valve closes is dependent on the pressure in the model aorta. The latter is an attempt to simulate the natural aorta. One may consider the natural aorta as an elastic vessel - a windkessel. Fluid entering this vessel encounters a pressure, which is determined by two elements, the static resistance $R = p/\dot{V}$ and the dynamic resistance $I = \dot{V}\rho c/F$; with R being the static resistance, p the pressure in the aorta, $\dot{V}$ the instantaneous flow into the aorta, I the impedance, ρ the fluid density, c the speed of the pulse wave and F the cross section of the aorta. The static resistance R can be modeled easily with a number of parallel tubes. Their diameter, length and number define the resistance. The dynamic resistance or impedance is usually modeled with an elastic vessel, such as a vessel partially filled with air. Both elements form a model of the natural aorta, but only as a first approximation: the natural aorta is more than a simple elastic vessel, it is an elongated vessel with pulse waves travelling at a speed of 4 m/s; these pulse waves reach branches of the aorta after a few milliseconds, then are reflected and superimpose the original pulse wave. However, the pulsewaves are not only reflected, they are also attenuated and in such a way the unique aortic pressure curve is generated. In addition to the fact that the flow curve through the valve is difficult to generate, it is also difficult to measure. Electromagnetic flow meters tend to drift and need frequent recalibration. Moreover, they are sensitive to an asymmetry of the velocity profile.

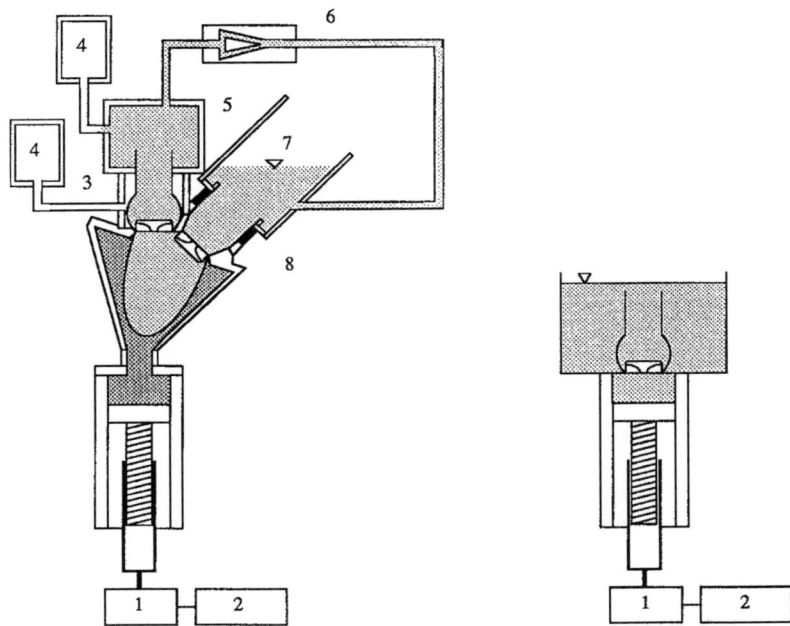

Figure 1. Left: schematic drawing of a conventional flow model, a mock circulation. An artificial ventricle generates a pressure or flow pulse, two valves direct the flow out of the reservoir into the artificial aorta and to the artificial resistance. A flow meter and two pressure transducers permit the measurement of flow and pressure difference as a function of time. On the right side the flow model is reduced to only a few elements: a piston to generate the flow pulse, the valve and two pressure transducers. A flow meter can be omitted, since the displacement of the piston is proportional to the flow through the valve.
1: Motor; 2: Control; 3: Compliant transparent aorta; 4: Supplementary compliance; 5: Characteristic resistance; 6: Adjustable peripheral resistance; 7: Left atrium; 8: Mitral flow transducer.

The same is true for ultrasonic flow meters. A flow rectifier to equalize the velocity profiles is unwelcome because it changes the impedance of the model aorta.

An improved flow model should avoid these problem areas and have the following features:

- defined flow curve,
- defined impedance,
- precise flow measurement.

Figure 2 shows in diagramatic form the hemodynamic basis of the new flow model: on top the ventricular pressure and the aortic pressure are depicted. The hatched area is the pressure difference over time across the aortic valve. The diagram below shows this pressure difference. Finally, the bottom diagram shows the flow through the valve. During systole the pressure difference across the valve is determined by the flow alone, and as is readily noted, no pulse wave or any other pressure difference has any influence on it. Therefore, it is sufficient to provide the flow in order to measure the pressure drop across the valve. An elastic vessel or a resistance is not required. Therefore, the first condition the flow model has to meet is the generation of a defined flow curve which during systole is independent of other conditions.

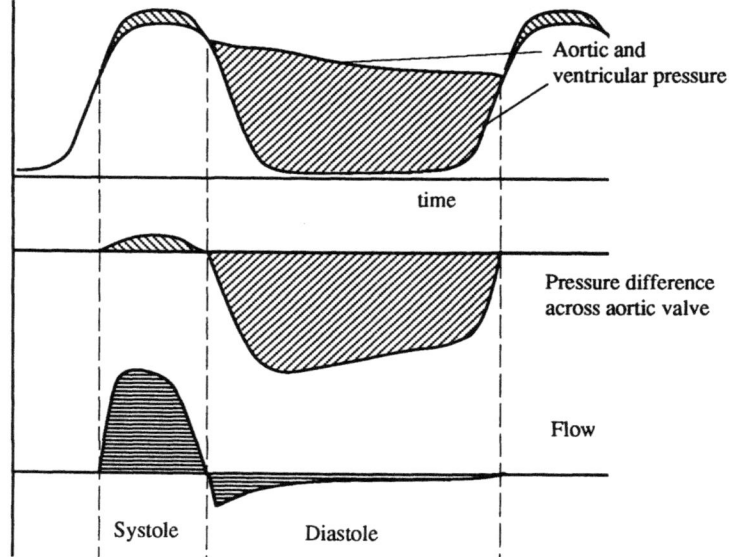

Figure 2. Hemodynamic basis of the new flow model. The top curve shows the aortic pressure and the ventricular pressure. The pressure difference - the hatched area - is caused by the aortic valve. During systole it causes a flow resistance, during diastole it is closed.

At the end of systole the flow approaches the baseline and even goes below it, because the valve is not yet closed. The closing action of the valve is, as the pressure drop, solely determined by the flow. However, as soon as the valve is closed the situation is different: the pressure difference now is determined by the aortic pressure and by the pressure in the ventricle. Only the pressure difference is important and no further parameters are required. This defines the second condition the flow model has to meet during diastole: the generation of an aortic-ventricular pressure difference.

If one designs the flow model to fulfill these conditions[4], one has to achieve a controlled flow during systole and a controlled pressure during diastole. A special controller has been designed to perform this action. It switches its mode from volume-control to pressure-control just when the valve is closed. Figure 1, right, shows the new principle in comparison with a conventional flow model. Many of the elements which were difficult to define are thus omitted. The new flow model is driven by a piston, which is computer controlled and is programmed to follow a specified curve. Figure 3 shows the ideal curve and the deviation of the piston

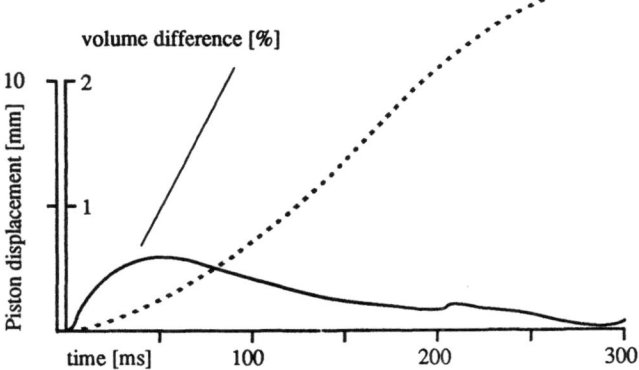

Figure 3. Movement of the piston in comparison with its ideal curve. The deviation is below one percent.

displacement from the ideal curve. The deviation can be kept below one percent. The design assures that the displacement of the piston is proportional to the volume, which flows through the valve. The time derivative of the piston displacement is the flow. Because there is only one valve, all the flow goes through it. As such, the displacement of the piston permits a volumetric measurement. When the piston comes close to the end of its systolic motion and is decelerated, comes to a stop and reverses its direction, the valve will close according to the generated flow curve. A measurement of the pressure is done upstream and downstream of the valve. These locations are shown in the schematic diagram in figure 5, which on the right side gives a schematic overview of the flow model, while on the left side the photograph shows its realization. The piston is driven by a ball screw with a lead of 5 mm inside a brass cylinder of 80 mm inner diameter. Special care has been taken for the design of the seal. It is required that the piston moves with little friction and still is leak-tight. Figure 4 shows the principle of the rolling seal[5]. It avoids a stick-slip-effect and is composed of a hollow vessel made of thin polyurethane foil with a thickness of 0.2 mm which is filled with air at a pressure of 2 bar. The outer vessel area is pressed against the cylinder wall, the inner one against the piston wall so that the only resistance is caused by the bending of the foil. Additional sealing is achieved with silicon grease. The ball screw is driven by a pulse width modulated pancake motor which transmits its power with a synchronous belt drive to the screw. An angular transducer with a resolution of 2000 pulses/revolution is directly attached to the screw. The length of the cylinder allows up to 7 heart cycles of 70 ml stroke volume. The cylinder leads to a water basin with a free surface. Because of the vertical construction of the flow model, venting problems are avoided. The valve is fixed in a rigid aortic root made of acrylic glass which can be easily mounted by screwing it into the bottom of the basin at the end of the cylinder.

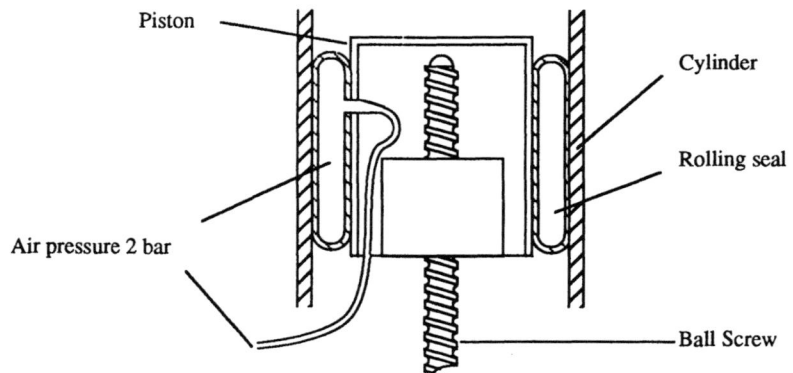

Figure 4. Drawing of the rolling seal.

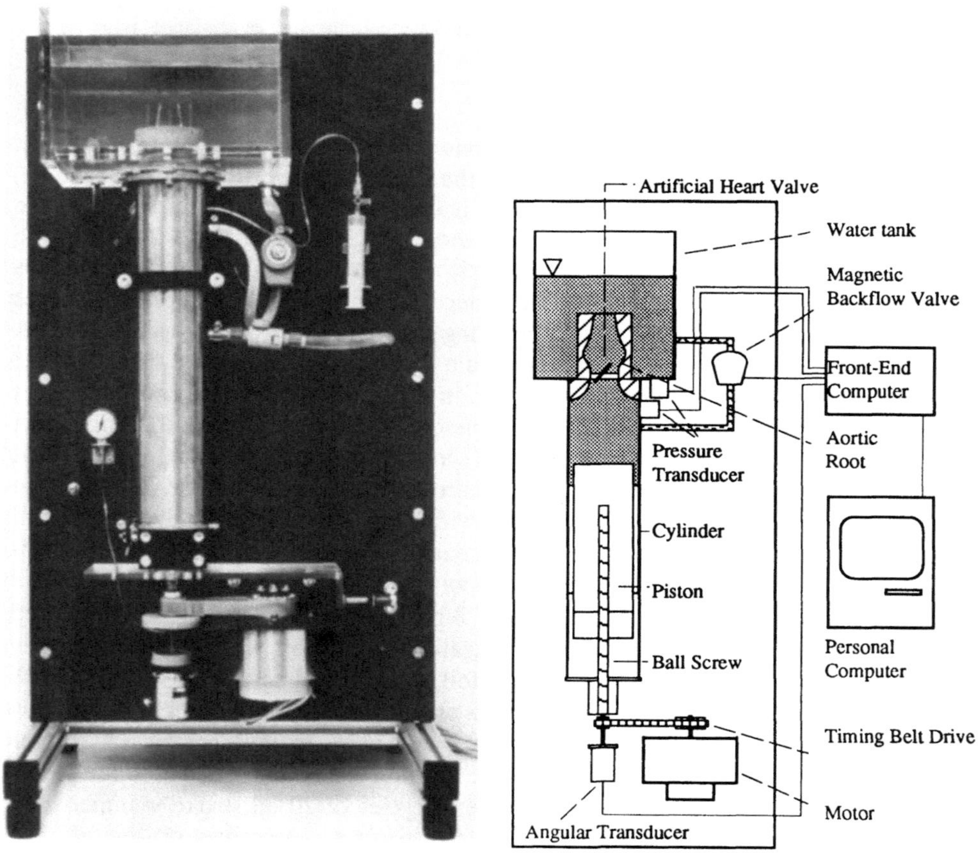

Figure 5. Photograph of the flow model (left) and schematic diagram (right).

As shown in figure 5 the heart valve tester is controlled by a special front end computer which has a control frequency of 5000 Hz. The user sets the test conditions with the host computer, an Apple Macintosh Classic®. For a test the software provides a menu as shown in figure 6, where the user can choose a cycle rate between 30 and 150 cycles per minute and a stroke volume between 30 and 100 ml just by moving the heart valve icon with the mouse. It is also possible to choose more than one cycle in order to obtain an ensemble mean average and smoother curves in the output plot. The host computer then generates the needed piston displacement curves and pressure difference curves in accordance to those found in any standard textbook of physiology. These curves are sent to the front end computer which starts and controls the electric motor. Calculation and graphics are made by the host computer. Figure 7 shows the output diagram with the curves for flow, pressure difference, energy loss, closing time, closing and leakage volume as well as the energy loss during systolic, closing and diastolic phase.

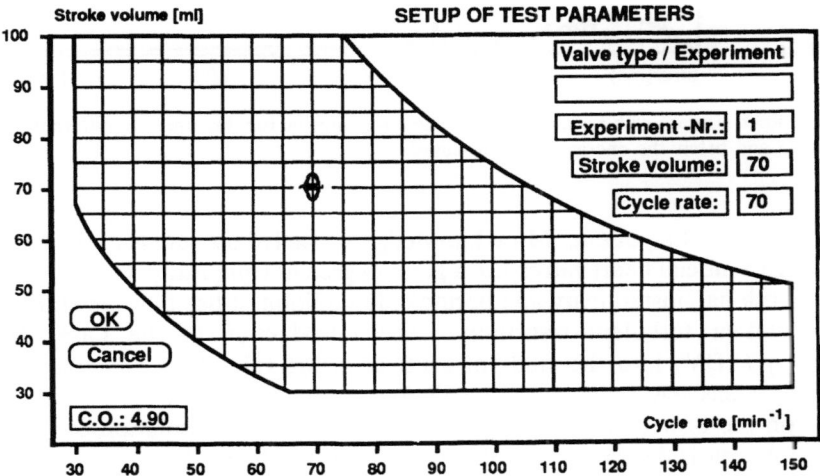

Figure 6. Screen menu to set the flow model and the test conditions.

Refinements of this type of model result in more consistent and more precise measurements. The obtained data are valuable, but also have only limited significance. The present authors feel, that the bulk hydrodynamic qualities of a valve have been overestimated up to now in their clinical relevance. If, for instance, a stenosed natural valve is excised and an artificial valve implanted instead, the heart still is trained to overcome a large pressure difference across the valve. The small energy loss caused by the artificial valve is easily supplied by the otherwise healthy heart. It is hardly conceivable, that the small energy loss of a modern artificial valve plays an important role in the long term compatibility of the valve. Other qualities are likely to come into account. These are researched in another type of flow model.

ENLARGED FLOW MODEL FOR THE RESEARCH OF THE THROMBOEMBOLIC QUALITIES

Despite the merits of the computer controlled flow model many questions remain open and cannot be adequately researched with this device. The reason is that the bulk hydraulic qualities of a valve are not sufficient to describe the performance of the

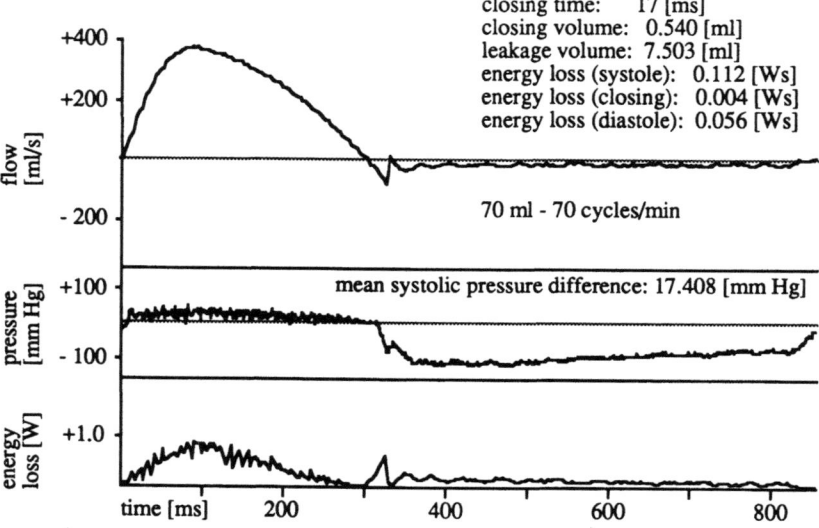

Figure 7. Output diagram with results: the flow curve is shown on top, the pressure difference is shown in the middle and below the energy loss is displayed.

valve in the body of a patient. More knowledge is needed on details of the flow. The flow through artificial heart valves has a great degree of complexity: the flow is usually three dimensional, instationary and periodic; in addition, it has flow separations, and often a laminar turbulent transition. The introduction of the Laser Doppler Velocimetry (LDV) into the study of artificial heart valve flow has enabled us to investigate extensively fluid mechanics[6,7]. With this instrument the velocity at one point within the flow field can be measured with a refined resolution of time. The spatial resolution is determined by the accuracy of the traversing unit. With several laserbeams it is possible to measure all three components of the velocity as reported by Yoganathan et al.[8]. From this, information on the turbulence such as the Reynolds shear stresses can be calculated. A great number of artificial heart valves have been systematically investigated in this way by Reul[9]. However, as successful as the LDV method is, it has the great disadvantage that the velocity can only be measured at one particular location at a time. The picture of the flow must be composed from many velocity vectors, which in addition are averaged vectors. Information is lost in this process. An overview of the flow can be obtained by the use of flow visualization, which has been used for heart valve flow studies since they were first available. Hitherto, however, they have lacked great detail, mainly due to the small size of the valves and the high velocity of the fluid. Thus, the methods to visualize the flow render an integrated picture but lack the precision of the LDV method. The methods of flow visualisation require low velocities - the ones we observe in an artificial heart valve have to be reduced. The desired reduction can be achieved with scaling - an enlarged model flow. In engineering sciences scaling is done often. The actual structure usually is not enlarged but is geometrically scaled down to a manageable size and subjected to the flow. Attendant to this procedure, there are difficulties in maintaining the similarity of the models in the correct relation. In contrast, in heart

valve flow one can scale up as opposed to scaling down, because our real flow is of small geometrical dimensions and, in addition, slow compared to most technical flows. So if one keeps the Reynolds number constant one can scale up the model and without having to sacrifice anything - indeed, one gains a number of advantages, namely:
- the time scale is greatly extended, i.e. the flow can be observed in slow motion,
- the lighting is much easier,
- powerful methods of flow visualisation can be applied.

By enlarging the valve it is possible to reduce the velocity and therefore to overcome the problems encountered in the measurement of original size flow.

Most methods of flow visualization require a low velocity of around 5 cm/s in water. At higher velocities dyes disperse too quickly and particles are difficult to light properly. Velocities of 150 cm/s which are typical for flow in an artificial heart valve are much too high for good flow visualization. A sufficiently low velocity can be achieved, however, with upscaling. The model flow will remain similar to the real flow, provided the Reynolds number is the same. A scale of 10 to 1 and the use of water as model fluid allow the reduction of the velocity in the model flow to 1/25 th of the real one, producing a maximum velocity of about 7 cm/s. In order to obtain such a similar flow the enlarged model must be similar to the real valve. For this purpose a water tunnel has been built which can be thought of as a mock circulation, figure 8. The model valve has a diameter of 220 mm and is mounted in a transparent

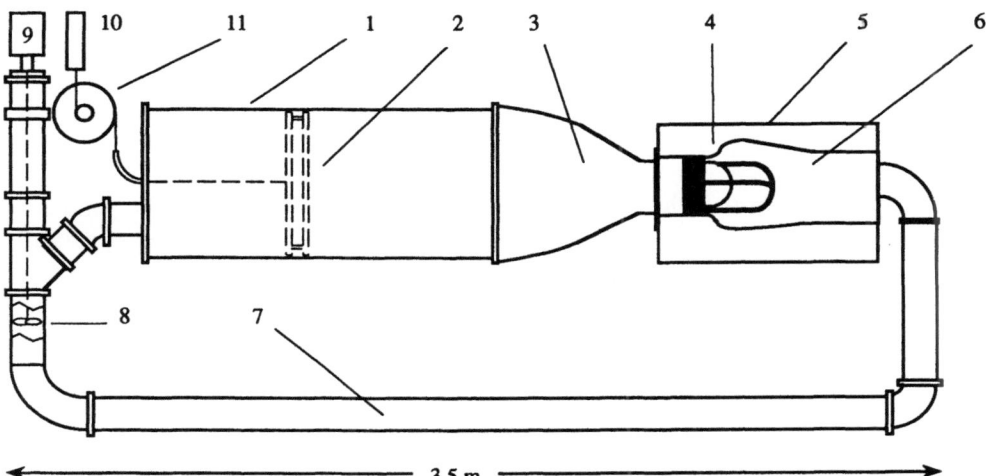

Figure 8. Enlarged flow model.
1: Settling chamber; 2: Floating piston; 3: Flow contraction; 4: Test valve; 5: Water basin; 6: Aortic root; 7: Backflow pipe; 8: Axial flow pump; 9: Electric motor; 10: Displacement transducer; 11: Reduction gear

model of the aortic root. Water is used as the fluid. The model weighs around 600 kg and holds some 400 liters of water. The total length is 3.7 m. Unlike a conventional watertunnel this one produces a pulsatile flow as is required for artificial heart valves. Another unique feature is the way an even and laminar flow to the model valve is established: still water from a settling chamber is separated from the disturbed fluid by a free floating wall. This wall in the form of a piston moves with the flow from one end of the settling chamber to the other. Once the piston has reached the far end, the flow is stopped and the piston reversed before the next run is started. The volume contained in the settling chamber limits the number of cycles to three, but this is sufficient for all methods of flow visualization. The floating piston also serves another purpose: since hardly any forces are acting upon it, it floats very precisely with the fluid and can be used to monitor the flow. The displacement of the piston is the same as the displacement of the water and so through differentiation the water flow can be calculated. The piston is connected to a displacement transducer by a thin string. The fluid is moved by a propeller housed in the return section of the tunnel and is driven by a computer controlled electric DC motor. By integrating the aortic flow curve taken from a textbook of physiology a displacement-time curve has been obtained. The electric motor is controlled so that the displacement of the piston follows the physiological curve. The proper speed of this displacement is calculated with the help of the similarity laws. Most important is the Reynolds number (Re) similarity. With u being the velocity, d the diameter of the valve and v the kinematic viscosity we obtain:

$$Re = \frac{u \cdot d}{v} \qquad (1)$$

With a model scale of 10:1 for the geometry and 1:2.54 for the kinematic viscosity - (water, 20°C: blood, 37°C) - we obtain a reduction of the velocity of 1:25.4 and a time expansion of 254:1. This means that a systole of 300 milliseconds lasts in the model more than 76 seconds thereby giving the flow a slow motion appearance. The second number to be observed is the Strouhal number (S):

$$S = \frac{d}{u \cdot t} \qquad (2)$$

This number assures that the stroke volume is in the proper relation to the geometry of the valve. However, this similarity is automatically kept, if the Reynolds-similarity is observed for each time step, i.e. if the flow curves of original valve and its model are properly scaled. Another number to be considered is the Archimedes number (Ar). With ρ_{fluid} being the specific density of the fluid, $\rho_{occluder}$ that of the valve occluder and g the gravity, the Archimedes number is defined as:

$$Ar = \frac{\rho_{occluder} - \rho_{fluid}}{\rho_{fluid} \, u^2} d \cdot g \qquad (3)$$

The Archimedes number takes into account the buoyancy of the occluder. In the real valve the specific weight of the occluder is about 1.6 and differs considerably from that of the blood. This has little influence on the valve's movements because the velocities are high and the dynamic pressure easily overcomes the influence of its buoyancy. In the enlarged model this cannot be said because the dynamic forces are so small that the buoyancy plays an important role. With the model scale inserted in the formula one obtains the specific weight of the model occluder as $\rho_{occluder} = 1.000079$ times the specific weight of water. This is practically equal to that of the water. To achieve this density the model occluder is made of polyethelene, which is lighter than water and weighted with lead to obtain a neutral buoyancy. The lead is positioned at the occluder's center of gravity to maintain its balance. The movement of the occluder is very sensitive to its specific weight, even some small air bubbles which attach easily when the tunnel is freshly filled will influence its movement.

Some valves are not made of a rigid material, but of flexible one. The natural heart valve, for instance, has three very thin and flexible leaflets, which are attached to the vessel wall and which float with the flow during opening, flutter during the systolic phase, and convert to a membrane structure when closed. To model such a valve, the fluid forces have to be in the proper relation to the elastic forces which resist the bending of the leaflet. As an example for this kind of similarity, imagine a tree which is bent in the wind. The model of this tree will have a similar shape only if the proper law is observed. From this example - a beam under the load of its dynamic flow resistance - a similarity law can be obtained:

$$K - \frac{\rho u^2 d^2}{E s^3} \qquad (4)$$

E is the modulus of elasticity of the material, d a characteristic length, u the fluid velocity, and s the thickness of the leaflet. The leaflets of the natural valve are extremly flexible as measured by Affeld et al.[10]. Jansen et al.[11] have described a valve with leaflets of polyurethane with a thickness of 0.15 to 0.22 mm. If the enlarged model valve is made of this material and the model law - equation 4 - is applied one obtains the same thickness for the model leaflet, a condition which obviously is easy to meet. This valve has leaflets that are bent in only one axis. However, if the leaflets have a double curvature a change of its shape can lead to folding and buckling, and in this case the relation of thickness and length has to be kept constant. One must then fabricate the leaflet from a material with a different modulus of elasticity. The enlarged model has to be much softer, for the model relation of ten to one obtains that the modulus of the model leaflet is only a thousandth of the original. Since the modulus of the natural valve is already low, this condition certainly is difficult to achieve. This clearly indicates the limitations of this model technique.

Unfortunately, however, this is not the only limitation of the enlarged model. Another one is: one cannot model the fluid precisely. Blood is a non-Newtonian fluid and there are succesful attempts to model it with water and polymeric additives, such as Separan® and Xanthem®[12]. Over a wide range of shear this model fluid imitates well the rheology of blood, showing a major deviation only in the low-shear range. Since the enlarged flow model allows for the discernment of fine details of the flow in small areas, where the shear is low, it is desirable to have a non-Newtonian fluid under these flow conditions. Modeling of non-Newtonian fluids in general is difficult and even may be impossible[13]. A simple calculation may illustrate this difficulty. If one assumes a simple law for a shear dependent viscosity such as:

$$\eta - a\dot{\gamma}^b \qquad (5)$$

η being the viscosity, a, b constants and $\dot{\gamma}$ the shear rate.
one obtains laws which model well blood and as well its model fluid water-Separan. However, if one scales the fluid to a ten times enlarged model flow one needs a fluid, which has a lower viscosity than water at larger shear rates. Thus water cannot be

applied as the model fluid. Obviously it is impractical to look for another fluid in place of water to run the model, because another fluid, which is likely to be an organic fluid, will have other undesired effects, as being toxic or hazardous, probably being expensive in addition - 600 liters are needed. Despite these limitations the enlarged model retains its merits still for the understanding of flow separations. For the precise modeling of flow separations in the very low shear range, however, one has to go back to the real size flow model and use a model fluid or even blood itself.

Once all the limitations are considered, the enlarged valve being fabricated and the proper movement of the occluder being established, the flow can be made visible. To this end a great variety of flow visualization techniques have been developed[14]. From these methods, the most appropriate ones have been selected and applied; as is described in detail in Affeld et al.[15,16].

One of these methods is the particle method. It provides a comprehensive picture of the flow, particularly the flow in a selected light plane. In this method small plastic (polystyrene) spheres with a diameter of 0.2 to 0.4 mm are added to the water. The flow is then lit with a light plane so that only a section of the flow becomes visible. The spheres move with the flow and show the path lines. In photographs taken with an extended exposure time, each particle produces a streak the length of which is proportional to the velocity. Figure 9 shows the center line section of the flow through the model of the Björk-Shiley Monostrut valve. The jet through the major orifice is visible as well as the flow separations behind the monostrut and at the valve ring. The flow is visualized in great detail, which one could not detect in a real size flow model.

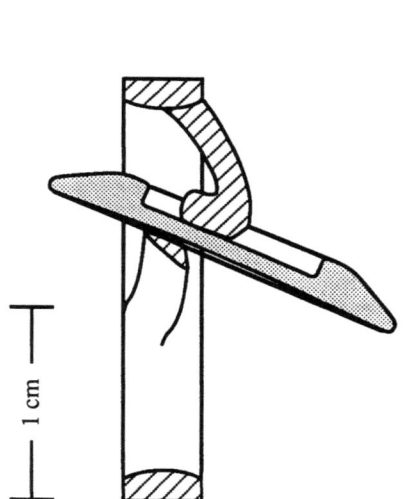 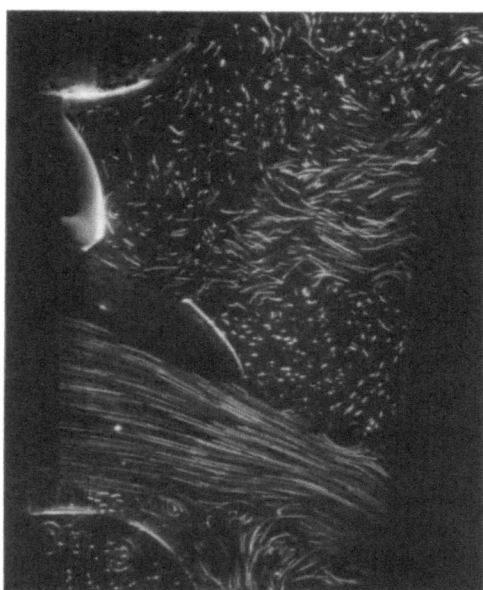

Figure 9. Still photograph of the flow through a Björk-Shiley Monostrut valve. The particles appear as streaks because of their movement during exposure. This provides an overview of the flow. The wake and the jet at the major orifice are clearly visible. The technical drawing on the left side provides orientation how the flow relates to the geometry. The light plane is in the centerline of the valve and identical with the dissecting plane in the drawing.

The use of a videorecorder adds a further feature - the slow motion effect permits the observation of the flow in its development and the repetition of the observation as often as needed. The timescale of 254 to 1 results in a camera speed which is 25 times 254 equal to 6350 frames per second, a speed, which corresponds to that of a high speed camera.

Using these video recordings it is possible to quantify the flow with methods of image processing. From the displacement of the particles between different video frames, the velocity field can be calculated, the time between frames is known from the above mentioned frame rate. The particle displacement has been obtained using the cross correlation method. Once the velocity is obtained, the stream function and stream lines can be calculated in order to make the flow patterns more understandable. The flow was filmed using a Panasonic F10 video camera and recorder. The video was digitized by a frame grabber board and software (Quick Capture from Data Translation) which was mounted in a Macintosh II computer from

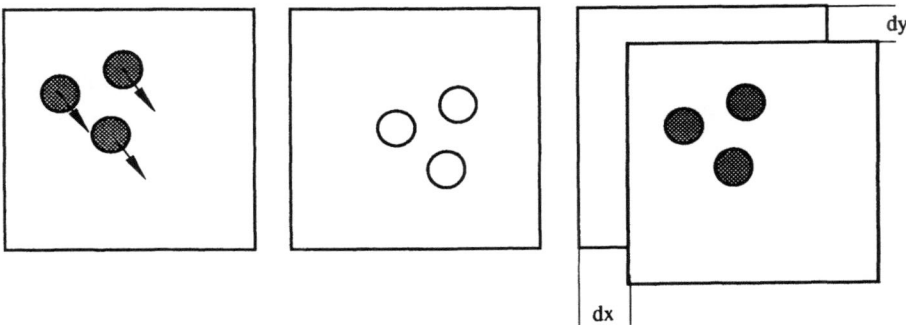

Figure 10. A pixel pattern in schematic drawing. Left is the pattern of the first frame, right the pattern of the second one. The characteristic pattern has moved one pixel to the left and one up, which is indicated by the frame.

Apple. The resolution of the digitizer is 768 lines by 512 pixels. Standard picture analysis was performed by the Quick Capture software and the specialized cross correlation analysis by in-house programs written in Z-Basic. This method uses standard pattern recognition procedures to match the patterns formed by the particles between pictures. Two consecutive video frames are used, each is divided up into a 19x12 grid, each grid square is 16 by 16 pixels large, figure 10. The gray scale value of the pixels range from 0 to 255, let these be represented by $f(x,y)$ for the first video frame and $g(x,y)$ for the second; x is the horizontal pixel number and y the vertical. Within each grid square the white particles upon the black background form a pattern, each grid square has a different pattern. However, corresponding grid squares in the two video frames will have the same pattern except that the position of the pattern within the grid will be different as the particles will have moved because of the velocity of the water. The cross correlation method involves moving the grid square from the first video frame over the corresponding grid square from the second video frame until the patterns of the particles match one another or at least until the best match is made. The difference in position of the pattern between the first and second frames then gives the particle displacement and hence the velocity vector. Intrinsic in this process is the assumption that the pattern undergoes a pure solid body translation and entails no rotation or shearing. This is, of course, in all but the simplest of motions untrue but the method can be used if the size of the grid squares are kept small and the time between pictures short, relatively speaking. The process of finding the best match between grid squares is mathematically equivalent to the cross correlation as defined in the equation (6):

$$f(x,y) \circ g(x,y) = \sum_{m=0}^{M-1} \sum_{n=0}^{N-1} f(m,n) \cdot g(x+m,y+n) \qquad (6)$$

This process is slow if performed using equation (6), but can be vastly accelerated if use is made of equation (7)[17]. That is, the cross correlation of two signals can be calculated by taking the Fourier transform of the two signals, one transform is then complex conjugated and the two transforms are multiplied together to give one complex signal, an inverse Fourier transform is then performed upon this signal to produce a real signal corresponding to the cross correlation.

$$f(x,y) \circ g(x,y) \iff F^*(u,v) \cdot G(u,v) \qquad (7)$$

Where f(x,y) and g(x,y) represent the gray scale values of the two pictures and F(u,v) and G(u,v) their Fourier transforms, the * is the complex conjugate and <=> the forward and inverse Fourier transform. The Fourier transforms are speeded up by using a fast Fourier transform (FFT) routine and long integer (32 bit) numbers. The first step in the formation of the cross correlation is to read in the gray values of the grid squares under consideration, that is to form f(x,y) and g(x,y). These arrays are then expanded up to 64 by 64 pixels to increase the accuracy of the result. A Fourier transform is now performed upon the two arrays to produce the fourier image of the two grid squares. The complex conjugate of one image is taken and the two images multiplied together, the product is then transformed back to the physical plane using an inverse Fourier transform. The resulting array is equal to the cross correlation of the two grid squares. Array element i, j corresponds to the correlation between the two pictures. If the first picture is displaced i pixels to the right and j pixels downwards, the largest element in the final cross correlation array therefore corresponds to the position of best correlation between the two grid squares. The displacement of the particles in the grid square is now known in terms of pixels which can be converted to actual displacement in meters and, by dividing by the time difference between the two video frames, the velocity can be calculated. By performing the above procedure upon all 228 (19x12) grid squares the velocity field over the entire frame at regular intervals is obtained. Figure 11 shows an example.

The particle method renders an overview of the flow and permits quantification of the flowfield as well. It also gives an indication of the numerous flow separations of the heart valve flow. These flow separations play an important role in the long term performance of the heart valve in the patient. A flow separation in the bloodstream is inevitably connected with a stagnant flow area and it is well known that such a stagnant flow is associated with the danger of thromboembolic complications. How can a stagnant flow be made visible and discriminated from the rest of the flow? The appropriate method is coloring a part of the fluid with a dye. The dye is infused into the model of the aortic root and thoroughly mixed. The aortic root is the space immediately downstream of the valve and the mixing is done to assure a homogeneous concentration of the dye. A sliding metal diaphragm next to the test valve separates the volume upstream and is taken out manually before the systole is initiated. This

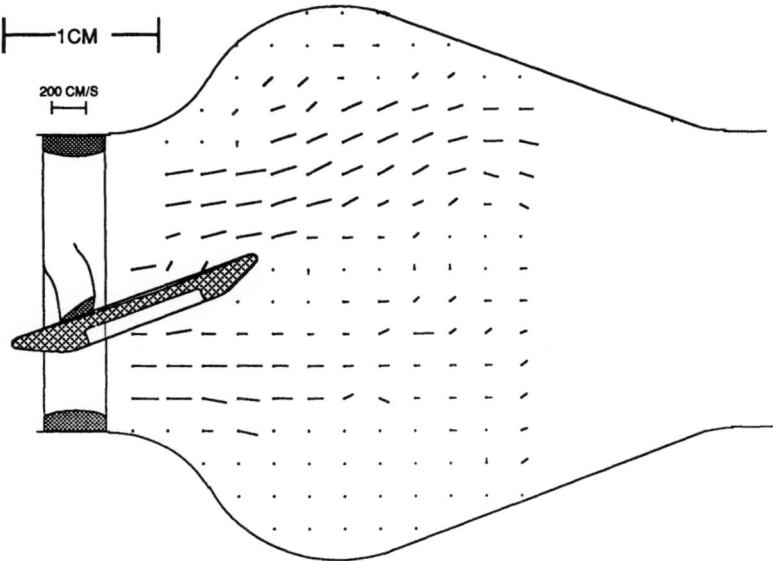

Figure 11. Vector field downstream of a valve.

initial separation of the clean portion from the dyed fluid provides an optical distinction between the fluid from the previous cycle and the fluid from the new cycle. If there is no flow separation, the new fluid will completely replace the old fluid forming a distinct front. However, in case of a flow separation the old fluid will remain in some areas and stay there for a while, the dye discriminating it from the new fluid. Uranin® is used as a dye. This substance is a watersoluble dye and flares up in yellow-green tinge when lighted up. Areas which are not lit remain transparent. This important property makes the unlit dye invisible, so that the dye between the observer and the light plane does not obscure the view. This permits us to look into the center of the flow which is undisturbed by the dye surrounding the center. Figure 12 shows the flow through a Björk-Shiley Monostrut valve. The fluid on the upstream side - coming from the left - thus contains no dye and appears black. The light plane is positioned parallel to the swivel axis of the occluder and goes right through the minor orifice, i.e., it is perpendicular to the light plane in figure 9. As the valve opens the clear fluid displaces the dyed fluid except in those regions where separations occur.

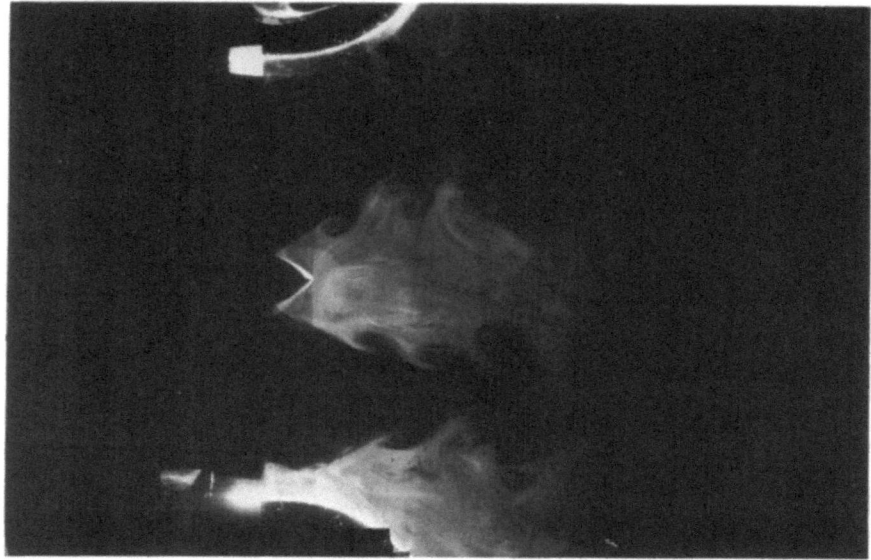

Figure 12. Flow through a Björk-Shiley Monostrut valve. The black fluid is clear water which displaces Uranin-colored water in the aortic root. Most prominent is the flow separation behind the strut.

We observe flow separations right at the ring, at the aortic sinuses and behind the monostrut. This explains, why the particle picture in figure 9 shows no flow behind the strut - the light plane goes right through the flow separation. Figure 12 also makes the turbulence of the mixing process visible, an observation, which has not yet been made in a real scale model. The main frequency of the turbulence in this case is 272 Hz, calculated to the real size flow.

The flow through the valve is instationary and so are the flow separations. The development of the flow separations can be observed and recorded with the video technique. Figure 13 shows 4 frames from the opening of a Björk-Shiley Monostrut

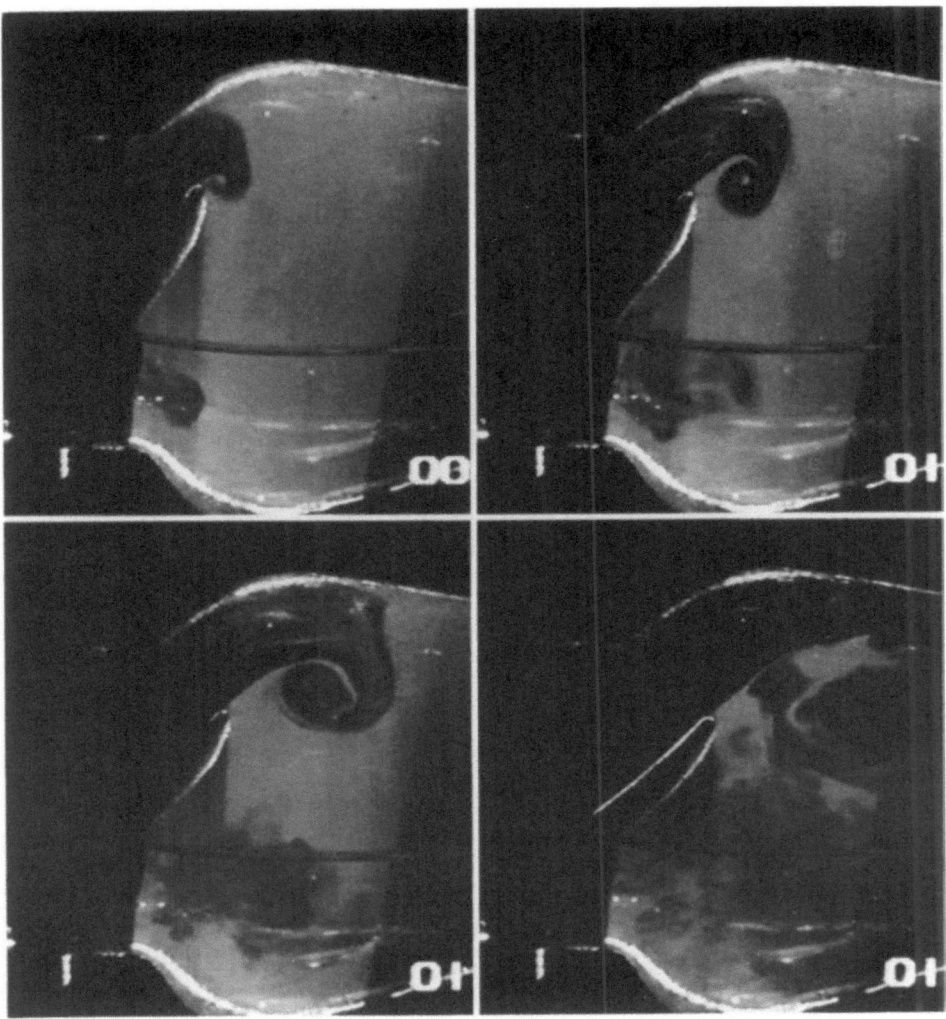

Figure 13. Opening phase of a Björk-Shiley Monostrut valve made visible with Uranin-colored water.

valve. The pictures are 40, 48, 55 and 69 milliseconds away from the beginning of systole (300 ms duration), calculated to real time scale. The fresh fluid - dark - comes from the left and a vortex is formed at the edge of the circular occluder already at the very beginning of the opening phase. It entrails dyed fluid and breaks up after 69 milliseconds, which is equivalent to 23% of systolic duration. The vortex sheet breaks up in a series of smaller vortices, which are 3 mm apart and would generate a turbulent oscillation of 200 Hz. The dark fluid does not replace the bright fluid at the lower end of the valve. The light section goes right through the centerline of the valve, and with this orientation also through the monostrut, which is responsible for the large flow separation, in this way we are looking into the wake of the monostrut. The area immediately behind the ring deserves special interest: it remains bright at the strut, but also close to the intense jet through the major orifice bright fluid is visible. There is a permanent influx at some portions of the ring and it creeps along the periphery of the ring and then enters the jet and mixes. It comes into the light plane with a movement which is perpendicular to the observer, see also the fourth frame at the top of the ring. It is interesting to compare this flow with a flow through a ball valve. An enlarged model of a Starr-Edwards heart valve has been fabricated and investigated. Figure 14 shows four frames, 40, 49, 64 and 82 milliseconds apart,

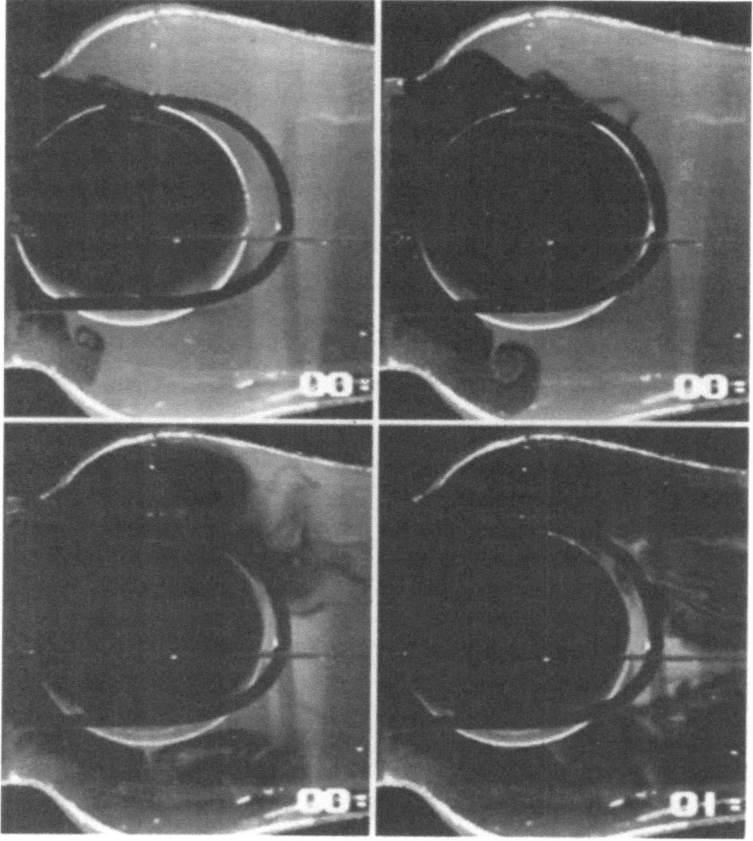

Figure 14. Opening phase of a Starr-Edwards ball valve made visible with Uranin-colored water.

which form the beginning of systole (300 ms duration). The dyed fluid is much faster replaced by the dark fluid than in the previous valve. One reason for this is that the ball has a much larger volume than the disk occluder of the Monostrut valve and consequently it displaces more fluid. As one would expect, there is an extended flow separation at the downstream side of the ball. This flow separation is greatly influenced by proper fitting of the voluminous ball into the aortic root. This proper fit often may have not been achieved, having the consequence of flow separations and thromboembolic complications. Thus the ball valve has slowly disappeared from the clinical practice. However, some cases are reported from patients with ball valves having an extraordinary longevity. On the basis of these flow studies we assume that these patients accidentally had a very good fit of vessel and ball. It is theorized that the small or favourably shaped flow separations were influential with respect to the lack of thrombus generation and thus helped the patient to sustain the implant well.

This leads to an important question: has the influence of the vessel geometry on flow separations sufficiently been considered? We think it has not. The flow around the occluder is usually looked at as a flow around an airfoil, which is surrounded by an infinitely extended fluid. The influence of the pressure along the x-axis is only little considered. In the valve flow, the structure which extends into the flow is subjected to a pressure which in general increases downstream, because the aortic root has a larger cross section in the vicinity of the valve. Thus we have a ducted flow with increasing static pressure, a flow separation is virtually inevitable. In engineering application one would avoid such flow; that is, one would decrease the cross section and accelerate the flow. We cannot change the human anatomy and it certainly is difficult surgically to change the shape of the aorta and thus create a better duct. However, there are applications where one has the freedom to design the duct - in artificial bloodpumps the occluder and the duct is subject to design freedom. Up to now little effort has been made to match the duct and the occluder of an artificial valve. Instead, artificial heart valves made originally for the replacement of the diseased natural valve are put into an artificial duct without considering the basic laws of ducted flow. Gentle[18] and Tanseley[19] have investigated a ball valve in a duct specifically designed for it in order to minimize the energy loss. Their ducted valve is intended to be used clinically as a conduit. However, the important issue of flow separation has not been adressed. In an attempt to create a valve with little or no flow separation a duct has been calculated and fabricated for a ball as occluder[20]. Flow visualization studies showed, that flow separation can be totally avoided thus giving hope for the design of a bloodpump with superior performance. Of course, flow separations can be avoided in the implantable ducted ball valve, the above mentioned conduit, as well. Such a device is guiding the bloodflow from the apex of the heart to the aorta descendens, bypassing the diseased natural aortic valve.

Despite the many efforts made so far, it has not been possible to quantify in a specific valve the flow conditions which lead to a thrombus formation. However, one knows from detailed in vitro experiments the flow conditions, which are relevant for the generation of a thrombus. The first condition is an area of high shear stress, either in a jet or at a wall, which exceeds $\tau = 100$ N/m^2 and lasts for more than 30 ms[21]. No information is yet available on the influence of a repetitive exposure to the shear stress. The second condition is the presence of an area of recirculation close to a wall of a foreign material. The recirculation is found in a flow separation and allows a concentration of platelets. If they are activated, accumulated densely enough and in addition placed in the vicinity of foreign material to which they can attach, then the formation of a thrombus is likely[22].

In order to check if the first condition is met - the exposure to high shear stress, an evaluation of the shear stress in the shear layer was made. If one evaluates the velocity profiles generated with the hydrogen bubbles method[23] the shear rate can be obtained. The derivative of these velocity curves is calculated and through the use of Newton's formula the resulting shear stress is found. The shear stress easily exceeds 100 N/m^2 and sometimes comes close to even 200 N/m$^{2[24]}$. These values agree with data from other authors[25]. The second condition is the presence of flow separations: we do find many of these in a valve, some with fast and some with slow mixing and concentration. How can one quantify the thrombogenicity of these flow separations? One attempt is to calculate the average time the platelets remain in the flow separation, which is defined as residence time. The platelets are modeled with dye. The flow separation is considered to be a vessel filled with a dyed fluid, while the inflowing fluid contains no dye. The residence time is considered to be a measure for the average time a dye particle remains at one place and is defined as:

$$T = \frac{1}{C_0} \int_{t=0}^{t=t_{end}} C(t)dt \qquad (8)$$

T being the residence time, C the concentration of the dye, the integration is performed during the systole. This definition requires the knowledge of the concentration of the dye. It has been assumed, that the intensity of the light is proportional to the concentration of the dye. However, this is not the case. The intensity of the light, which is the gray value of the video picture, depends on the width of the illuminating light plane and of the dye concentration. This has been measured and plotted, as figure 15 shows. These curves can be approximated by polynoms and with the help of these functions the gray values can be corrected to be proportional to the dye concentration. Figure 16 shows the concentration at two selected spots. The concentration varies with time because mixing takes place and vortex structures which are larger than the integrating frame are passing by. An integration is required and consequently 12 consecutive video frames were digitized. Each individual frame was divided into a number of small frames, the gray level in these frames was averaged using equation (7)[26]. Figure 17 shows the residence time derived in such manner in a plane just behind the ring of a Björk-Shiley Standard valve. The black half circle is a part of the occluder, which protrudes into the light plane. Long residence times are plotted dark. Below the occluder at the minor orifice dark areas appear, i.e., areas with high residence time. As mentioned above, the dye represents platelets which have been sheared in their previous history and which are now activated and ready to attach themselves to a natural or artificial

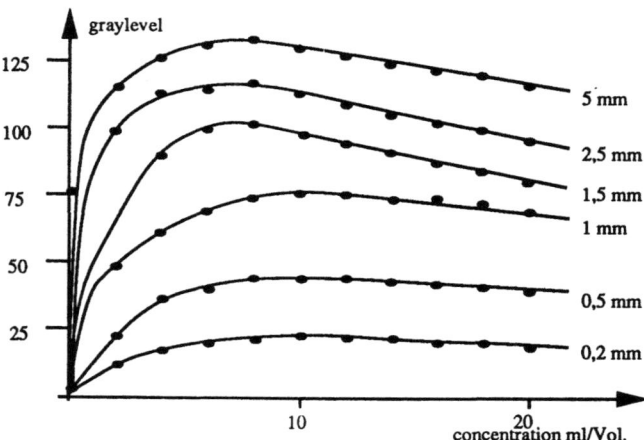

Figure 15. Gray value as a function of light plane thickness and dye concentration. These curves show, that the intensity of fluorescence cannot be taken as the concentration of the dye. Instead one has to compute the concentration.

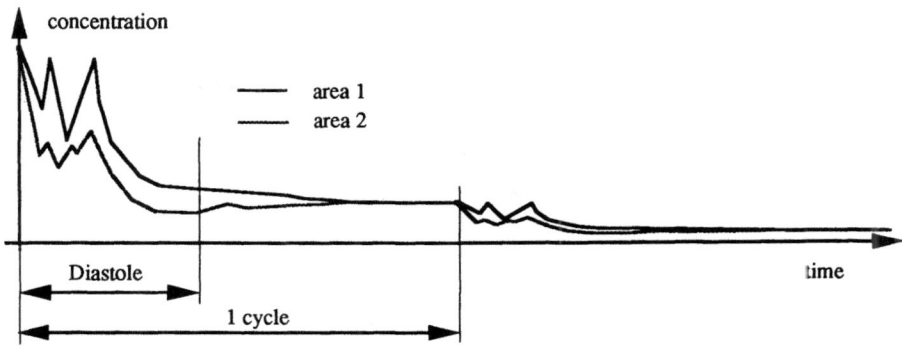

Figure 16. Concentration during two cycles in two different observation areas.

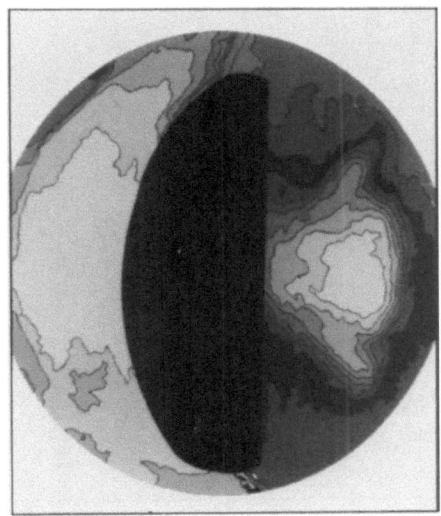

Figure 17. Mean residence time in a plane parallel to the ring and shortly downstream. The occluder is protruding the light plane. The calculation has been made over one systole. Dark areas are areas of long residence time.

surface. A long residence time of the dye thus corresponds to a long residence time of platelets. Consequently the dark areas are sites where thrombi are likely to form. See for comparison figure 18, which shows two photographs taken from valves at the post mortem. Both show thrombi which have developed in areas corresponding with those with an extended residence time in figure 17. These areas also correspond to low shear stress areas which have been assessed with other experimental methods[27].

As convincing the correspondence of residence time and thrombus generation may appear, these calculations of the platelet's residence time are not yet a quantification of the thrombogenicity of an artificial valve. Still, it would be very valuable if one could predict its thrombogenicity before a newly developed valve leaves the laboratory and prior to any animal experiments. In additon to the residence time we need to know the average stress history of a platelet. This means that for our flow studies we will have leave the Eulerian way of looking at the flow in favour to the Lagrangian way, i.e., the observer does not look at the flow from the outside, but looks at an individual particle and so to speak travels with the fluid, as done experimentally by Goldsmith and Turitto[28].

But also we need to know more on the flow conditions a thrombus needs to form. There are experiments to study the thrombus formation in vitro and relate it to the shear rate[22]. Blood was drawn and subjected to shear in a cone-plate viscometer. The resulting thrombus was plotted as a function of the shear rate. Figure 19 shows the results. At zero shear rate a maximum of thrombus generation is reported. However,

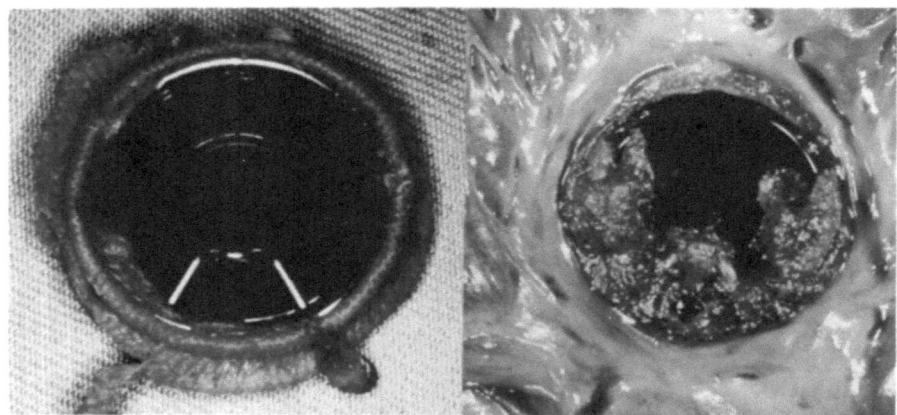

Figure 18. Two Björk-Shiley Standard valves taken out at post mortem. Clots are visible at the lower orifice of the valves and attached to the ring. Compare these sites with the dark areas in figure 17.

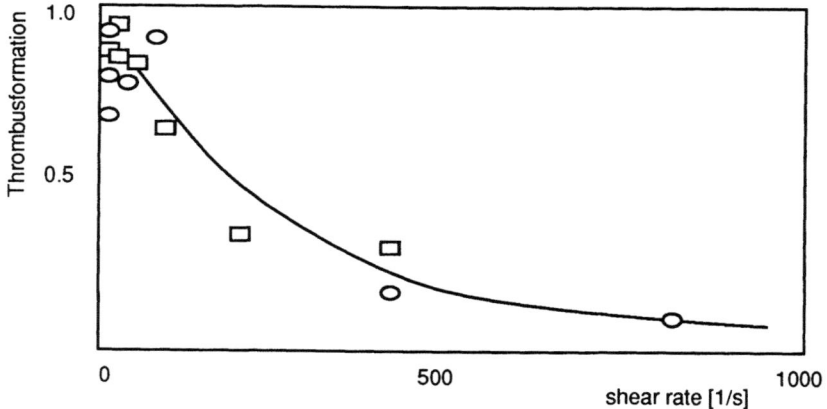

Figure 19. Thrombus formation as a function of shear rate - results from an in vitro experiment (redrawn from Hashimoto et al.[22]). This curve indicates that the formation of thrombi is maximal at zero shear rate. Experience contradicts this and indicates, that at zero shear rate the thrombus generation should be zero as well.

this appears to be an artifact, since it is known, that blood will hardly clot if left undisturbed. Therefore, the thrombus formation at zero shear rate has to be zero instead, increases with higher shear rate and decreases only after reaching a maximum at a certain critical shear rate. Which is the shear rate which is optimal for the formation of a thrombus? A first order model may bring some clarification: The first assumption is that the platelets have been sheared above a certain limit[21] and are releasing thrombin which diffuses around the platelet with its specific diffusive speed. This is expressed with the diffusion constant for thrombin of $D = 10^{-6}$ $[cm^2/s]^{(29)}$. The second assumption is that, at a critical shear rate, the thromboactive agent is concentrated rather than diluted. Due to the shear flow the platelets will pass each other at close distances, but react only if a certain time for a reaction is permitted, i.e., at this shear rate thrombin is concentrated rather than diluted. We find 3×10^5 platelets in the cubicmillimeter of blood. From this an average platelet distance of 13 microns results. This distance has to be bridged by the fluid movement, the shearing motion. This leads to the third assumption, which is, that the velocity of the thrombin diffusion should be equal to the speed of approach. The shear flow brings the platelets to a close distance and their diffusive fields superimpose and a higher concentration of thrombin will result, which eventually will lead to an adherence of the platelets. It is assumed that this effect takes place, when the convective and the diffusive velocity both have an equal value or are at least of equal order. Figure 20 shows an assumed arrangement of two platelets. If the shearing takes place only parallel to the x-axis, the platelets will pass each other and not come to a close approach. When a y-component of the shear is added, however, we will find the platelets approaching each other. With d being the diameter of the platelets, D the diffusion coefficient, and $\dot{\gamma}$ the shear rate we obtain for the velocity of approach u_p

$$u_p - \dot{\gamma} d \qquad (9)$$

The diffusive speed depends on the distance from the source, which is the platelet and is defined by the diffusion coefficient of thrombin divided by the distance, in this case

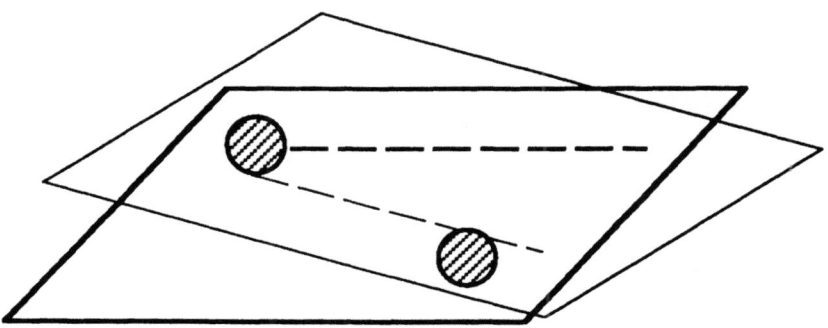

Figure 20. From the concentration of platelet follows a mean distance at which the platelets are spaced and sheared. If the shear is performed at a certain tilt, the platelets will approach and pass each other at a close distance. If in addition the speed of approximations is set equal to the diffusive speed of thrombin, a condition for the evaluation of the shear rate is found.

the diameter of the platelet:

$$u_d \sim \frac{D}{d} \qquad (10)$$

Equation (9) and (10) combined lead to the critical shear rate $\dot{\gamma}$

$$\dot{\gamma}_c \sim \frac{D}{d^2} \sim 10 \qquad (11)$$

At this shear rate we expect a maximum concentration of thrombin and thatfore a maximum likelihood of a platelet interaction-coagulation. In heart valve flow this means a velocity difference of 10 mm/s over a distance of 1 mm. This would be typical for the flow separation in the vicinity of a valve ring, while in the main stream we find shear rates of two order of magnitude higher. It also is a shear rate at which we already find a distinct difference of the blood's viscosity to the Newtonion viscosity. This gives an outlook for further studies and quantifications of flow separations.

DISCUSSION

The flow through an artificial heart valve is very complex. It is non-stationary-periodic, three-dimensional, with moving boundaries, with a laminar turbulent transition and in addition the fluid is non-Newtonian. Thus heart valve flow is much more complex than many typical engineering flows. No wonder it has so far resisted a complete clarification, despite the sophisticated methods which have been applied. However, with new approaches one can clarify some of the still unsolved questions. A mechanically simplified flow model permits a more defined flow and more precise measurements of the bulk hydrodynamic qualities of a heart valve. Another flow model makes use of an enlarged geometry and of similarity laws. This flow model permits the application of many flow visualization methods using particles and dye. The resulting still-pictures and video-pictures show a high resolution in space and time, which previously was unobtainable. With this model it is possible to locate and quantifiy the areas of flow separations, which are likely sites for a thrombus to grow. Pathologists have indeed found thrombi at these loci through the examination of artificial heart valves which have been implanted in patients or which have been subjected to bloodflow in artificial bloodpumps. This gives an indirect proof that areas of thrombus generation and areas of flow separation coincide. This is an encouraging result, because there are many ways to modify the flow. With the enlarged model we have a way to objectively check the modifications and to predict the thrombogenicity of a valve.

REFERENCES

1. International Standard Organization (ISO) 5840: (E), Cardiovascular Implants - Cardiac Valve Prostheses, Geneva, Switzerland, 1989.
2. Reul H: In vitro evaluation of artificial heart valves, In: Advances in Cardiovascular Physics, 5:17-30, Karger Pub. Co., 1983.
3. Scotten LN, Walker DK, Smith DW, Brownlee RT: A versatile pump for simulating physiological fluid flows, Proc. AAMI, 18 Ann. Meeting, 1983.

4. Affeld K, Spiegelberg A, Schichl K, Mohnhaupt A: Design of a new tester for artificial heart valves, Life Support Systems, 4(2):142-144, 1986.
5. Affeld K: Dichtung zum Abdichten eines pneumatisch oder hydraulisch betriebenen Kolbens, Offenlegungsschrift DE 3325179 A 1, Deutsches Patentamt, 1985.
6. Affeld K, Pszolla H, Lehmann B, Mohnhaupt R: Measurement of the flow field behind artificial heart valves with the help of the laser-doppler-effect, Proc. ISAO, II:439-441, 1979.
7. Pszolla H, Affeld K, Lehmann B, Mohnhaupt A: Messung des Geschwindigkeitsfeldes hinter einer künstlichen Herzklappe (Björk-Shiley-Ventil) mit dem Laser-Doppler Anemometer, Biomed. Technik 24, Erg.-Band, 1979.
8. Yoganathan AP, Woo YR, Sung HW: Turbulent shear stress measurements in the vicinity of aortic valve prostheses, J. Biomech., 19(6):433-442, 1986.
9. Reul H, Giersiepen M, Knott E: In vitro testing of bioprostheses, Trans. ASAIO, 34:1033-1039, 1988.
10. Affeld K, Schmidt S, Bücherl ES: Form- und Festigkeitsuntersuchungen von Aorten- und Pulmonalklappen des Rindes, Jahrestagung der Deutschen Gesellschaft für Biomedizinische Technik, 85-88, 1973.
11. Jansen J, Willeke S, Reiners B, Harbott P, Reul H, Rau G: New J-3 flexible-leaflet polyurethane heart valve prosthesis with improved hydrodynamic performance, The International J. of Artificial Organs, 14(10):655-660, 1991.
12. Liepsch DW, Levesque M, Nerem RM, Moravec ST: Correlation of Laser Doppler Velocity Measurements and Endothelial Cell Shape in a Stenosed Dog Aorta, Vascular Endothelium in Health and Disease, Plenum Publ. Corp., 43-50, 1988.
13. Astarita G: Scale-up problems arising with non-Newtonian fluids, J. of Non-Newtonian Fluid Mechanics, 4:285-298, 1978.
14. Gad-el-Hak M: Visualization techniques for unsteady flows: An overview, J. of Fluids Engineering, 110:231-243, 1988.
15. Affeld K, Walker P, Schichl K: Novel Flow Visualization to Detect Sites of Thrombus Formation at Artificial Heart Valves, Proc. ESAO, Brno:91-100, 1988.
16. Affeld K, Walker P, Schichl K: The use of image processing for the investigation of artificial heart valve flow, Proc. ASAIO, 294-298, 1989.
17. Gonzalez RC, Wintz P: Digital Image Processing, Addison-Weseley Publishing Company, Reading, UK, 1987.
18. Gentle R: Minimizing of pressure drop across heart valve conduits: A preliminary study, Life Support Systems, 1(4):263-270, 1983.
19. Tansley GD: Numerical Analysis of Turbulent, Non-Newtonian Fluid Flow through Heart Valve Conduits, Ph.D. thesis, Trent Polytechnic, Nottingham, U.K., 1988.
20. Vondran T: Untersuchung der Umströmung einer Starr-Edwards-Kugelherzklappe im Modell, Diplomarbeit, Technische Universität Berlin, Unpublished M.A. - thesis, 1991.
21. Wurzinger L, Opitz R, Wolf M, Schmid-Schönbein H: Shear induced platelet activation, Biorheology, 1985.
22. Hashimoto S, Maeda H, Sasada T: Effect of shear rate on clot growth at foreign surfaces, Artificial Organs, 9(4):345-350, 1985.
23. Schraub FA, Kline SJ, Henry J, Runstadler PW, Littell A: Use of hydrogen bubbles for quantitative determination of time-dependent velocity fields in low-speed water flows, J. of Basic Engineering, Transactions of the ASME, 429-444, 1965.

24. Affeld K, Walker P, Schichl K: Upscaling as a tool in biofluidmechanics - demonstrated at the artificial heart valve flow, In: Biomechanical Transport Processes (Ed., Mosora F et al.), Series A: Life Sciences, 193, Plenum Press, New York, 1990.
25. Woo YR, Yoganathan A: In vitro pulsatile flow velocity and turbulent shear stress measurement in the vicinity of mechanical aortic heart valve prostheses, Life Support Systems, 3:283-312, 1985.
26. Taenzer L: Konvektiver Fluidaustausch bei künstlichen Herzklappen, Diplomarbeit, Technische Universität Berlin, unpublished M.A. thesis, 1991.
27. Brami B: Current trends in prosthetic heart valves, In: Frontiers in Cardiovascular Engineering (Eds., Hwang NHC, Turitto VT, Yen MRT), Plenum Publishing Corp., New York, 1992.
28. Goldsmith H, Turitto VT: Rheological aspects of thrombosis and haemostasis: Basic principles and applications, Thrombosis and Haemostasis, 55(3):415-435, 1986.
29. Schmid-Schönbein H: Thrombose als ein Vorgang in "strömendem Blut". Wechselwirkung fluiddynamischer, rheologischer und enzymologischer Ereignisse beim Ablauf von Thrombozytenaggregation und Fibrinpolimerisation, Hämostaseologie, 8:149-173, 1988.

HEART TRANSPLANTATION: THE PRESENT AND FUTURE – THE "REINA SOFIA" HOSPITAL (CORDOBA, SPAIN) EXPERIENCE

Manuel Concha, Manuel Anguita, Anastasio Montero,
José M. Arizón, Federico Vallés, José M. Latre,
Amelia Jiménez, and Fernando López-Rubio

Heart Transplantation Unit
Hospital Regional "Reina Sofía"
University of Córdoba
14004 Córdoba (Spain)

INTRODUCTION

Although the first experimental studies on heart transplantation were carried out in the first decade of this century by Carrel and Guthrie[1], it was not until December 2, 1967 that the first heart transplant from a human donor to a recipient with end-stage cardiopathy was performed, by the team led by Dr. Barnard in Capetown, South Africa[2]. The technique followed had been developed in Stanford University by Lower and Shumway[3], and still forms the basis of the methods used today by most teams. Since then, advances made in various fields related to transplantation, mainly in immunosuppression and in the diagnosis and treatment of rejection, have given rise to a rapid increase in the number of transplants performed in the world. According to figures provided by the ISHT (International Society for Heart and Lung Transplantation) for 1991[4], by the end of 1990 there had been over 15600 orthotopic and over 300 heterotopic heart transplants in over 150 hospitals all over the world. Allowing for the fact that this Registry is voluntary and that not all transplant centers supply the Registry with information, the total is probably even higher.

In our center we have performed 83 orthotopic heart transplants during the period May 1986 to October 1991. The purpose of this study is to examine our experience with these patients and look at the results obtained, concentrating mainly on the following points: 1) evaluation of patients considered for transplant (clinical and hemodynamic characteristics, short- and medium-term prognostic factors, risk of sudden death and contraindications for transplant); 2) criteria for selection and characteristics of donors, including study of thyroid function; 3) influence of donor and recipient characteristics on posttransplant evolution; 4) analysis of early death and morbidity (during the first 30 days posttransplant); 5) role played by the graft preservation temperature in the possible onset of bradyarrhythmia and the need for a pacemaker after transplant; 6) noninvasive diagnosis of acute rejection using radionuclide and doppler echocardiography techniques; and 7) analysis of late complications and death.

EVALUATION OF RECIPIENTS

The prognosis of patients having severe cardiac failure secondary to myocardial dysfunction of any etiology has improved greatly in the last decade, mainly due to heart transplantation. However, innovations in medical treatment have also led to the prolongation and improvement of these patients' lives[5,6], although in the long term a transplant is still required in most cases[7]. As the number of transplants is mainly limited by the shortage of donors[8], it would seem important to be able to classify patients by risk in order to establish indication and the correct timing for transplant. Some studies[9-12] have identified clinical, analytical and hemodynamic parameters as predictors for survival in series of patients who had cardiac failure in various stages of development, belonged to various functional classes and had different etiologies (Table 1), but the prognostic value of these indicators in a more homogenous group of patients, such as those referred for evaluation for heart transplantation, seems somewhat limited.

Table 1. Classic prognostic markers in patients with heart failure.

Functional class
Ventricular arrhythmia
Exercise capacity
Serum sodium
Left ventricular ejection fraction
Left ventricular end-diastolic pressure
% of fibrosis (endomyocardial biopsy)

For this reason, other more suitable parameters had to be found, such as norepinephrine and atrial natriuretic peptide blood levels[13] or myocardial gammography with meta-iodo-benzyl-guanidine (as an expression of the beta-sympathetic innervation of the myocardium)[14]. These and other techniques are complex, however, and not available to many centers, so it would be useful to find simpler clinical and hemodynamic parameters to assist in determining the prognosis of different subgroups of patients.

General clinical and hemodynamic features

To this end, we have analyzed our experience of 155 consecutive patients referred to our center over the last 5 years for evaluation for possible heart transplant (15-17). Mean age was 45±12 years (12-65). 89% were male. The etiology of cardiac failure was idiopathic dilated cardiomyopathy in 52% of cases, ischemic cardiomyopathy in 34%, and myocardial dysfunction secondary to valvulopathy in the remaining 14%. The mean interval since the onset of symptoms (Class II dyspnea) was 23±27 months.

All patients were examined according to a strict diagnostic protocol in order to evaluate their functional and hemodynamic status, as well as to check for possible contraindications for transplant. The main clinical and hemodynamic characteristics of these patients are shown in Tables 2 and 3. It is worth noting that 93% of the patients were in Functional Class IV of the NYHA and that the left ventricular ejection fraction and cardiac index were very low and telediastolic pressures were very high in both ventricles.

Table 2. Baseline clinical features in our patients.

Age		45±12 years
Sex:	Male:	89%
	Female:	11%
Etiology:	IDCM:	52%
	ICM:	34%
	VHD:	14%
NYHA class:	III:	7%
	IV:	93%
Prior cardiac surgery:		20%
Time from onset of symptoms:		23±27 months
Severe ventricular arrhythmias:		27%
Serum sodium:		136±6 mE/L
Creatinine clearance:		69±30 mil/min

IDCM: idiopatic dilated cardiomyopathy. ICM: ischemic cardiomyopathy. VHD: valvular heart disease.

Table 3. Baseline hemodynamic features in our patients

Left ventricular ejection fraction:	22±7%
Left ventricular end-diastolic pressure:	27±7 mmHg
Right ventricular end-diastolic pressure:	11±6 mmHg
Systolic aortic pressure:	107±16 mmHg
Cardiac index:	2.2±0.6 ml/m/m2
Pulmonary vascular resistance:	4.5±4.1 wood U/m2
Left ventricular end-diastolic volume:	173±63 ml/m2
Left ventricular end-systolic volume:	135±49 ml/m2

All patients received digoxin and diuretics, and 97% tolerated the administration of the angiotensin converting enzyme inhibitors (ACEI) captopril or enalapril. 42% required inotropics (dopamine, dobutamine, amrinone) administered intravenously for a mean period of 13±14 days. Medication enabled clinical stabilization of 76% of cases (15).

Patient evolution

After completion of the entire clinical and hemodynamic examination, and evaluation of response to intensive medication, 56% of patients were considered to be indicated for a heart transplant. In another 20% of cases, contraindications prevented the operation. Finally, in the remaining 24%, transplantation was not considered to be indicated at that point in time due to the acceptable clinical situation of the patients[15]. All patients were followed up for a mean time of 12±11 months, or until death or transplant. Up to April 1991, 53 patients had undergone transplantation (35%), 53 had died without receiving transplant and 41 were still alive without transplant[15].

Actuarial probability of survival was significantly greater for patients without initial indication for transplant than for those who were indicated for transplant but could not be operated on due to contraindications or death while on the waiting list (80% vs 39% the first year and 69% vs 27% at two years, $p<0.001$). It is worth pointing out, however, that 20% of patients initially considered to be "too well" for transplant had died or needed transplant by the end of the first year of follow-up, and the same was true for 31% at two years. Most of these deaths were sudden, and this will be discussed later.

Prognostic factors

Follow-up was greater than 6 months for 30 of the 41 patients surviving without a transplant. Patients who had not received a transplant were divided into two groups: those who died during follow-up (n=53) and those survivors who, when this study was made, had a follow-up time of more than 6 months (n=30)[15,17]. An univariate and multivariate comparison (stepwise logistic regression) of more than 30 clinical and hemodynamic variables of possible prognostic value was made between the two groups. On univariate analysis, the group of survivors, when compared to the group of patients who died, showed lesser need of intravenous inotropics ($p<0.001$), responded in greater numbers to medication ($p<0.001$), tolerated a higher dosage of captopril ($p<0.01$), and showed a longer time for the evolution of symptoms ($p<0.01$). Survivors also had higher systolic arterial pressure and serum sodium levels ($p<0.001$). Left ventricular ejection fraction was slightly greater in live patients than in those who died during follow-up ($p<0.05$). The remaining parameters studied (telediastolic pressure of the left ventricle, ischemic or idiopathic etology, ventricular or supraventricular arrhythmia, age, sex, amiodarone therapy, cardiac index, pulmonary vascular resistance etc.) showed no significant differences[15-17].

On multivariate analysis, only three variables were independent predictors of prognosis: need for intravenous inotropics, systolic arterial pressure and captopril dosage. When transplanted patients were included in the figures for dead patients and comparison between the two groups was repeated, there was practically no change in the results of the univariate analysis, although on multivariate study the response to medical therapy showed as an independent predictor (Table 4).

Table 4. Independent factors predicting prognosis in our patients with severe heart failure.

	Coefficient	Standard error	p value
Need for inotropics	1.8304	0.8419	<0.001
Systolic blood pressure	-0.5029	0.1938	0.003
Captopril dosage	-0.2209	0.9369	0.043
Response to therapy	-0.3090	0.1765	0.008
Constant	7.8160	2.3110	

Stepwise logistic regression

To conclude, the "classic" indicators of prognosis in cardiac failure, such as left ventricular ejection fraction and end-diastolic pressure, or the presence of ventricular arrhythmia are unable to discriminate in a homogenous group of patients with very severe clinical and hemodynamic impairment. However, some simple parameters, such as need for inotropics, systolic arterial pressure, tolerated captopril dosage or response to intensive medical therapy, may be of great use in such circumstances, enabling the identification of the patients with the poorest prognosis and thus the greatest priority for heart transplantation.

Sudden death in patients with severe heart failure

Incidence of sudden death in patients with severe cardiac failure is around 5-10%[13,18], so we should try to identify this group of high-risk patients and thus be able to determine if and when transplantation is appropriate. To this end we have conducted a survey of the 155 patients previously mentioned[15]. During the follow-up period of 12±11 months there were 52 deaths: 34 deaths were due to cardiac failure and 18 were sudden[19]. Four of these 18 patients were resuscitated and they later successfully underwent transplantation. Overall incidence of sudden death in our series was 12%, one third of these occurring in the patients group initially not considered suitable for transplant because of their acceptable clinical condition. In this group, 16% of all patients died suddenly, compared to 10% in the group initially indicated for transplant. The ratio of sudden deaths to total deaths was 86% for the first group and only 26% for the second (p<0.05), where death due to cardiac failure was clearly predominant. Moreover, the greatest number of sudden deaths in the group indicated for transplant occurred in those patients whose situation was most stable[19].

50% of all sudden deaths happened during the first two months of follow-up, with only 15% occurring after one year. The mechanism behind sudden death could be clearly identified in 9 cases: 6 were due to ventricular tachychardia/fibrillation and 3 due to bradycardia; the other 9 deaths took place in the patients' homes.

We have also performed univariate and multivariate studies of sudden deaths and deaths due to cardiac insufficiency, in order to identify possible predictors. In summary, patients who died due to cardiac failure, when compared to those who died suddenly, showed worse response to initial medical therapy (p<0.01), greater need for intravenous inotropic drugs ($p < 0.01$), lower serum sodium ($p < 0.01$), higher transpulmonary gradient ($p < 0.01$) and higher pulmonary resistance and pressures ($p < 0.05$). On the multivariate study, the need for inotropics and the response to treatment were again seen to be independent predictors[19].

When sudden death patients were compared to survivors without transplant (the group with better evolution), the former presented a higher percentage of ventricular arrhythmia ($p < 0.05$), lower tolerance to captopril ($p < 0.01$) and lower left ventricular ejection fraction, systolic blood pressure and time of exercise (ergometry) ($p < 0.05$)[19].

Thus, it can be concluded that, using clinical and hemodynamic parameters, it seems possible to identify within a group of patients with severe cardiac failure a subgroup of patients with a tendency for sudden death, distinguishing them not only from those who die from cardiac failure but also from patients showing good evolution with no need for transplantation. This could be of great help in indicating this type of patient for transplantation.

EVALUATION OF DONORS

The correct selection of donors is crucial to the success of any heart transplant program. Until just a few years ago, the universally-accepted characteristics for a heart donor were: age under 40 years; donor/recipient weight ratio not less than 0.85 (in other words, the recipient should not be more than 15% heavier than the donor); total ischemic time less than 4 hours; and absence of moderate inotropic support. The ideal donor should also satisfy the following conditions: no prolonged cardiac arrest; no intense thoracic traumatism; no previous cardiopathy; absence of active infection; and no neoplastic processes (except for primary cerebral tumors). He or she should display hemodynamic stability and not require inotropics or only low-dose catecholamine support.

However, given the limited number of donors and the widening of the selection criteria for recipients, it has been necessary to use donors who fail to satisfy some - or many - of these criteria. Several authors have reported good results from such organs[20,21]. In the present day, implantations are being made of hearts with a weight difference with respect to the recipient of up to 25-30%, and also of organs from donors with thoracic traumatism, aged over 45-50 years (although in this case previous coronariography is recommended in order to check for coronary lesions, especially if the donor has a high coronary risk factor), or even of donors suspected of infection. Logically, any decision to accept such a donor would also depend on the clinical and general situation of the recipient.

ABO compatibility is absolutely essential, as is the absence of cytotoxic antibodies against the donor antigens. The donor must have a normal physical and cardiological examination, and a normal chest x-ray film and electrocardiogram (although brain death can cause alterations in the electrocardiogram). Where doubt exists, echocardiography of the donor is of great importance, and this should be carried out systematically. In our experience, we have rejected 5 hearts from questionable donors after finding significant disorders of regional and global contractility. In one case, death had been caused by carbon monoxide poisoning, in another the donor was addicted to cocaine and in yet another there were several coronary risk factors. Two other hearts were rejected during explantation when the surgeon observed the presence of mitral stenosis in one case and atrial septal defect with dilation of the right cavities in another. These lesions could have been detected by echocardiography. Mean age of our donors was 26±8 years (15-49). 81% were male and 19% female. Mean donor weight was 73±12 kg (45-100), with a donor/recipient weight ratio of 1.1±0.2 (0.7-2.1). No significant statistical association was found between any of the features of the donors and the incidence of perioperative complications or early or late death, nor with post-transplant hemodynamic condition.

Study of thyroid function in potential donors

Previous studies have documented the existence of hyperthyroidism in patients after brain death[22]. This thyroid disorder could have harmful effects on the myocardium. We have studied thyroid function before and after brain death, as well as the density of T3 receptors in the myocardium[23,24]. Results showed T3 levels before brain death to be lower than those of the control group, whereas rT3 levels were higher, response to TSH being normal after stimulation by TRH. After brain death, however, there was no change in T3 levels, but rT3 and T4 levels fell significantly and the normal response of TSH to TRH was abolished. Additionally, occupied myocardial T3 receptors were less abundant in donor hearts (3 receptors / 3 micron2) than in healthy myocardia (15 receptors / 3 micron2)[23,24]. So it would seem that while before brain death a "low T3 syndrome" exists, after brain death (when the patient is now a potential donor) there is an obvious hypothyroidism. The first experiments on the treatment of these patients with T3 produced no satisfactory results, however, and this point remains controversial.

INFLUENCE OF DONOR AND RECIPIENT CHARACTERISTICS ON POST-TRANSPLANT EVOLUTION

We stated earlier that there was no correlation between donor characteristics and early or late post-transplant evolution, nor with mortality or graft function, although it should be noted that, generally speaking, we have been working within rather strict limits where the acceptance of donors is concerned, and this could obviously influence results.

With respect to the pre-transplant characteristics of recipients, it is widely accepted that the initial selection criteria for candidates for heart transplant comprised a very strict list of contraindications (Table 5). Time, however, has produced advances in immunosupression and improved results, so these contraindications have become more and more relative and patients with one or even several of these features can now be considered for transplant. Various studies have reported good results in patients over 55 or even 60 years old[25,26], patients with diabetes mellitus[27], and indeed in patients with a history of malignancies[28].

We have made a survey of our first 57 transplant patients, evaluating the influence of six factors theoretically unfavorable to survival after transplant[29]. These were: 1) age over 55 years (21% of cases); 2) diabetes mellitus requiring administration of insulin or oral hypoglycemic agents (7%); 3) active infection during the week before transplant (9%); 4) significant pulmonary hypertension, defined by the presence of a pulmonary vascular resistance greater than 5 Wood units/m2 and/or a transpulmonary

Table 5. "Classic" contraindications for heart transplantation

Pulmonary vascular resistance > 8 Wood units
Active infection
Diabetes mellitus
Malignancy
Active peptic ulcus
Drug or alcohol addiction
Pulmonary embolism (< 2 months)
Severe psichiatric disturbs
Severe obstructive or restrictive pulmonary disease
Severe, irreversible hepatic or renal failure
Age over 50 years

gradient greater than 12 mmHg (26%); 5) renal failure, defined by serum creatinine levels over 2 mg/dl and/or creatinine clearance under 35 ml/min before transplant (11%); and 6) critical or unstable clinical condition at the moment of operation (urgent transplantation), most of these cases being in cardiogenic shock (25%). We have also evaluated the influence of an overall risk score which was calculated for each patient by adding one point for each of the above characteristics. 38% of patients had zero score (the "ideal" recipient); 25% had a score of 1; 23% scored 2; and 14% scored over 2.

Actuarial probability of survival at one month, one year and at eighteen months for patients with or without each of the characteristics studied are shown in the Table 6.

It should be noted that only pulmonary hypertension seemed to influence early death, whereas the other factors seemed to come into play in the longer term. Probability for survival was significantly lower for patients over 55 years of age ($p<0.05$), for those with higher pulmonary vascular resistance ($p<0.01$) and for those with serious renal failure ($p<0.01$). On the other hand, infection, diabetes and pretransplant clinical condition did not seem to have much influence.

By risk score, actuarial probability of survival was significantly higher for patients with lower scores (0 or 1) than for patients with scores of 2 or more (74% vs 31% at 18 months, $p<0.001$). There was no significant difference between patients with a score of 0 or 1 (87% vs 67% at 18 months) and between patients with a score of 2 or more (31% vs 39% at 18 months). On multivariate analysis (stepwise logistic regression), global risk score ($p=0.003$) and renal failure ($p=0.074$) were the only independent predictors for death after transplant.

So although it appears quite possible to perform a heart transplant with a guaranteed outcome on patients with theoretically unfavorable characteristics, the association of several of these factors in one patient considerably raises the risk involved. Accordingly, given the limited number of donors, teams should be more selective when deciding whether a patient with "relative" contraindications should be indicated for transplant.

Table 6. Actuarial survival (Kaplan-Meier method) of our patients according to pretransplant recipient features.

	1 month	1 year	18 months
Age > 55 years	83%	60%	45%
Age < 55 years	91%	78%	68%
Elevated PVR	73%	38%	38%
Lower PVR	95%	87%	72%
Diabetes mellitus	100%	75%	50%
No diabetes	88%	73%	63%
Renal failure	83%	33%	16%
No renal failure	90%	80%	70%
Active infection	80%	60%	60%
No infection	90%	76%	63%
Unstable pre-HT	88%	60%	45%
Stable pre-HT	92%	78%	69%

PVR: pulmonary vascular resistance. HT: heart transplantation

ANALYSIS OF EARLY MORBIDITY AND MORTALITY

By early morbidity and mortality we normally mean that which occurs during the first 30 days after transplant. Despite recent advances, rejection and infection are still the leading causes of morbimortality when we examine the global results after heart transplantation. Over the years there has been a gradual decrease in early mortality, but figures are still relatively high - between 10 and 14 percent, according to the Registry of the ISHT[4,30]. In the Spanish Registry figures for early death corresponding to the same period are somewhat higher[31]. The causes of death in more than 40% of cases were basically of a cardiac nature, including technical complications arising during operation (bleeding, poor preservation, etc.) and acute right cardiac failure. Many of these complications could be avoided or reduced by suitable handling of the donor heart, meticulous technique and constant respect for myocardial preservation procedure during the transportation and implantation of the graft.

Here we present a retrospective study analyzing early morbidity and mortality (< 30 days) in our total experience of 83 patients undergoing heart transplantation, with the aim of determining the incidence of different complications and the predictive value of the different variables on which they might depend.

Material and methods

From May 1986 to October 1991, a total of 83 patients underwent heart transplantation in Reina Sofia Hospital, Córdoba, (Spain). From 1986 through 1989, 26 transplants were performed, while the remaining 57 were done in 1990 and 1991. The underlying cardiopathy was idiopathic dilated cardiomyopathy in 34 cases, ischemic cardiomyopathy in 33, severe secondary myocardial dysfunction to valvulopathy in 13, and miscellaneous in the 3 remaining cases. Age ranged from 12 to 62 years, the majority (58 patients, 69%) being between 40 and 60 years old. Eleven were female (14%) and 72 male (86%).

Of the more important preoperative data it is worth noting that 16 patients (20%) had undergone previous operation, 6 had endocavitary pacemakers fitted, 38 required anticoagulants (48%) and 31 were given amiodarone before operation (39%). Inotropic support (dopamine, dobutamine and/or amrinone) was required by 24 patients (30%) just before transplant. Overall preoperative clinical status was considered to be less-than-ideal in 38 cases (46%); recipients were considered to be thus if they needed inotropics at the moment of transplant, were in cardiac shock, and/or pulmonary vascular resistance was greater than 3.5 Wood units x m2, or when the recipient was over 55 years of age and had a serum creatinine level of over 2 mg/dl. Twelve patients in cardiac shock received emergency transplant.

Pulmonary vascular resistance varied between 0.5 and 8.7 Wood units x m2, with a mean value of 2.5±1.4. Total ischemic time was 136±38 minutes (50-255), and cardiopulmonary bypass time was 121±53 minutes. In every case myocardial temperature was systematically controlled both during extraction of the organ and on its arrival at the hospital, prior to implantation; pretransplant graft temperature was 6±2.6 °C (1.9-14.3). Temperature of the organ preservation liquid during transport was 7.1±2.1 °C (1.3-10.8). At the time of extraction, every heart was given the same cardioplegic dose of one liter of St. Thomas II solution, immediately after aortic clamping. Left ventricular distension was carefully avoided. Extracorporeal circulation was maintained by a membrane oxygenator, with general hypothermia at 28°C. Superficial cardiac irrigation was associated using Ringer's lactate at 4°C. After placing sutures on the left atrium, pulmonary artery and aorta and once suture of the right atrium was started, the aorta was unclamped and controlled perfusion was maintained at an arterial blood temperature of 37°C, perfusing the myocardium at a maximum pressure in the aorta of 50 mmHg until cardiac rhythm was reestablished.

Hematic cardioplegia at 8°C was employed in the most recent sixteen consecutive cases when the heart reached the theater; ischemic time in these cases was 161±44 minutes, significantly higher than that of transplants performed using traditional cardioplegia (131±37, p<0.05). Preimplantation temperature (8±2°C vs 5.4±2.5°C, p<0.05) was also significantly higher.

In the 36 consecutive transplants performed since October 1990, a protocol of routine administration of aprotinine (Trasylol-R) has been followed.

Details of the immunosuppression protocol have already been published[32]. Initial myocardial biopsy was performed at 7-10 days posttransplant, and repeated weekly for the first month. We have previously described a protocol for noninvasive follow-up and detection using radionuclide methods, and this has been applied to all patients (33).

Statistical analysis was performed using Student's t-test and the chi-square test.

Results: complications

During the immediate postoperative period there was severe pleural effusion in 8 patients (10%) and reoperation was necessary in 4 cases (5%) due to bleeding (but there was no occurrence of this after the application of the aprotinine protocol). 5 patients (6%) presented some type of neurological complication, but never with sequelae; 12 suffered from temporary renal failure (15%), and one patient required dialysis. There were 9 cases of sustained fever in the first month (11%).

Postoperative bleeding during the first 24 hours was 733±537 cc in the first 47 cases, and reoperation was required in 4 of these (8%). In the other 36 cases, to whom the aprotinine protocol had been applied at operation, blood loss was only 268±109 cc on the first day, and no patient required reoperation.

Total intubation time for respiratory assistance was 36±25 hours (7-168), and patients spent an average of 20±13 days (5-62) in the Heart Transplantation Unit. In the first month, six patients (8%) experienced an episode of severe rejection and another six (8%) suffered moderate rejection.

Low cardiac output was in evidence in 12 patients (16%); 6 of these cases (8%) were severe and required intensive inotropic support. There was no clinical evidence of low cardiac output in 84% of patients. Statistical analysis showed no association between low output and any of the parameters studied (age of recipient, pretransplant serum creatinine, inotropics, pulmonary resistance, non-optimal recipient, etc.).

The incidence of bradyarrhythmia and the need for pacing is discussed later. There was an overall postoperative incidence of bundle branch block of 67% (almost always, 65%, of the right branch).

In spite of increased ischemic time in the last 16 transplants, in which hematic cardioplegia was employed, there was no evidence of either complete atrioventricular block or sinus dysfunction, the incidence of low output was nul and the incidence of bundle branch block was reduced from 70% to 25% ($p<0.05$). It should also be borne in mind that myocardial preservation temperature was also higher in these 16 cases, as mentioned previously.

Results: mortality

There were seven in-hospital deaths (8.7%) during the first month. When this mortality was examined by periods there were no significant differences. In 1986-1989, of 26 transplanted patients three (10%) died in the first month; in 1990 and 1991, during the same initial posttransplant period, there were 4 deaths (7%) from 57 patients who underwent transplantation. No significant statistical association was found between early death and the clinical history of the recipient (underlying cardiopathy, pulmonary resistance, previous operation, etc.), need for preoperative inotropics, urgent or elective operation, ischemic and cardiopulmonary bypass times or donor characteristics.

Of the 7 hospital deaths, 3 were due to cardiac causes (primary failure of graft); 2 were caused by sepsis (one of these cases had had pneumonia previous to transplant); 1 death was due to a hematological disorder with widespread intravascular coagulation); the remaining case was due to anaphylactic shock brought on by the administration of ATG.

Conclusions

There has been a remarkably low rate of morbidity and early death in our patients, and this has been true since the first transplants were performed. This must be connected with the scrupulous care exercised during every phase of the transplantation process (donor selection, preservation, surgical technique and postoperative care). It is also worth pointing out the good results obtained by the aprotinine protocol, and also the preliminary experience with the use of hematic cardioplegia in the 16 most recent patients, whose early mortality has been non-existent and in whom there has been an extremely low incidence of complication.

INFLUENCE OF MYOCARDIAL PRESERVATION TEMPERATURE ON THE DEVELOPMENT OF POSTTRANSPLANT BRADYARRHYTHMIA

4-7% of transplant patients require permanent pacing, according to most reports[34,35], though around 25% of patients need temporary electrical stimulation because of severe bradyarrhythmia occurring during the first few days posttransplant. While complete atrioventricular block may be produced in some cases, the mechanism most often responsible is sinus dysfunction. The mechanism of this bradyarrhythmia is probably multifactorial, attributable to acute or chronic rejection, ischemia of the organ, previous ingestion of amiodarone, coronary anomalies, surgical trauma, etc.[34-36]. Occasionally, the development of sinus dysfunction or atrioventricular block is clearly due to rejection, although in these cases bradyarrhythmia does not usually appear so early.

We have studied the incidence of bradyarrhythmia and the need for pacing in our first 52 consecutive transplants. Other factors possibly connected to these disorders have also been studied[37]. The overall incidence of bradyarrythmia requiring at least 24 hours of pacing during the first posttransplant days was 27% (14 patients). In all but four patients, normal sinus rhythm was recovered in less than 3 weeks, and so only those 4 patients (7.6 of the 52 patients and 28% of those requiring temporary pacing) underwent permanent pacemaker. Another two patients developed late atrioventricular block in the second posttransplant month, associated to episodes of acute rejection, and they also needed to be fitted with a pacemaker. Thus, the overall incidence of permanent pacemakers was almost 11%.

We carried out a study to compare various variables between groups of patients with or without need of temporary pacing during the first days after transplant. Age and sex of donors and recipients were similar in both groups, as were the percentage of patients who had been receiving amiodarone before transplantation, total ischemic time and the time of cardiopulmonary bypass. No moderate or severe rejection was seen in the first biopsy (performed at 7-10 days after transplant) of the 14 patients who required pacing. Only the temperature of myocardial preservation was found to be significantly different. The initial temperature of the cardioplegic solution was $5.3 \pm 1.7 °C$ for patients with bradyarrhythmia and $6.5 \pm 1.5 °C$ for those without ($p < 0.05$). In addition, myocardial temperature before implantation was $3.9 \pm 1.6 °C$ for the first group and $5.7 \pm 2.6 °C$ for the second ($p < 0.01$). As previously mentioned, myocardial preservation temperature in the 16 most recent cases - not included in this study - was higher ($8 \pm 2 °C$), and no type of bradyarrhythmia has been observed in these patients.

To conclude, although the mechanism of early bradyarrhythmia occurring after transplant is probably multifactorial, low preservation temperature may well play an important part, and should be avoided. No difference was found either in graft functionality or mortality by raising myocardial preservation temperatures to the limits described above.

NONINVASIVE DIAGNOSIS OF REJECTION

Cardiac rejection is a basically anatomopathological concept, characterized by the presence of inflammatory (mainly lymphocyte) infiltration and necrosis of the transplanted myocardium. For this reason the "gold standard" for diagnosis of rejection is endomyocardial biopsy - a cruent, invasive process which, in addition to being of great discomfort to the patient, has to be repeated at certain intervals. So other noninvasive methods which might give a reliable diagnosis of rejection have been sought. Noteworthy amongst the numerous techniques which have been employed are doppler-echocardiography and radioisotopic studies of ventricular function and antimyosin antibodies.

Doppler-echocardiography

Echocardiography is an accessible, reliable, noninvasive technique for the study of systolic and diastolic ventricular function and which should therefore be of great use in the determination of secondary alterations occuring as a result of rejection. We carried out a study of our patients, performing both echocardiography and endomyocardial biopsy on the same day, and correlated various echocardiographic parameters with the biopsy results[38]. In our experience, the presence of severe damage to ventricular systolic function is highly indicative of a serious degree of rejection. The practising of echocardiography on these patients permitted evaluation of the effectiveness of the anti-rejection treatment used. Fortunately, cases of severe rejection with deterioration of myocardial contraction are very rare these days, probably owing to prophylactic immunosupression therapy.

The study of diastolic function could, theoretically, detect less severe degrees of rejection, as the first thing to deteriorate on cardiac rejection is ventricular distensibility. In our experience, however, the usual parameters of diastolic function determined by doppler-echocardiography are incapable of differenciating between patients with or without rejection, nor can they distinguish between treatable- and non-treatable cases[38]. The only significance found was for the shortening of isovolumetric relaxation time in patients with moderate rejection compared to those with light or zero rejection ($p < 0.05$), but sensitivity and specificity were not very high. This lack of sensitivity may be explained by the changes in atrial contractility present in transplant patients, due to the existence of two "atria" (one from the donor and one from the recipient). Transplant patients also display a high variability in atrioventricular filling patterns[38].

Antimyosin antibodies. Study of left ventricular systolic and diastolic function by radionuclide methods.

We investigated the usefulness of myocardial uptake of antimyosin antibodies labeled with indium 111 and of left ventricular systolic and diastolic function determined by gated ventriculography in the diagnosis of acute rejection[39].

A prospective protocol which included 160 studies on 36 consecutive patients was made. This included gated left ventriculography, myocardial gammagraphy with antimyosin antibodies and endomyocardial biopsy, all performed within 48 hours. Isotope ventriculography was carried out by the administration of 740 MBq of 99mTc-albumin and taken at 32 frames per cycle. Analysis was made of overall ejection fraction, mean and peak emptying rates (systolic parameters) and mean and peak filling rates (diastolic parameters). Filling and emptying rates were obtained from the first derivative of the ventricular activity/time curve, and normalized as end-diastolic volume/sec. Immediately after ventriculography, 74 MBq of indium-111 antimyosin were injected intravenously; 48 hours after injection, the myocardial/pulmonary uptake ratio was calculated, in counts/pixel. Endomyocardial biopsy was performed by following the usual technique of percutaneous puncture of the right internal jugular vein. Rejection was classified according to the standards of the University of Stanford.

The results of this protocol showed statistically significant differences ($p < 0.001$) between mean values for the myocardial/pulmonary antimyosin uptake index, and the mean and peak emptying rates of cases with or without rejection. When patients were divided into those who had treatable rejection (moderate or severe) and untreatable (light or non-existent), systolic functional parameters, including overall ejection fraction, were also statistically significant ($p < 0.05$). Antimyosin index, diastolic rates and ejection fraction were significantly correlated ($p < 0.05$) to the extent of rejection[40,41].

The sensitivity of an antimyosin uptake index > 1.75 in the diagnosis of rejection was over 90%[40,41].

To conclude, these results demonstrate the immense usefulness of radioisotope studies in the noninvasive determination of rejection, and this has led to the continued reduction of the use of routine endomyocardial biopsy on our patients in examinations made one month or more after transplant, with no detrimental changes to results.

Radionuclide study of right ventricular function

Due to the morphological and functional characteristics of the right ventricle, changes brought on by rejection will predictably show up here sooner than in the left ventricle. For this reason we have also begun a study of the usefulness of right ventricular function, as determined by radioisotope techniques, in the diagnosis of rejection[42]. 98 studies were carried out on 52 patients. Right ventricular function was studied by right ventriculography using the first pass and equilibrium technique with 99mTc-albumin, and results were correlated to those of endomyocardial biopsy. Preliminary results have shown significant differences in the ejection fraction obtained both in first pass and in equilibrium between patients with or without rejection and those with or without treatable rejection (as defined above)[42].

Conclusions

The present results point to how useful radioisotope studies can be in the noninvasive diagnosis of rejection. Future studies must determine which of the techniques and parameters, or combinations thereof, yield the highest sensitivity and specificity. These techniques may also have a role to play in following-up episodes of rejection and in monitoring the effectiveness of their treatment. Finally, the long-term evolution of post-transplant ventricular function must be studied, as should the factors which could influence this, and here again radioisotope studies have much to offer.

LONG-TERM RESULTS

Heart transplantation involves a high rate of morbidity, especially in the first year posttransplant, primarily due to rejection, infections and complications attributable to immunosuppressive drugs. In our experience, the latter complications are frequent but not too serious. In only one case did death - by anaphylactic shock - result, as the result of administration of ATG during the first few days posttransplant.

The incidence of moderate rejection determined by endomyocardial biopsy was 18% in our patients, whereas severe rejection was only detected in 2.5% of biopsy specimens.

There were 34 episodes of serious infection (0.4/patient), and these are listed in the table 7. Bacterial infections were the most common, followed by those caused by protozoans (Toxoplasma gondii and Pneumocystis carinii), viruses (cytomegalovirus, herpes zoster and hepatitis C) and fungi (Candida albicans and Aspergillus).

Table 7. Incidence and causes of infections in our patients

Bacterial	17(3*)	50%
Protozoans	7	20%
Viral	5 (2*)	15%
Fungal	2 (1*)	6%
Unknown	3	9%
Overall	34 (6*)	

* Cause of death (27% of all deaths)

Mortality

There were 7 cases (8.4%) of hospital death (<30 days). Total mortality was 23 patients (27%). Most deaths were due to acute rejection (5 patients, 22% of total mortality) and infection (6 cases, 27%). The table 8 lists causes of death, related to posttransplant time. Only one death was due to cancer, where a patient with an adenocarcinoma of the colon died a few months after its extirpation because of recidivation of the tumor. Early mortality has been notably low, whereas late death has perhaps been somewhat higher, generally owing to rejection and sudden death due to unknown causes.

Overall actuarial probability of survival for our patients was 80% at 6 months and 72% at one year. These figures were slightly higher for patients undergoing operation in 1990-1991. The great majority of survivors are in excellent functional condition and have successfully returned to their pretransplant occupations.

PERSPECTIVE

Heart transplantation represents one of the greatest recent advances in the field of cardiac pathology, and has become a safe, effective option for the treatment of a large group of patients who previously had nothing to look forward to but imminent death. The introduction of cyclosporine has meant a great leap forward in the control of rejection. There are still some problems to be solved, though most of these are currently the subject of research. One of the main problems is the limited number of donors, which in turn restricts the widening of criteria for possible recipients; this is particularly serious in the pediatric sector. The answer lies in increasing public awareness and in the use of other sources (artificial hearts and xenografts). Great importance should also be attached to the development of new preservation techniques for donor hearts, which would allow augmentation of ischemic time. Currently existing palliative procedures, such as cardiomyoplasty, may also have an important role to play

Table 8. Causes of death in our patients

In-hospital (one month)	7 (8.7%)
Cardiac	3
Infection	2
Intravascular coagulation	1
Anaphylactic shock	1
Medium-term (1 month - 1 year)	9 (10.8%)
Infection	3
Arrhythmia	2
Acute rejection	2
Colon carcinoma	1
Sudden death	1
Long-term (> 1 year)	7 (8,7%)
Acute rejection	3
Sudden death	2
Viral hepatitis	2

in certain cases. All these factors, together with both the development of immunosuppressive drugs with increased selectivity but fewer side-effects and the creation of new techniques for the early diagnosis of rejection, should help widen even further the exciting prospects already offered by heart transplantation.

REFERENCES

1. Carrell A, Guthrie CC: The transplantation of veins and organs, Am. Med. (Philadephia), 10:1101, 1905.
2. Barnard CN: The operation: a human cardiac transplantation. An interin report of the successfull operation performed at Groot Schur Hospital, Cape Town, S. Afr. Med. J., 41:1271, 1967.

3. Lower RR, Shumway NE: Studies on orthotopic transplantation of the canine heart, Surg. Forum., 11:18, 1960.
4. Kaye MP: Heart Registry Report 1991, Eleventh Annual Meeting of the International Society for Heart and Lung Transplantation, Paris, 1991.
5. The CONSENSUS Trial Study Group, Effects of enalapril on mortality in severe congestive heart failure, N. Eng. J. Med., 316:1429, 1987.
6. Stevenson LW, Dracup KA, Tillisch JH, et al: Efficacy of medical therapy tailored for severe congestive heart failure in patients transferred for urgent cardiac transplantation, Am. J. Cardiol., 63:461, 1989.
7. Stevenson LW, Fowler MB, Schroeder JS, Stevenson WG, Dracup KA, Fond V: Poor survival of patients with idiopathic dilated cardiomyopathy considered too well for transplantation, Am. J. Med., 83:871, 1987.
8. Evans RW, Mannihen DL, Garrison LP, Maier AM: Donor availability as the primary determinant of the future of heart transplantation, JAMA, 255:1892, 1986.
9. Cleland J, Dargie H, Ford I: Mortality in heart failure: clinical variables of prognostic value, Br. Heart J., 58:572, 1987.
10. Franciosa J, Wilen M, Ziesche S, Cohn J: Survival in men with severe chronic left ventricular failure due to either coronary heart disease or idiopathic dilated cardiomyopathy, Am. J. Cardiol., 51:831, 1983.
11. Fuster V, Gersh B, Giuliani E, Tajik A, Brandenburg R, Frye R: The natural history of idiopathic dilated cardiomyopathy, Am. J. Cardiol., 47:525, 1981.
12. Lampert S, Lown B, Graboys T, Podrid P, Blatt C: Determinants of survival in patients with malignant ventricular arrhythmia associated with coronary artery disease, Am. J. Cardiol., 61:791, 1988.
13. Keogh AM, Baron DW, Hickie JB: Prognostic guides in patients with idiopathic or ischemic dilated cardiomyopathy assessed for heart transplantation, Am. J. Cardiol., 65:903, 1990.
14. Benvenuti C, Merlet P, Loisance D, et al: Optimal timing of heart transplantation in dilated cardiomyopathy using radionuclide I-123 metaiodobenzylguanidine (MIBG) cardiac imaging, J. Heart Lung Transplant, 10:174, 1991.
15. Vallés F, Anguita M, Arizón JM, et al: Factores predictores de supervivencia a medio plazo en pacientes con insuficiencia cardiaca severa, Rev. Esp. Cardiol., 44 (Supl.1):23, 1991.
16. Anguita M, Vallés F, Arizón JM, et al: Indication of heart transplantation in severe left ventricular dysfunction and class IV cardiac failure patients, Eur. Heart J., 11 (Suppl):289, 1990.
17. Vallés F, Anguita M, Arizón JM, et al: Prognostic factors in patients with severe left ventricular dysfunction and NYHA class IV, Philippine J. Cardiol., 19:I-320, 1990.
18. Luu M, Stevenson WG, Stevenson LW, Baron K, Walden J: Diverse mechanism of unexpected cardiac arrest in advanced heart failure, Circulation, 80:1675, 1989.
19. Anguita M, Vallés F, Arizón JM, et al: Incidencia y factores predictivos de muerte súbita en pacientes con insuficiencia cardiaca severa, Rev. Esp. Cardiol., 44 (Supl. 1):112, 1991.
20. Emery RW, Cork RC, Levinson MM, et al: The cardiac donor: a six-year experience, Ann. Thorac. Surg., 41:356, 1986.

21. Schuler S, Parnt R, Warnecke H, Matheis G, Hetzer R: Extended donor criteria for heart transplantation, J. Heart Transplant, 7:326, 1988.
22. Novitzky D, Cooper DKC, Reichart B: Hemodynamic and metabolic responses to hormonal therapy in brain-death potential organ donors, Transplantation, 43:852, 1987.
23. Montero JA, Mallol J, Alvarez F, Benito P, Concha M, Blanco A: Biochemical hypothyroidism and myocardial damage in organ donors: are they related?, Transplant. Proc., 20:746, 1988.
24. Benito P, Montero JA, Avila L, et al: Thiroid function in potential organ donors. Influence on myocardial damage, Clin. Transplantation, 3:355, 1989.
25. Olivari MT, Antolick A, Kaye MP, Jamieson SW, Ring S: Heart transplantation in elderly patients, J. Heart Transplant, 7:258, 1988.
26. Renlund DJ, O'Connell JB, et al: Exclusion of the older heart transplant candidates from consideration: lack of medical justification, J. Heart Transplant, 9:56, 1990.
27. Rhenman MJ, Rhenman B, Icenogle T, Christensen R, Copeland J: Diabetes and heart transplantation, J. Heart Transplant, 7:356, 1988.
28. Edwards BS, Hunt SA, Fowler MB, Valantine HA, Stinson EB, Schroeder JS: Cardiac transplantation in patients with preexisting neoplastic diseases, Am. J. Cardiol., 65:501, 1990.
29. Anguita M, Arizón JM, Vallés F, et al: Influence of "unfavourable" features of recipients on survival after heart transplantation, J. Heart Lung Transplant, (in press).
30. Heck CF, Shumway SJ, Kaye MP: The Registry of the International Society for Heart Transplantation. Sixth official report, J. Heart Transplant, 8:271, 1989.
31. Vázquez de Prada JA: Registro Nacional de Trasplante Cardiaco, Rev. Esp. Cardiol., 44:293, 1991.
32. Anguita M, Arizón JM, Vallés F, et al: Mixed fungal (Candida and Aspergillus) sepsis after heart transplantation, Trasplante, 2:25, 1991.
33. Latre JM, Arizón JM, Anguita M, et al: Diferenciación del rechazo "tratable" en el trasplante cardiaco mediante métodos radioisotópicos, Rev. Esp. Cardiol., 43 (Supl. 3):122, 1990.
34. Leonelli F, Boahene A, O'Neill G, et al: Temporary and permanent pacing requirements in patients after orthotopic heart transplantation, Eur. Heart J., 11:289, 1990.
35. DiBiase A, Tse CT, Schnittger I, Stinson EB, Valantine HA: Indications for permanent pacemaker implantation in cardiac transplant patients, Circulation, 80:II-527, 1989.
36. Loria K, Salinger M, McDonough T, Frohlich T, Arentzen C: Activitrax AAIR pacing for sinus node dysfunction after orthotopic heart transplantation: an initial report, J. Heart Transplant, 7:380, 1988.
37. Montero JA, Anguita M, Concha M, et al: Pacing requirements after orthotopic heart transplantation. Incidence and related factors, J. Heart Lung Transplant, (in press).
38. Vivancos R, Anguita M, Segura J, et al: Ecocardiografía-doppler en la valoración del rechazo cardiaco, Bol. Soc. and Cardiol., 17:18, 1991.
39. Arizón JM, Latre JM, Anguita M, et al: Diagnosis of acute heart rejection by radioisotopic left ventricular function and 111-In.antimyosin myocardial uptake, Eur. Heart J., 11 (Suppl):186, 1990.

40. Latre JM, Arizón JM, Jiménez A, et al: Non-invasive radioisotopic diagnosis of acute cardiac rejection (abstract), J. Heart Transplant, 10:173, 1991.
41. Latre JM, Arizón JM, Jiménez A, et al: Non-invasive radioisotopic diagnosis of acute cardiac rejection, J. Heart Lung Transplant, (in press).
42. Latre JM, González F, Jiménez-Heffernan A, et al: Diagnóstico no invasivo del rechazo cardiaco mediante el estudio isotópico de la función ventricular derecha, Rev. Esp. Cardiol., 44 (Supl 1):114, 1991.

CURRENT ADVANCES IN CARDIAC ASSIST DEVICES

Jean-Raoul Monties, Patrick Havlik, Thierry Mesana

Laboratoire de Recherches Chirurgicales
University of Aix-Marseille II
France

Under the heading of circulatory assist devices, we include systems designed for complete or partial cardiac replacement. We exclude agents acting on myocardial contraction and peripheral circulation as well as so-called counter-pulsation procedures using intra-aortic balloons which modify pressure rather than the flow rate.

Despite early reports about an "artificial heart" by Carrel and Lindberg in 1932, it was not until the first extracorporal circulation procedures in 1954 (Gerbode, Lillehei) that circulatory assistance became a clinical reality. Development of the artificial heart began at the same time and, as early as 1959, Kolff and Akutsu were able to keep a dog alive for one hour with an orthotopic artificial heart.

Heart transplantation and artificial heart implantation have been studied for decades. The first successful transplantation was carried out in 1967 by Chris Barnard, based on the work of N.C. Shumway. The first prolonged survival with an artificial heart was achieved much later in 1982 by De Vries with the Jarvik heart designed by Dr. Kolff in Salt Lake City. This "first" was followed by 5 other procedures in 1984 but the conditions of survival with the Jarvik heart were deemed unacceptable for permanent implantation.

Temporary assistance was proposed by Jarvik (total intrathoracic artificial heart) and Pierce-Donachy (external pneumatic ventricles) and since 1985 has been widely used as a bridge to transplantation. In this regard the work of Griffith and Pennington in the U.S.A., of Cabrol in France, and Hetzer in Germany is notable.

In addition to these highly invasive, labor-intensive procedures, simpler and less expensive assist devices have been developed including centrifugal pumps and turbines. As a result, clinical applications have been extended. In particular, circulatory assistance is used to relieve the natural heart after myocardial infarction or heart surgery. With experience, this technique, which was first used as a last resort after failure of all other methods, has been refined and thanks to better patient and device selection, timing, and monitoring, results have improved greatly.

In this article we will review:
- the devices that are currently available or being tested;
- the obstacles to the development of artificial ventricles and hearts;
- the various known research projects;
- the current status of our own research program.

CURRENT CIRCULATORY ASSIST DEVICES

With the Jarvik heart out of production, no artificial heart is presently in the clinical test phase.

Artificial hearts can be divided into two groups, i.e. reciprocating artificial ventricles with a pulsatile flow and rotary pumps with a continuous flow.

Reciprocating ventricles

Reciprocating devices provide pulsatile flow very similar to that of the natural heart. They consist of a chamber of variable capacity determined either by membrane movement or wall compression; these hearts always have inlet and outlet valves. Valves may be mechanical (Bjork, Medtronic Hall), polyurethane (Abiomed) or biological (bioprosthesis). Movement of the membrane or compression of the sack is usually achieved pneumatically. A more and more widespread design feature for compression is the use of one or two pusher plates. Ejection volume is relatively constant.

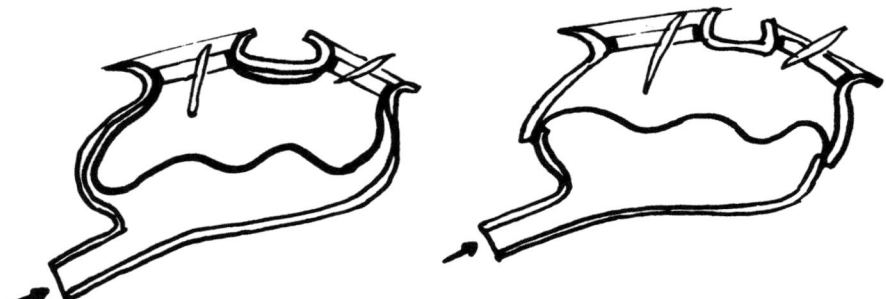

Figure 1. Pneumatic ventricle: sac type on the left, diaphragm type on the right

Two electrically powered reciprocating ventricles, the Novacor and TCI, are implantable. Presently used as a bridge to heart transplantation, these devices are destined for prolonged implantation. In the Novacor system, the walls of the ventricle are compressed by pusher plates actuated by electromagnets. In the TCI, compression is achieved using a cam driven by an electric motor.

One or two control systems are used, i.e.: the constant frequency system and the variable frequency system. In the constant-frequency system, flow is determined by the volume aspirated during filling. In the variable-frequency system, the ventricle is emptied as soon as it is full ("fill to empty mode"). Emptying may be asynchronous with natural heart beats or governed by ventricular diastolic pressure.

Figure 2. Ventricle activated by a pusher plate

An essential difference between pneumatic and pusher plate ventricles should be emphasized. Pneumatic ventricles are connected by cannulation between the atrium and the aorta. Filling is passive, assisted only by a depression in the pneumatic circuit of 20 to 30 mm Hg. Pusher plate ventricles are connected by cannulation between the left ventricle and the aorta. Filling is facilitated by ventricular contraction and there is no depression. Thus unlike pneumatic ventricles which can be disconnected after a weaning period, the two pusher-plate ventricles are suitable only for assistance pending heart transplant or as a permanent replacement since implantation results in irreparable damage to the left ventricle.

It is also noteworthy that the two pusher-plate ventricles can only be used for assistance of the left ventricle. Pneumatic ventricles can and often are used in pairs: one between the left atrium and the aorta as a left ventricular assist device (L.V.A.D.) and the other between the right atrium and the pulmonary artery as a right ventricular assist device (R.V.A.D.).

Mechanical failure has been extremely rare. The duration of assistance ranges from a few days for the Abiomed system to a few weeks for the Thoratec and Symbion systems to a few months for the Novacor system. Prices are in relation with these durations. Not including purchase of the power and control unit, the Abiomed system costs around $16,000, the Thoratec and Symbion around $60,000, and the Novacor around $200,000.

Rotary pumps

Under this heading, we include:
- peristaltic pumps,
- centrifugal pumps,
- spiral pumps.

Peristaltic pumps are identical to those used during heart-lung bypass for open-heart surgery. They are traumatic for vessels and above all for blood components. Because they result in high grade hemolysis, peristaltic pumps are rarely used for cardiac assistance and, when they are, the duration of assistance rarely exceeds a few hours. In general, the system includes an oxygenator and assistance is achieved by peripheral cannulation of the femoral vein and artery (ECMO).

Centrifugal pumps use impellers (3M, St Jude Medical) or a vortex mechanism (Biomedicus). The speed of rotation ranges from 1,500 to 4,000 rpms with flows of up to 5 to 6 liters/minute. Thanks to aspiration and continuous flow, relatively thin tubing can be used and as a result high flow assistance can be achieved by peripheral cannulation. This is especially advantageous for assistance in infants and newborns. However, to minimize loading loss, it is necessary that connecting tubes be kept as short as possible.

The hemolysis rate with these pumps is acceptable. They can be used for partial or total right and/or left ventricular assistance for durations up to two to three weeks provided the pumpheads are changed regularly.

Since they are inexpensive, these systems are used in many centers to gain time for recovery of the natural heart. Clinical experience has shown that patients adapt well to a continuous flow and that a totally pulseless flow does not have any severe adverse effects. When the natural heart retains some degree of effectiveness, it is important to divert only part of the flow (40 to 60%) so that the natural ventricle continues to pump. This precaution is necessary to facilitate weaning and to avoid intracardiac thrombosis.

Spiral pumps or turbines consist of an Archimedean screw enclosed in a catheter (Hemopump). The catheter is introduced into the aorta via the femoral route. The tip is placed in the left ventricle. The turbine, which is located at the other end of the catheter, at the level of the aortic isthmus aspirates blood from the left ventricle into the aorta. The screw which is activated by a cable connected to an external motor rotates at a speed of 25,000 rpm. Flow is continuous and is strictly dependent on aortic pressure. Outputs of up to 3.5 liters/minute can be achieved.

The useful life of these devices is less than 6 days after which the risk of rupture of the drive cable is high. Although this system provides effective relief of the left ventricle, it is relatively expensive ($5,000). Placement and especially removal are delicate. With improvement, this device might be very useful in cardiac intensive care to manage cardiogenic shock secondary to extensive myocardial infarction.

Table I lists current systems and their characteristics.

Table 1. Circulatory assist devices: current devices and (in progress devices)

Centrifugal Pumps	Pneumatic V.A.D.	Electric V.A.D.
Biomedicus	Thoratec	Novacor
Sarns-3 M	Symbion	T.C.I.
S.J.M.	Berlin heart	(Cora)
(Spindle Pump)	Toyobo	
	Nippon Zeon	
	Abiomed	
Axial Pump		**Electric TAH**
Hemopump		(Abiomed)
(Nimbus)		(Pennstate)
		(Utah)
		(IRCV)

Use of circulatory assistance has become more and more widespread throughout the world since 1985. World statistics allow assessment of the results:

Table 2. Circulatory assistance: overall result.
(According to W. E. Pae - International Registry for all Years - ASAIO 1991)[1]

VAD/TAH	No. Patients	Staged cardiac transplantation	
		Transplanted	Discharged
LVAD	122	87 (71.3%)	76 (62.3%)
RVAD	4	1 (25.0%)	1 (25.0%)
BVAD	161	105 (65.2%)	73 (45.3%)
TAH	169	135 (39.6%)	67 (39.6%)
Total	476	328 (68.9%)	217 (45.5%)

VAD/TAH	No. Patients	Post-cardiotomy cardiogenic shock	
		Weaned	Discharged
LVAD	494	254 (51.4%)	137 (27.7%)
RVAD	121	47 (33.8%)	31 (25.6%)
BVAD	350	132 (37.7%)	69 (19.7%)
Total	965	433 (44.8%)	237 (24.5%)

The following table summarizes the criteria of device selection for the following indications: bridge to heart transplantation, recovery from acute peroperative cardiac insufficiency, and treatment of cardiogenic shock after myocardial infarction.

Table 3. Choice of systems - Indications

	Recovery	A.M.I.	Bridge	Long Term
Centrifugal Pump	+++	+	++	-
Axial Pump (Hemopump)	+	++	-	-
Extracorporeal pneumatic VAD	+	+/-	+++	-
Implantable Electromechanical VAD	-	-	+	+++

Experience has shown the need for two types of assist devices to meet two different therapeutic applications. Short-term assistance lasting between a few hours and 2 weeks requires a simple external system that can be rapidly placed preferably via a peripheral vessel. Prolonged or permanent assistance requires a more durable, implantable device. Short-term devices should be reasonably priced, i.e. between $2,000 and $10,000 per unit. The potential market for such devices is estimated to be 50,000 to 100,000 cases per year.

Research and development of temporary external assist devices, artificial ventricles and implantable artificial hearts would thus seem worthwhile. Recently in a reversal of its 5-year-old grant policy giving priority to implantable artificial ventricles, the National Institute of Health decided to follow the recommendations of the United States Academy of Sciences and to provide funding for an artificial heart. This agency predicts that a permanent artificial ventricle will be available on the market by the end of the decade and an artificial heart by about the year 2005. Although this may seem somewhat distant, it should be remembered that the first predictions were for 1985 then 1990. These postponements are a good indication of the obstacles that have been encountered.

OBSTACLES TO THE DEVELOPMENT OF ARTIFICIAL VENTRICLES AND HEARTS

This section describes the objectives, possible solutions, and obstacles in the quest to develop artificial ventricles and hearts.

Objectives

The goal of maintaining blood circulation can be divided into two categories: assistance and replacement. Assistance implies that the natural ventricle is still capable of furnishing some of the flow. Assistance involves the left ventricle much more often than the right and in a few cases biventricular assistance is necessary. Pumps designed for assistance do not need to have a high volume output. It must simply be sufficient to insure part of the flow or eventually all of the flow under basal conditions. These pumps can be connected by cannulation, applied externally, or implanted. Replacement implies total failure of the natural heart. This is the goal of orthotopic artificial hearts designed for implantation in the place of the heart.

Pulsatile versus non-pulsatile flow: it was first assumed that pumps must provide a pulsatile flow as close as possible to the normal heart. This assumption provided the basis for the development of reciprocating pumps simulating the output of the natural heart with a diastole ratio in the magnitude of 1:3.

Research with continuous flow devices such as centrifugal or spiral pumps was a later development. The work of Nose[2] demonstrated that the body is able to adapt to continuous circulation. This adaptation can be divided into three phases:

- the "compressing stage" (first 6 hours) which requires a flow 20% higher than normal and is associated with edema;
- the fighting stage (3 weeks) during which the body reacts, edema disappears and hemodynamic status improves.
- the "live together stage" (after 3 weeks) in which an idioperipheral vascular pulsatility appears with cyclic pressure variations in the magnitude of 6 mmHg occurring at a slow frequency of about 20 cycles/mn independently of respiration.

Clinical application has borne out these findings since good adaptation to continuous pulseless flow has been reported in patients for periods up to one month with no adverse effect on any organ system.

Regulation of supply and demand: flow provided by the device must adapt automatically to the requirements of the body. The overall circulatory flow is the sum total of local requirements in each organ system and muscle. Each part of the network is self-regulated so that it can adapt supply to demand during different phases of activity. The heart which maintains this flow depends on the filling pressure (preload) and total vascular resistances (afterload). According to the law of Starling, end-systolic ejection fraction can vary 20% in either direction. Heart rate can vary from 1 to 3 fold depending on the age and physical condition of the subject.

A pump designed for replacement must be capable of automatically adapting to changes in flow requirements within a range of 1 to 3. For assistance the natural ventricle insures regulation.

Possible solutions

Temporary assist devices include one or two pumps, an activator, and a control system to monitor the device through sensors and supply power from an external source which in most cases is electric. For short-term assistance in hospitalized patients the system need not be implantable.

For prolonged or permanent assistance the pump, activator and control system should be implantable. Since the power supply remains outside the body, a transfer system is necessary.

Pumps: three types of pumps have been tested to fulfil these objectives:
- reciprocating pumps with membranes, diaphragms, or pouches.
- rotary pumps.
- rotary piston pumps.

The theoretical advantages and disadvantages of these three pumps are listed in the following table:

Table 4. The pumps: three different concepts

Membrane Pump	Rotating Piston Pump	Centrifugal Pump
Pulsatile flow	Quite pulsatile flow	Continuous flow
Alternative motion	Continuous rotation	Continuous rotation
	(Low speed)	(High speed)
Soft material	Solid material	Solid material
Valves	No valve	No valve

Activation: Until now membrane ventricles have been driven by pneumatic power. This power supply system cannot be miniaturized and thus cannot be implanted. Electric power is the only feasible source for implantation at the present

time. An energy convertor is indispensable. Several solutions have been proposed:
- a pump or a turbine to provide electrohydraulic power (Kolff, Olsen, Abiomed.);
- an electromagnetic field applied to the pusher plates (Novacor);
- an electric motor to provide reciprocating movement (linear motor - Akutsu) or cam system (Pierce-Donachy) or screw (I.R.V. Sion). Rotary pumps such as centrifugal and spiral pumps can be powered simply by a high speed electric motor. Rotary piston pumps require a low speed, high torque electric motor.

All these motors are brushless, direct current motors with permanent magnets and an electronic commutator. For implantation in the body, protection from humidity is one of the most difficult problems to solve.

Other motors have been used. For example, thermic motors were tested with turbines at the University of Washington (Stirling) but miniaturization and implantation are difficult. In the future, electrohydraulic or eletromechanical convertors will probably be used.

A control system is needed to adapt supply by the ventricle to the demands of the body. All systems have been based on the same principles: filling dependent feedback and safety system avoiding harmfully high arterial pressure. These systems thus take into account both preload and afterload. The objective is to prevent preloading from exceeding 10 mm Hg and maintain a mean arterial pressure between 80 and 100 mmHg.

Filling is passive in almost all membrane ventricles and insured only by the natural flow. When pneumatic power from an external source is used, filling can be promoted by gentle aspiration (not more than 30 mmHg). This is not possible with electrohydraulic and electromechanical systems.

With rotary pumps, filling is active and it is necessary to avoid overpumping which leads to depression, cavitation, and blood gas emboli.

Reciprocating pumps can be regulated according to two modes:
- constant frequency, according to the Starling principle. When filling pressure increases, the amount of blood entering the ventricle during diastole increases and as a result ejection volume increases. This method is best suited to pneumatic power and at best can achieve a 10 to 15% variation in systolic volume. Because fluids cannot be compressed, this method is more difficult to use with electrohydraulic or electromechanical power;
- variable frequency. The filling volume is always the same and ejection occurs as soon as the ventricle is full. Filling is assessed by a simple system. A magnet attached to the membrane activates a Hall effect sensor when contact is made. Measuring impedance variations have also been tested.

Sensors especially stress gauges can be used to measure filling pressure. The control system modifies pump output in order to maintain upstream pressure within preset limits. These sensors can also be used to estimate downstream pressure which varies in function of flow and resistances. However in most cases control is achieved by presetting the maximum pump power not to exceed certain pressure levels.

The power requirements of an artificial ventricle is equal to the sum of the energy needed to activate each of its subsystems and therefore depends on the efficiency of these subsystems.

The power supplied by a natural left ventricle is between 2 and 3 watts, 3 to 5 times less by the right ventricle, under basal conditions. During exertion this power can increase two fold.

The most efficient of currently available systems require 20 watts, which represents an overall efficiency of 8 to 12%.

At the present time, the only portable power source able to provide sufficient autonomy for normal activity is an external electric battery pack. The general consensus is that the life of the battery pack must be at least ten hours. This is why great efforts are being devoted to improving the efficiency of all subsystems, i.e. pumps, activators, convertors, sensors, control and regulation systems, and energy delivery systems.

Energy delivery from the external battery to the orthotopic ventricle is achieved by means of a low-voltage electric wire. Although this method is restrictive for the patient, it has the advantage of being highly reliable and efficient. Another method of energy transfer that has been used is percutaneous high-frequency electromagnetic induction. The efficiency of this method is 80% without risk of burning the skin.

Between 1975 and 1980, some research with nuclear isotope power sources was carried out in the United States and the Soviet Union. Conversion was achieved using a turbine, thermocouple or a thermic motor. However because of the prohibitive cost and the inherent danger of spreading dangerous materials, this avenue of investigation was abandoned.

Obstacles

The problems that have been encountered are all related to the interface between the living organism and the artificial device:
- bio- and hemo-compatibility;
- regulation;
- in vivo durability and reliability;
- tolerance and autonomy.

Compatibility with the organism: an implanted foreign body must be perfectly tolerated by the body. It must be inert. This means that it must not lead to adverse toxic or immunologic reactions and that it must not be altered by the body. In the body construction materials are subjected to forces as hostile and aggressive as in the sea. Rapid deterioration has been observed in a number of polymers under in vivo conditions. Some subcutaneous implants lose half their flexibility and elasticity after only a few months. Hairline cracks appear as a result of invasion by body substances. Microthrombi forming in these cracks promote thrombosis and calcification.

Whether flexible or rigid materials are used, the objective is to encourage the formation of indigenous protein coating which serves as an interface between the living organism and the artificial device. Attempts have been made to promote binding of heparin molecules which have an anti-coagulant effect.

However activation of platelets is observed during the first moments of contact between the material and blood. Ideally it would be necessary either to create a perfectly smooth surface with no irregularities or cracks or to promote and control coating of the surface with a solid stable layer of blood components.

A number of studies have allowed selection of polyurethanes better able to resist aggression in the body. Surfaces have been specially treated to minimize consequences on the blood and coagulation.

New materials covered with ultrathin layers of hemocompatible materials (derived from carbon and titanium) are now being tested.

Control System: adapting supply and demand is another obstacle to the development of artificial ventricles. The main difficulty is evaluating requirements. This requires a sensor device for data acquisition. However, because sensors are inaccurate and unreliable and must be frequently recalibrated, they are not suitable for implantation. Moreover a number of factors associated with normal daily life such magnetic fields, variations in atmospheric pressure, forces of acceleration and deceleration, changes in altitude, etc..., can interfere with the performance of sensors.

It is for this reason that efforts are being directed at developing a feedback system using data from physiologic sensors. Such data is difficult not only to collect but also to interpret since they are subject to a number of variables. Experience in the use of so-called physiologic pacemakers which adapt the frequency of stimulation to prevailing requirements, attests to the difficulty in solving these problems.

At the present time, two approaches are being used:

- approximate regulation and ventricular assistance through a bypass, leaving the natural ventricle in place to insure fine tuning through its own physiologic controls.
- evaluation of variations in preload and afterload is the basis for any control system, completing the true regulation insured at the local circulation level.

Durability and reliability: though this problem mainly involves technology, it is complicated by the rapid deterioration of materials in the body. A natural heart accomplishes a considerable amount of work in a lifetime. The artificial ventricle that replaces it must conform to strict rules of safety and reliability. It must be able to insure for one year 40 million pulsations and pump 3 million liters of blood for a period of 8,760 hours without a major failure. In comparison, an automobile is built to last only 1,500 to 2,000 hours.

The present requirements of the N.I.H. for a "permanent" device is failure-free performance for at least 2 years, i.e. 17,520 hours.

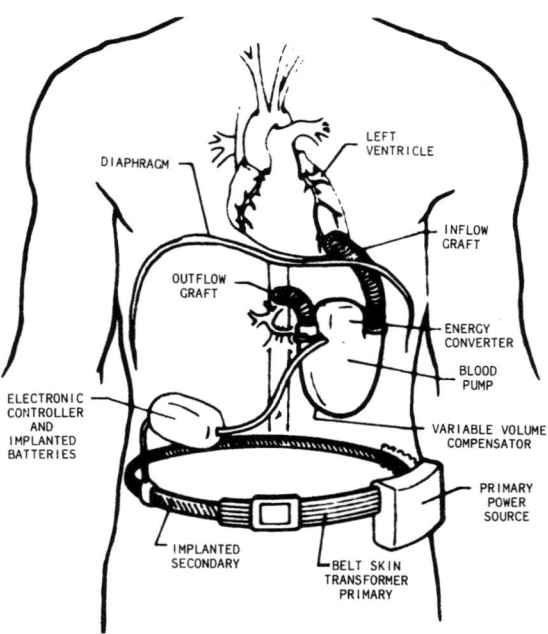

Figure 3. Novacor LVAD: the total project

Another requirement for an artificial ventricle or artificial heart is to provide an acceptably independent lifestyle. It is thus absolutely necessary that the energy source and activation system be portable with a minimum autonomy of 10 hours.

The noise and vibration levels caused by the device as well as local heat production must be within tolerable limits.

Finally the form, volume, and weight of the device must be compatible with implantation in the body[4,5].

All of these requirements and difficulties encountered account for the delays in the development of an artificial ventricle, the scarcity of projects now under way, and the search for innovative ideas.

PRINCIPLE RESEARCH PROJECTS UNDER WAY IN THE WORLD

In 1985 the N.I.H. gave priority to the development of artificial ventricles designed for permanent implantation and capable of performing reliably for at least 2 years. Research grants for total artificial hearts were drastically reduced. Recently the N.I.H. reversed this policy and accepted financing of an artificial heart program.

A few such projects are under way. Two left ventricular assistance devices are in the clinical testing phase, namely:
- the Novacor device,
- the Thermo-cardiosystem.

The **Novacor artificial ventricle** (Baxter) developed by Portner[5] in California is a very elaborate system that has been extensively tested in vitro and in vivo and used in more than 80 patients as a bridge to heart transplantation. In some cases the duration of assistance was over 1 year.

This device includes:

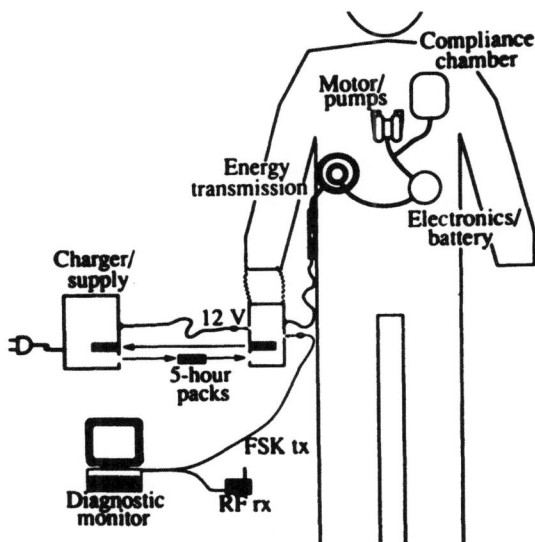

Figure 4. Penn-State project (Pierce)

- a pump consisting of a polyurethane pouch fitted with either porcine bioprosthesis or mechanical valves;

- the pouch is compressed by a double pusher plate system driven by electromagnets. Filling is passive, the ventricle being placed connected between the tip of the left ventricle and aorta. The artificial ventricle is placed in the abdomen. Because of variations in pump volume, a compliance compartment is placed in the pleural cavity.

- a control system operating on the "fill to empty" principle using the pouch filling sensors as electromagnets. The control and power supply unit is implantable.

Energy is supplied by induction from a belt on the outside to a subcutaneous belt.

This system has been approved by the F.D.A. for clinical evaluation in two centers in the U.S., one in Pittsburgh, PA, and the other in St. Louis, MO. The first implantations are imminent.

The **Thermo-Cardiosystem (TCI) artificial ventricle system** was developed in Houston[6]. It consists of a membrane pump made in part of titanium. The valves used are mechanical. The surface in contact with the blood is covered with microbeads. The membrane is compressed by a pusher plate and cam mechanism driven by a brushless, alternative rotation electric motor. The control system reportedly operates on the full-to-empty principle and is implantable. Energy from external batteries is transmitted by a percutaneous wire. This pump has been successful used as a bridge to heart transplantation in 40 patients. Implantation of the complete unit, which was attempted twice, was successful in one patient.

The Novacor and TCI systems are the only two devices to have reached the clinical test phase. According to the N.I.H., if all goes well, these devices should receive F.D.A. approval before the end of the decade.

Other pouch or membrane artificial ventricles with electromechanical activation are in various stages of design. Although these projects are confidential, some appear to be in advanced stages of development.

One approach is the pouch-type double ventricle with alternative compression using pusher plates driven by an electric motor/cam system[7], (figure 4), a screw (I.R.C.V. Sion), or a linear motor[8]. Another is the electrohydraulically controlled double ventricle driven by a turbine[9], or a micro-turbine placed between the two ventricles[10].

Two other avenues of research for left ventricular assistance are the implantable centrifugal pump[11,12] and the spiral micropump (Wampler-Nimbus). Components in contact with blood rotate at high speed (25,000 rpm) resulting in high shearing forces for blood cells and harmful heating.

Two other noteworthy projects are the "Tea Spoon pump" and the double-action centrifugal pump[13].

The fact that the prototypes for Novacor and TCS systems now in the clinical phase of testing were designed and built ten years ago attests to long research and development phase needed.

THE CORA PROGRAM

Our project constitutes a novel approach to the problem. As early as 1976, while other groups were studying ventricles that simulated the pumping action of the natural heart[14], we opted for a system that maintained circulation without necessarily

mimicking the natural heart. We turned our attention to a rotary piston system based on the Wenkel compressor system. This system has several advantages over conventional cardiac pumps[15]:
- slow rotation,
- pulsatile output at low speed,
- constant volume,
- valveless operation.

Since this pump can be made of solid materials, the premature aging of supple polymers in the body can be avoided. Activation can be achieved directly by an electric motor. This technology was already well-known to engineers and did not require special design study.

Our pump is a hypocycloidal pump with an eliptically shaped rotor. Pump movement is guided by an excentric gear mechanism. The stator has exactly the form defined by the turning rotor. Guidance is achieved by a cog and planetary wheel arrangement.

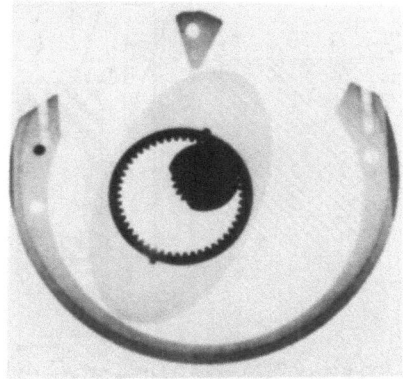

Figure 5. CORA rotating piston pump: an X-ray view

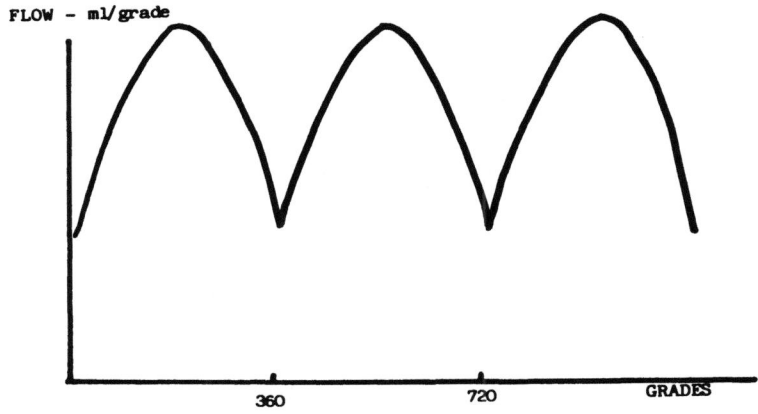

Figure 6. CORA Pump: pulsatile Flow

At each rotation of the rotor inside the stator, two cavities open and close carrying a constant volume of liquid from the inlet to the outlet of the pump. Filling and injection occur simultaneously. As shown in figure 6, the flow pattern is the sum of the two flow components: continuous or pulsatile.

This valveless pump which was extensively tested on a mock circulatory circuit can provide a pulsatile flow without regurgitation. Following these successful tests, three questions remained to be resolved:

1) what were the effects of this pump on blood components in general and the hemolysis rate in particular?

2) what construction material was most suitable in terms of bio- and hemo-compatibility?

3) what was the best means of driving and above all controlling the pump?

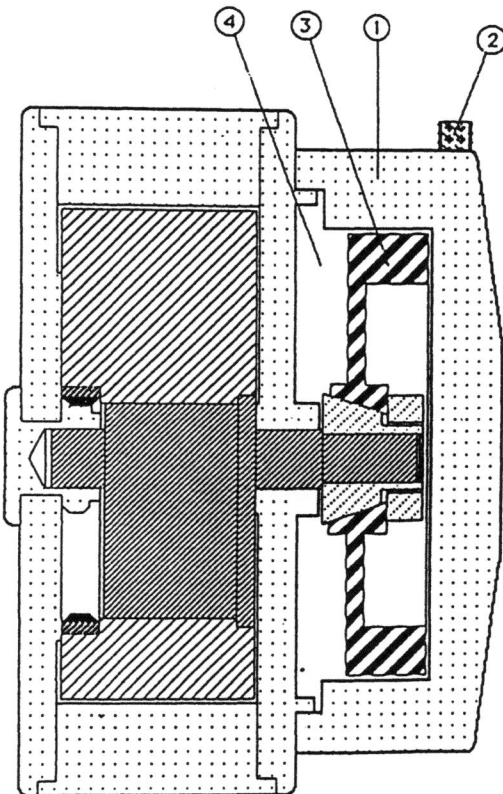

Figure 7. CORA ventricle: cross-section of the motor-pump unit
1 - Stator of the motor, 2 - Electrical sealed connection, 3 - Rotor of the motor, 4 - Sealed cavtity of the motor

Hemolysis

A study of pump operation was performed at the Center of Rheology in Toulouse[16]. A transparent pump was tested on a mock circulatory system. Turbulence and shearing stress inside the pump were measured by laser velocimetry. These measurements were recorded at different rotation speeds under various hemodynamic conditions within normal limits. Rotors of different sizes were used in order to study different clearance parameters.

Calculations based on the data obtained showed that, with clearances between 80 and 100 microns, shearing forces were maintained at values below 1,000 Pascals with no adverse effect on output. Under these conditions of operation hydraulic efficiency always exceeded 80%. The current pump has an ejection volume of 60 ml with 80 to 100 micron clearances. Output is function of rotation speed and pressure.

Construction material

Because manufacturing precision within a few microns is necessary for certain components, polymers cannot be used. The current model is made of titanium and alumine ceramic. The latter material is used for parts submitted to wear.

The search for a suitable construction material has been delayed because of the lack of data concerning the hemocompatibility of solid materials. Pyrolitic carbon, the only previously studied material, was not strong enough for our requirements which are stringent. The material must be as light as possible but hard enough to obtain a perfectly smooth surface that resists abrasion associated with constant operation.

At the present time we are focusing on surface coating obtained by chemical vaporization deposition on a graphite frame. Various materials have been tested to determine their physical properties (tribology, surface, resistance, durability, and stability), hemocompatibility and biocompatiblity. So far, our results show that amorphous diamond or, better still, crystallin diamond coatings are far better than titanium nitride and Borum carbide coatings.

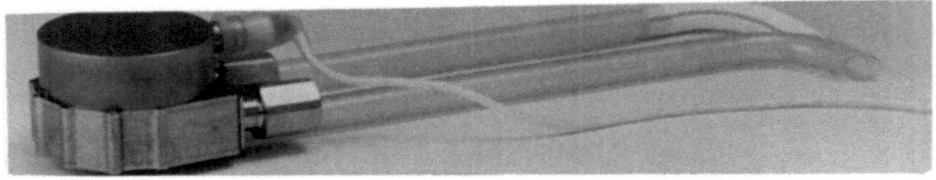

Figures 8. Views of the CORA ventricle showing the two pieces of the motor and ready for implantation

Activation and control

An electric motor was designed and developed especially for our artificial ventricle (Artus). This highly efficient brushless motor with permanent Neodymium iron magnets rotates a slow speed (50 to 180 rpm) and produces high torque (0,39 Newtons/meter). Heat production is low and easily dissipated. There is no hot spot on its surface.

It is composed of two distinct parts which are enclosed in a totally sealed titanium case. The rotor/magnet assembly is attached to the pump shaft. The stator is placed over it and then secured to the pump case, thus forming a single piece, sealed, implantable pump/motor unit (figure 7).

Pump operation is silent and vibration free with no detectable torque effect. The pump/motor unit is 9.1 cm in diameter and 5.2 cm high and weighs 900 g. With further study and the use of new materials, it should be possible to reduce these dimensions by about 10 to 20% (figure 8 and figure 8bis).

The control system must adapt pump output to the volume of incoming blood. It must also maximize flow while avoiding overpumping. The rotary pump being volumetric, it is necessary to prevent depressions at the inlet. Since filling is active, rotation speed must depend on preload. By monitoring the form of the motor intensity curve, it is possible to predict and prevent depressions[18]. Control is based on this principle (patent no. 90.01.850, February 2, 1990 and patent no. 90.01.851, February 2, 1990). A control calculator is now being developed as an external control unit. Clinical testing will soon be undertaken.

The CORA research and development program includes three phases. The first phase will be aimed at the development of an artificial ventricle for temporary assistance. The ventricle and motor will be connected by cannulation between the left atrium and the aorta. The external power and control unit will be connected to the pump by a percutaneous electric wire. This first phase is necessary to study the control system. The objective of the second phase is development of a permanent artificial ventricle with an implantable miniaturized electronic control unit. The third phase will be aimed at developing a permanent implantable biventricular artificial heart.

In vivo tests[19] have been performed on sheep for 10 years and on calves for 4 years. The artificial ventricle is connected between the left atrium and the thoracic aorta and placed either in a subcutaneous pocket against the rib cage or under the abdominal wall. In these experiments we have observed the total hemodynamic effectiveness of the ventricle for durations up to 13 days. The pump is able to maintain all or part of the arterial circulation at mean pressures of 100 to 140 mmHg.

Laboratory tests have shown no harmful effects on blood biochemistry or components. The hemolysis rate is low and decreases after two or three days. Notably, after an initial decrease, platelet count returns to a satisfactory level.

Clinically the animal appears normal with an excellent appetite. Urine is clear and no diuretic treatment is required. Moderate anticoagulant therapy with heparin has been administered throughout these experiments. We have not yet tested coumarin agents.

Research is under way to improve the pump. In particular, lighter and sturdier materials are being tested to increase the pump reliability. Though many problems remain, this project constitutes an original and promising solution to an important problem.

Credible studies have estimated that between 20,000 to 50,000 patients a year could benefit from ventricular assistance or artificial heart replacement. Heart transplantation can cover only about a tenth of this demand. A prosthetic device cannot be used unless it provides results comparable to those of heart transplantation, i.e. a long, active, independent live. The permanent artificial heart remains one of the major technological challenges of our time.

ACKNOWLEDGMENTS

Our research was subsidized by grants of "ANVAR," Ministery of research, Region PACA and supported by "Lions Club de France" and "Lions Club International Foundation."

REFERENCES

1. Miller CA, Pae WE, Pierce WS: Combined registry for the clinical use of mechanical ventricular assist pump and the total artificial heart in conjunction with heart transplantation, Fourth Official Report 1989, J. of Heart Transplantation, 9:453, 1990.
2. Nose Y, Golding L: Is a pulsatile pump necessary for artificial heart? Proc. ESAO, 11, 1980.
3. Affeld K, et al.: The use of a computer graphics to find an optimal fit for a human artificial heart, Tran. ASAIO, 103, 1983.
4. Shiono M, Shah AS, Sasaki T, et al.: Anatomic fit study for development of a one piece total artificial heart, Trans. Primer, 11(3):57, 1988.
5. Portner PM, Jassawallla JS, Oyer PE, et al.: A totally implantable ventricular assist for terminal heart failure, ASAIO Primer, 11(3):57, 1988.
6. McGee MG, Myers TJ, Frazier OH, et al.: Extended support with a left ventricular assist device as a bridge to heart transplantation, Trans. ASAIO, 37:425, 1991.
7. Snyders A, Rosenberg G, Pierce W, et al.: A completely implantable total artificial heart system, Trans. ASAIO, 37:237, 1991.
8. Yagura A, Taruaka UY, Akutsu T, et al.: An electrohydraulic ventricular assist system with a linear actuator, Trans. ASAIO, 35:447, 1989.
9. Diegel PD, Mussivand T, Olsen DB, et al.: Electrohydraulic ventricular assist device development, Trans. ASAIO, 37:206, 1991.
10. Sheng YL, Bi Y, Kolff WJ, et al.: New polyurethane valves in new soft artificial heart, Trans. ASAIO, 35:301, 1989.
11. Golding LAR, Stewart RW, Sinkewitch M, et al.: Non pulsatile ventricular assist bridging tot transplantation, Trans. ASAIO, 34, 1988.
12. Hager J, Brandshaetter F, Klima G, et al.: Functionnal heart replacement with the spindle pump: First results, Trans. ASAIO, 36, 1990.
13. Imachi K, Chinzei T, Abe Y, et al.: A new pulsatile total artificial heart using a single centrifugal pump, Trans. ASAIO, 37:242, 1991.
14. Montiès JR, Havlik P, Mesana T, et al.: Rotary pump artificial heart: a new system for circulatory assistance or heart replacement, Life Support Systems, 5:251, 1987.
15. Montiès JR, Mesana T, Havlik P, et al.: Another way of pumping blood with a rotary but non centrifugal pump for an artificial heart, Trans. ASAIO, 36:258, 1990.
16. Bellet D, Sengelin M, Henge JC: Ecoulements au sein d'une prothèse cardiaque hypocycloïdale sans clapet, Innov. et Technologie en Biologie et Médecine, 7:3, 1986.
17. Dion I, Baquey C, Candelon B, Montiès JR: Haemocompatibility of titanium nitride, XIIe Congress of ESAO, Sept. 19, 1990, to be published in Artificial Organs.
18. Trinkl J, Mesana T, Havlik P, et al.: Control of pulsatile rotary pumps without pressure sensors, Trans. ASAIO, 37:208, 1991.
19. Mesana T, Mitsui N, Trinkl J, et al.: First significant animal survival with a Wankel-type left ventricular assist device, Trans. ASAIO, 37:166, 1991.

MECHANICAL SUPPORT FOR THE FAILING HEART

Peter P. McKeown and Stephen G. Kovacs

The University of South Florida
Division of Cardiovascular Surgery
Department of Surgery, MDC 0016
12901 Bruce B. Downs Blvd.
Tampa, FL USA

ABSTRACT

For an increasing number of cardiac patients mechanical support devices have meant the difference between life and death. Mechanical support encompasses cardiopulmonary bypass, intra-aortic balloon pumping, ventricular assist devices including pulsatile and nonpulsatile systems, and the total artificial heart (TAH). Circulatory support can be considered temporary (pending recovery of the natural heart or transplantation) or permanent. While the early experience with the Jarvik-7 TAH was disappointing, the severe shortage of donor hearts is stimulating attempts to develop other permanent implantable support systems. The technical and engineering challenges to produce a successful permanent artificial heart are significant but provide an exciting field of research for what we believe is an achievable goal.

INTRODUCTION

Development of a successful mechanical heart has long been a matter of speculation and ambition. For over thirty years researchers have aggressively pursued attempts to develop artificial devices to support the circulation, either on a temporary or permanent basis. Much of the early work on the artificial heart and support devices was initiated by Willem Kolff[1,2]. In 1964 the artificial heart program (AHP) was established as a mission oriented project within the then National Heart Institute. As many as 35,000 patients per year under the age of 70 in the United States could be candidates for a total artificial heart or permanent left ventricular assist device[3].

Use of artificial hearts and left ventricular assist devices for circulatory support as a bridge to transplant and for recovery has now become widespread[4,5]. Patient selection and management are critical in determining longterm outcome and assessing the usefulness of circulatory support systems for the failing heart. Choosing the right device for the right patient may pose ethical as well as clinical dilemmas[6-9]. A wide spectrum of mechanical support devices are now available or in development[10-12]. While the initial clinical results of the first total permanent artificial heart were disappointing, use of the same device (Jarvik-7, Symbion Inc., Salt Lake City, Utah, USA) as a temporary support for bridging to transplantation, provides encouragement for those who believe a permanent device is a technical possibility[13-16].

Cardiopulmonary Bypass (CPB)

Temporary mechanical substitution for cardiac (and pulmonary) function became possible with the development of the heart-lung machine in 1951[17]. Initially introduced for application in cardiac surgery cardiopulmonary bypass was subsequently used in a series of patients with pump failure associated with acute myocardial infarction[18,19]. In 1965, Spencer, Eisman and Trinkle reported using cardiopulmonary bypass to successfully support patients with postoperative cardiogenic shock[20]. "Resting the heart" on cardiopulmonary bypass remains one of the most common forms of mechanical support for the failing heart. Cardiopulmonary bypass techniques have been modified for extracorporeal membrane oxygenation (ECMO) with percutaneous femoral artery/femoral vein cannulas, membrane oxygenator and a centrifugal pump (Bard Cardiopulmonary Bypass System (CPS) Bard, Inc., Billerica, MA, USA)[11]. This system provides a means of rapid resuscitation and ventricular support. Heparinization is necessary and hematological complications limit the usefulness of cardiopulmonary bypass as a ventricular support system. If recovery does not occur in a relatively short period of time (<6 hr) alternative support systems need to be considered.

Intra-Aortic Balloon Pump

The IABP is a readily available, inexpensive, simple device for circulatory support. Moulopoulas and colleagues implemented the concept of diastolic intra-aortic balloon augmentation in 1962[21]. Kantrowitz refined these techniques and reported the first clinical success in 1968[22]. Because of the relative simplicity, the IABP has been used extensively for circulatory support[23,24]. Original clinical application was for failure to wean from cardiopulmonary bypass after cardiac surgery. Its use has expanded to include use in cardiogenic shock from acute myocardial infarction, pre- and post-infarction angina, acute ischemia complicating a PTCA, post infarction ventricular septal defect and mitral regurgitation, ventricular tachyarrhythmias and as a mechanical support prior to transplantation[12,25]. IABP related complications occur in about 10% of patients. These are usually vascular injuries related to insertion. Improved techniques of insertion include the use of a percutaneous guide wire[26,27]. Peripheral vascular disease will contribute to the complication rate and may preclude the use of this device. Between 1.5 to 8% of adult cardiac surgery patients require IABP postoperatively for temporary cardiac support[8]. Overall, it is used in about 30,000 adult patients per year in the United States. IABP is not suitable for use in patients with aortic dissection or with significant aortic regurgitation. The application in pediatric patients has been very limited because of size constraints and increased aortic compliance[23,24]. While the device is pulsatile, it has limited abilities to support the circulation. The IABP requires an intrinsic patient ECG or arterial wave form for appropriate augmentation. The IABP has proved useful in many patients requiring circulatory support, however the effectiveness is limited and in some patients augmentation is insufficient to match circulatory demands.

Pulmonary Artery Balloon Counterpulsation (PABCP)

Isolated clinical and experimental reports have also demonstrated the efficacy of pulmonary artery balloon counterpulsation (PABCP) for RV support[28-30]. Successful clinical results were achieved by anastomosing a Dacron or polytetrafluoroethylene (PTFE) graft to the main pulmonary artery and inserting a standard intra-aortic balloon (usually 40 cc) into the graft. The graft acts as a reservoir conduit. The device is triggered from the ECG signal or the arterial wave form for diastolic augmentation. These techniques require an open surgical approach for insertion and removal[28-30].

Experimental studies have suggested a role for PABCP using a balloon placed within the pulmonary artery[31,32]. If future studies document the efficacy of this approach, efforts should be directed toward developing a percutaneous nonsurgical insertion technique. Augmentation by an intrapulmonary balloon will be limited by the balloon's size and has yet to become a clinical reality. Pulmonary artery dimensions place finite limits on balloon length and volume. Oversizing could cause obstruction of pulmonary blood flow and/or damage to the pulmonary artery. Smaller sizes limit appropriate augmentation.

Hemopump

The hemopump is an intravascular device in the fashion of an Archimedean screw turned by an electrical motor. It is inserted through a small graft conduit anastomosed to the femoral or the external iliac artery via a cutdown approach. The screw is incorporated in the terminal (21 cm) 21 French portion of the device. It is passed retrograde across the aortic valve into the left ventricle. A 9 French sheath is connected to the terminal portion of the system and contains a flexible drive cable which then exits to be connected to the drive motor. The screw is rotated up to 25,000 rpm and can provide flows up to 3.5 liters/minute. Flushing the system with a 40% dextrose solution provides lubrication to the drive cable and hydrodynamic components and prevents entry of blood into the system. As the screw is rotated, blood is pumped from the left ventricle to the aorta[33].

Advantages of the system include low cost, simplicity in insertion and ease of operation. Flow rates are reasonable. Augmentation is provided without the need for an arterial pressure tracing or ECG synchronization. Disadvantages of the device include the fact that it produces a nonpulsatile flow and there is a potential for hemolysis. There have been reports of complications citing fracture of the drive cable. After modifying insertion techniques and redesigning and replacing the drive cable, fracture rates have been reduced from 60% to 5.2%[34]. Peripheral vascular disease and small femoral arteries as well as inability to cross the aortic valve have precluded insertion of the device in some patients[34]. In preliminary clinical trials, 41 patients with acute MI or post cardiotomy shock were supported with the hemopump for a

Table 1. Implantable (Corporeal) Ventricular Assist Devices

**DEVICE	POSITIONS	ENERGY/ DRIVER	CHAMBER (BLADDER)/ DIAPHRAGM	VALVES	TRIGGER	STROKE VOL	FLOW	COMMENTS
Novacor	L+	Electrical Solenoid	Biomer (outer shell epoxy impregnated Kevlar)	Baxter porcine pericardial (25mm/21mm)	Fixed/Fill Rate	70 / 90 ml	9	L+
Thermedics (HeartMate)	L+	Pneumatic (Electrical)*	Sintered Titanium/ textured Biomer	Porcine (25 mm)	Fixed/Automatic fill ECG	80 ml	9 L	Electrical version currently undergoing clinical trials
UNDER DEVELOPMENT								
Abiomed VAS	L+	electro-hydraulic		Abiomed Tri-leaflet	ECG		10+L	2 pumping chambers in series serve as internal compliance

+ LV Apex to Thoracic or Abdominal Aorta

PARACORPOREAL PULSATILE VENTRICULAR ASSIST DEVICES

**DEVICE	POSITIONS	ENERGY/ DRIVER	CHAMBER (BLADDER)/ DIAPHRAGM	VALVES	TRIGGER	STROKE VOL	FLOW	COMMENTS
Thoratec	Paracorporeal L, R, B	Pneumatic	Thoratec (BPS-215m) Biomer/polyurethane	Bjork-Shiley monostrut (Delrin)	Asynchronous (fixed) Volume (Fill to empty)/ External ECG	65 ml	6.5 L	
Symbion AVAD	Paracorporeal L, R, B	Pneumatic	Pelethane & Biomer polyurethane	Medtronic Hall 25 mm	Asynchronous Fill volume External ECG	70 ml	<5.5 L	unavailable
ABIOMed BVS 5000	Extracorporeal L, R, B	Pneumatic	Angioflex polyurethane	Abiomed trileaflet angioflex polyurethane	Fill to empty	80 ml	<5.0 L	2 chambers configured ventrically First chamber acts as compliance reservoir
UNDER DEVELOPMENT								
USF Kovacs/ McKeown	L, R, B	electro-magnetic	Vivathane copolymer	Medtronic Hall/ Omniscience (25 mm)	Asynchronous ECG	65 ml	7 L	

** Biomer (Ethicon Inc., Somerville, NJ, USA), Thermedics (Thermo Cardiosystems), Woburn, MA, USA, Novacor (Division of Baxter, Deerfield, IL, USA), Thoratec (Berkley, CA, USA), Medtronic (Minneapolis, MN, USA), Edwards CVS (Baxter Healthcare Corp, Oakland CA, USA), ABIOmed (Danvers, MA, USA), Symbion (Symbion, Inc., Salt Lake City, Utah, USA), Bjork-Shiley (Shiley, Inc., Irvine, CA, USA)

L = Left Ventricular Assist   R = Right Ventricular Assist   B = Biventricular Assist mean of 52.8 hours (1-194 hours). Fifteen of the 41 patients (36.5%) were successfully weaned. In 12 other patients, insertion of the hemopump was not achieved. Survival in this group was only 16.6%[34]. The hemopump has a potential to fulfill a role between intra-aortic balloon pump and the more complex ventricular assist devices.

A totally implantable Nimbus system (Nimbus, Inc., Rancho Cordova, CA, USA) based on the hemopump design is currently being developed. This device is external to the heart but placed intrathoracically and pumps blood from the left ventricular apex to the ascending aorta[10,35]. It is powered electrically and provides nonpulsatile flow.

VENTRICULAR ASSIST DEVICES

Ventricular assist devices (VADs) provide mechanical circulatory support superior to that of the IABP or hemopump. In some of the devices flows of up to 9 liters/minute are possible (Table 1). The VADs can be considered under several classifications; a) pulsatile versus nonpulsatile, b) temporary versus permanent, c) implantable (corporeal) versus para-corporeal, and d) electrical versus pneumatic (Table 1,2).

Table 2. Ventricular Assist Devices

Device	Non-Pulsatile	Pulsatile	Para-corporeal	Implantable	Permanent	Temporary	Electrical	Pneumatic
Biopump	X		X			X		
Centri-med	X		X			X		
Thoratec		X	X			X		X
Thermedics (Pneumatic)		X		X		X		X
Thermedics (Electric)		X		X	X	X	X	
Novacor		X		X	X	X	X	
ABIOMed BVS		X	X			X	X	X
Symbion AVAD		X	X			X		X

Figure 1. Biomedicus. Dual consoles for Biomedicus, providing right and left heart assist.

Nonpulsatile

The nonpulsatile centrifugal pumps are the simplest, lest expensive and most readily available VAD. Two types of vortex centrifugal pumps are commercially available, the Biopump (Biomedicus, Minneapolis, MN, USA) and the Centrimed (Centrimed Corp, Hopkins, MN, USA). In these similar devices, an acrylic impeller is mounted on a circular magnet enclosed in a plastic chamber (Figure 1). A drive motor turns a second magnet which turns the impeller without direct coupling. In some centers, centrifugal pumps are used instead of the traditional roller pumps in routine cardiac surgery cases. Special FDA approval is not required for these devices. There are no inlet and outlet valves and standard cannulation techniques and cannulas are used. The devices can be configured for left heart assist (LVAD) with a left atrial inflow and aortic outflow, right heart assist (RVAD) with a right atrial inflow and pulmonary artery outflow, or as a combination to deliver biventricular assistance (BiVAD). After insertion the cannulas are externalized via separate skin incisions to enable closure of the chest. The pump head is then placed in an extracorporeal position. Removal of the device requires a return to the operating room and repeat median sternotomy. Systemic anticoagulation is recommended with these pumps and the heads are changed every 24-26 hours to prevent potential thrombo-embolic complications. Flow rates in excess of 4 liters/minute are possible. These devices are used in conjunction with the IABP to provide a modicum of pulsatile flow. Circulatory support with a centrifugal pump for over 30 days has been documented[36], however better results and fewer complications are present when used for period of less than 1 week[4,37]. Karl (1991), reports encouraging results using a Biomedicus VAS in 12 children with congenital heart disease (age 6 days-12 years)[38]. Ten of the 12 patients were successfully weaned and six were eventually discharged from hospital. The disadvantages of the centrifugal pump include potential for hemolysis and micro-emboli, and the lack of pulsatile flow.

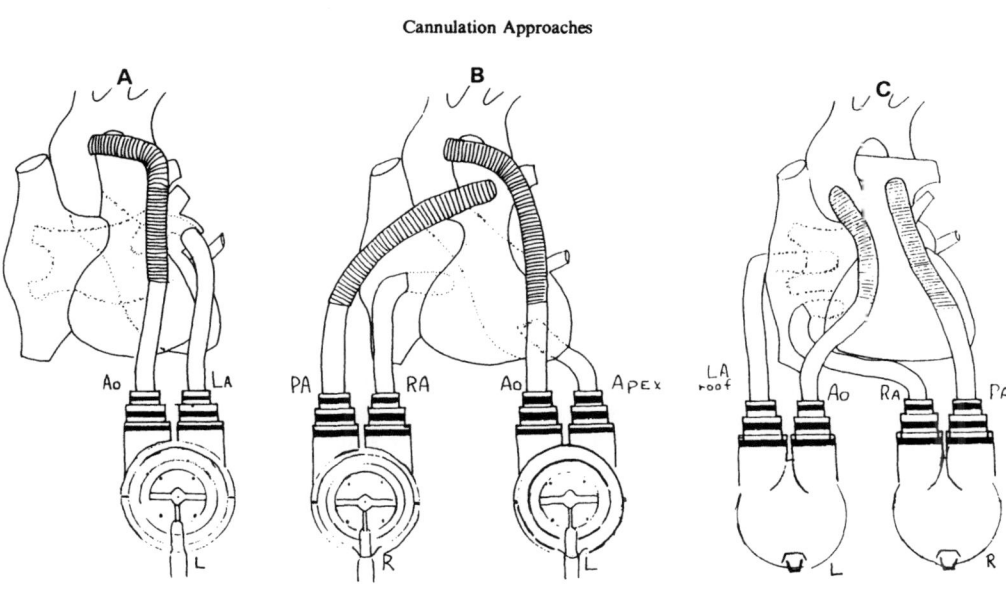

Adapted from Thoratec Ventricular Assist Device System
Clinical Investigational Plan, 1988
Thoratec Laboratories, Berkeley, CA

Figure 2. Configurations for the Pierce-Donachy Ventricular Assist Device and Cannulation Techniques

Pulsatile Left Ventricular Assist Devices

Pulsatile ventricular assist devices have been used extensively in clinical trials for over 10 years. The majority of the implants have used pneumatic sac type pumps with inflow and outflow valves to provide pulsatile, unidirectional flow. There has been a recent trend towards electrical, electro-hydraulic and electromagnetic systems. Initially these pumps were designed for a specific purpose, to provide temporary mechanical support for the heart for short term periods of 7 to 21 days. They have proven successful for periods of up to 232 days[4]. Totally implantable devices such as the Novacor (Baxter, Deerfield, IL, USA) and Thermedics "HeartMate" (Thermo Cardiosystems, Woburn, MA, USA) are already being used as a bridge to cardiac transplant and plans for clinical trials as a permanent implant are under way. Totally implantable devices allow improved patient mobility when compared with paracorporeal or extra-corporeal devices. However, there is a higher degree of technical difficulty when implanting or removing the device. In the event of a technical malfunction in parts of the drive mechanism or valve failure accessibility to the device is difficult.

The Pierce-Donachy (Thoratec, Inc., Berkeley, CA) ventricular assist device developed at Pennsylvania State University was a prototype for VADs and has been used extensively both as a bridge to transplant and in weaning patients from bypass after cardiac surgery[39-45]. Configurations include LVAD, RVAD and BiVAD (Figure 2,3). While the number of survivors in the post-cardiotomy support group is small (19 of 93 patients = 20.4%), most if not all of these patients would have succumbed without mechanical circulatory support provided by the device. Results in the bridge to transplant group are very encouraging; of 154 patients in the bridge to transplant group, 98 have received a heart transplant. Eighty-two of the 98 patients transplanted (83.7%) survived (Table 3).

THORATEC VAD CLINICAL SUMMARY

Bridge to Cardiac Transplantation
154 Patients

34 LVADs	120 BiVads
2 Waiting	1 Waiting
23 Transplanted	75 Transplanted
20 Survivors	62 Survivors

Weaning From Bypass
93 Patients

55 LVADS	10 RVADs	28 BiVADs
0 Waiting	0 Waiting	0 Waiting
19 Weaned	5 Weaned	10 Weaned
11 Survivors	3 Survivors	5 Survivors

Thoratec's Heartbeat 5:4 (1991)

Table 3. Thoratec Clinical Summary as of October 30, 1991 in implantation as a bridge to cardiac transplantation and weaning from cardiopulmonary bypass.

The Pierce-Donachy pump is a pneumatically driven device with a segmented polyurethane sac enclosed in a rigid polysulfone case. Inlet and outlet valves are monostrut Bjork-Shiley (Shiley, Inc., Irvine, CA, USA) Delrin valves providing unidirectional flow. The pump can be set at a fixed rate or triggered automatically by a full-to-empty mode or by an external synchronous signal using the R-wave from the ECG. Special cannulas have been designed and fabricated with proprietary BPS215M polyurethane to connect the device to the circulation. Cannulation

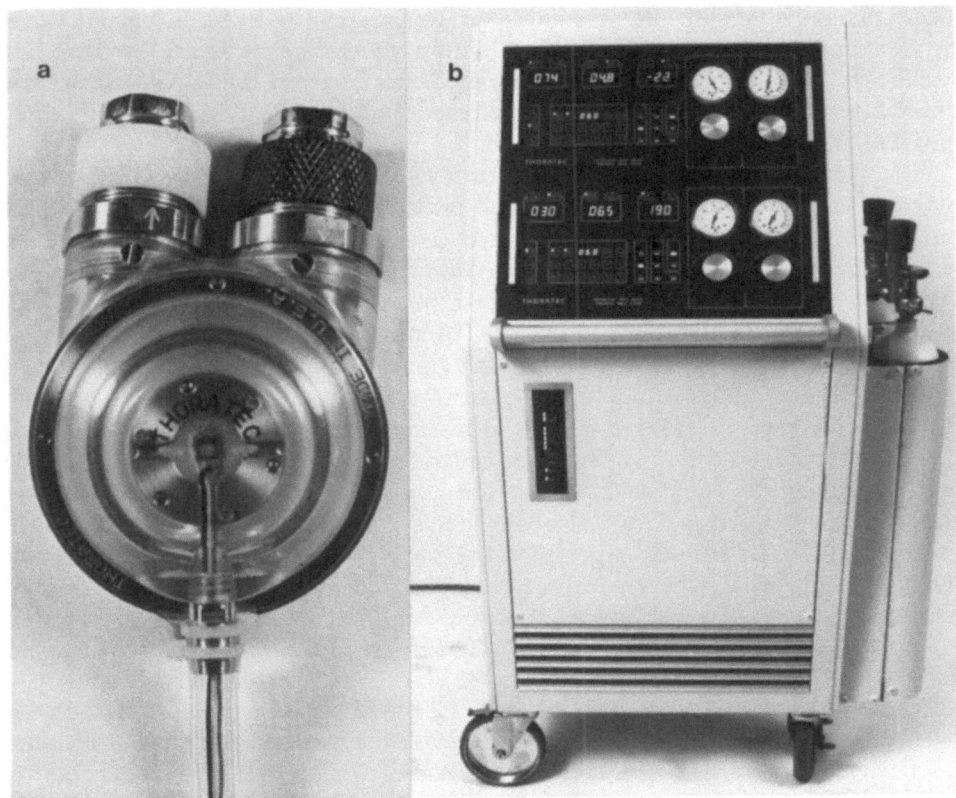

Figure 3. Thoratec. a) Thoratec (Pierce-Donachy) pneumatic VAD demonstrating pump pneumatic drive line. b) Thoratec device console, capable of controlling 1 or 2 pump systems.

techniques include inflow using the left atrial appendage or the left atrial roof, occasional left ventricular apex with outflow to the ascending or thoracic aorta for left ventricular assist configuration. For right ventricular assist, inflow is via the right atrium to the pulmonary artery (Figure 2). Cannulae exit the chest through small incisions and connect to the device which is positioned para-corporeally on the anterior abdominal wall. While unique cannulas have been developed for this device, the cannulation techniques are fairly standard for most of the pulsatile and non-pulsatile temporary ventricular assist devices. Anticoagulation is necessary and thrombo-embolic problems remain a major concern[46].

The ABIOMed BVS System 5000 (ABIOMed Incorporated, Danvers, MA, MSA) is a pneumatically powered pulsatile assist device which can be configured for univentricular or biventricular support. It is mounted extra-corporeally on a stand which limits the patient's mobility. The device fills passively by gravity and has two chambers: an inflow reservoir chamber and a pumping chamber. Both the pumping chambers, the inflow and outlet valves are made from smooth Angioflex polyurethane. The pumps run in a full-to empty trigger mode and have proven successful when used as a bridge to transplantation[47]. A permanent implantable electro-hydraulic version of the ABIOMed LVAD system (VAS) is currently under development[11].

The Symbion (Symbion Inc., Salt Lake City, UT, USA) acute ventricular assist device (AVAD) is a pneumatically driven VAD suitable for uni- or biventricular support. The device consists of a pelethane injection molded housing with a polyurethane diaphragm. Blood contact surfaces are covered with Biomer (Ethicon Inc., Somerville, NJ, USA). Inflow and outflow valves are 25 mm Medtronic-Hall (Medtronic Inc., Minneapolis, MN, USA) mechanical valves. Initial clinical trials as a bridge to transplant were moderately successful. Use of the device was halted by the GDA in early 1991[10].

The Thermedics "HeartMate" (Thermedics/TCI, Woburn, MA, USA) is an implantable pneumatic device with several distinctive features. It is designed to be implantable and has been limited to LVAD configuration, with inflow from the left ventricular apex cia a short cannula and outflow to the ascending or thoracic aorta. The pump is inserted below the diaphragm, in the abdominal cavity, and a single pneumatic drive line exits through the skin. Another unique feature of this device is the use of textured surfaces in blood contact areas within the pump. The diaphragm is textured Biomer polyurethane and the chamber is constructed in sintered titanium. The use of textured surfaces is a concept to provide a suitable surface for development of a smooth pseudo-neointima. Porcine xenograft valves (25 mm) are used for inflow and outflow and the combination of the textured surface and tissue valves has been successful in reducing thromboembolic rates[48,49]. Patients are not routinely anticoagulated after the initial perioperative period but, are placed on a long-term antiplatelet regimen. The device has been inserted in 50 patients as a bridge to cardiac transplantation with average implant time of 75 days (1-324). Of the fifty patients, 43 have a completed outcome. Twenty-six of the 43 patients (61%) were transplanted and 22 of the 26 transplanted (85%) were discharged from hospital. An electrical implantable version of the TCI HeartMate has now been used in two patients[50]. The electrical and pneumatic devices are essentially the same, except that the electrical device has a low speed torque motor which is used to directly drive a pusher plate to compress the diaphragm (Figure 4a, 4b).

The Novacor (Division of Baxter, Deerfield, IL, USA) ventricular assist system was designed to be an implantable electrical device for a bridge to transplant and as a permanent LVAD. The inflow cannula is connected to the apex of the left ventricle and outflow cannula is anastomosed to the ascending or descending thoracic or abdominal aorta. The pump is positioned in the left upper abdomen and placed in the pre-peritoneal tissues. Electrical energy is used to power a pulse solenoid which drives dual power pusher plates together causing contraction of the polyurethane Biomer bladder. Porcine pericardial tissue valves (Edwards CVS, Baxter Healthcare Corp, Oakland, CA, USA) are used for inflow (25 mm) and outflow (21 mm). Epoxy impregnated Kevlar is used to fabricate the outer shell of the device. Currently, percutaneous vent, control and power leads are required. Future plans incorporate an internal compliance chamber and transcutaneous energy transfer systems (TETS). As of May 1990, the device has been used in 68 patients with duration up to 90 days prior to successful transplantation[51-53].

The Thermedics HeartMate and Novacor devices, while implantable, are limited to left ventricular assistance only. Unfortunately, biventricular failure is not uncommon[54,55]. Biventricular infarction, potentially lethal arrhythmias or elevated pulmonary vascular resistance may affect the function of these devices. One experimental study demonstrated reduction in left ventricular pressure by use of an LVAD had minimal effect on right ventricle function by induced ischemia[56,57]. Biventricular failure and ventricular fibrillation may limit the usefulness of a permanent left ventricular assist device. Both the Thermedics HeartMate and Novacor use tissue valves which seems to decrease the risk of thromboembolic

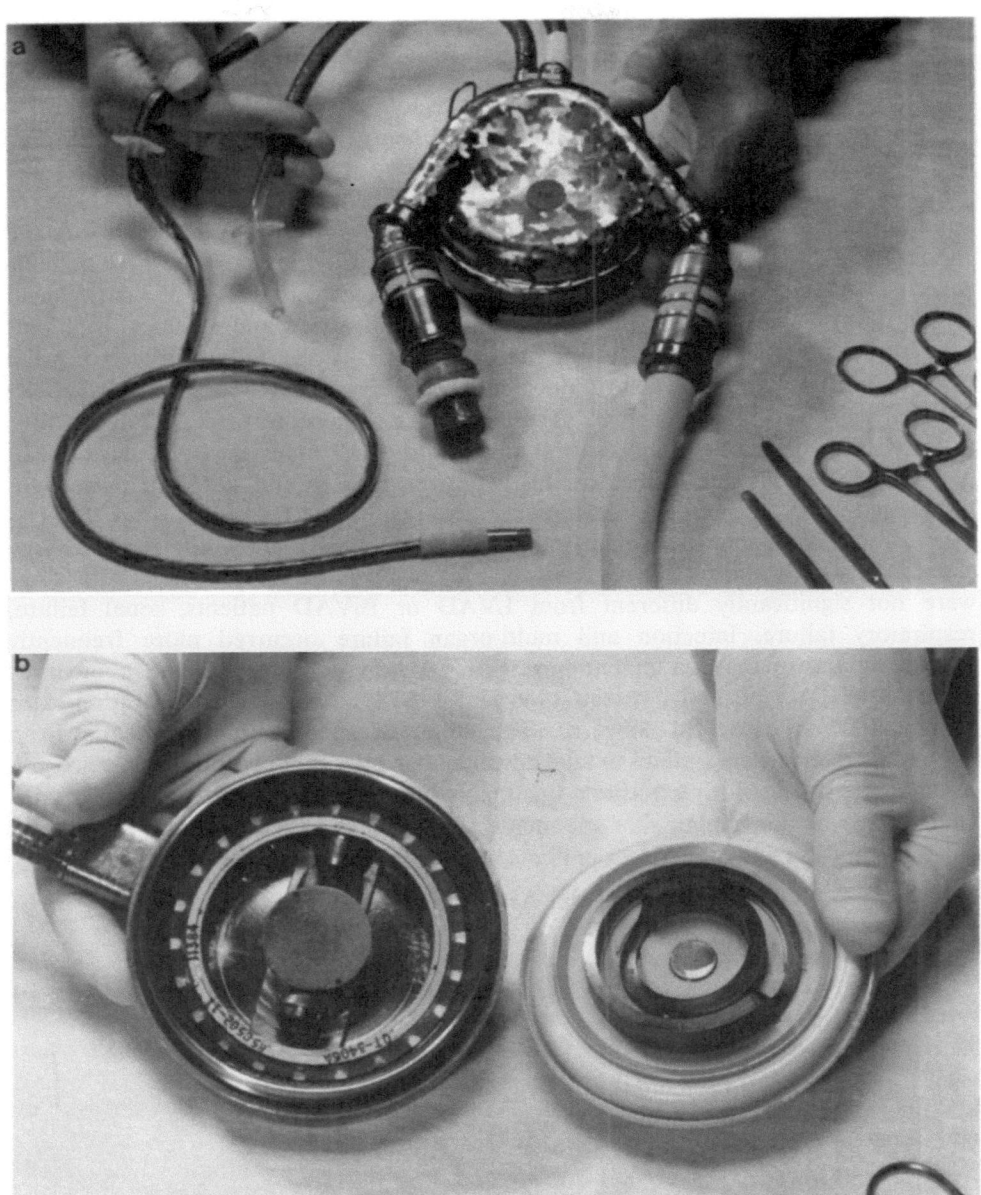

Figure 4. a) External view of electrical Thermedics HeartMate implantable LVAD showing connections and cannulae. b) Internal view demonstrates electrical drive mechanism and pusher plate/diaphragm.

complication, but the durability of tissue valves in implantable permanent LVADs is still uncertain[58]. Criteria for implantable ventricular assist systems (VAS) have now been well established and development of a tether-free implantable VAS remains a major goal of the circulatory support program of the National Heart, Lung and Blood Institute (NHLBI)[59].

TOTAL ARTIFICIAL HEART

The first clinical implant of a total artificial heart (TAH) was performed in 1969 by Cooley using the Liotta heart. Clinical experience with TAH now includes 194 implants in about 30 centers worldwide. While over a dozen different TAHs have been developed (Table 4) the majority of the clinical experience has been with the Jarvik 7 and Jarvik 7-70[16,60] (Symbion Inc., Salt Lake City, UT, USA). Five patients received a TAH as a permanent implant and survived from 10 days to 552 days (mean of 274 days). Unfortunately thromboembolic complications and technical problems have aborted further attempts at long term implants. Valve failure occurred in the first implant. A conversion from Bjork-Shiley (convexo-concave) valves to Medtronic-Hall valves solved this problem[13]. The device has been more successful as a bridge to transplant. It has been inserted for this purpose 189 times for 0-438 days (mean 23.7 days). Seventy-one of the patients were transplanted and 49.6% of those were discharged from hospital. Unfortunately, the results for transplant after TAH is significantly worse than transplant following LVAD or BiVAD where the discharge rates were 87.4% and 69.5% respectively. While bleeding complications with TAH were not significantly different from LVAD or BiVAD patients, renal failure, respiratory failure, infection and multi-organ failure occurred more frequently precluding transplantation in patients with TAH as a bridge[4]. The number of implants of TAH peaked between 1986 and 1988[4]. Costs and complication rates remain a major cause for concern. The emphasis appears to be switching from pneumatic to electrical systems to take advantage of TETS energy transfer systems and to eliminate the need for a pulsatile pneumatic line exiting from the body. It remains to be seen whether implantable pneumatic LVADs such as Novacor and HeartMate will supplant the use of TAH in patients with irreversible cardiac failure.

Table 4. Total Artificial Heart Devices

Akutsu	Milwaukee Heart MHTAH
Berlin	Moscow
BRNO/Czech	Penn State - Electric
Ellipsoid - Unger	Penn State - Pneumatic
Jarvick 7 (Symbion)	Phoenix
Jarvick 7-70 (Symbion)	Poisk
Lepeyre/Aerospatiale	Rostock
Liotta	

THE FUTURE

The ideal permanent circulatory assist device should be small, extremely efficient and totally implantable. It should be non-thrombogenic, cost-effective and have a power source and control system that provides patient mobility.

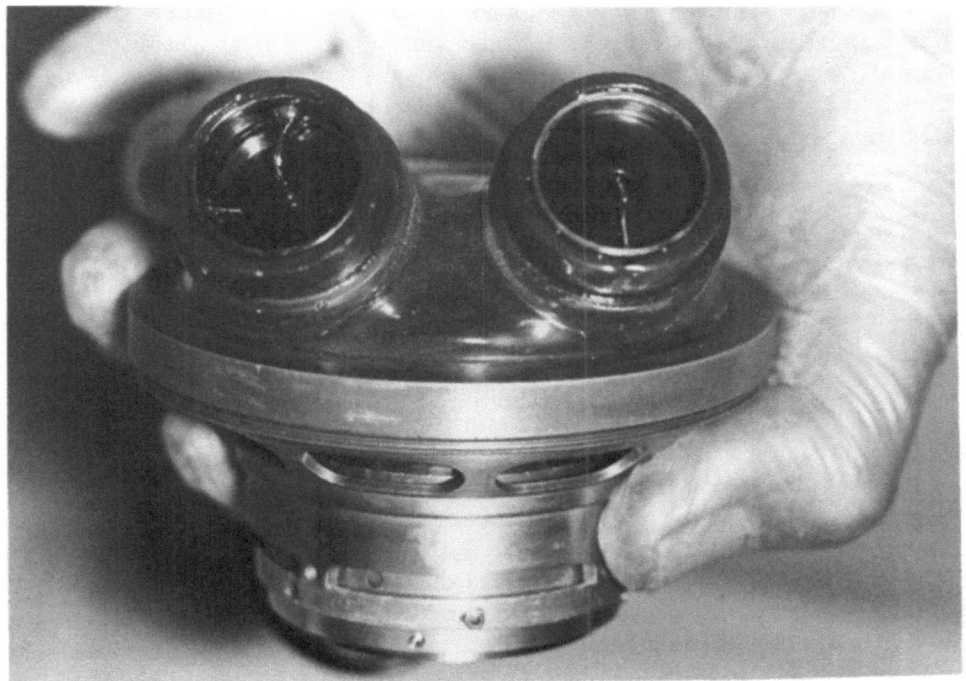

Figure 5. Magnetically actuated ventricular assist device (Kovacs/McKeown pump) demonstrating inflow and outflow Medtronic-Hall valves, pump chamber and magnetic actuator.

Technical challenges to be surmounted in the development of the ideal ventricular assist/artificial heart will focus on (1) power source/drive mechanism, (2) size and weight, (3) thrombogenicity/hemolysis, (4) compliance chamber, and (5) valves.

Power Source/Drive Mechanism

The early successful heart pumps were generally pneumatically driven by an external compressor. The ideal actuator and drive mechanisms should be small, energy efficient and light-weight. Current emphasis has focussed on electrical and electro-magnetic devices[12,50,61-64]. While pneumatic systems require an external compressor, electrical systems can be configured with the drive mechanism incorporated within the device itself. Novacor can be configured with the drive mechanism incorporated within the device itself. Novacor and TCI HeartMate are electrical left ventricular assist devices currently undergoing clinical evaluation as totally implantable devices. Efforts at the University of South Florida in Tampa have concentrated on an electromagnetic assist device (Kovacs/McKeown pump) (Figure 5). This pump is unique in its sophisticated simplicity. There are no moving mechanical parts. The actuator is magnetically coupled to an integrally mounted diaphragm/pusher plate magnet in repulsion mode. With power requirements of 7.8 to 11.4 watts (60-100 bpm) this system has an efficiency near 30% with an average dynamic stroke volume of 65 ml. Early bench testing and acute animal studies of this device have been very encouraging[61,62].

All four groups currently commissioned by the National Heart Lung and Blood Institute (NHLBI) to develop totally implantable artificial hearts are working on electrical devices. The Texas Heart Institute, the University of Utah, and the Cleveland Clinic are investigating electrohydraulic devices in which electrical energy will be used to pump silicone hydraulic fluid to power a pusher plate. The Penn State electric heart uses electrical energy to drive a low speed, reversing DC motor to turn a screw moving pusher plates back and forth. These four devices produce asynchronous, alternating right and left ventricular systole. Electrical device systems are better able to utilize the revolutionary developments in micro-circuitry. Implantable pacemakers and AICDs have evolved dramatically within the past 15 years. A similar course can be predicted with internal, micrologic physiological controllers and telemetry monitoring of artificial heart devices.

Current electrical pumps may produce a lower DP/DT with reduced closing velocity of the valves when compared with the pneumatic devices. This has the potential to reduce the amount of hemolysis. Electrical system are able to best utilize the portable power systems such as the NIH developed transcutaneous energy transfer system (TETS)[65-68]. Using coils, energy can be inductively transmitted across the skin to charge an internal implantable battery. Battery technology still lags behind other scientific developments in this field and remains a limiting factor. Current weight and size of battery packs, while significantly improved, still only provides a limited scope. For improved lifestyle and ease of mobility, the systems must eliminate tethering to the control and power source. Besides the thrombo-embolic complications, the Jarvik-7, being pneumatic, had significant limitations in portability. Although a portable compressor system was designed to provide partial, very temporary (4-6 hours) untethered support of the patient, electrical systems will offer a significant improvement. Alternative energy systems such as nuclear or chemical energy may play a role in future developments. A comparison of size and weight of some of the LVAD energy convertors is presented in Table 5.

Size and Weight

Human anatomic considerations raise significant size constraints; e.g., for abdominal implantation, a device 10.5 cm in diameter will be acceptable to only 73% of the adult population. Reduction to 8.5 cm diameter will allow universal use. Intrathoracic implant requires device geometry considerations to avoid lung compression, and enable chest wall attachment[69-71]. Most implantable devices currently in use or under development have size constraints and are not suitable for smaller adult or pediatric patients. Pivotal technological advances will be required to produce efficient devices useful in smaller patients. Many of the changes center on miniaturization of the current devices, although, it is likely entire new systems of drive mechanism, valve types and pump configuration will be required to meet the needs of pediatric patients. Calcification of tissue valves may prevent long term use of some devices in pediatric patients, necessitating further investigation and advances in valve technology including development of copolymer heart valves.

Thrombogenicity/Hemolysis

Hemolysis and thrombo-embolic complications associated with artificial heart devices are profound problems. Clot formation can occur on the surface of the

Table 5. Comparison of LVAD Energy Convertors

Developer	Description	Designation	Weight (g)	Volume (cc)	Project Time (Yrs)
Thermedics	Low-speed torque motor, rotor with center bearing, and 2 cam followers	MK IV	620	220	16
Gould	A high-speed (4,000 rpm), brushless DC motor drives face cam through gear reduction mechanism. Face cam actuates levers that drive pusher-plate	Mod. V	490	147	12
Kovacs/ McKeown	Single-stage direct magnetic energy converter using mutual repulsion mode pulsed magnetic fields to produce nonmechanical displacement of magnetic pusher-plate	MK V	312	32.5	7
Novacor	Fast-acting, short-stroke solenoid energizes spring that when unlatched drives pump	MK 22C	675	250	16
AVCO/Univ. of Utah	High-speed motor drives axial flow pump	-	-	-	-
Univ. of Washington	Modified Stirling engine	System 7	970	516	14
Aerojet	Stirling cycle thermocompressor	MK 8	990	480	15

bladder/diaphragm, on the valves, or on the cannulae. Design of the pump chamber is critical in preventing areas of stasis and reduced turbulent flow[72]. Finding the ideal substance for blood material interface has posed major technical challenges[73].

Two almost diametrically opposed philosophies have dictated research into surface materials for pump chamber and diaphragm. While many researchers have explored smooth thrombo-resistant materials, others have concentrated on textured blood contact surfaces. Ethicon's Biomer segmented polyurethane has been used extensively for fabrication of blood sacs and cannulae. Thoratec's BPS-215M, a surface modifying copolymer blend, seems to be an improved version of polyurethane[74]. The USF research team is investigating an ultrasmooth copolymer (Vivathane) for fabrication of the pump chamber, diaphragm and cannulas.

In addition to its ultrasmooth character, Vivathane has proven to have excellent flex strength endurance. In acute 1 day animal studies (5 sheep, 18 calves) thrombus formation has not been a significant problem. In these studies however the animals were heparinized and long term trials are obviously essential to confirm low hemolysis and low thrombogenicity rates.

In contradistinction to the smooth surface approach, TCI researchers have elected to use textured surfaces. This can be created using integrally textured polyurethane (ITP) for the diaphragm and Sintered Titanium Microspheres (STM) for the chamber. The blood contacting surfaces in the HeartMate are essentially all textured. These are designed to promote development of a natural adherent, pseudoneo-intima. While there was an initial concern the neo-intimal surface might shear off in pieces and embolize, this does not appear to occur and the early clinical results for this approach are encouraging[49,50,75].

Compliance Chamber

Concomitant with a switch from pneumatic to implantable electrical pump systems is the need for a compliance chamber (CC). Air driven pneumatic systems do not require internal CCs because evacuation of air was intrinsic to the mechanism of action. Implantable air-tight enclosed systems with moving diaphragms require a vent/compliance chamber to permit evacuation of air during diastole. Fabrication of a sealed pump with the pusher place in the end diastolic position creates a vacuum during systole therefore decreasing the efficiency of the device and increasing power requirements. The solution lies in fabricating an implantable compliance chamber connected to the pump or venting through a skin port which creates a portal for infection. Although some progress has been made in the development of compliance chambers including the use of bi-lamina chambers and two-phase fluid compensators, further research efforts in this area are required.

Valves

A choice of heart valves for an LVAD or total artificial heart poses several dilemmas and needs to be individualized for each device. Thrombo-embolic complications of mechanical valves are well documented[76,77]. The use of four such artificial valves in TAH significantly magnifies the problem. In one clinical series, results in routine (non-device related) use of mechanical and tissue valves was compared analyzing total morbidity and valve related mortality[76]. The risk of thrombo-embolism was significantly greater in the mechanical valve group. While there was no significant difference in total morbidity and valve related mortality over

Figure 6. Prototype copolymer valve fabricated from Vivathane.

a ten year period, the results favored tissue valves in the first five years and mechanical valves in the second five years, reflecting the superior durability of the mechanical valves. Calcification and stress tears are the major cause of failure in bioprosthetic (tissue) valves[78-81]. Attempts have been made to decrease calcification in tissue valves with some improvement in porcine aortic valvular prostheses but not in pericardial bioprostheses[78,79]. The lack of uniformity in bovine pericardium is likely to prevent fabrication of a pericardial valve with predictable durability[81,82]. In the clinical devices currently using tissue valves (HeartMate and Novacor) the short and intermediate results have been encouraging with regards to both thrombo-embolism and valve function. Experimental data using tissue valves in an LVAD however have been discouraging enough to raise significant concern[58]. Mechanical valves tend to be more durable, but the in vitro durability data in both mechanical and tissue valves does not necessarily correlate with the clinical implant results[83]. High flow rates and increased stress may precipitate structural failures in tissue or mechanical valves that under other circumstances might not become evident. Such was the case with the Bjork-Shiley convexo-concave valve in the first Jarvik artificial heart implant[13]. In comparing mechanical valves using hydrodynamic in vitro performance assessments, wide variations may occur in such parameters as mean systolic pressure difference, regurgitant volumes and energy losses[84]. Clinical results from mechanical valves also demonstrate wide variability[77]. Choice of valves for VAD/TAH may also significantly influence the amount of hemolysis and thrombo-embolism. Prosthetic valve selection for these devices however encompasses many features. In addition to thrombo-embolic rates, regurgitant volumes, DP/DT, closing velocities, and hemodynamic variables, durability, hemolysis and fabrication parameters may all influence a final choice of valve prosthesis[85].

In the future, successful prosthetic valves may combine the durability and predictability of nonbiologic prostheses with the shape of biological prostheses. Copolymer fabrication makes this possible (Figure 6).

SUMMARY

Mechanical support for the failing heart has now become a technical and clinical reality. Multiple devices ranging from the intra-aortic balloon pump to the total artificial heart are widely available. Patient selection is becoming more refined and results are steadily improving. In most transplant programs, availability of temporary bridging support devices has become essential. The deficiency in donor organs has prompted a renewed interest in implantable permanent circulatory support systems, either LVAD or TAH. Further technological challenges remain to be overcome to produce successful permanent systems. Nonthrombogenic surfaces, miniaturized power drive systems and energy sources, compliance chambers and improved heart valves are all areas of research requiring scientific solutions to clinical problems associated with LVAD and TAH.

ACKNOWLEDGEMENT

Very special thanks to Pat Conant, MS ARNP and Bonnie Heath for their research and editing efforts, Jim Dozier for artwork, and the American Heart Association, Florida Affiliate for their research grant support.

REFERENCES

1. A pioneer in artificial organs, Heart Trans., 1:158, 1982.
2. Kolff WJ, Akutsu T, Dreyer B, et al: Artificial heart inside the chest and the use of polyurethane for making valves and aortas, Trans. Am. Soc. Artif. Intern. Organs, 7:198, 1959.
3. The working group on mechanical circulatory support, National Heart, Lung and Blood Institute. Artificial heart and assist devices: Directions, needs, costs, societal and ethical issues, U.S. Dept. of Health and Human Services, No. (NIH) 85-2723, Public Health Service, 1985.
4. Oakes TE, Pae WE, Miller CA, Pierce WS: Combined registry for the clinical use of mechanical ventricular assist pumps and the total artificial heart in conjunction with heart transplantation: Fifth official report-1990, J. Heart and Lung Trans., 10:621, 1991.
5. Hill JD: Bridging to transplantation, Ann. Thorac. Surg., 47:167, 1989.
6. Ott RA, Mills TC, Eugene J, Gazzania AB: Clinical choices for circulatory assist devices, Trans. Am. Soc. Artif. Intern. Organs, 36:792, 1990.
7. Pennington DG, Swartz MT: Management: By circulatory assist devices, Cardiology Clinics, 7:195, 1989.
8. Pennington DG, Joyce LD, Pae WE, Bucholder JA: Patient selection (proceedings from Circulation Support 1988, Topical Meeting of the Society of Thoracic Surgeons), Ann. Thorac. Surg., 47:77, 1989.
9. Pennington DG, Termuhlen DF: Mechanical circulatory support: Device selection, Cardiac Surgery: State of the art reviews, 3:407, 1989.
10. Emery RW, Joyce LD: Directions in cardiac assistance, J. Cardiac Surg., 6:400, 1991.
11. Ott RA, Mills TC, Eugene J: Current concepts in the use of ventricular assist devices, Cardiac Surgery: State of the art reviews, 3:521, 1989.

12. Pae WE, Pierce WS: Intra-aortic balloon counter-pulsation, ventricular assist pumping and the artificial heart, In: Glenn's Thoracic and Cardiovascular Surgery (Ed., Bane AE), Appleton and Lange, Norwalk, CT, 1990.
13. DeVries WC, Anderson JL, Joyce LD, et al: Clinical use of the total artificial heart, New Engl. J. Med., 310:274, 1984.
14. DeVries WC: The permanent artificial heart; four case reports, JAMA, 259:849, 1988.
15. Pierce WS: Permanent heart substitution: better solutions lie ahead, JAMA, 259:891, 1988.
16. Joyce LD, Johnson KE, Cabrol C, Griffith BP, et al: Trans. Am. Soc. Artif. Intern. Organs, 34:703, 1988.
17. Dennis C, Spring DS, Nelson GE, et al: Development of a pump oxygenator to replace the heart and lungs; an apparatus applicable to human patients, and application as one case, Ann. Surg., 709:721, 1951.
18. Stuckey JH, Newman MM, Dennis C, et al: The use of the heart lung machine in selected cases of acute myocardial infarction, Surg. Forum, 8:342, 1957.
19. Gibbon JH, Jr.: Application of a mechanical heart and lung apparatus to cardiac surgery, Minn. Med., 37:171, 1954.
20. Spencer FC, Eisman B, Trinkle KJ, Rossi NP: Assisted circulation for cardiac failure following intracardiac surgery with cardiopulmonary bypass, J. Thorac. and Cardiovasc. Surg., 49:56, 1965.
21. Moulopoulas SD, Topaz S, Kolff WJ: Diastolic balloon pumping (with carbon dioxide) in the aorta: Mechanical assistance to the failing circulation, Am. Heart J., 63:669, 1962.
22. Kantrowitz A, Tionneland S, Krakauer JS, et al: Mechanical intra-aortic cardiac assistance in cardiogenic shock: Hemodynamic effects, Arch. Surg., 97:1000, 1968.
23. Sanfilippo PM, Baker NH, Ewy HG, Moore PJ, et al: Experience with intra-aortic balloon counter-pulsation, Ann. Thorac. Surg., 41:36, 1986.
24. Weintraub RM, Thurer RL: The intra-aortic balloon pump-a ten year experience, Heart Transplantation, 3:8, 1983.
25. Reemstsma K, Drusin R, Edie R, et al: Cardiac transplantation for patients requiring mechanical circulatory support, N. Engl. J. Med., 298:670, 1978.
26. Bregman D, Casarella WJ: Percutaneous intra-aortic balloon pumping: Initial clinical experiences, Ann. Thorac. Surg., 29:153, 1980.
27. Subramanian VA, Goldstein JE, Sos TA, et al: Preliminary clinical experience with percutaneous intra-aortic balloon pumping, Circulation, 62(supplement I:I), 1980.
28. Miller DC, Moreno-Cabral RJ, Stinson EB, et al: Pulmonary artery balloon counterpulsation for acute right ventricular failure, J. Thorac. Cardiovasc. Surg., 80:760, 1980.
29. Symbas PN, McKeown PP, Santora AH, Vlasis SE, Hatcher CR: Pulmonary artery balloon counterpulsation for treatment of intraoperative right ventricular failure, Ann. Thorac. Surg., 39:437, 1985.
30. Moran JM, Opravil M, Gorman AJ, et al: Pulmonary artery balloon counterpulsation for right ventricular failure II. Clinical experience, Ann. Thorac. Surg., 38:254, 1984.
31. Spence PA, Weisel RD, Easdown J, et al: Pulmonary artery balloon counterpulsation in the management of right heart failure during left heart bypass, J. Thorac. Cardiovasc. Surg., 38:242, 1984.
32. Spence PA, Weisel D, Easdown J, et al: The hemodynamic effects and mechanism of action of pulmonary artery balloon counterpulsation in the treatment of right ventricular failure, Ann. Thorac. Surg., 39:437, 1984.

33. Wampler RK, Moise JC, Frazier OH, Olsen DP: In vivo evaluation of a peripheral vascular access axial flow blood pump, Trans. Am. Soc. Artif. Intern. Organs, 34:450, 1988.
34. Wampler RK, Frazier OH, Lansing AM, Smalling RW, et al: Treatment of cardiogenic shock with the hemopump left ventricular assist device, Ann. Thorac. Surg., 52:506, 1991.
35. Duncan JM, Frazier OH, Radovancevic B, Velebit V: Implantation techniques for the hemopump, Ann. Thorac. Surg., 48:733, 1989.
36. Golding LAR, Stewart RW, Sinhowich M, et al: Non-pulsatile ventricular assist bridging to transplantation, Trans. Am. Soc. Artif. Organs, 34:476, 1988.
37. Zumbro GL, Shearer G, Kitchens WR, Galloway RF: Mechanical assistance for biventricular failure following coronary bypass operation and heart transplantation, Heart Transplantation, 4:348, 1985.
38. Karl TR, Sano S, Horton S, Mee RBB: Centrifugal pump left heart assist in pediatric cardiac operations, J. Thorac. Cardiovasc. Surg., 102:624, 1991.
39. Gray LA, Ganzel BL, Mavroudis M, Slater AD: The Pierce-Donachy ventricular assist device as a bridge to cardiac transplantation, Ann. Thorac. Surg., 48:222, 1989.
40. Farrar DJ, Lawson JH, Letwak P, Cedarwall G: Thoratec VAD system as a bridge to transplantation, J. Heart Transplant. 9:415, 1990.
41. Farrar DJ, Hill JD, Gray LA, Pennington DJ, et al: Heterotopic prosthetic ventricles as a bridge to cardiac transplantation, N. Engl. J. Med., 318:333, 1988.
42. Pennington DG, Samuels LD, Williams G, Palmer D, et al: Experience with the Pierce-Donachy ventricular assist device in post-cardiotomy patients with cardiogenic shock, World J. Surg., 9:37, 1985.
43. Pierce WS, Parr GVS, Myers JL, et al: Ventricular assist pumping in patients with cardiogenic shock after cardiac operations, N. Engl. J. Med., 305:1606, 1981.
44. Lawson JH, Cederwall G: Clinical experience with the Thoratec ventricular assist device, In: Assisted Circulation 3 (Ed., Unger F), Springer-Verlag, Berlin, 1989.
45. Kanter KR, McBride LR, Pennington DG, Swartz MT, et al: Bridging to cardiac transplantation with pulsatile ventricular assist devices, Ann. Thorac Surg., 46:134, 1988.
46. Termuhlen DF, Swartz MT, Pennington DG, et al: Thromboembolic complications with the Pierce-Donachy ventricular assist device, Trans. Am. Soc. Artif. Intern. Organ., 35:616-618, 1989.
47. Champsaur G, Ninet J, Vigneror M, Cochet P, Neidedur J, Boissonnat P: The use of the ABIOMed BVS System 5000 as a bridge to cardiac transplantation, J. Thorac. Cardiovasc. Surg., 100:122, 1990.
48. McGee MG, Parnis SM, Nakatani T, Myers T, et al: Extended clinical support with an implantable left ventricular assist device, Trans. Am. Soc. Artif. Intern. Organs, 35:614, 1989.
49. Goldsmith MF: Portable heart pump recipient recovering well; evaluation begins of role for this technology, JAMA, 266:2666, 1991.
50. Dasse K: Personal communication, 10/29/1991.
51. Starnes VA, Oyer PE, Portner PM, et al: Isolated left ventricular assist as a bridge to cardiac transplantation, J. Thorac. Cardiovasc. Surg., 96:62, 1988.
52. Kormos RL, Borovetz HS, Grasior T, Antaki JF, et al: Experience with univentricular support in mortally ill cardiac transplant candidates, Ann. Thorac. Surg., 49:261, 1990.

53. McCarthy PM, Portner PM, Tobler HS, Starnes VA, et al: Clinical experience with Novacor ventricular assist system, J. Thorac. Cardiovasc. Surg., 102:578, 1991.
54. Portner DM, Oyer PE, Pennington DG, Baumgartner WA, Griffith BP, et al: Implantable electrical left ventricular assist system; bridge to transplantation and the future, Ann. Thorac. Surg., 47:142, 1989.
55. Termuhlen DF, Swartz MT, Pennington DG, et al: Predictors for weaning from ventricular assist devices (VADs), Trans. Am. Soc. Artif. Intern. Organs, 33:683, 1987.
56. Farrar DJ, Chow E, Compton PG, Foppiano L, Woodard J, Hill JD: Effectors of acute right ventricular ischemia on ventricular interactions during prosthetic left ventricular support, J. Thorac. Cardiovasc. Surg., 102:588-595, 1991.
57. Fukuda S, Takano H, Tanenaka Y, et al: Chronic effect of left ventricular assist pumping on right ventricular function, Trans. Am. Soc. Artif. Intern. Organs, 34:712, 1988.
58. Ramasamy N, Chen H, Miller PJ, Jassawalla JS, et al: Bioprosthetic value calcification and pseudo-neointimal proliferation in bovine and ovine model, Trans. Am. Soc. Artif. Organs, 34:696, 1988.
59. Altieri FD, Watson JT: Implantable ventricular assist systems, Artificial Organs, 11:237, 1987.
60. Olsen DB, Burnes GL: Total artificial heart: The Salt Lake perspective, In: Cardiovascular Science and Technology: Basic and Applied II, Oxymoron Press, Boston, 1990.
61. McKeown PP, Kovacs SG, Reynolds DG, Wassell JA, et al: Magnetically circulated LVAD: Initial animal test results, In: Cardiovascular Science and Technology: Basic and Applied II (Ed., Norman JC), Louisville, KY, USA, 1990.
62. McKeown PP, Kovacs SG, Ondrovic L, et al: Acute animal studies using the Kovacs/McKeown MALVAD Electro-magnetic Heart Pump, AAMI Proceedings, Dec. 2-4, Washington, DC (abstract), 1991.
63. Kovacs SG, Reynolds DG, McKeown PP, et al: A magnetically actuated left ventricular assist device (MALVAD), Trans. Am. Soc. Artif. Intern. Organs, in press, 1991.
64. Koroly MV, Ida N, Roemer LF: An electromagnetic actuator using recently developed rate earth permanent magnets, In: Recent Advances in Cardiovascular Surgery (Ed., Reichart B), R. S. Schultz, Seehang, Germany, 1989.
65. Miller PJ, Green GF, Chen H, et al: In vivo evaluation of a compact implantable left ventricular assist system (LVAS), Trans. Am. Soc. Artif. Intern. Organs, 29:551, 1983.
66. Portner PM, Oyer PE, Jassawalla JS, et al: A totally implantable ventricular assist system for end-stage heart disease, In: Assisted Circulation 2 (Ed., Unger F), Springer-Verlag, Berlin, 1984.
67. Portner PM, Oyer PE, Jassawalla, Miller PJ, et al: An implantable permanent left ventricular assist system for man, Trans. Am. Soc. Artif. Intern. Organs, 24:98, 1978.
68. Weiss WJ, Rosenberg G, Snyder AJ, Cleary TJ, et al: Permanent circulatory support systems at Penn State University, Intern Electronic and Elec. Sciences Transaction in Biomedical Engineering, 37:138, 1990.
69. Igo SR, Hibbs CW, Fuqua JM, et al: Theoretic design considerations and physiologic performance criteria for an improved intracorporeal (abdominal electrically-actuated long-term left ventricular assist device (E-type ALVAD) or partial artificial heart, Cardiovasc. Dis. (Bull. THI), 5:172, 1978.

70. O'Bannon W, Donachy HH, Brighton JA, et al: A comparison of three ventricles used for left ventricular bypass in the calf, Trans. Amer. Soc. Artif. Int. Organs, 22:450, 1976.
71. Fuqua JM, Igo SR, Edmonds CH, Hibbs CW, Normal JC: Evaluations of left ventricular apex orientation in patients with congestive cardiomyopathy: Implications for long-term left ventricular assist devices, Clin. Res., 26:646A, 1978.
72. Baldwin JT, Tarbell JM, Deutsch S, Geselowitz DB: Mean flow velocity patterns within a ventricular assist device, Trans. Am. Soc. Artif. Intern. Organs, 35:429, 1989.
73. Portner PM, Green GF, Ramasamy N: The blood interface at artificial surfaces within a left ventricular assist system, Ann. NY Acad. Science, 471-499, 1983.
74. Farrar DJ, Litwak P, Lawson JH, et al: In vivo evaluation of a new thromboresistance polyurethane for artificial heart pump, J. Thorac. Cardiovasc. Surg., 95:191, 1988.
75. Dasse KA, Chapman SD, Sherman CN, Devine AH, Frazier OH: Clinical experience with textured blood contacting surfaces in ventricular assist devices, Trans. Am. Soc. Artif. Intern. Organs, 33:418, 1987.
76. Hammond GL, Geha AS, Kopf GS, Hashim SW: Biological versus mechanical valves, J. Thorac. Cardiovasc. Surg., 93:182, 1987.
77. Cortina JM, Martinell J, Artiz V, et al: Comparative clinical results with Omniscience (STM1), Medtronic Hall and Bjork-Shiley convexo-concave (70 degrees) prostheses in mitral valve replacement, J. Thorac. Cardiovasc. Surg., 91:174, 1986.
78. Schoen FJ: The future of bioprosthetic valves, Trans. Am. Soc. Artif. Intern. Organs, 34:1040, 1988.
79. Jones M, Eibo EE, Hilbert SL, Ferrans VJ, Clark RE: The effects of anticalcification treatments on bioprosthetic heart valves implanted in sheep, Trans. Am. soc. Artif. Intern. Organs, 34:1027, 1988.
80. Garcia Paez JM, San Martin AC, Garcia Sestafe JV, et al: Is cutting stress responsible for limited durability of heart valve bioprostheses? J. Thorac. Cardiovasc. Surg., 100:580, 1990.
81. Clark RE: The future of bioprosthetic valves, Trans. Am. soc. Artif. Intern. Organs, 34:1021, 1988.
82. Gabbay S, Welch H: Reducing the variability in durability of heart valve prostheses, Trans. Am. soc. Artif. Intern. Organs, 34:1022, 1988.
83. Clark RE, Swanson WM, Kardow JL, et al: Durability of prosthetic heart valves, Ann. Thorac. Surg., 26:323, 1978.
84. Knott E, Reul H, Knoch M, et al: In vitro comparison of aortic heart valve prostheses, J. Thorac. Cardiovasc. Surg., 96:952, 1988.
85. Kovacs SG, McKeown PP: Prosthetic valve selection for a pulsatable LVAD, J. Clin. Engineering, Nov/Dec, in press, 1991.

VASCULAR GRAFTS:

CLINICAL AND HEMODYNAMIC APPLICATIONS

[1]Travis J. Phifer, and [2]Ned H. C. Hwang

[1]Department of Surgery, Louisiana State University School of Medicine, Shreveport, LA 71130, USA; [2]Department of Biomedical Engineering, Memphis State University, Memphis, TN 38152, USA

HISTORY OF VASCULAR GRAFTING

Early experiments in the development of vascular grafts began with simple tubes of impervious nonbiologic material placed into the arterial circulation of animals. These experiments were largely unsuccessful, however, with high rates of thromboembolism and hemorrhage related to lack of coverage of the graft lumen by a biologic surface and failure of incorporation of the synthetic materials into the perivascular tissues. In 1906, Carrel reported experiments with homologous and heterologous artery and vein grafts placed in the canine arterial circulation[1]. In that same year, Goyanes replaced a segment of human artery with autologous vein[2]. Despite this early work with vascular grafts, treatment of most arterial injuries as late as World War II was by ligation[3]. In 1952, however, Dubost made a landmark contribution by repair of an aortic aneurysm with an arterial homograft[4]. The subsequent introduction of Vinyon-N as a synthetic graft material[5] ushered in a new era of elective arterial reconstruction. Repair of arterial injuries, also, soon became commonplace.

Repair of venous injuries remains controversial even today, however, with reconstruction of venous occlusive lesions and correction of valvular incompetence still something of a frontier. Delay in development of venous surgery is in part because of lack of understanding of the basic hemodynamics of this circulation, but also because of lack of availability of a satisfactory prosthetic venous conduit with uncertainty regarding feasibility for maintenance of graft patency and risk of pulmonary thromboembolism. The history of vascular surgery, therefore, is largely the history of graft development.

CLASSIFICATION OF GRAFTS

Tissue Grafts

1. Autologous vein

Autologous vein is the most common tissue graft utilized for peripheral vascular reconstruction, suitable for placement in both the arterial as well as the venous circulation. the vein of choice for this application and the standard of comparison for other grafts is the greater saphenous, despite suitability for grafting of other expendable veins such as the lesser saphenous in the posterior calf[6] as well as the cephalic and basilic in the arm[7]. Mechanical properties of the autogenous vein conduit are similar to native artery, with maintenance of compliance match even after arterialization of the graft[8,9]. Also, the endothelial cell lining of the autogenous vein graft is antithrombotic with production of substances such as prostacyclin and plasminogen activator as well as antithrombin III. Mediators released from these endothelial cells likely affect not only vascular tone in the conduit but also function of smooth muscle cells in the media of the graft.

A primary goal in grafting of autologous vein, therefore, is preservation of the endothelial lining in the conduit. Controversy exists, however, as to the optimal technique for accomplishment of this goal. Dependent upon application desired and vein utilized, different techniques of graft preparation are applicable. In the infrainguinal application, both in situ techniques[10] as well as translocated reversed[11] and translocated nonreversed[12] utilizations of the vein are applicable.

The translocated reversed technique is a classic method for preparation of vein grafts. The translocated nonreversed technique offers the advantage of better size matching at the anastomotic level, but requires incision of the biscusped valves in the conduit lumen. Both techniques for preparation of a translocated vein graft require complete dissection of the vein with ligation of all tributaries and disruption of the vasa vasora. Capillary ingrowth from the bed of a vein graft re-establishes circulation in the vasa vasora by 72 hours[13], however, with normal histologic appearance in experimental arteries and veins examined four months after adventitial stripping[14]. Also, translocated vein grafts prepared with good surgical technique perhaps offer equal preservation of the endothelium as compared to in situ techniques[11]. Minimization of endothelial damage during preparation of a graft for translocation is possible by careful handling of the graft[15,16], limiting the period of warm ischemia by immediate flushing after harvest with cold (4°C) and heparinized autologous blood[17] or balanced salt solution, avoidance of intraluminal distention pressure beyond 200 mmHg[18-21], and prevention of vasospasm[22].

As with the translocated nonreversed vein graft, the in situ technique of vein graft preparation requires incision of bicusped valves in the conduit lumen but offers advantages of good size matching between the graft and host vessels at the proximal and distal anastomoses. Minimal handling with avoidance of complete dissection and preservation of the vasa vasora further characterize the in situ technique of vein graft preparation.

Some believe the in situ technique not only limits damage to the endothelial lining of the graft, but also prevents ischemia to the outer media by maintaining integrity of the vasa vasora[10]. Perfusion of the outer media depends upon intact vasa vasora rather than lumenal flow, in contrast to the inner media and endothelium. Disruption of the vasa vasora produces ischemia in the outer media with acute swelling and patchy necrosis of medial myocytes at three days despite little endothelial damage[23]. A fallacy exists, however, in the argument favoring in situ grafting for preservation of the vasa vasora. Technical nuances necessary for placement of the in situ graft actually require complete mobilization of perhaps as much as 10 cm of the vein at both the proximal as well as the distal anastomoses, with myointimal hyperplasia most common at the distal anastomosis rather than in the body of the graft with nondisrupted vasa vasora. Also, some clinical[24] as well as experimental studies[25,26] show development of intimal hyperplasia despite preservation of the vasa vasora with the in situ technique of vein graft preparation.

Despite desirability of autogenous venous conduits for vascular reconstruction, the supply of this valuable commodity is finite. In 10% to 20% of cases dependent upon application and technique, saphenous vein is unavailable either because of previous utilization or because of inadequacy related to size or damage from previous thrombophlebitis[27-29]. In other situations, this otherwise suitable but unmodified venous tributary is too small for the particular application needed. Splicing together of short segments of usable vein offers an option for preparation of a longer conduit in situations with no adequate single tributary[30-32]. Another option particularly in the infrainguinal location is a composite graft[33-37] to a distal target, with prosthetic material in perhaps the above-knee segment combined with a graft of autologous vein crossing the knee to a small diameter target vessel. Formation of a large diameter spiral graft[38] from a smaller diameter vein such as the greater saphenous is a tedious technique described for preparation of a conduit suitable for a superior vena cava graft.

2. Autologous artery

Autologous artery is not only suitable but perhaps ideal for many situations confronting the vascular surgeon[39-42]. Replacement of a diseased artery with a functional artery from the same person certainly seems logical. Advantages cited for the autologous artery graft includes retention of viability, proportional growth in children, maintenance of maximal flexibility across joints, and optimal compliance matching of graft to vessel[43]. Size matching of donor graft to host vessel diameter is frequently possible by proper choice of donor artery (internal iliac = 7.0 mm to 8.0 mm, external iliac artery = 6.0 mm to 8.0 mm), superficial femoral artery = 5.0 mm to 6.0 mm). Resistance to infection makes these conduits particularly desirable for utilization in infected fields[44]. A particularly useful application for these grafts is in renovascular reconstruction. Aneurysmal dilation of autologous artery grafts in the aortorenal location is uncommon, in contrast to autologous vein placed in this location in a young host.

Despite utility of these conduits in varied applications, availability of expendable autologous artery is quite limited. A single internal iliac artery, however, is usually expendable. Harvesting of an autologous arterial graft from the external iliac, common iliac, or superficial femoral location followed by reconstruction with a prosthetic graft is also sometimes justified. In other case, endarterectomy of an already occluded superficial femoral artery provides a vascular conduit for placement in another location.

3. Homologous tissue

The length and diameter of unmodified human umbilical vein combined with the natural absence of valves and branches makes this structure attractive as a possible conduit for infrainguinal arterial reconstruction. Glutaraldehyde tanning stabilizes the basic architecture of the graft for up to five years. A subsequent substantive increase in biodegradation occurs[45,46], however, with aneurysm formation at or beyond 36% by five years[47,48] despite reinforcement of the graft with an external polyester mesh. Indications for utilization of this graft, therefore, are limited[49].

Early work with cryopreserved vascular conduits obtained from cadaveric donors suggested, perhaps, some role for utilization of these grafts in peripheral vascular reconstruction[50,51]. A role for this graft in peripheral vascular reconstruction, however, is not yet established.

4. Heterologous tissue

Heterologous conduits for arterial reconstruction are at this time mostly of historical interest. Although limited by premature thrombosis and especially the potential for aneurysm formation, these grafts perhaps ushered in a new generation of vascular grafts (the tanned collagen tube) and served as the prototype for the human umbilical cord vein allograft. In cardiac valve replacement, however, xenogeneic tissue functions quite well. Also, some recent experimental work suggests a possible role for xenogeneic tissue in arteries[52] venous valve replacements[53].

Prosthetic Grafts

1. The "ideal" prosthetic graft

The near ideal prosthetic graft[54] is readily available in a variety of sizes at reasonable cost, with preimplantation shelf stability as well as capability of repeated sterilization. Handling characteristics are good, with an uncomplicated preparation procedure. Tissue reactivity is low, but with good incorporation of the external surface into surrounding tissue. Implantation durability is good with maintenance of compliance as well as a biologically renewable and healthy lining, but with no toxic or allergic side effects. The graft produces minimum alteration of fluid energy or flow kinetics with practically no change in cellular components of blood. The graft is resistant to infection, and thrombogenesis.

2. Graft materials and clinical utility

Dacron and polytetrafluroethylene are the materials commonly used in fabrication of most modern prosthetic vascular grafts. Utilization of these grafts is primarily for elective arterial reconstruction in clean fields. Despite superb function of Dacron grafts in the aortic as well as other central arterial locations, infrainguinal placement of this material certainly limits patency results as compared the venous autograft[55,56]. Polytetrafluorethylene grafts enjoy limited application at the infrainguinal level particularly in the above knee location[57]. Patency data with venous application of either Dacron or Polytetrafluorethylene protheses are generally unsatisfactory, with the exception of limited application for large diameter and relatively high flow conduits perhaps with external support placed in central locations[38]. Utilization of prosthetic materials in contaminated wounds is generally contraindicated because of the risk of graft infection. Polytetrafluoroethylene grafts, however, seem relatively resistant to infection[58-61].

a. Dacron Grafts

The fabrication of most modern Dacron grafts is generally in either a woven or knitted configuration. Each configuration has specific indications and advantages as well as disadvantages. Porosity is a major factor in the choice of graft configuration for a specific application. In general, the porosity of knitted grafts exceeds that of woven grafts. The standard for measurement of prosthetic graft porosity is in cubic centimeters of water per minute per square centimeter at 120 mm Hg pressure[62]. High porosity grafts have superior healing characteristics, but with an increased rate of bleeding complications. Woven grafts are useful, therefore, in situations associated with coagulopathy and platelet dysfunction such as leaking aortic aneurysms and cardiopulmonary bypass. Knitted grafts are best in elective extrathoracic reconstructive procedures with no associated platelet or coagulation derangements.

Some knitted Dacron grafts also incorporate a velour fabrication such that fibers of Dacron are perpendicular to the surface plane of the graft. This velour results in a "velvety" feel to the graft, perhaps with improved handling characteristics. Also, the velour probably enhances healing characteristics related to fibroblast attachment and movement along the interstices of the fabric.

Crimping of the Dacron tube is now a common practice in the manufacture of vascular grafts. This perhaps reduces kinking of the graft at the level of obligatory bends, such as the level of joint crossing. A disadvantage of the crimping, however, is the trapping of fibrin in lumenal ridges produced by the crimp itself. The effects of crimping are substantially reduced, however, by the standard surgical practice of sizing the graft with stretching of the fabric at the time of graft placement such as to remove most crimps.

Most knitted grafts require preclotting, thus sealing interstices of the porous tube with thrombus. A new design of knitted graft, however, requires no preclotting consequent to pretreatment of the graft with collagen during the manufacturing process resulting in temporary occlusion of the graft interstices. Preclotting of the less porous woven graft is not essential. A common practice with woven grafts, however, involves soaking the graft in concentrated albumen followed by autoclave heating of the albumen soaked prothesis thus sealing the graft with congealed protein immediately prior to implantation. Subsequent removal by phagocytosis of biologic plugs such as thrombus or albumen ultimately returns the Dacron fabric to the original level of porosity, thus realizing optimal healing potential for the graft with reduced porosity at a time of placement. Despite wide acceptance of these practices and proven durability of most modern Dacron grafts, early prototypes of "compound" grafts were unpredictable and unreliable with increased complications related to late bleeding and perigraft hematoma formation[63-65].

b. Polytetrafluroethylene (PTFE) Grafts

The first clinical application of the polytetrafluoroethylene vascular graft came in 1971[66]. Polytetrafluoroethylene (PTFE) is a chemically inert but electronegative and hydrophobic expanded polymer of Teflon. The material requires no preclotting, but is highly porous in regards to fibroblast ingrowth. The determinant of porosity is the length of fine fibrils connecting solid nodes producing a pore with the standard size of 30 microns. Graft healing by fibroblast ingrowth and formation of a neointima (pseudointima) varies with pore size[67,68]. Thrombogenicity of PTFE grafts is greater than autogenous saphenous vein, but less than Dacron[69-71]. Also, declotting of PTFE grafts is more simple than with Dacron grafts, in part because of the smooth lumenal surface of the PTFE grafts. These characteristics as well as graft requirements in excess of autogenous supply and reasonable secondary patency after primary thrombotic failure makes PTFE a popular material for construction of grafts such as the extraanatomic axillofemoral and in creation of hemodialysis access fistulas. Covering the PTFE graft with a thin external reinforcement layer or increasing the wall thickness eliminated the problem reported in early grafts of aneurysm formation in the body of the graft in the absence of injury or infection[72]. Also, utilization of external rings for graft support perhaps enhances patency in some situations.

HEALING OF VASCULAR GRAFTS

Tissue Grafts

Endothelial injury, sustained from either mechanical trauma or ischemic damage during preparation of venous autografts, undergoes subsequent repair after reperfusion. Endothelial denudation or even separation of intercellular junctions between endothelial cells with exposure of basement membrane results in platelet coverage of the exposed elastin and collagen with subendothelial infiltration by polymorphonuclear leucocytes[72]. The inflammatory response becomes transmural by seven days, with fragmentation of medial myocytes and progressive necrosis[74]. Both endothelial as well as smooth muscle cells begin proliferating[75], with mitotic activity apparently peaking at about one week. Smooth muscle cells also develop secretory attributes similar to fibroblasts in soft tissue wounds, secreting a matrix of collagen and glycosaminoglycans. Collagen increases nearly 50% by four to eight weeks[76]. Platelet-derived growth factor is a likely link between endothelial injury and proliferation of modulated smooth muscle cells[77,78], despite inclusive clinical data[79,83]. A linear relation exists, however, between development of intimal hyperplasia and graft injury, such as that sustained from surgical trauma or mural ischemia[84].

Prosthetics

After implantation, the flow surface of a prosthetic vascular graft also undergoes a process of healing[85]. Platelets and early-phase coagulation factors adhere to the surface of prosthetic grafts immediately after reflow, with synthesis of thromboxane A2 and platelet aggregation as well as release of platelet-derived growth factor with platelet factor IV and B-thromboglobulin. Thrombin production results in further platelet aggregation with production of fibrin. Complement activation embellishes the process with C3a stimulating platelet aggregation and release, and C5a stimulating polymorphonuclear neutrophil release of tissue thromboplastin. Thrombus formed on the surface of the graft lumen subsequently undergoes organization by ingrowth through graft interstices of fibroblasts from surrounding tissues.

Increased porosity and velour fabrication facilitate fibroblast movement and incorporation in the interstices of the graft. This basic biologic process leaves connective tissue covering part if not all of the graft surface, effectively excluding the foreign material of the prosthetic graft from the vascular lumen. Failure of complete coverage of the graft surface perpetuates the process of blood component interaction with the graft, however, potentiating graft thrombosis as well as progression of distal occlusive disease consequent to microembolization of thrombotic debris and endothelial injury with continued exposure to mitogens and vasospastic agents.

Thickness of the layer of lumenal thrombus formed on the surface of prosthetic vascular grafts depends upon several factors including diameter of the graft as well as local velocity of blood flow and separation of the flow boundary layer[85]. Thickness of the surface thrombus in arterial grafts varies inversely with mean blood velocity and mean wall shear stress. Mean blood velocity varies inversely to the square of the radius of the conduit. Mean wall shear stress varies inversely to the cube of the radius of the conduit. Grafts oversized to the lumen of the recipient vessel as well as grafts with low flow, therefore, form a thicker layer of lumenal surface thrombus. With reduction of wall shear, the thickness of thrombus deposited on the graft surface increases with reduction of the effective graft lumen until wall shear stress in the conduit and recipient vessel are somewhat equalized. Thick lumenal surface thrombus, as well as limited graft porosity with inadequate fibroblast ingrowth and incomplete surface thrombus organization, potentiate fragmentation of thrombus, however, with peripheral embolization and/or graft occlusion. Proper sizing of the graft to the host vessel, therefore, is important. The ideal graft diameter provides adequate flow without a pressure decrease both at rest as well as during exercise, with flow limitation consequent to resistance of the peripheral vascular bed rather than resistance of the graft itself.

Perianastomotic factors

Preservation of endothelium on the lumenal surface and limitation of injury to smooth muscle in the media are major concerns with the preparation of autogenous vein grafts. In these venous autografts, complications consequent to endothelial injury occur both in the body of the graft as well as at the perianastomotic level. Prosthetic vascular conduits in the human host, however, heal without endothelialization beyond the perianastomotic level[87-91]. Even attempts at implantation of endothelium on the surface of prosthetic vascular conduits[92], either prior to implantation or at the time of implantation, are as yet unsuccessful in terms of pragmatic clinical application. Complications in prosthetic grafts related to endothelial injury, therefore, occur only at the perianastomotic level in the autogenous host artery. Injury to the host artery occurs during vascular occlusion consequent to both clamp trauma as well as lumenal ischemia, with further injury caused by arteriotomy and placement of anastomotic sutures.

After revascularization, compliance mismatch between host vessel and prosthetic graft, as well as unfavorable geometrical design of the prosthesis-to-host anastomosis, potentiate additional endothelial injury. The response to endothelial injury regardless of origin is similar in both the autogenous vein autograft and the recipient host artery. The response of vascular endothelium and smooth muscle to injury, therefore, is relevant not only to vascular autografts but also to vascular prostheses.

The response of the arterial wall to injury is complex and multifactorial[93-109]. In the rat carotid artery balloon injury model, repair of the denuded endothelial surface begins with deposition of a monolayer of platelets on the injured surface. Soluble factors such as fibronectin, von Willebrand factor, fibrin, and thrombospondin modulate this attachment of glycoprotein receptors on the platelet membrane to the subendothelial matrix. Growth factors released with platelet degranulation (platelet derived growth factor, fibroblast growth factor, epidermal growth factor, and transforming factor B) possibly act as mediators signaling cellular events such as medial proliferation. Both endothelium as well as smooth muscle begin proliferating within 48 hours, with mitotic activity peaking at about one week. Advancement of endothelium as a continuous sheet from adjacent uninjured vessel displaces the platelets in an attempt at regeneration of a complete monolayer at the level of denudation. This monolayer is sometimes incomplete, however, leaving areas of denuded surface with increased permeability to large molecules.

Smooth muscle proliferation begins by 24 hours after injury with migration of these cells into the intima some days later. Endothelial regeneration limits this process. In the absence of surface endothelium, smooth muscle proliferation continues for two to four weeks. Intimal thickening occurs both as a result of this proliferation and migration of smooth muscle, and because of the subsequent deposition in the intima of a connective tissue matrix consisting of collagen and elastic fibers as well as proteoglycans synthesized by the smooth muscle. Compensatory dilation of the vessel only partially mitigates lumenal narrowing caused by this process. Failure of endothelial regeneration, however, sometimes results in abnormal vasoactive responses with development of vasospasm.

All modulating factors in this complex process regulating smooth muscle and endothelial growth, as well as matrix deposition, are not completely clear. Not only platelets but perhaps also smooth muscle and leukocytes as well as the endothelium produce factors capable of either self-modulation or modulation of adjacent cells. The endothelium, therefore, has a potential role not only in maintenance of a nonthrombotic surface and regulation of vessel diameter but also in modulation of cellular events related to repair of vascular injury. In the rat carotid artery balloon injury model, but not in the vascular graft, regeneration of surface endothelium curtails the process of smooth muscle proliferation and migration. The presence or absence of smooth muscle proliferation, then, perhaps reflects the relative production of promotors or inhibitors from either the overlying endothelium or perhaps the smooth muscle cells themselves as modulated by either biochemical or biomechanical factors.

Compliance mismatch related to either the graft material or the anastomotic technique[43,110], as well as turbulent blood flow or local stasis resulting from slow boundary layer flow[111,112], are mechanical factors of likely relevance to the development of myointimal hyperplasia.

Energy loss and flow disturbance occur at any change in direction of blood between a vascular graft and host vessel, with both parameters proportional to the angle subtended between the graft and vessel. The end-to-end anastomosis is the optimal choice of anastomotic configuration[113]. The end-to-side and side-to-end configurations are sometimes best, however, because of either clinical or technical reasons. Despite energy loss, blood transmits in anastomoses even at obtuse anastomotic angles[114]. Flow separation and stagnation as well as turbulence and distorted velocity vectors consequent to flow disturbance with suboptimal anastomotic configurations, however, potentiate intimal thickening and thrombus formation with early graft failure[115-119]. Substantive compliance mismatch between the vascular graft and host vessel, likewise, potentiate injury with a healing response[120]. Hemodynamic parameters specific to the graft material and anastomotic configuration at a particular site, therefore, probably impact significantly upon the response to healing of perianastomotic tissues. These factors will be discussed in details in the following sections.

HEMODYNAMIC EVENTS IN VASCULAR GRAFTING

The hemodynamic factors that may influence vascular graft healing include turbulence, stasis and the distributions of wall shear stresses. These factors are unequivocally related to the anastomotic angles, the geometrical design and the physical properties of the graft.

Surgical grafting of a newcomer conduit, whether biological or synthetic, to a host vessel imposes a new hemodynamic environment to the host. The so-called "healing" is accomplished only when the host and the graft are completely adapted to the new hemodynamic environment. It has long been suspected that hemodynamical factors play an important role in the initiation and localization of various types of vascular lesions and the deposition of platelet thrombi which can eventually lead to closure of the vessel lumen[111-120]. Recent research to further define the role of hemodynamics in healing and failure of vascular grafts are discussed according to the design of each grafting configuration.

End-to-End Grafting

End-to-end arterial graftings are usually made to replace or repair larger size arteries (e.g., aortic aneurysm). Hemodynamic parameters that are of clinical concern in this type of grafting are usually different from that of the smaller arterial graftings. For patients on heparinization undergoing aortic replacement surgery, excessive blood loss may result from bleeding through the interstices of the large diameter grafts. Blood forced through the graft fabric probably undergoes defibrinization leading to bleeding diathesis or coagulopathy. The need of blood replacement by multiple transfusions also aggravates the condition. Therefore, graft of low-porosity construction with Dacron threads is the basic material of choice in surgery[121]. If the graft is made to replace a significant portion of the ascending aorta and/or the aortic arch, control of the after-load of the heart and preservation of circulation to the distal body may become a consequential concern. In this case, the compliance of the graft conduit will be an important factor for considerations.

For cylindrical blood vessel of diameter D and length L, a change of pressure Δp will cause a corresponding change of vessel diameter ΔD. Assuming no change in length when the cylinder is distended, the vessel compliance can be defined as:

$$C = \left(\frac{\Delta V}{V}\right)/\Delta p = \left(\frac{2\Delta D}{D}\right)/\Delta p \qquad (1)$$

By tracking diameter axially across end-to-end arterial anastomosis created using 6-0 Prolene sutures on transected femoral arteries in dogs, Hansson, et al[122,123] were able to generate longitudinal profiles of compliance which demonstrated: 1) Compliance of the anastomosis (suture line) is significantly lower than that of the reference artery; 2) Diameter decreased monotonically from artery to suture line to produce an average 8 to 18% diameter stenosis, and this effect was greater with continuous suture compared to interrupted suture; and 3) the compliance between the reference artery and suture line first increased to a peak value, approximately 50% higher than that of the reference artery, then decrease to its anastomic nadir. The term "para-anastomotic hypercompliance" was introduced by these authors to describe peak compliance on the arterial side (Fig. 1). The observations were made to compare continuous suture vs interrupted suture, but the peak compliance value and its location were found independent of the suture technique. While the influence of the axially varied vessel compliance on the formation of vascular thrombosis and/or initial hyperplasia is yet to be established quantitatively, Duncan et al[124] was able to show that the outer wall compliance reduces shear rate at the wall.

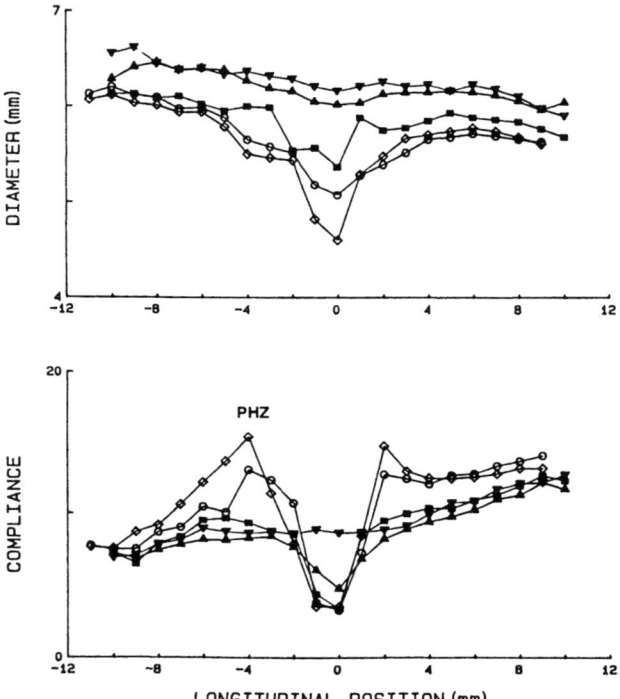

Figure 1. Typical longitudinal profile of compliance (measured on day of surgery) in a canine femoral artery showing Para-anastomotic Hypercompliance Zone (PHZ) (from Mergerman and Abbott, 1989, with permission).

Side-to-End Grafting

Side-to-end anastomosis is surgically constructed as the proximal anastomosis of a vascular bypass graft. At the anastomosis, blood flow from the upstream vessel is

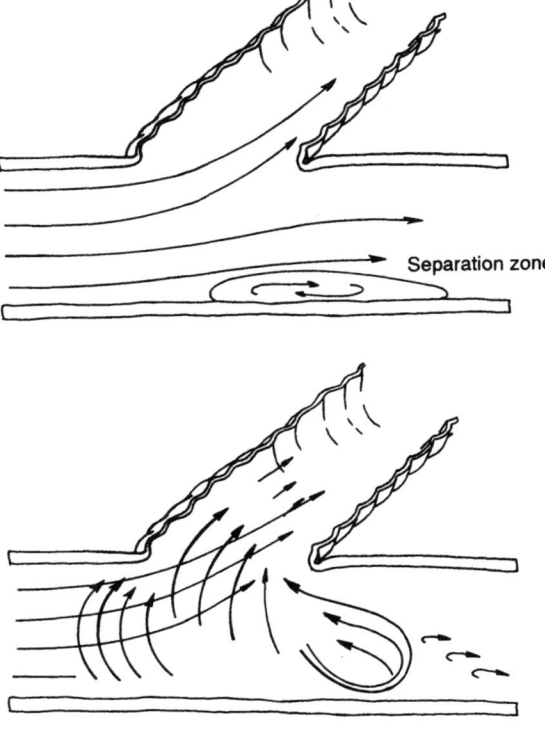

Figure 2. (a) Side-to-end anastomosis with patent host artery; (b) side-end anastomosis with significant arterial occlusion.

usually divided into two streams. One of which continues to flow into the distal host vessel, while the other is diverted into the graft conduit as shown in Figure 2.

Under steady flow conditions, LoGerfo et al[116] performed a carefully planned hydraulic model study by monitoring the pressure distribution from 15 flush-mounted wall probes located around the anastomotic region of an acrylic conduit. The experiment was carried out in an 8 mm acrylic tube with the tube Reynolds number valued between 300 and 5000. The testing fluids include both water (at 4°C) and blood (at 37°C). Their results demonstrated that the presence of a side-to-end anastomosis causes little loss of fluid energy with either the smooth wall graft or the Dacron graft. Their results however showed that there is a linear pressure gradient with distance in the inlet tube (due to friction). As the fluid enters the anastomosis, there is a decrease in pressure along the outside wall in the host vessel. This is followed by a sharp pressure recovery. The pressure gradient again became linear within 20 diameters of the anastomosis (Fig. 3).

Although the local adverse pressure gradient (i.e., pressure increasing in the direction of flow) is small in magnitude, it provides the force necessary to cause flow separation from the wall and to change the direction of the flow within the separation zone. The cine studies made with dye injection demonstrated that the separation at the bifurcation formed a three-dimensional shell-like flow pattern as shown in Fig. 2.

The studies also demonstrated that flow within the separation is unstable but not turbulent, at Reynolds numbers within the clinical range. Flow within the separation

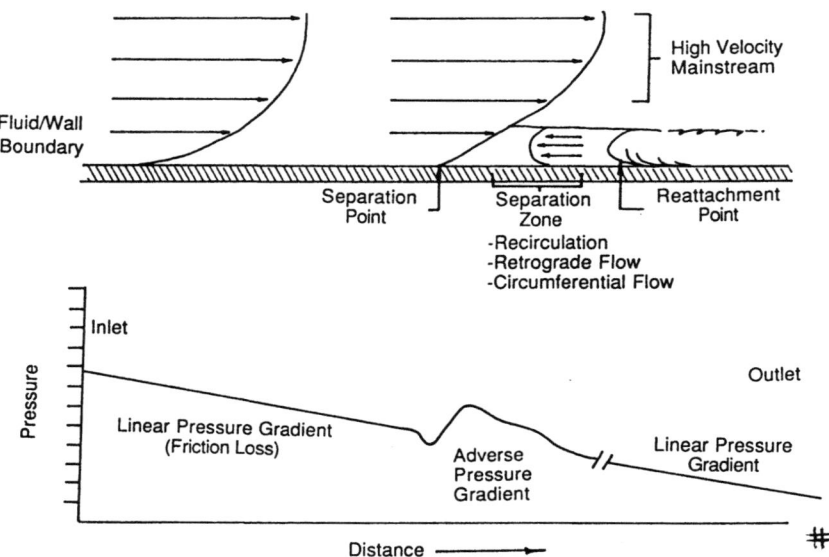

Figure 3. Pressure gradients at an arterial side-to-end anastomosis (from LoGerfo et al, 1979, with permission).

zone is slow compared to that in the mainstream. Therefore, this is a region of relatively low shear stress, where fluid in contact with the wall moving slowly. In contrast, wall shear is high at the flow divider where the mainstream comes into proximate with the wall. The authors suggested that these findings are pertinent to the clinical problems of intimal or neointimal hyperplasia that occurs at the anastomosis, and that the adherence of platelets to the subendothelium within the separation may be one of the mechanisms involved in development of anastomotic hyperplasia.

From a hydrodynamic point of view, one would notice that both the location and the magnitude of this pressure reversal (hence the location of the separation) depend on the ratio of the flow rates into the graft to that into the distal artery. A common application of the side-to-end grafting would be that of a proximal anastomosis for a graft bypass over a totally occluded or near occluded artery. In such case, the development of intimal or neointimal hyperplasia in the distal host vessel would no longer be a clinical concern.

On the other hand, when the side-to-end anastomosis is applied to a high flow situation (e.g., hemodialysis angioaccess arteriovenous loop graft, AVLG), the flow field in the anastomotic region is highly pulsatile. We found in this case, the mainstream flow into the distal artery may not be always unidirectional. Retrograde flow has been measured during certain phase of the pulse cycle (Fig. 4). This is possibly due to the significant development of collateral circulation after the AVLG implantation (Fig. 5). Consequently, the point of separation was found to oscillate up- and downstream along the opposite wall in the axial direction[125]. To quantify these observations with respect to the development of anastomotic intimal hyperplasia and/or formation of atherosclerotic plaque, however, requires further investigations. Clinical observations have established that most hemodialysis angioaccess loop grafts fail due to lesion development at the distal (end-to-side) anastomosis[126-128].

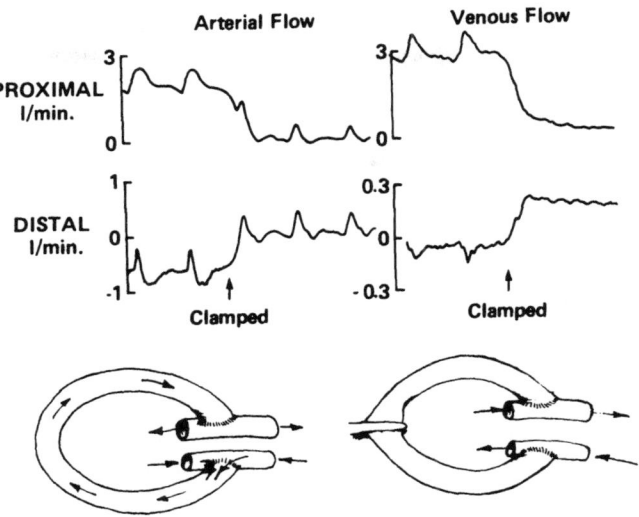

Figure 4. Antigrade and retrograde flows at an arterial end-to-side AVLG anastomosis.

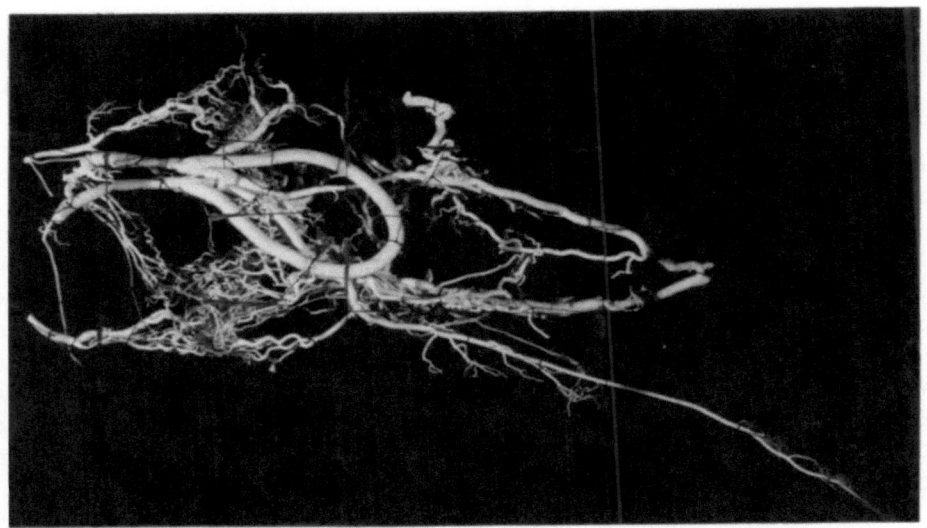

Figure 5. Collatarol circulation development in a 24-week canine femoral-to-femoral AVLG implantation.

End-to-Side Grafting

End-to-side anastomosis is surgically created to bring high energy arterial blood to an otherwise low or no flow distal vessel. The longevity of these grafts is usually limited by the development of intimal hyperplasia which leads to stenosis and thrombosis. The occlusive lesions are predominantly initiated from a location in the host vessel at or in the vicinity of the anastomosis[126,129-131]. The frequently mentioned contributing factors include: (1) the hemodynamics of blood flow at the anastomosis,

(2) the rheologic behavior of the blood and the blood vessel, (3) the characteristics of the graft luminal surface, (4) the adhesiveness of platelets, and (5) the activation of the clotting and fibrolytic mechanisms. Viewing it comprehensively, each of these factors may, in one degree or another, be related to the hemodynamics of the graft flow in the anastomotic region.

The surgically constructed end-to-side anastomosis of an arterial bypass graft brings high energy arterial blood to an otherwise occluded or nearly occluded distal host blood vessel. This may result in significant hemodynamic changes in the host vessel. Thickening of the intima is one of the typical reactions associated with the elevated intravascular pressure found in larger arteries.

Quantitation of the detailed local hemodynamic events in the anastomotic region may be accomplished by measurements made in full-size or up-scaled flow models using a blood analog fluid[119,128]. The flow models are designed and constructed to preserve the dynamic similarity of the prototype vessels. This is realized by maintaining the same Reynolds number for steady flow models, and maintaining both the Reynolds number and the Wormersley parameter values for pulsatile flow models. The Reynolds number (Re) and the Wormersley parameter (α) are respectively defined by:

$$Re = \frac{VD}{\nu} \qquad (2)$$

$$\alpha = \frac{D}{2}\sqrt{\frac{2\pi f}{\nu}} \qquad (3)$$

where D is the vessel diameter, V is the mean velocity, ν is the kinematic viscosity and f is the frequency of the heart beat.

A typical example of modeling a 4mm (I.D.) artery with maximum flow velocity of 100cm/sec at heart rate 70 beats/min requires that Re=1000 and α=3.5. When the system is modelled with a 10mm (I.D.) tube using an analog fluid having the same viscosity as the blood, the corresponding mean velocity in the model conduit should be 15.7cm/sec, and the corresponding pulse rate is 4.72 beats/min. Since the pulse wave at 4.72 beats/min can not be physiological, nor the maximum arterial flow velocity of 15.7cm/sec, careful plans must be drawn before the measurements made in the model can be interpreted to quantify the hemodynamic parameters in the in vivo graft-host anastomosis.

Most fluid mechanicians recognize that the motion of the vessel wall can significantly alter the flow structure inside the blood vessels, nevertheless, a significant number of the in vitro vascular flow model studies reported in literature were performed in rigid conduits, using either full-size or scaled models[131-134]. Rather than quantifying any particular hemodynamic parameters, these rigid models are usually used to explore a certain biochemical mechanism. Using an otherwise identical 90° bifurcation, the difference in flow patterns between a flexible wall model and that derived from a rigid wall model was clearly demonstrated by Liepsch[135]. The effects of non-Newtonian viscoelasticity on flow at junctions were investigated[136]. Recently, several groups have performed anastomotic flow studies using elastic transparent models[137-139].

We constructed a series of three elastic transparent flow models based on the silicone rubber casts obtained from the femoral end-to-side graft anastomoses implanted in dogs. These casts were obtained at reoperation after the grafts had been implanted for different periods of time up to 62 weeks. Typically, they represent the luminal geometries at different stages of lesion development at the anastomoses (Fig. 6). Detailed flow dynamic studies were made using transparent Silastic models derived from these casts[137]. The three flow model conduits used in the study may be considered as the representations of the host vessel in three different stages of lesion development: (A) with no lesion development, (B) with moderate development, and (C) with severe lesion development. In order to create a flow model that is similar to the in vivo host vessel, several elastic transparent Silastic flow model conduits were made from each of the RTV casts. Based on the measured wall compliance (Eq. 1), one of the conduits was selected as the flow model.

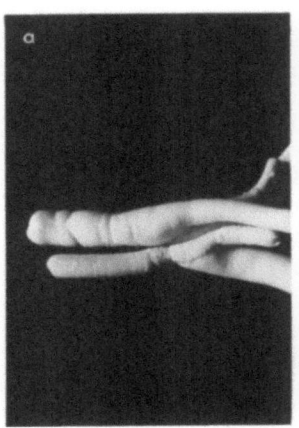
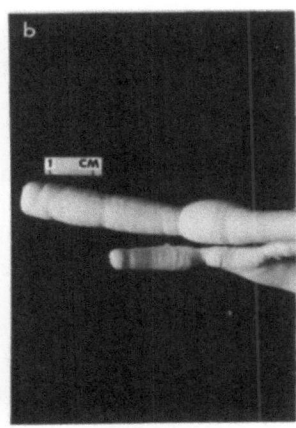
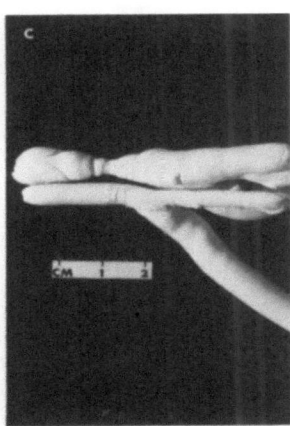

Figure 6. Silicone RTV casts showing: (a) minor (b) moderate (c) severe lesion development in the venous anastomotic region.

The in vivo hemodynamic conditions were replicated by using a mock flow loop system. The system (Figure 7) included a steady and a pulsatile energy sources, a flow laminator and an afterload system with adjustable resistor and compliance units on both the arterial and venous sides. The Silastic model was installed in an open-top box which was connected to the mock loop system. Physiological pressure transducers and electromagnetic flowmeters were installed at each end of the flow model conduits, while the preload (energy sources) and the afterloads (resistor and compliance units) were carefully adjusted until both the pressure and flow waveforms similar to that of the in vivo measurements made at the reoperations were obtained.

The arterial flow model conduit was connected to the venous conduit by using a uniform 25 cm long, 6 mm expanded polytetrafluoroethylene (ePTFE) Gore-Tex graft (Gore Inc., Flagstaff, AZ) in a fashion similar to that created for the hemodialysis angioaccess loop graft (ALG). Under pulsatile flow, the diameter change in each flow model conduit was continuously monitored by a pair of laboratory constructed ultrasonic dimension gauges. Since one of the main interests was to study the effects of wall motions on the hemodynamics in the anastomotic region, the conduit compliances were carefully compared to that measured from the in vivo host vessels during the reoperation.

Flow visualizations were made at both the proximal and distal anastomoses prior

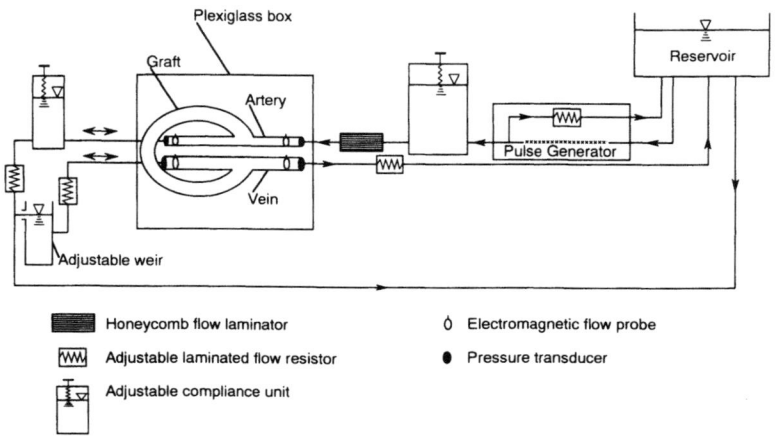

Figure 7. Mock flow loop system for AVLG studies.

to detailed mapping of the patterns of the anastomotic flow field. The flow visualizations, made with electrically generated hydrogen bubbles in a laser (15 mW HeNe) illuminated field, were made to provide a comprehensive view of the flow field and to identify areas of unusual flow activities (e.g., boundary layer separation, area of stasis, etc.).

To facilitate laser Doppler anemometer (LDA) measurements, the flow model was mounted on a motorized two-dimensional transverse table which had an accuracy of 16 μm per dial division in both X and Y directions. The LDA used in the experiment was a TSI (Thermal Systems, Inc., St. Paul, MN 55164) Model 1090 tracker system with Bragg cells. Using an achromat focusing lens of 120 mm focal length, the elliptical measuring volume has major and minor axes of 0.4 mm and 0.75 mm, respectively. The elliptical measuring volume consists of 64 fringes, spaced 1.38 μm apart.

Velocity surveys were made at seven different sections in the venous and arterial host vessels in the flow models (Fig. 8). The sections were spaced 7 mm (approximately one diameter) apart.

At each cross-section surveyed, the starting position at the wall was manually adjusted to the upper side of the arterial wall and at the lower side of the venous wall. The 'at the wall' locations were determined without flow in the loop. Near the wall, the first three sampling points along the diameter were spaced 100 μm from each other, and the fourth sampling point was 400 μm from the third. For all the remaining points along the same diameter up to the opposite wall, 800 μm spacings were maintained. At this time, the same measuring procedure was repeated from the opposite wall. Certain overlap of the positions surveyed was necessary to check the consistency of the measurements.

To assess the value of wall shear rate, the distensible host vessel was instrumented with ultrasonic dimension gauges to measure changes in the segment diameter[140]. The wall motions in the flow models can also be monitored by the laser technique described in Chapter   . The details of this new technique are described elsewhere[141]. The velocity signals, pressure signals and the wall motion signals were simultaneously recorded on a magnetic analogue tape using a FM record (Hewlett-Packard Model 3968A, band-width 0-10 kHz) at a tape speed of 19.05 mm s^{-1}. The signals were recorded for a minimum of 2 min at each sampling point and later played back for digitization using a PDP 11/23 digital computer (16 bits per word) at a sampling rate of 2500 points/s. According to the Nyquist criterion for signal processing, this would guarantee a frequency response of 1000 Hz in the flow measurement. The digitized data were processed on a Micro Vax 3.4 computer. Wall movement as a function of time was calculated from the variation of the recorded ultrasonic signal.

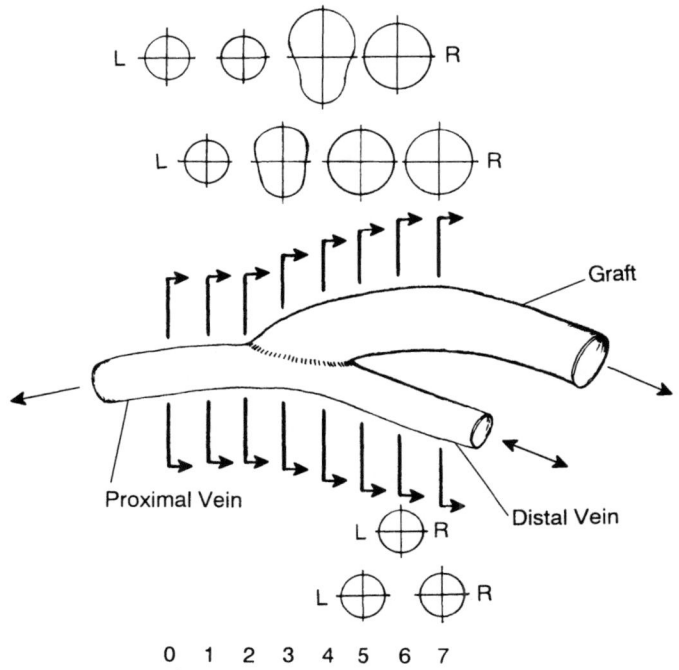

Figure 8. Sections of LDA velocity survey in the venous anastomotic region.

The velocity signals obtained at each point could be expressed as:

$$U_j(t) = \frac{1}{N}\sum_{n=1}^{N} u_j(t+nT) - \frac{1}{N}\sum_{n=1}^{N}[u_j(t+nT) - C_j] \qquad j = 1,2,...,k \qquad (4)$$

where j indicates the surveyed point at each cross-section, k is the number of measured points at each section, C_j is the constant frequency shift value of the jth point surveyed, T is the period of a cycle and N is the number of ensembling cycles. The pressure waveform at the proximal artery was used as a time trigger to identify the beginning of a cycle. The digitized velocity signals were 'ensemble averaged' for 30 cycles of continuous pulses at each measuring point. The wall shear stress distribution was calculated from the slope of the measured velocity profiles 'at the wall,' as defined above. Since the wall motion at each cross-section surveyed was recorded simultaneously with the velocity profiles, the distance $dr_i(t)$ between the vessel wall and the nearest point to the wall is a function of time. Thus, the wall shear stress can be calculated as:

$$\tau_i(t) = \frac{\mu}{N}\sum_{n=1}^{N}\left[\frac{du(t+nT)}{dr_i(t)}\right]_{\gamma_R} \qquad (5)$$

where $i = 0, 1, 2, 3, 4, 5, 6$, and μ is the viscosity of the testing fluid.

The venous anastomotic conduit of AVLG provides a well-defined flow field for studying the end-to-side anastomotic hemodynamics and their influences on the development of an occlusive lesion. Flow in the AVLG venous anastomotic region remained pulsatile in all three model stages studied. It was particularly advantageous as the hemodynamic parameters of an end-to-side anastomosis were generally accentuated. The experimental results from the flow study in these models may be discussed based on data displayed in the following figures.

Figures 9a, b, c, d, and e display a sequence of venous flow patterns taken a different phases of a pulse cycle. The stagnation point (S) on the opposite wall, the two large areas of recirculation (C) near the stagnation point and downstream of the anastomotic toe, respectively, can be clearly identified. Of particular interest is the oscillation of the stagnation point on the opposite wall during the pulse cycle.

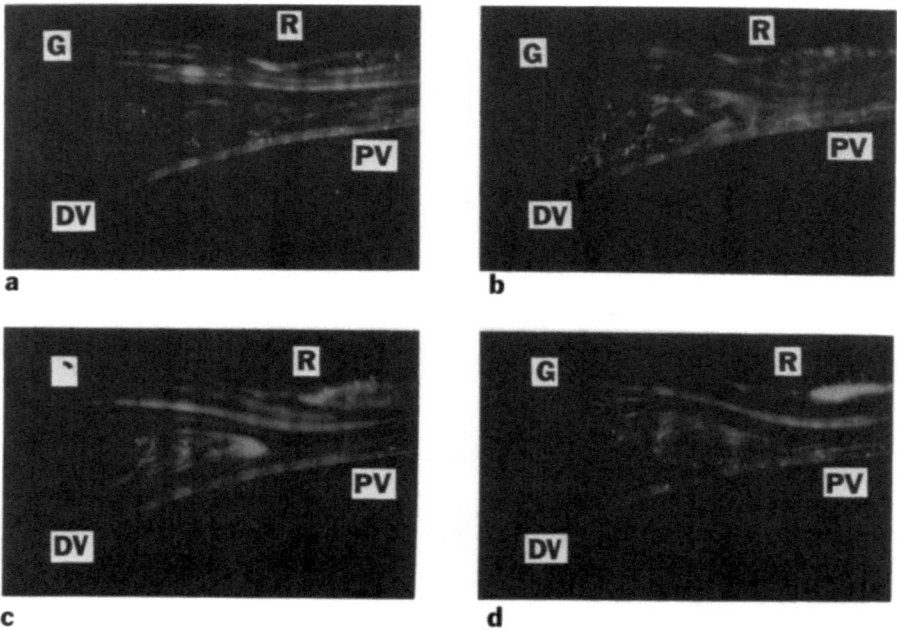

Figure 9. Pulsatile flow patterns at the venous anastomosis.

Figures 10 shows the typical LDA velocity profiles measured at different anastomotic sections as indicated in Figure 8. At each section surveyed, the section diameter was traversed to obtain the velocity along the diameter assuming quasi-steady conditions. Eight velocity profiles were obtained at each cross section to determine the phasic variations of the velocity. The velocity profiles were taken by dividing a pulse cycle into eight (8) equal phase angle intervals (45° each). The abscissa represents the diametric position, the ordinates are velocities in units of cm/sec. The origins of the coordinates are placed along the inside of the opposite wall of the anastomoses. Figure 10 shows a set of skewed velocity profiles taken at a short distance downstream of the anastomotic toe. The separation region, as indicated by the velocity reversal, is clearly demonstrated. Figure 10 was taken at a section further downstream where the jet stream is more organized as the separation region is reduced. Figure 10 shows the velocity profiles taken at a section inside the distal host vessel (Sec. 5) where the flow activities are rather complex in nature; however, the large recirculation region can be seen.

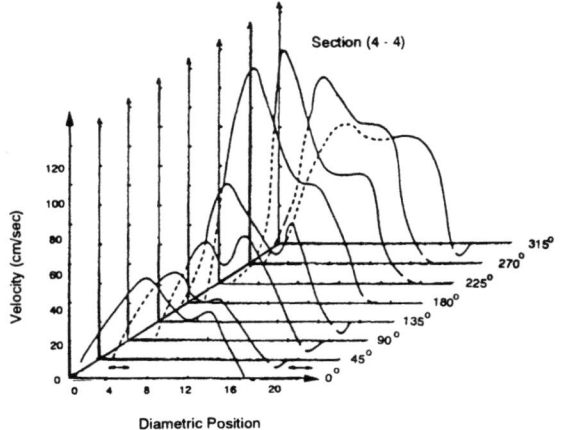

Figure 10. Typical velocity profiles at an AVLG venous anastomotic cross section.

Figure 11 shows the distribution of wall shear stress (WSS) along the opposite (bottom) wall and the wall of anastomosis (top). The values of WSS were calculated from Equation (2) using the measured velocity profiles and wall motions as inputs. The oscillation of WSS near the flow stagnation point on the opposite wall (Figure 11), and that on the top wall (Figure 11) are believed to be a causative factor of the lesion development, as discussed by Ku and Giddons[131].

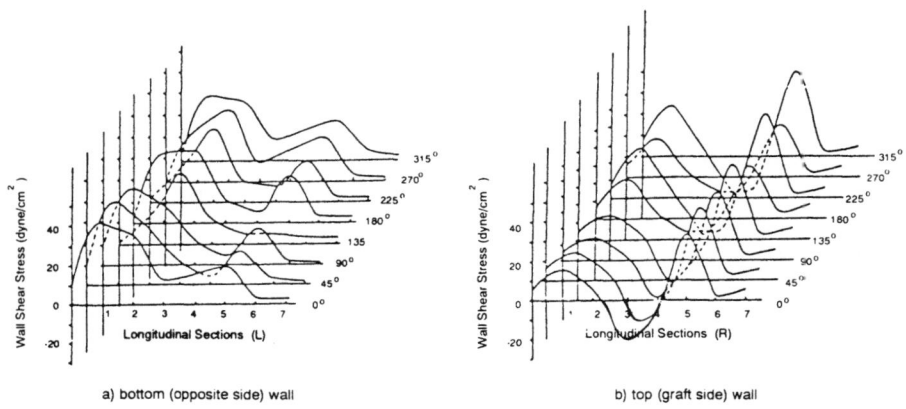

Figure 11. Typical wall shear stress distributions at an AVLG venous anastomosis.

From the above measurements, we were able to make the following observations on an end-to-side anastomoses:

Generally, the flows at an end-to-side anastomosis remains pulsatile in nature. Boundary layer separations occurred at both the toe and the heel of the anastomoses. Respective regions of flow recirculations can be clearly identified (e.g., Figure 9's).

The flow visualizations showed that the hydrogen bubble tracers accumulated around the stagnation region, and in the separation region near the toe during the systolic phase. Bubble accumulation in the stagnation region seems to indicate that bubbles are trapped to the wall by the force of the impinging jet. The accumulation at the separation region was thought to be a combined effect of the recirculation in the separation zone and the inward motion of the wall.

The surgical construction of the venous anastomosis creates inner tension and stress mismatch at the toe. Vascular compliance at this site is reduced. A small "dead" flow zone is formed immediately downstream from the toe and upstream of the recirculation region. Comparatively low shear rates and limited convective mass exchange may provide a favorable condition for activated platelets to aggregate.

The momentum created by the oscillatory impinging jet transfers into a hydraulic force acting on the endothelial layer of the host vessel. In the meantime, WSS generated at the vicinity of the stagnation point acts in opposite directions. Under the combined action of impinging force and oscillating WSS, the endothelial cells may be stretched in the shear stress direction[142] and the endothelial layer in the vicinity of the stagnation region can be injured. In addition, the jet stream may transport activated platelets or other thrombotic agents from the upstream graft to the stagnation region. Initiation of the vascular lesion may take place when the activated blood elements are brought into contact with the damaged endothelial layer. Blasberg et al.[143] have shown that even maximally activated platelets do not seem to adhere to walls, unless transportation toward the wall is provided by hydrodynamic forces.

In the present study, we found that the hydrodynamic conditions favor lesion development at the stagnation point opposite to the venous anastomotic toe and at the boundary separation point near the venous anastomotic toe. The lesion development at these regions was also found by other groups. Lesion seems to develop circumferentially, instead of starting along the anastomotic surgical ring as suggested by early studies which claimed that compliance mismatch at the surgical ring might stimulate the growth of the lesion along the ring[122,123]. Compliance mismatch along the anastomotic surgical ring between the graft and the host vessel did not seem to be a major factor in stimulating the lesion at the venous anastomosis in this study.

Frontiers in Vascular Grafts

The primary goal for vascular grafting is healing of the graft in man[144]. Current existing vascular graft prostheses can endothelialize and heal in dogs and many other animal models, but not in man. Large diameter (6 mm I.D. and larger) arterial graft prostheses function reasonably well in all species. The small bore arterial replacements and venous replacement prostheses do not[145]. These are the two major frontiers in vascular grafts research and development in the coming decade. While new vascular graft materials with improved thromboresistance and physical properties that promote healing remains the main challenge in the field[52,146], better understanding of the graft-host hemodynamics continues to be beneficial. This is probably even more important in the development of a clinically applicable venous graft prosthesis.

Our knowledge of venous hemodynamics are extremely limited at present[148]. This is particularly true in regards to the venous valves, their behavior and operation in given venous hemodynamic environments. In general, we know that the valves in the veins of the lower limb are unevenly distributed. Their main function is to prevent blood flow towards the periphery. In the communicating veins, they prevent flow from the deep to the superficial veins. Venous blood flow in the lower limbs returns to the heart primarily by the musculovenous pump action during normal rhythmic exercise, such as walking. In a state with completely relaxed and inactive muscles, the pressure drop driving the blood from the ankle towards the heart is very small (5 to 10 mmHg remnant arterial blood pressure). In a completely relaxed, standing person, it is believed that this pressure is the only energy available to drive the venous blood to the heart. For an average adult, the distance between the feet and the heart represents a liquid column of more than 1 meter height. In this state, the numerous venous valves along the entire length of the lower limb are assumed to be completely opened to provide minimum resistance to the returning blood flow. Venous valves are supposed to be extremely sensitive to pressure gradients, and can operate under slight change in the direction of flow.

In a recent article, van Bemmelen et al[147] reported the results of their in vivo measurements performed in a series of patients and healthy volunteers. From these measurements, they conclude that "closure of the (venous) valves does not simply respond to the cessation of antigrade flow, but requires reversal of flow at a certain level (>30cm/sec) before it will occur."

Using a 10mm (dia.) glutaraldehyde fixed bovine jugular vein valve, we performed a series of carefully planned in vitro experiments in a hydraulic flow loop (Fig. 12) allowing systematic variations of both oscillatory pressure levels and discharge flow rates.

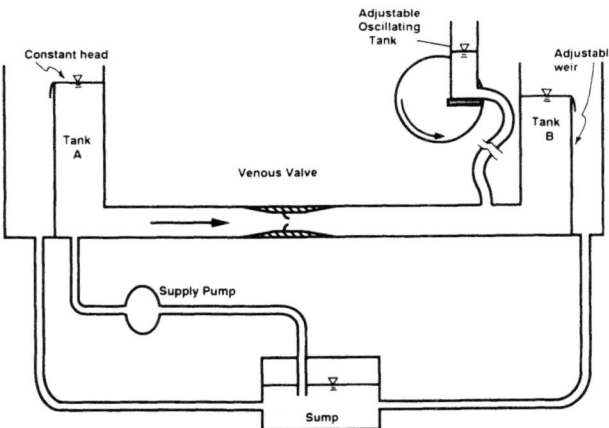

Figure 12. Mock flow loop system for testing venous valves.

The testing valve is installed in the mid-section of a horizontal tube connecting reservoirs A and B. A 20°C saline solution (0.9% NaCl by volume) is pumped to fill both tanks A and B until the elevations in the tanks reaches the centerline of the wheel. The experiment begins by adjusting the platform so that the water elevation in tank C oscillates at 5 cm (peak-to-peak). The elevation in the tail tank B is reduced gradually until the valve begins to operate, as defined by explicit cyclic open and closure (contact) of the leaflet as observed through an inclined mirror installed on the bottom of tail tank B. The discharge through the valve is measured by the volume/time method.

The magnitude of the oscillation is then increased by a small amount, while the water elevation in tank B is reduced to increase the discharge through the valve. The measurement is repeated at the point where the valve begins to close-and-open again. The experimental run continues until the oscillatory pressure reaches the maximum 13.5 cm level. The experimental series with the increasing discharge is termed "close-to-open" (C-to-O) operation.

The entire experimental run is then reversed by repeating the same procedures at a gradually increased elevation in tank B, thus, reducing the elevation difference between the head tank A and the tail tank B. In this reversed run, the discharge rate Q is gradually reduced as the elevation difference decreases. The valve remains opened throughout the entire cycle at the high discharge rate. The discharge at which the two leaflets begin to meet at a given oscillatory pressure (H) defines the stage of valve open and close functions. The reversed run is termed "open-to-close" (O-to-C) operation. The same C-to-O and O-to-C experiments are performed on all three approaching conduits using the same venous valve.

The flow fields immediately upstream and downstream of the venous valve are visualized by using micron-size mica chips as tracers in laser illuminated fields[2]. The flow fields both upstream and downstream of the valve are recorded on the same screen with the simultaneous leaflet motions (viewed through the mirror) using synchronized video cameras.

Experiments were performed on each of the three different approaching conduits (Fig. 13) using the same procedures as described above. The O-to-C and C-to-O data are individually plotted in Figs. 14a, 14b and 14c, respectively.

In each of these figures, there are two sets of data points presented. The open dots represents the maximum discharge rate at which the leaflets still can make contact to close, during the course of increasing discharge (C-to-O). The closed dots, on the other hand, represent the discharge rate at which the leaflets begin to make contact to close, during the course of decreasing discharge (O-to-C).

For all three approach conduits tested, there is a clear trend showing that the leaflets behave quite differently during increased or decreased flow rates. The courses of increasing discharge and decreasing discharge each followed a different path that they formed a "loop" similar to that of the hysteresis loop[3]. We believe that the shape of this loop can be correlated to the behavior (stiffness) of the valve leaflets, and the hydrodynamic effects of the approaching flow on the performance of the valve leaflets.

Compared with van Bemmelen et al[1,4,7], our experiment was made on a much larger (10 mm dia.) glutaradhyde fixed bovine jugular vein valve, which was much stiffer than the natural valves in vivo. Our results from all three different types of approaching conduits showed that very little reflex velocity is needed to accomplish complete closure of the venous valve within the physiological ranges of pressure and flow rate. There is practically no valvular regurgitation as confirmed by both high resolution video recording of the flow fields and by actual flow rate measurements.

The need for a replacement prosthesis that will functionally replace the natural vessel and heal in man is clear, but the specifics for designing such a prosthesis is not yet available. While large diameter (>6 mm) prostheses perform satisfactorily in spite of deposition of fibrous material in the lumen, thrombogenicity and intimal hyperplasia, these factors do contribute to rapid occlusion of small vessel grafts.

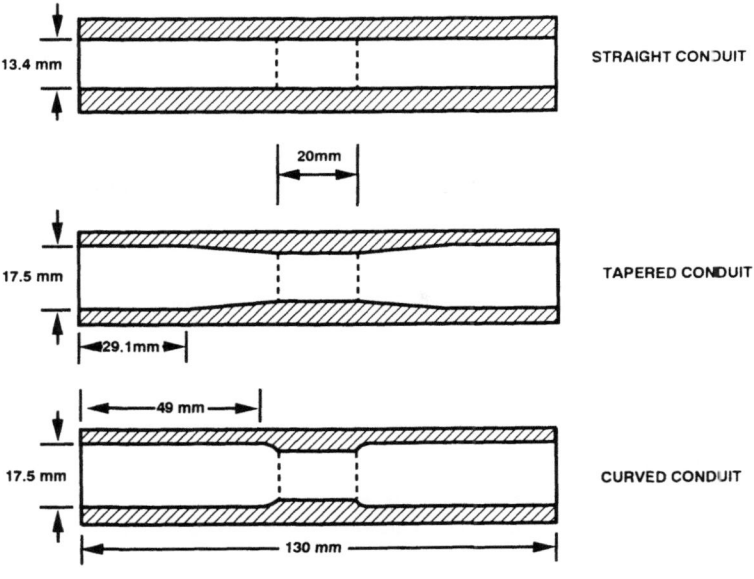

Figure 13. Geometric configurations of three different approaching venous valve conduits.

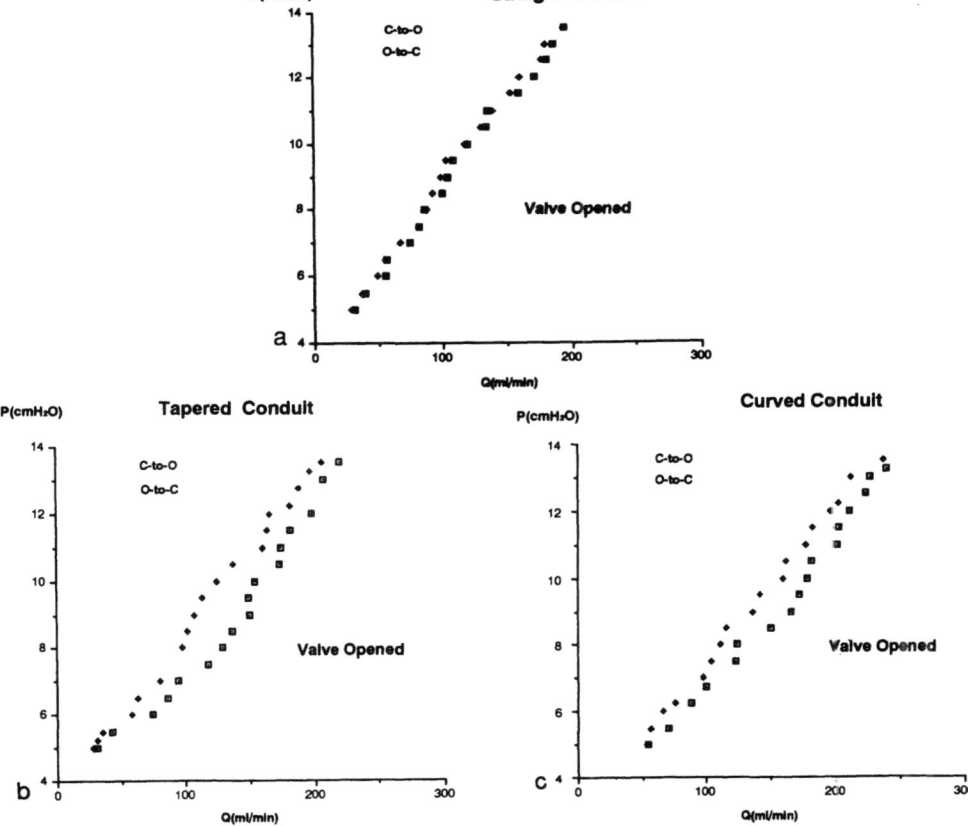

Figure 14. Hydraulic behavior of a venous valve in: (a) a straight conduit (b) a tapered conduit (c) curved conduit.

Numerous studies have been performed in experimental animal models. Unfortunately such studies in experimental models are not directly transferrable to man. With more than 200,000 of bypass surgeries performed every year in man, it is unfortunate that too few of these grafts are recovered at autopsy for study of healing process in man[144].

While continued efforts are made to develop new thromboresistant, nontoxic, blood compatible graft materials, better understanding of local hemodynamic parameters, blood flow, and mechanical properties of the grafts and their relationship to the healing process in man becomes an even more important issue. Continued research and development in venous replacements and small bore arterial grafts remain as frontier areas in vascular grafts for the next decade.

ACKNOWLEDGEMENTS

The glutaraldehyde fixed bovine jugular vein valve and the three approaching conduits used in the hydraulic study experiment were provided by the Baxter Edwards CVS Division, Irvine, CA 92714, USA. The authors acknowledge Lorna Lee Horishny for her artwork presented in this chapter.

REFERENCES

1. Carrel A, Guthrie CG: Uniterminal and biterminal venous transplantations, Surg. Gynecol. Obstet., 2:266, 1906.
2. Goyanes DJ: Substitution plastica de las arterias por las venas, o arterioplastia venosa, aplicada, como nuevo metodo, al tratamiento de los aneurismas, El Siglo Medico, September 1:346, September, 8:561, 1906.
3. Debakey ME, Simeone FA: Battle injuries of the arteries in World War II, Ann. Surg., 123:534, 1946.
4. Dubost C, Allary M, Oeconomos N: Resection of an aneurysm of the abdominal aorta: Reestablishment of the continuity by a preserved human arterial graft, with results after five months, Arch. Surg., 64:405, 1952.
5. Voorhees AB Jr, Jaretzki A III, Blakemore AH: The use of tubes constructed from Vinyon "N" cloth in bridging arterial defects, Ann. Surg., 135:332, 1952.
6. Chang BB, Paty PSK, Shah DM, et al: The lesser saphenous vein: An underappreciated source of autogenous vein, J. Vasc. Surg., 15:152, 1992.
7. Vellar IDA, Doyle JC: The use of cephalic and basilic veins as peripheral vascular grafts, Aust. NZ J. Surg., 40:52, 1970.
8. Cambria RP, Megerman J, Brewster DC, et al: The evolution of morphologic and biomechanical changes in reversed and in situ vein grafts, Ann. Surg., 205:167, 1987.
9. Lye CR, Sumner DS, Strandness DE, The transcutaneous measurement of the elastic properties of the human saphenous vein femoropopliteal bypass graft, Surg. Gynecol. Obstet., 141:891, 1975.
10. Leather RP, Shah DM, Karmody AM, Infrapopoiteal arterial bypass for limb salvage: Increased patency and utilization of the saphenous venin used in situ, Surgery, 190:1000, 1981.
11. Taylor LM, Phinney ES, Porter JM: Present status of reversed vein bypass for lower extremity revascularization, J. Vasc. Surg., 3:288, 1986.

12. Batson RC, Sottiurai VS: Nonreversed and in situ vein grafts: Clinical and experimental observations, Ann. Surg., 201:771, 1985.
13. Wyatt AP, Rothnie NG, Taylor GW: The vascularization of vein grafts, Br. J. Surg., 51:378, 1964.
14. McCann RL, Hagen PO, Fuchs JCF: Aspirin and dipyridamole decrease intimal hyperplasia in experimental vein grafts, Ann. Surg., 191:238, 1980.
15. Haudenschild C, Gould KE, Quist WC, et al: Protection of endothelium in vessel segments excised for grafting, Circulation, 64(Suppl II):101, 1981.
16. Abbott WM, Wieland S, Austen WG: Structural changes during preparation of autogenous venous grafts, Surgery, 76:1031, 1974.
17. Gundry SR, Jones M, Ishihara T, et al: Optimal preparation techniques for human saphenous vein grafts, Surgery, 88:785, 1980.
18. Cambria RP, Megerman J, Abbott WM: Endothelial preservation in reversed and in situ autogenous vein grafts: A quantitative, experimental study, Ann. Surg., 202:50, 1985.
19. Bonchek LI: Prevention of endothelial damage during preparation of saphenous veins for bypass grafting, J. Thorac. Cardiovasc. Surg., 79:911, 1989.
20. Kuruz M, Christman EW, Derrick JR, et al: Use of cold cardioplegia solution for vein graft distention and preservation: A light and scanning electron microscopic study, Ann. Thorac. Surg., 32:68, 1981.
21. Malone JM, Kischer CW, Moore WS: Changes in venous endothelial fibrinolytic activity and histology with in vitro distention and arterial implantation, Am. J. Surg., 142:178, 1981.
22. Baumann FG, Catinella FP, Cunningham JN, et al: Vein contraction and smooth muscle cell extensions as causes of endothelial damage during graft preparation, Ann. Surg., 194:199, 1981.
23. Brody WR, Angell WW, Kosek JC: Histologic fate of the venous coronary artery bypass in dogs, Am. J. Pathol., 66:111, 1972.
24. Bandyk DF, Kaebrick HW, Stewart GW, et al: Durability of the in situ saphenous vein arterial bypass: A comparison of primary and secondary patency, J. Vasc. Surg., 5:256, 1987.
25. Batson RC, Sottiurai VS: Nonreversed and in situ vein grafts: Clinical and experimental observations, Ann. Surg., 201:771, 1985.
26. Cambria RP, Megerman J, Brewster DC, et al: The evolution of morphologic and biomechanical changes in reversed and in situ vein grafts, Ann. Surg., 205:167, 1987.
27. Leather RP, Shah DM, Karmody AM: Infrapopliteal arterial bypass for limb salvage: Increased patency and utilization of saphenous vein used "in situ," Surgery, 90:1000, 1981.
28. Leather RP, Shah DM, Buchbinder D, et al: Further experience with the saphenous vein used in situ for arterial bypass, Am. J. Surg., 142:506, 1981.
29. Veith FJ, Moss CM, Sprayregen S, et al: Preoperative saphenous venography in arterial reconstructive surgery of the lower extremity, Surgery, 85:253, 1979.
30. Harris RW, Andros G, Salles-Cunha SX, et al: Totally autogenous venovenous composite bypass grafts, 121:1128, 1986.
31. Graham JW, Lusby RJ: Infrapopliteal bypass grafting: use of upper limb vein alone and in autogenous composite grafts, Surgery, 91:646, 1982.
32. Taylor LM Jr, Edwards JM, Brant B, et al: Autogenous reversed vein bypass for lower extremity ischemia in patients with absent or inadequate greater saphenous vein, Am. J. Surg., 153:505, 1987.

33. Snyder SO Jr, Gregory RT, Wheeler JR, et al: Composite grafts utilizing polytetrafluorethylene-autogenous tissue for lower extremity arterial reconstructions, Surgery, 90:881, 1981.
34. Wheeler JR, Gregory RT, Snyder SO Jr, et al: Gore-Tex autogenous vein composite grafts for tibial reconstruction, J. Vasc. Surg., 1:914, 1984.
35. Hall RG, Coupland GAE, Lane R, et al: Vein, Gore-Tex or a composite graft for femoropopliteal bypass, Surg. Gynecol. Obstet., 161:308, 1985.
36. Scribner RG, Beare JP, Harris EJ, et al: Polytetrafluoroethylene vein composite grafts across the knee, Surg. Gynecol. Obstet., 157:237, 1983.
37. Verta MJ: Composite sequential bypasses to the ankle and beyond for limb salvage, J. Vasc. Surg., 1:381, 1984.
38. Glovicski P, Pairolero PC, Cherry KJ, et al: Reconstruction of the vena cava and its primary tributaries: A preliminary report, J. Vasc. Surg., 11:373, 1990.
39. Stoney RJ, Wylie EJ: Arterial autografts, Surgery, 67:18, 1970.
40. Lye CR, String ST, Wylie EJ, et al: Aortorenal arterial autografts, Arch. Surg., 110:1321, 1975.
41. Stoney RJ, DeLuccia N, Ehrenfeld WK, et al: Aortorenal arterial autografts, Arch. Surg., 116:1416, 1981.
42. Ehrenfeld WK, Stoney RJ, Wylie EJ: Autogenous arterial grafts, In: Biologic and Synthetic Vascular Prostheses (Eds. Stanley JC, et al), Grune and Stratton, New York, 1982.
43. Abbott WM, Megerman J, Hasson JE, et al: Effect of compliance mismatch on vascular graft patency, J. Vasc. Surg., 5:376, 1987.
44. Qvarfordt PG, Reilly LM, Ehrenfeld WK, et al: Surgical management of vascular graft infections-local treatment, graft excision, and methods of revascularization, In: Complications in Vascular Surgery (Eds. Bernhard VM, Towne JB), Grune and Stratton, New York, 1985.
45. Dardik H, Hessler K, Ibrahim IM, et al: Arteriovenous fistulas: Preliminary clinical experience employing glutaraldehyde tanned human umbilical cord vein, Trans. Am. Soc. Artif. Int. Organs., 4:64, 1981.
46. Dardik H, Baier RE, Weinberg S, et al: Morphologic and biophysical assessment of long-term human umbilical cord vein implants employed as vascular conduits, Surg. Gynecol. Obstet., 154:17, 1982.
47. Dardik H: Modified human umbilical vein allograft, In: Vascular Surgery (Ed. Rutherford RB), WB Saunders, Philadelphia, 1989.
48. Dardik H, Ibrahim IM, Sussman B, et al: Biodegradation and aneurysm formation in umbilical vein grafts: observation and a realistic strategy, Ann. Surg., 199:61, 1984.
49. Dardik H, Miller N, Dardik A, et al: A decade of experience with the glutaraldehyde-tanned human umbilical cord vein graft for revascularization of the lower limb, J. Vasc. Surg., 7:336, 1988.
50. Brockbank KGM, Donovan TJ, Rub ST, et al: Functional analysis of cryopreserved veins, J. Vasc. Surg., 11:94, 1990.
51. Elmore JR, Glovicski P, Brockbank KGM, et al: Cryopreservation affects endothelial and smooth muscle function of canine autogenous saphenous vein grafts, J. Vasc. Surg., 13:584, 1991.
52. Noishiki Y.
53. Phifer TJ, Gerlock AJ, Grafton WD, et al: Valvular xenografts in the inferior vena cava: An animal study, Am. J. Surg., 157:588, 1989.

54. Scales JT: Tissue reactions to synthetic materials, Proc. R. Soc. Med., 46:647, 1953.
55. Reichle FA: Criteria for evaluation of new arterial prostheses by comparing vein with Dacron femoropopliteal bypasses, Surg. Gynecol. Obstet., 146:714, 1978.
56. Stephen M, Lowenthal J, Little JM, et al: Autogenous veins and velour Dacron in femoropopliteal arterial bypass, Surgery, 81:314, 1977.
57. Veith FJ, Gupta SK, Ascer E, et al: Six year prospective multicenter randomized comparison of autologous saphenous vein and expanded polytetrafluoroethylene grafts in infrainguinal arterial reconstructions, J. Vasc. Surg., 3:104, 1986.
58. Shah PM, Ito K, Clauss RH, et al: Expanded microporous polytetrafluoroethylene (PTFE) grafts in contaminated wounds: Experimental and clinical study, J. Trauma, 23:1030, 1983.
59. Stone KS, Walshaw R, Sugiyama GT, et al: Polytetrafluoroethylene versus autogenous vein grafts for vascular reconstruction in contaminated wounds, Curr. Surg., 41:267, 1984.
60. Shah DM, Leather RP, Corson JD, et al: Polytetrafluoroethylene grafts in the rapid reconstruction of acute contaminated peripheral vascular injuries, Am. J. Surg., 148:229, 1984.
61. Feliciano DV, Mattox KL, Graham JM, et al: Five-year experience with PTFE grafts in vascular wounds, J. Trauma, 25:71, 1985.
62. Wesolowski SA, Dennis C: Fundamentals of Vascular Grafting, McGraw-Hill, New York, 1963.
63. Jordan GL, Stump MM, Allen J, et al: Gelatin-impregnated Dacron prosthesis implanted into procine thoracic aorta, Surgery, 53:45, 1963.
64. Krajicek M, Zastava V, Chvapil M: Collagen-fabric vascular prostheses: Biological and morphological experience, J. Surg. Res., 4:290, 1964.
65. Wesolowski SA, Fries CC, Domingo RT, et al: The compound prosthetic vascular graft: A pathologic survey, Surgery, 53:19, 1963.
66. Norton L, Eiseman B: Replacement of portal vein during pancreatectomy for carcinoma, Surgery, 77:280, 1975.
67. Campbell CD, Goldfarb D, Roe R: A small arterial substitute: Expanded microporous polytetrafluoroethylene: Patency versus porosity, Ann. Surg., 182:138, 1975.
68. Florian A, Cohn LH, Dammin GJ, et al: Small vessel replacement with Gore-Tex (expanded polytetrafluoroethylene), Arch. Surg., 111:267, 1976.
69. Sauvage LR, Walker MW, Berger K, et al: Current arterial prostheses. Experimental evaluation by implantation in the carotid and circumflex coronary arteries of the dog, Arch. Surg., 114:687, 1979.
70. Hamlin GW, Rajah SM, Crow MJ, et al: Evaluation of the thrombogenic potential of three types of arterial graft studied in an artificial circulation, Br. J. Surg., 65:272, 1978.
71. Goldman M, Hall C, Dykes J, et al: Does 111-indium-platelet deposition predict patency in prosthetic arterial grafts? Br. J. Surg., 70:635, 1983.
72. Campbell CD, Brooks DH, Webster MW, et al: Aneurysm formation in expanded polytetrafluoroethylene prostheses, Surgery, 79:491, 1976.
73. Pearce JE, Dujovny M, Ho KL, et al: Acute inflammation and endothelial injury in vein grafts, Neurosurgery, 17:626, 1985.
74. Brody WR, Angell WW, Kosek JC: Histologic fate of the venous coronary artery bypass in dogs, Am. J. Pathol., 66:111, 1972.

75. Zwolak RM, Adams MD, Clowes AW: Kinetics of vein graft hyperplasia: Association with tangential stress, J. Vasc. Surg., 5:126, 1987.
76. Seidel CL, Lewis RM, Bowers R, et al: Adaptation of canine saphenous veins to grafting: Correlation of contractility and contractile protein content, Circ. Res., 55:102, 1984.
77. Boerboom LE, Olinger GN, Bonchek LI, et al: Aspirin or dipyridamole individually prevents lipid accumulation in primate vein bypass grafts, Am. J. Cardiol., 55:556, 1985.
78. McCann RL, Hagen PO, Fuchs JCF: Aspirin and dipyridamole decrease intimal hyperplasia in experimental vein grafts, Ann. Surg., 191:238, 1980.
79. Fuster V, Chesebro JH: Role of platelets and platelet inhibitors in aortocoronary artery vein-graft disease, Circulation, 73:227, 1986.
80. Sharma GV, Khuri SF, Josa M, et al: The effect of antiplatelet therapy on saphenous vein coronary artery bypass graft patency, Circulation, 68(Suppl II):218, 1983.
81. Cahill PD, Sarris GE, Cooper AD, et al: Inhibition of vein graft arteriosclerosis by eicosapentanoic acid: Correlation with reduced platelet thromboxane production but no change in lipoproteins or LDL receptor density, J. Vasc. Sug., 7:108, 1988.
82. Landymore RW, MacAulay M, Sheridan B, et al: Comparison of cod-liver oil and aspirin-dipyridamole for the prevention of intimal hyperplasia in autologous vein grafts, Ann. Thorac. Surg., 41(1):54, 1986.
83. Kohler T, Kaufman J, Kakiayanos G, et al: Effect of aspirin and dipyridamole on the patency of lower extremity bypass grafts, Surgery, 96:462, 1984.
84. Gunstensen J, Smith RC, El-Maraghi N, et al: Intimal hyperplasia in autogenus veins used for arterial replacement, Can. J. Surg., 25:158, 1982.
84. Berguer R, Higgins RF, Reddy DJ: Intimal hyperplasia: An experimental study, Arch. Surg., 115:332, 1980.
85. Morinaga K, Okahome K, Kuroki M, et al: Effect of wall shear stress on intimal thickening of arterially transplanted autogenous veins in dogs, J. Vasc. Surg., 2:430, 1985.
86. Karayannacos P, Rittgers SE, Kakos GS, et al: Potential role of velocity and wall tension in vein graft failure, J. Cardiovasc. Surg., 21:171, 1980.
88. Morinaga K, Eguchi H, Miyazaki T, et al: Development and regression of intimal thickening of arterially transplanted autologous vein grafts in dogs, J. Vasc. Surg., 5:719, 1987.
85. Bush Jr HL: Mechanisms of graft failure (Special Communication), J. Vasc. Surg., 9:392, 1989.
86. Sanders RJ, Kempczinski RF, Hammond W, et al: The significance of graft diameter, Surgery, 88:856, 1980.
87. Berger K, Sauvage LR, Rao AM, et al: Healing of arterial prostheses in man: Its incompleteness, Ann. Surg., 175:118, 1972.
88. DeBakey ME, Jordan GL, Abbot JP, et al: The fate of Dacron vascular graft, Arch. Surg., 89:757, 1964.
89. Reichle FA, Stewart GJ, Essa N: A transmission and scanning electron microscope study of luminal surfaces in Dacron and autogenous vein bypasses in man and dog, Surgery, 74:945, 1973.
90. Warren R, McCoombs HL: Morphologic studies on plastic arterial prostheses in humans, Ann. Surg., 161:73, 1965.

91. Wesolowski SA, Fries CC, Henningar G, et al: Factors contributing to long-term failures in human vascular prosthetic grafts, J. Cardiovasc. Surg., 5:544, 1964.
92. Herring MB, Compton RS, LeGrand DR, et al: Endothelial seeding of polytetrafluoroethylene popliteal bypasses. A preliminary report, J. Vasc. Surg., 6:114, 1987.
93. Clowes AW: Arterial wall response to injury and healing (Special Communication), J. Vasc. Surg., 9:373, 1989.
94. Clowes AW, Reidy MA, Clowes MM: Kinetics of cellular proliferation after arterial injury. I. Smooth muscle growth in the absence of endothelium, Lab. Invest., 49:327, 1983.
95. Clowes AW, Reidy MA, Clowes MM: Mechanisms of stenosis after arterial injury, Lab. Invest., 49:208, 1983.
96. Clowes AW, Schwartz SM: Significance of quiescent smooth muscle migration in the injured rat carotid artery, Circ. Res., 56:139, 1985.
97. Reidy MA: Biology of disease: a reassessment of endothelial injury and arterial lesion formation, Lab. Invest., 52:513, 1985.
98. Owens GK, Reidy MA: Hyperplastic growth response of vascular smooth muscle cells following induction of acute hypertension in rats by coarctation, Circ. Res., 57:695, 1985.
99. Reidy MA, Chao SS, Kirkman TR, et al: Endothelial regeneration. VI. Chronic nondenuding injury in baboon vascular grafts, Am. J. Pathol., 123:432, 1986.
100. Clowes AW, Kirkman TR, Reidy MA: Mechanisms of arterial graft healing: rapid transmural capillary ingrowth provides a source of endothelium and smooth muscle in porous PTFE prostheses, Am. J. Pathol., 123:220, 1986.
101. Ross R: The pathogenesis of arteriosclerosis-an update, N. Engl. J. Med., 314:488, 1986.
102. Schwartz SM, Campbell GR, Campbell JH: Replication of smooth muscle cells in vascular disease, Circ. Res., 58:427, 1986.
103. Sporn MB, Roberts AB, Wakefield LM, et al: Some recent advances in the chemistry and biology of transforming growth factor-beta, J. Cell. Biol., 105:1039, 1987.
104. Ross R, Raines EW, Bowen-Pope DF: The biology of platelet-derived growth factor, Cell, 46:155, 1986.
105. Carpenter G, Cohen S: Epidermal growth factor, Ann. Rev. Biochem., 48:193, 1979.
106. Gospodarowicz D, Neufeld G, Schweigerer L: Molecular and biological characterization of fibroblast growth factor, and angiogenic factor which also controls the proliferation and differentiation of mesoderm- and neuroectoderm-derived cells, Cell. Diff., 19:1, 1986.
107. Warner SJC, Auger KR, Libby P: Human interleukin I induces interleukin I gene expression in human vascular smooth muscle cells, J. Exp. Med., 165:1316, 1987.
108. Esmon CT: The regulation of natural anticoagulant pathways, Science, 235:1348, 1987.
109. Clouse LH, Comp PC: The regulation of hemostasis: the protein C system, N. Engl. J. Med., 314:1298, 1986.
110. Abbott, WM, Megerman J: Adaptive responses of arteries to grafting (Special Communication), J. Vasc. Surg., 9:377, 1989.
111. Bush HL Jr: Mechanisms of graft failure (Special Communication), J. Vasc. Surg., 9:392, 1989.

112. Lo Gerfo FW: Hemodynamics and the arterial wall (Special Communication), J. Vasc. Surg., 9:380, 1989.
113. Strandness DE Jr, Sumner DS: Hemodynamics for Surgeons, Grune and Stratton, New York, 1975.
114. Lye CR, Sumner DS, Strandness DE Jr: Hemodynamics of the retrograde crosspublic anastomosis, Surg. Forum., 26:298, 1975.
115. Crawshaw HM, Quist WC, Sarrallach E, et al: Flow disturbance at the distal end-to-side anastomosis. Effect of patency of the proximal outflow segment and angle of anastomosis, Arch. Surg., 115:1280, 1980.
116. LoGerfo FW, Soncrant T, Teel T, et al: Boundary layer separation in models of side-to-end arterial anastomoses, Arch. Surg., 114:1369, 1979.
117. McMillan DE: Blood flow and the localization of atherosclerotic plaques, Stroke, 16:582, 1985.
118. Zarins CK, Giddens DP, Bharadvaj BK, et al: Carotid bifurcation atherosclerosis. Quantitative correlation of plaque localization with flow velocity profiles and wall shear stress, Circ. Res., 53:502, 1983.
119. LoGerfo FW, Quist WC, Nowak MD, et al: Downstream anastomotic hyperplasia. A mechanism of failure of Dacron arterial grafts, Ann. Surg., 197:479, 1983.
120. Edwards WS: Arterial grafts. Past, present and future, Arch. Surg., 113:1225, 1978.
121. Cooley DA: Surgical treatment of aortic arch aneurysms, In: International Practice in Cardiothoracic Surgery (Eds. Wu YK, Peters RM), Science Press, Beijing, 1985.
122. Hansson JE, Megerman J, Abbott WM: Increased compliance near vascular anastomoses, J. Vasc. Surg., 2:419, 1985.
123. Hansson JE, Megerman J, Abbott WM: Suture technique and para-anastomotic compliance, J. Vasc. Surg., 3:196, 1986.
124. Duncan DD, Bergeron CB, Borchardt SE, et al: The effect of compliance on wall shear in casts of a human aortic bifurcation, J. Biomech. Engr., 112:183.
125. Shu MC, Hita CE, Hwang NHC: Hemodynamic models in vascular grafting, in: Vascular Dynamics-Physiological Perspectives (Eds. Westerhof N, Gross DR), Plenum Press, New York, 1989.
126. Rapaport A, Noon GP, McCollum CH: Polytetrafluoroethylene (PTFE) grafts for hemodialysis in chronic renal failure: Assessment of durability and function at three years, Aust. NZ J. Surg., 51:561, 1981.
127. Schwab SJ, Raymond JR, Saeed M, et al: Prevention of hemodialysis fistula thrombosis; Early detection of venous thrombosis, Kidney Int., 36:707, 1989.
128. Glagov S, Giddens DP, Bassiouny H, et al: Hemodynamic effects and tissue reactions at graft to vein anastomosis for vascular access, In: Vascular Access for Hemodialysis II (Eds. Sommer B, Michell H), W. L. Gore and Assoc., 1991.
129. Clowes AW, Gown AM, Hasson SR, Reidy MA: Mechanisms of arterial graft failure: Role of cellular proliferation in early healing of PTFE prostheses, Am. J. Path., 118:43, 1985.
130. Fillinger MF, Kerns DB, Schwartz RA: Hemodynamics and intimal hyperplasia, In: Vascular Access for Hemodialysis II (Eds., Sommer B, Michell B), W. L. Gore and Assoc., 1991.
131. Ku DN, Giddens DP: Laser Doppler anemometer measurements of pulsatile flow in a model carotid bifurcation, J. Biomech., 20:407, 1987.

132. Lutz RJ, Hsu L, Menawat A, Zrubek J, Edwards K: Comparison of steady and pulsatile flow in a double branching arterial model, J. Biomech., 16:753, 1983.
133. Einav S, Avidor J, Viden B: Hemodynamics of coronary artery-saphenous vein bypass, J. Biomed. Engr., 7:305, 1985.
134. Pei H, Xi BS, Hwang NHC: Wall shear stress distribution in a model human aortic arch: assessment by an electrochemical technique, J. Biomech., 18:645, 1985.
135. Liepsch D: Flow in tubes and arteries, a comparison, J. Biorheology, 23:395, 1986.
136. Ku DN, Liepsch D: The effects of non-Newtonian viscoelasticity on flow at 90 degree bifurcation, Biorheology, 23:359, 1986.
137. Shu MSC, Hwang NHC: Hemodynamics of angioaccess venous anastomoses, J. Biomed. Engr., 13:103, 1991.
138. Deters OJ, Bargeron CB, Mark FF, Friedman MH: Measurement of wall motion and wall shear stress in a compliant arterial cast, J. Biomech. Engr., 108:355, 1986.
139. Liepsch D, Moravec S: Pulsatile flow on non-Newtonian fluids in distensible models of human artery, Biorheology, 21:571, 1984.
140. Gentile BJ, Gross DR, Chuong CTJ, Hwang NHC: Segmental volume distensibility of the canine thoracic aorta in vivo, Cardiovas. Res., 22:385, 1988.
141. Zhou JS, Wahab SA, Guo XM, et al: Monitoring vascular wall motions with a laser optic system, In: 1990 Advances in Bioengineering (Ed., Goldstein SA), ASME BED, 17:193, 1990.
142. Therat DP, Levesque MJ, Sato M, Nerem RM, Wheeler LT: The application of a homogeneous half-space model in the analysis of endothelial cell micropipette measurements, ASME J. of Biomech. Engr., 110:190, August 1988.
143. Blasberg P, Wurzinger LJ, Musler K, Myrenne H, Schmid-Schonbein H: A platelet aggregometer with automatic data processing, Thromb. Haemostas., 46:132, 1981.
144. Wesolow A: The healing of arterial prostheses - the state of the art, Thorac. Cardiovas. Surg., 30:196, 1982.
145. Haubold A, Borovetz, HS: Stress-strain characteristics of vascular prostheses: Is there a relationship to healing and graft patency? In: Vascular Dynamics: Physiology Perspectives (Eds., Westerhof N, Gross DR), Plenum Press, New York, 277, 1989.
146. Mori Y: Vascular graft materials and their structure, In: Vascular Dynamics: Physiological Perspectives (Eds., Westerhof N, Gross Dr), Plenum Press, New York, 287, 1989.
147. Van Bemmelen PS, Beach K, Bedford G, Strandness DE Jr: The mechanism of venous closure: Its relationship to the velocity of reverse flow, Arch. Surg., 125:617, 1990.
148. Eklof B, Gjores JE, Thulesius O, Bergqvist D: Controversies in Management of Venous Disorders, Butterworths, London, 1989.

NEW CONCEPTS AND DEVELOPMENT OF VASCULAR GRAFT PROSTHESES

Yasuharu Noishiki

First Department of Surgery
Yokohama City University
School of Medicine
3-9, Fukuura, Kanazawa-ku
Yokohama, 236, Japan

INTRODUCTION

There are several kinds of vascular graft prostheses, such as fabric Dacron prostheses, biological grafts, EPTFE grafts, cell seeding grafts, chemically treated connective tissue tube grafts, etc. They have been used safely in clinic, although with advantages and disadvantages. The biggest disadvantages are the poor healing ability of the neointima and lack of antithrombogenicity. Therefore, they cannot be used as venous and small diameter arterial grafts. For example, fabric Dacron prostheses have no natural antithrombogenicity. The surface is covered with fresh thrombi for a long period of time after implantation. Endothelialization of the grafts is limited to areas near the anastomotic sites. Conversely, EPTFE grafts have been expected to prevent thrombus formation. The grafts are hydrophobic and have less adhesive property. They can prevent the hydrophilic substances, but, hydrophobic substances can adhere to it followed by thrombi. Biological grafts are also not antithrombogenic and have a thrombus layer on the lumenal surface. If these grafts have an excellent healing ability of the neointima, they can maintain their patency for long periods of time with a powerful anticoagulant therapy just after implantation.

To accelerate the endothelialization, or to maintain the antithrombogenicity of the grafts before the endothelialization, we developed our own methods which showed satisfactory results in animal studies. In this communication, four types of vascular prostheses will be displayed. One is a heparinized hydrophilic polymer graft which has an antithrombogenic property[1]. The graft showed problems in long-term implantation. Other three are new types of grafts, i.e., a temporally antithrombogenic biological graft which can be reconstructed by host cells[2], a fabric prostheses transplanted with autologous tissue fragments[3], and a fabric prosthesis fabricated with ultrafine polyester fibers[4].

VASCULAR PROSTHESIS MADE OF AN ANTITHROMBOGENIC POLYMER

We have developed a heparinized hydrophilic polymer utilizing an ionically-bound heparin in order to render excellent antithrombogenicity for a long period[5]. This material has been successfully applied to medical devices, such as catheters, drainage tubes, ascites tubes and chambers of blood pumps[6-8]. As previously reported, its thromboresistance is caused by the continuous release of a certain amount of heparin from the surface into the blood stream. In this experiment we have applied this heparinized polymer to vascular prostheses by making it porous in order to render pliability and ease of suturing.

Materials and methods

Preparation of heparinized hydrophilic polyurethane (H-USD): The cationic hydrophilic polyurethane (USD) was synthesized by blending N,N-dimethylacetamide (DMAC) solutions of segmented polyurethane (Lyra T-127, DuPont, polymer concentration: 14 wt%) and DMAC solution of a cationic copolymer (SD^+) was synthesized by radical copolymerization of methoxypolyethyleneglycol methacrylate (SM) of 60 wt% and N,N-dimethylaminoethyl methacrylate (DAEM) of 40 wt% in DMAC at 45°C for 30 hours using azodimethylvaleronitrile as an initiator and subsequently by quarternizing dimethylaminoethyl groups at 55°C for two hours using ethylbromide. Heparinization was achieved by ionically binding heparin to the quarternized nitrogen groups of USD polymer after fabrication.

Preparation of the vascular prosthesis: The cylindrical glass mandrils, ranging from 3 to 8 mm in diameter, were dipped into the DAMC solution of USD polymer (polymer concentration: 1 wt%), dried under varying conditions and subsequently dipped into the N,N-dimethylformamide (DMF) solution of the segmented polyurethane (polymer concentration: 10 wt%). For the purpose of reinforcement, polyester mesh of fiber was incorporated in the coating layer of the segmented polyurethane. Without drying the coated segmented polyurethane layer, the mandril was immediately soaked in distilled water to precipitate the polymer by substitution of water for the solvents (DMF and DMAC) involved in the polymer matrix. Then the polymer coating layer was slipped over the mandril and soaked in distilled water at room temperature for one day in order to extract the residual solvents and unreacted monomers. Heparinization of the USD coating layer was carried out in 3 wt% sodium heparin aqueous solution at 60°C for two days. The average heparin concentration in the H-USD layer measured by electron probe x-ray microanalyzer (EMX) was 15 wt%.

In vivo experiments: Seventy-three healthy mongrel dogs of both sexes weighing 6 to 12 kg were used as the test animals. Forty prostheses were implanted in the aortas of 40 dogs. A 5.5 cm segment of the thoracic aorta was resected and replaced by the prosthesis (8 mm in internal diameter and 5.7 cm in length). The implantation periods ranged from one to 575 days.

Twenty-eight prostheses were implanted in both the external iliac arteries of 14 dogs. The segments of both the external iliac arteries (3 to 5 cm long) were excised and replaced by the prostheses (3 mm in internal diameter and 4 to 7 cm in length). The implantation periods ranged from one to 98 days.

Nineteen prostheses were implanted in the inferior vena cavae of 19 dogs. A 2 cm segment of the inferior vena cava was resected and replaced by the prosthesis (8 mm in internal diameter and one to 3 mm in length). The implantation periods ranged from one to 309 days. In all experiments, antibiotics were used at the time of operation, but no anticoagulants were used at any time.

Angiographical examination: In the thoracic descending aorta and the inferior vena cava, the occlusion of the implanted prosthesis was confirmed by death of the animal. In the external iliac artery, however, the occlusion of the implanted prosthesis did not cause death, so that translumbar aortographic studies were performed in the animals with the prostheses implanted for more than one month in order to assess the patency.

Observations: The specimens for light microscopy were fixed with 10% formaldehyde aqueous solution and then embedded in paraffin and examined using Weigert elastic fiber stain, as well as hematoxylin and eosin. The specimens for scanning electron microscopy were fixed with 2.5% glutaraldehyde in phosphate buffered saline solution (PBS) and stained with 1.0% osmium tetroxide aqueous solution at 4°C. The fixed specimens were dehydrated in a graded series of ethylalcohol and in amyl acetate, then critical-point dried with carbon dioxide, and coated with gold palladium. The examination was performed with a JSM-50A scanning electron microscope. The specimens for transmission electron microscopy were dehydrated in a graded series of ethyalcohol and propylene oxide and then embedded in epoxy resin (Epon 812). Ultrathin sections were cut and stained with uranyl acetate and lead citrate. These stained sections were examined with a JSM-100C electron microscope.

Results

All the prostheses implanted in the thoracic descending aorta were patent at the time of removal. The average implantation period was 146 days. Twenty-one prostheses out of those implanted in the external iliac arteries were patent and the residual seven prostheses were occluded. The patency rate and the average implantation period were 75% and 33 days respectively. In 19 prostheses implanted in the inferior vena cavae, 14 prostheses were patent and five prostheses were occluded. The patency rate and the average implantation period were 74% and 56 days, respectively.

Autopsy findings demonstrated that the occlusion of three prostheses implanted in the external iliac arteries for 28, 60 and 98 days was caused by pannus formation in the anastomotic lines. In addition, in the prostheses implanted in the inferior vena cavae for 23, 72 and 111 days, the occlusion was caused by the similar pannus formation. The occlusion was observed at the early stage (5 to 7 day-implantation) in the external iliac arteries; in the inferior vena cavae failure was due to thrombi formed in the anastomotic lines by technical failure.

One translumbar aortographic observation of a prostheses implanted in both external iliac arteries for 98 days, indicated that the right prosthesis was occluded and the left prosthesis was patent. At the distal area of the anastomotic site a small pannus was noticed in the photography.

Macroscopical observation revealed that the inner surface of the prostheses implanted in the iliac arteries, in the inferior vena cavae, and descending aortae (grafts at 91, 200 and 365 days respectively) were whitish and glistening, and no thrombus was observed (Fig. 1). Light microscopical observation revealed that numerous pores from one to 3 μm in size were seen in the H-USD layer and no deposit was microscopically observed on the H-USD surface. At the anastomotic site, pannus was observed (Fig. 2). Although the layer of a paste-like substance was observed on the surfaces by means of scanning electron microscopy, neither thrombus formation nor endothelial cells were seen. By means of transmission electron microscopy, the isolated domain of the SD^+ polymer and the continuous domain of the segmented polyurethane were clearly distinguished in H-USD layer. The nonspecific structured proteineous layer of 100 to 250 nm thickness was observed on the inner surface of the prostheses. The heparin concentration in the graft removed at 575 days was about 50% compared with that before implantation.

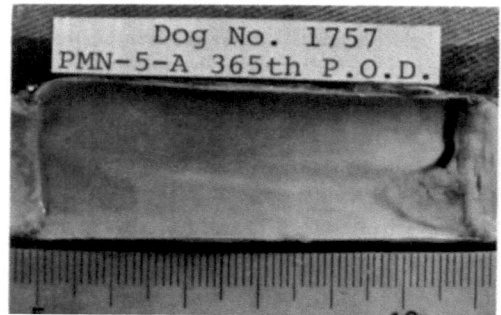

Figure 1. Luminal aspect of a heparinized graft implanted in the descending aorta at 365 days. No thrombus adheres on it, large pannus is observed at the distal anastomotic site.

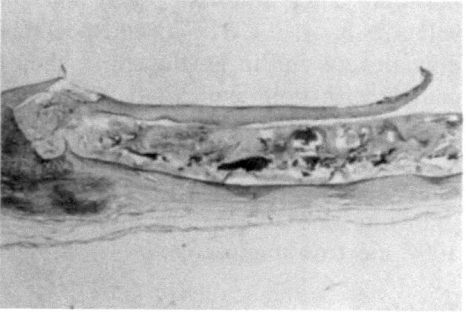

Figure 2. Photomicrograph of a cross-section of a heparinized graft near anastomotic site at 39th day. A pannus developed from the anastomotic site. The graft surface prevent the adhesion of the pannus. X 20. H.E.

Discussion

The scanning and transmission electron microscopic examination indicated that neither pseudoneointima nor endothelial cells were observed, but nonspecific structured proteineous layer was noted irrespective of implantation period (up to 1-1/2 yrs) and sites. In addition, the histological examinations showed that the pannus formation in the anastomotic lines indicated on the Tevdeck suture surface and extended to the H-USD surface of the prostheses; the pannus whose surface was surrounded by endothelial cells was not adhesive to the H-USD surface. These findings show that the H-USD surface completely prevents the deposition of blood cells and endothelial cells. This shows an excellent antithrombogenicity, but also a problem of the graft. The major cause of the graft occlusion was a pannus formation at the anastomotic site. This is the biggest problem with the graft. Endothelial cell lining is a natural activity of the wound healing in the blood vessels.

Antithrombogenic polymer prevents not only the thrombus deposition but also pannus adhesion. The graft cannot control the pannus growth. Therefore, small diameter graft and venous graft will be occluded easily with the pannus growth.

TEMPORALLY ANTITHROMBOGENIC BIOLOGICAL GRAFT WHICH CAN BE RECONSTRUCTED BY HOST CELLS

Biological materials have unique, fine structural and mechanical properties which cannot be simulated by any current technologies. For example, arteries have ideal hemodynamic shaped ramifications. Inside the arteries, they have uniquely and fine structures specially suitable for cell inhabitation. They also have suitable mechanical properties to accept and to pass the pulsatile blood pressure and flow. If we could use these special properties and structures for biomedical materials, we could make excellent artificial organs. One of the problems is antigenicity except for those of autologous origin. Biodegradability of the materials is also one of the problems. To reduce the biodegradability and antigenicity of the materials, chemical modifications by glutaraldehyde, dialdehyde starch, formaldehyde, and hexamethylene diisocianate have been used; however, these treatments make the materials hydrophobic and stiff. Glutaraldehyde is the most frequently used treatment, but it has cytotoxicity and prevents cell infiltration inside the graft wall. To overcome these problems, we introduced a new crosslinking reagent[9]. Another technology we have developed is the heparinization of biological materials[10]. Collagen is one of the major components of the biological materials. It has unique properties for the host cells migration and proliferation, and also for platelets adhesion and accumulation[11]. Therefore, this biological material is thrombogenic if used for cardiovascular artificial organs. To reduce the thrombogenicity, we developed a heparinization method. With the combined use of heparinization and the crosslinking method, a small diameter biological graft was developed.

Materials and Methods

Cross-linking reagent used: The cross-linking reagents used were polyepoxy compounds (PC), such as polyethylene glycol diglycidyl ether, glycerol polyglycidyl ether, polyglycerol polyglycidyl ether, and sorbitol polyglycidyl ether (Nagase Chemical, Ltd., Osaka, Japan). In this study, polyglycerol polyglycidyl ether (PGPGE) was used. Its molecular structure and cross-linking reaction with collagen molecules are illustrated in Fig. 3. The cross-linking reaction can be performed at room temperature, and the specimens cross-linked with PCs become hydrophilic.

Preparation of the graft: A fresh carotid artery with an inner diameter of 2.5 to 3.0 mm was obtained from dogs. The artery was soaked in distilled water for one hour and submitted to ultrasonic waves of 28 kilocycles for 20 seconds to produce cell destruction. Cell debris was then removed by washing with distilled water. In this way, a natural tissue tube composed of collagen and elastic laminae was obtained. A 2% protamine sulfate solution at pH 5.9 was poured into the natural tissue tube graft lumen, and the graft was inflated with air at a pressure of 80 to 100 mmHg for 30 min to force the protamine into the graft wall. The graft inflated with air pressure was treated with a 5% PC solution in 50% ethanol and 0.1 M Na_2CO_3 at pH 10.0 for five hours to cross-link the tissue and covalently immobilize protamine impregnated into the wall. The graft was then washed with distilled water. The graft was soaked in a 1% heparin solution at pH 7.0 for five hours at 45°C and repeatedly washed with distilled water. The graft was then preserved and sterilized in a 70% ethanol solution.

Control graft: For the control experiment, nonheparinized grafts treated with either GA or PC were prepared. Carotid arteries obtained from dogs were cross-linked with 1% GA solution, inflated at 80 to 100 mmHg air pressure for five hours after sonication to remove the cell components and used for mechanical property measurements and animal studies. The PC-treated control was prepared by the method described in the preceding section without using protamine or heparin. This control was used for mechanical property measurements.

Mechanical properties: Fresh canine, GA cross-linked, PC cross-linked and PC cross-linked and heparinized carotid arteries were used. Cylindrical specimens were fixed longitudinally and tensile strength and elongation measurements were performed. Each cylindrically shaped specimen was placed in an evaluation system developed by Hayashi et al[12].

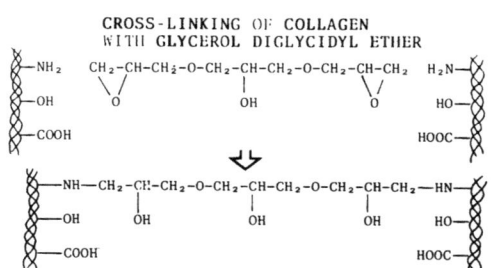

Figure 3. Molecular structure of a ethylene, polyethylene glycol diglycidyl ether and its reaction with collagen molecules.

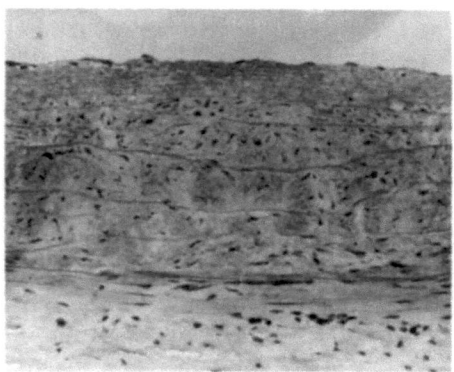

Figure 4. Photomicrographs of cross-sections of heparinized grafts at 389 day. X 100, N.E.

Animal experiment: Fifty-seven mongrel dogs weighing 8 to 12 kg were used for the experiment. About 6 cm of both carotid arteries was harvested and a 6 cm long by 2.5 to 3.0 mm internal diameter segment of heparinized graft was implanted end to end. Penicillin (500 mg) was given, but no anticoagulants were used at any time. Eighty-two heparinized grafts were implanted, as well as 16 GA treated control grafts. The animals were euthanized electively at selected time intervals. The implantation periods ranged from one hour to 429 days. To examine the patency of the grafts, angiographic studies were performed on those animals with the grafts in place for more than one month. The right brachial artery was exposed and an angiography catheter was inserted, contrast media was injected, and X-rays were taken when the tip of the catheter reached the brachycephalic artery.

Observations: All specimens excised from the experimental animals were subjected to macroscopic, microscopic, and scanning electron microscopic observations after routine handing. Heparin concentration in the whole graft wall was measured using the method of Lagnoff and Warren.

Results

Mechanical properties of the graft: Using the method of Hayashi et al., quantitative comprehensive wall stiffness of each specimen was calculated. From these results, the vascular compliance of each specimen at 90 mmHg was also calculated. The strength and elongation of each specimen was measured. The compliance of native, GA treated and PC treated artery were 2.33, 0.80, 1.70 percent radial change per mmHg $\times 10^{-3}$, respectively. Elongation rates were 11.6, 57, 127 %, respectively. Tensile strengths were 207, 127, and 199 g/mm^2, respectively.

Animal experiment: Heparinized grafts were white, pliable, and more elastic than the yellowish controls. The inner surfaces of both grafts were shiny and smooth, but the heparinized grafts were easier to suture and match to the host arterial wall. There was no blood leakage through the grafts wall on implantation, and no kinking occurred even when the grafts were bent.

Implanted grafts: All the grafts were patent at the time of the angiographic examination. The inner surface of the grafts was smooth throughout its length, and no stenosis or aneurysmal dilatation was observed in any of the grafts. At the graft explantation, the dogs were anesthetized with intravenous sodium pentobarbital, and heparin 2 mg/kg was given intravenously to prevent clot formation. In the heparinized graft, 79 were patent and three were occluded. In one dog, killed at 172 days after implantation, a graft implanted in the right carotid artery was patent, but the graft in the left artery was occluded. The occluded graft was soft and white, but anastomotic lines were hard. An angiographic examination of the dogs performed at 40 days after implantation, however, showed both grafts to be patent. Consequently, it was considered that the graft occluded a certain period of time after the angiogram. In another dog, killed at 11 days, the cervical wound was infected and the grafts implanted in both carotid arteries were occluded. As these were the only grafts occluded, the patency rate of the heparinized graft was 96%. All the patent grafts were still as soft and pliable as the native artery. Within 100 days after implantation, the inner surfaces were completely free from thrombus deposit. The surfaces were as shiny, white, smooth, and glistening as those of the host arterial intima. In the case of those grafts that remained in place for more than 100 days, slightly yellowish and semitransparent small spots were sporadically observed on the surface.

The control grafts were occluded within one week after implantation, and they were very hard and dark brown in color.

Scanning electron microscopic observations: Scanning electron microscopy (SEM) revealed the inner surface of the heparinized grafts before implantation not to be smooth, but to have a naked elastic lamina with many holes and wrinkles on the surface. No endothelial cells were seen. After implantation, the surface was covered with a layer of protein that was so thin that the structure of the elastic lamina could be observed throughout it. On the surface of the grafts removed at less than 100 days, there was neither fibrin deposition nor platelet aggregates, and the surfaces were rough due to the wrinkles in the elastic lamina. At the anastomotic line, pannus was first observed at 37 days and was noticed at each anastomotic line in all the grafts left in place for more than 37 days. The size of the pannus was not longer than 1 mm beyond the anastomotic line. The pannus was completely covered with endothelial cells and adhered on the graft surface. After 106 days, there were small fibrin deposits on the graft surface at the lines formed by the elastic lamina. Endothelial cells were not observed on the inner surface of the center areas of any grafts after periods as long as 172 days. However, in the cases of the grafts which remained in situ for 389 and 429 days, the whole entire surfaces were covered with endothelial cells.

Microscopic observations: Microscopic observations confirmed that there was neither thrombus nor fibrin deposit on any graft before 81 days. The luminal surface was composed of the internal elastic membrane, with no endothelial cells on the surface. At the early stage, there was no foreign body reaction such as giant cell infiltration on the outer surface of the graft. A small number of plasma cells were observed a short period of time after implantation. At the anastomotic line of the grafts implanted for more than 37 days, there were small panni covered with endothelial cells seen by SEM to be adherent to the graft surface. The size of the pannus grew with time. At more than 30 days, some macrophages were noticed in the inter elastic luminal spaces near the luminal surface. Before implantation, these spaces were occupied by disrupted smooth muscles cells, collagen, and elastic fibrils. After implantation, macrophages gradually phagocytized the smooth muscle cells. After more than two months, spaces containing only collagen and elastic fibrils were observed. After more than six months, smooth muscle-like cells infiltrated these spaces from the adventitial sides (Fig. 4). In the wall of long-term grafts, cells filled spaces completely. In 75% of the spaces near the luminal surface, these elongated cells were arranged circumferentially. The rest were oriented longitudinally. After 389 days, the central part of the inner surface was covered with endothelial cells, which impinged directly on the surface of the elastic lamina. The structure of the graft following long-term implantation closely resembled that of the native arterial wall, and near the anastomotic line at 389 days a thick layer of pannus with an endothelial cell lining covered the surface. The pannus adhered on the graft surface. The thickness of the pannus was about 30 μm. There was no foreign body reaction in the long-term specimens and no degenerative changes such as hyalinization, calcification, or arteriosclerosis.

Concentration of heparin in graft: Before and after the implantation, the total amount of heparin in the graft was measured. The results indicated that the amount before implantation was about 7.0 units/cm^2, but in specimens in place for more than 80 days, there was no heparin in the graft wall. This was confirmed for grafts in place for 106, 153, 172 and 389 days.

Discussion

Antithrombogenicity and compliance are both important factors in small-caliber vascular grafts. We previously developed a method that would afford antithrombogenic properties to collagenous biomaterials, such as vascular grafts made from carotid arteries[2] and ureters[13]. This method was very effective in preventing thrombus formation for both small caliber arteries and large vein vascular grafts. The mechanism is as follows. Heparin is bound ionically to protamine that has been previously covalently linked to the materials, so that the heparin is slowly released following implantation. This slow release of heparin can prevent fibrin formation on the graft surface. As the heparin is gradually desorbed, the graft becomes naturally antithrombogenic because endothelial cells begin to cover the graft surface. Consequently, the graft can remain permanently antithrombogenic by endothelialization. Animal experiments revealed that this method produces stable antithrombogenicity in small-caliber arterial grafts. In previous preparations of this heparinized graft, the protamine impregnated into the graft wall was cross-linked with GA under conditions of graft inflation, and although the graft showed no thrombus formation on its surface following implantation, the graft became less pliable and yellow with time. These changes occurred because when the graft was treated with GA to cross-link the protamine to its wall, it also cross-linked the protamine to the collagen molecules inside the graft wall. While GA cross-linking makes the materials less biodegradable, insoluble, and less antigenic, it makes the materials less flexible. Recently, other adverse effects of GA treatment have been reported.

To overcome these difficulties, a new cross-linking method was introduced. A marked difference in appearance between materials treated with GA and PC is their color: GA treatment makes the materials yellow, but PC makes them white. There are also marked differences in their softness and elasticity. In long-term animal experiments, those treated with PC maintained their elasticity. This natural compliance during implantation seems to be very important in obtaining permanent patency of small-caliber vascular grafts. Another characteristics of the PC treatment is the hydrophilic property imparted to the material, which is important for its affinity with the host tissue and makes the graft more nonthrombogenic.

Reconstruction of an arterial wall with the graft was most successful with smooth muscle-like cells infiltrating the graft wall. We observed the healing process of an implanted fabric Dacron vascular prosthesis and note that the smooth muscle cells infiltrated the neointima of the graft and were arranged parallel to the direction of the tensile stress upon the graft wall[14]. If the tensile stress is not present, smooth muscle cells seldom appear. The appearance of such cells in the graft treated with PC suggested that there is enough compliance of the graft to induce the migration of these cells. In case of GA-treated grafts, the environmental condition in the graft wall is considered to be insufficient for the infiltration of the smooth muscle-like cells. The grafts cross-linked with GA become yellowish and lose their elastic characteristics. By contrast, PC cross-linked grafts maintain their natural vessel compliance and are stronger than the original vessel, thus providing excellent suturability and compliance match. Furthermore, the PC cross-linked grafts are hydrophilic because of hydroxyls in the molecular structure, while the GA cross-linked grafts are hydrophobic. In this report, the PC cross-linked graft has superior antithrombogenic characteristics because the high hydrophilicity may give the material antithrombogenicity[15]. It has another merit with regard to antithrombogenicity in that it becomes weak-negatively charged after the cross-linking. This was because e-NH_2 groups in the collagen molecules were used for the cross-linking, and these increased the carboxyl groups relatively. The weak charge in negativity contributes to the antithrombogenicity of the materials because of the prevention of platelet aggregation on the negatively charged surface.

From these results, we can conclude that the combination of short-term antithrombogenicity of slow released heparin followed by the permanent antithrombogenicity of endothelialization, together with the natural tissue compliance of these grafts, were the major reasons for their success as small caliber vascular grafts.

A FABRIC VASCULAR PROSTHESIS TRANSPLANTED WITH VENOUS TISSUE FRAGMENTS

Endothelialization of vascular grafts in human is extremely delayed[15]. Most of the grafts implanted are not endothelialized, and are covered with fresh thrombi for a long time after implantation except for those areas near anastomotic sites. Endothelial cell seeding methods have been attempted for the past decade[17]. Some of them have produced satisfactory experimental results, but they remain unavailable for general use, as they require special cell culture techniques and facilities. They are also not available for emergency use, since the cell culture requires an extended period of time. Recently, we developed a new method to seed tissue fragments which contained endothelial cells, smooth muscle cells and fibroblasts[18]. This mixed cells seeding was very effective in making a new arterial wall in vivo. With this method, we made a fabric vascular prosthesis transplanted with autologous venous tissue fragments into the wall.

Materials and methods

Preparation of the graft: A highly porous fabric vascular prosthesis (Microknit, Golaski Laboratories, Inc., Philadelphia, PA; Water porosity: 4,000 ml) was used as the framework of the graft. The prosthesis was connected with a syringe through a three way stopcock, and was enveloped by a transparent bag connected to the three way stopcock through a tube. A piece of peripheral vein weighing about 0.3 g was obtained, cut into tiny fragments, and stirred into 20 ml of saline, thereby creating a tissue suspension. This suspension was sieved through the wall of the prosthesis by strong injection with a syringe. The residual suspension that passed through the prosthetic wall was suctioned again through the connective tube. The prosthesis was then implanted as a vascular graft into the same dog from which the vein used had been resected.

Implantation of the prostheses: Forty adult mongrel dogs, weighing 8-12 kg were used. The seeded grafts, 7 mm ID and 5.7 cm in length, were implanted into the descending aortae of 25 dogs. Fifteen preclotted prostheses (Microknit) were also implanted as controls.

Graft harvesting: Implanted grafts were harvested at from 1 to 61 days after implantation. Before harvesting, sodium heparin (100 IU/kg) was intravenously administered to prevent clotting. After removing the prostheses, the kidneys of each animal were resected and examined for fresh, trapped microemboli.

Histologic examination: The specimens were fixed with 1% glutaraldehyde in phosphate buffer, pH 7.4, and embedded in hydrophilic resin (JB-4, Polyscience Inc., Warrington, PA). Sections were stained with hematoxylin and eosin, PAP method, and Van Kossa stains.

Results

Implantation of the graft: Grafts seeded with tissue fragments were soft and pliable (Fig. 5), and there were no bleeding complications postoperatively, while in the control grafts, problems occurred after surgery. Three of 15 control animals (20%) bled into the pleural cavity and died within 24 hours after implantation.

Graft harvesting: Both the seeded and control grafts were patent at the time of harvesting. In the tissue fragments graft, fresh red thrombi adhered to the luminal surface just after implantation. The next day, the luminal surface became white with small spotted thrombi. On the third to fifth day, the luminal surface became whiter, and at 14 days, the surface was completely white, without any thrombus. No macroscopic changes were observed up to the two months observation period. In the control graft, the luminal surface was covered with fresh thrombi, except at anastomotic sites, at two weeks after implantation. After two months, the center area was still covered with fresh thrombi. Macroscopic examination of all the kidneys resected from the animals showed no trapped microemboli.

Light microscopic observation: The graft wall before implantation had many embedded tissue fragments that contained smooth muscle cells and endothelial cells. After implantation, fresh thrombus covered the luminal surface. At the fourth day, obvious migration of numerous fibroblasts, smooth muscle cells, and endothelial cells into the thrombus layer was noticed (Fig. 6a). Some of the endothelial cells reached the luminal surface to make a colony. At the 14th day, endothelial cells lined the whole luminal surface, and multilayers of smooth muscle cells were underneath; neointima formation was virtually completed at two weeks (Fig. 6b). In control grafts, endothelialization was seen at the anastomotic sites at 14 days, but the center area was not endothelialized after two months.

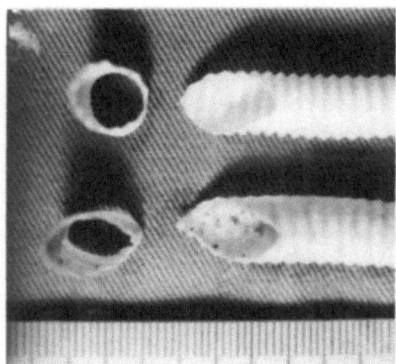

Figure 5. An original fabric vascular prosthesis (top) and a tissue fragments seeded graft (bottom).

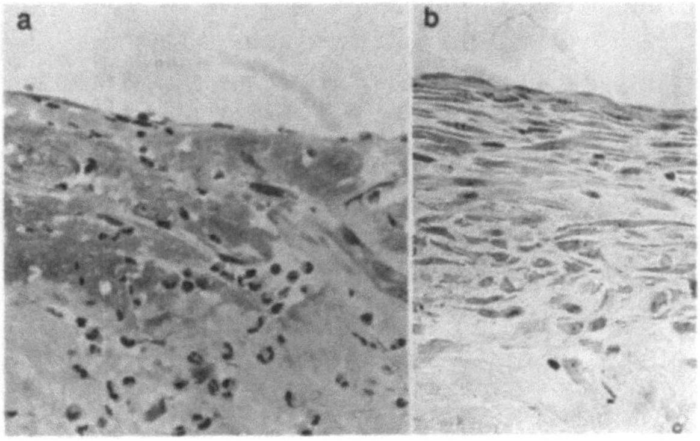

Figure 6. Photomicrographs of cross-sections of the tissue fragments seeded grafts at 4 days (a) and 8 days (b). X 100. H.E.

Discussion

The tissue fragments seeded graft showed extremely rapid healing along its entire length due to the venous fragments transplanted. Neointima formation started by the fourth day, and was completed at two weeks. This is the fastest healing of any grafts reported. The healing proceeded equally at all locations, with no difference in healing between the central area and the anastomotic sites. In the control graft, however, healing did not start until 14th day, and had not finished in the central area at two months.

This method was simple and easy to do, and can be available in any operation room without special techniques, instruments or facilities. In this experiment, the result showed excellent efficacy in seeding of the tissue fragments, and complete endothelialization within a short period of time with no negative side effects in the long-term specimens.

Transplantation of autologous tissue fragments has already been used in orthopedic and plastic surgery, as osteoblasts migrate very rapidly from the cut edges of transplanted bone fragments, and epidermal cells from skin fragments. Multiple tiny fragments have large areas of cut edges from which cell can migrate and proliferate very quickly under physiologic conditions.

In this experiment, three kinds of cells - fibroblasts, smooth muscle cells, and endothelial cells - migrated and proliferated at the same time from the fragments. This is a very unique phenomenon, that is not observed previously in cell culture. When fibroblasts and endothelial cells are culture in a petri dish, endothelial cells are suppressed and fibroblasts proliferated to form a confluent layer. However, in this in vivo experiment, we noticed the three cells migrated and proliferated together. Endothelial cells produced capillaries, and rose to the inner surface of the graft to face the blood stream. Accordingly, smooth muscle cells made a multilayer beneath the endothelial cells, and fibroblasts crawled down under the smooth muscle cells layer around the polyester fibers. This phenomenon suggests that growth of these cells was controlled by their physiological environment.

Mixed cell culture of different types have been reported to behave uniquely. For example, in the case of an experiment of a skin equivalent substitute[19], epidermal cells cannot make a satisfactory membrane by themselves. Fibroblasts are necessary as a feeder cells underneath the epidermal cells. In case of the neointima, endothelial cells cannot produce a stable neointima by themselves. With the smooth muscle cells or fibroblasts underneath, endothelial cells can maintain a more stable condition. Also, smooth muscle cells and fibroblasts cannot maintain stability without the protection of the endothelial cells.

These results in this experiment show the efficacy of transplantation of venous tissue fragments into vascular prostheses, and the possibility of overcoming delayed neointimal healing in humans.

FABRIC PROSTHESIS WITH HIGH CELL AFFINITY FABRICATED WITH ULTRAFINE POLYESTER FIBERS

After the detailed animal and clinical experiments of Wesolowski, a highly porous fabric vascular prosthesis is recommended because of the high healing ability of the neointima[20,21]. For this purpose, many preclotting methods have been developed to seal the highly porous grafts before implantation[22-24]. In clinic situations however, such a highly porous graft cannot be used due to bleeding through the graft wall[25]. Even after a perfect seal using these reliable preclotting techniques, the grafts have the possibility of bleeding due to the fibrinolysis. In the clinic, middle or low porous fabric grafts have been used to prevent postoperative bleeding. But these grafts show little endothelialization. Special methods such as an immobilization of growth factors have been tried, without success. For these reasons, a fabric vascular prosthesis with the high healing capability of the neointima has been desired. Recently we developed a new graft fabricated from ultrafine polyester fibers[4]. Our animal experiment revealed that fibroblasts adhere onto the ultrafine polyester fibers very avidly. We believe that these phenomena are from the basic affinity of the cells to the graft, i.e., the cells want to adhere to the sharp edge or fine fibers. We term this behavior "contact guidance." The vascular prosthesis fabricated from ultrafine polyester fibers can accumulate many host cells, which will require their nutrition. Therefore after migration and proliferation of these cells, capillary blood vessels will follow, resulting in a natural angiogenicity. Using the behavior of these cells, we developed a special vascular prosthesis.

Materials and methods

Preparation of the graft: The graft was prepared by the following procedures:

1) Tube formation: Ultrafine polyester fibers (UFPF), thickness: about 3 µm and ordinary polyester fibers (OFP), thickness: about 16 µm were used as the front and back yarn in the weft-backed woven method. The preparation and fabrication of UFPF have been reported in detail elsewhere[26]. Napping of the UFPF: The front surface of the tube was napped by a nap-raising machine. 3) Ravel proofing: The front surface of the tube was treated by a water jet machine. 4) Exchanges of the front (outer) and back (inner) surfaces: The tube was turned inside out and crimped. After these procedures, the graft has an inner surface composed of napped UFPF.

In vivo experiments: Forty-five healthy mongrel dogs of both sexes, weighing 8 to 12 kg were used as test animals. The grafts (8 mm in internal diameter, and 5.7 cm in length) were implanted in the thoracic descending aortae of 30 dogs for up to 375 days. As controls, Cooley Veri-Soft grafts (Cooley graft, Meadox Inc., indicated water porosity: less than 130 ml) were implanted in the remaining 15 dogs in the same manner. All specimens excised were submitted to light and scanning electron microscopy. The specimens for light microscopy were fixed with 1% glutaraldehyde in 0.2 M phosphate buffer, pH 7.4, and embedded in a hydrophilic resin (JB-4, Polyscience, Inc.). The cross sections were examined with hematoxylin and eosin stain, as well as Weigert's elastic fiber stain. Specimens for scanning electron microscopic observation were fixed with 1% glutaraldehyde in 0.2 M phosphate buffer, pH 7.4, and stained with a 1% OsO_4 solution. Specimens were dehydrated in a graded series of ethanols and in amilacetate, then critical point dried with carbon dioxide and spattered with gold palladium. The examination was performed with a JSM-50-A scanning electron microscope (JEOL, accerelating voltage:15kV).

Results

Preparation of the graft: SEM observations of the cross-sectional and the inner surfaces of the graft and control Cooley graft are shown in Fig. 7a and 7b, respectively. It should be noted that the difference in fiber diameter between UFPF and OPF is very large. In the graft, the napped and entangled UFPF's were seen on the inner surface, while in cross section, the tightly woven and entangled UFPF and OFP for reinforcement on the outer side were observed. The cut edge shows no raveling. On the other hand, the Cooley graft was constituted of tightly woven plain OPF alone, without napped fibers which ravels easily. Water permeabilitys of the graft and Cooley graft were 92.6 and 134 ml/min./cm^2, respectively.

In vivo experiment: Because the graft was soft and pliable, fitting and suturing it to the aortic wall became significantly easier. Although no preclotting was performed, blood leakage through the wall of the graft was minimal because of the dense structure. Blood infiltrated no further than into the whole cloth, which turned reddish but did not leak. On the other hand, the rigidity of the control graft caused difficulties in fitting and suturing and, in several cases, the surgical needle became bent. All the grafts were patent at the time of harvesting. The following healing processes were macroscopically noted in the graft. The red color was seen two hours after implantation, but the surface turned white in three days. during a period of seven to 37 days after implantation, the suture turned gradually reddish again, followed by a final change to yellow. After 42 days of implantation, white, glistening areas developed as in growths of the host vessel at both ends of the graft, and occasionally similar areas were noticed more centrally. These areas gradually grew to cover the whole of the surface in three months. The surface remained white and glistening for a long period of time. In the control grafts, however, severe clotting occurred immediately after implantation. At the anastomotic sites, healing of the control graft was similar to that of the graft, but in the middle of the control graft, a considerable amount of fresh clot still remained even after 360 days.

Observations: The light microscopic and scanning electron microscope (SEM) studies revealed that the thickness of the initial clot layer on the graft surface ranged from 10 to 100 µm and was very thin compared with that of the control graft. The clot was composed of fibrin, erythrocytes, leucocytes, and platelets, and the content of fibrin gradually increased with time (3 to 7 days). At about 15 days, endothelial cells appeared at the anastomoses and then migrated toward the middle part of the graft.

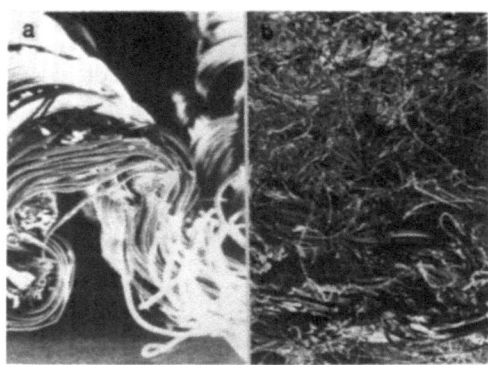

Figure 7. Scanning electron micrographs of a control graft (a) and a vascular prosthesis fabricated with ultrafine polyester fibers (b). polyester fibers (b). X 100.

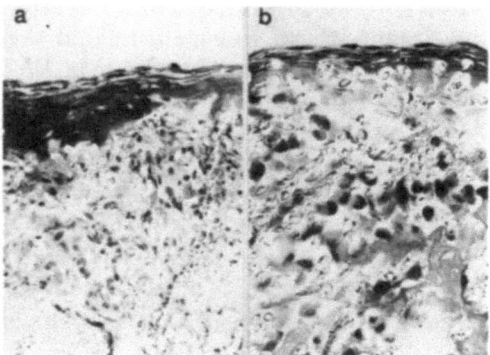

Figure 8. Photomicrographs of center areas of the grafts fabricated with ultrafine polyester fibers at 21 days (a) and 61 days (b). Fibroblasts are noticed among the ultrafine polyester fibers. The luminal surface are already endothelialized. X 100. H.E.

In addition, fibroblasts and a few plasma cells infiltrated the fibrin layer over the brushed UFPF (Fig. 8a), but this did not occur in those without the UFPF. SEM studies after 42 days of implantation showed either sporadic deposits or colonies of endothelial cells on the mid-surface of the grafts. Light microscopic observation of the central areas of the grafts demonstrated that by postoperative day 42, endothelial cells completely covered the surface and fibroblasts and other cells appeared beneath them. By 85 days, the middle parts of the grafts were completely covered with cells (Fig. 8b), the fibrin was gone, and a neointima ranging from 10 to 50 µm in thickness had formed. After long-term implantation, the formation of a vasa vasorum was clearly seen, and degenerative changes such as calcification, or arteriosclerosis were present, even at the center of the grafts. On the other hand, in the control grafts, the initial clot was markedly thick and uneven (50 to 1000 µm). Light microscopic and SEM studies showed that a fresh clot still remained in the middle part of the control graft after 99 and 360 days of implantation. Even at 360 days, endothelialization was not complete at the center of the graft. Microscopic cross-sectional views demonstrated little cell infiltration into the interstices of OPF, even after long-term implantation.

Discussion

Ultrafine polyester fibers (UFPF, about 3 μm in thickness) are used for making the suede-like fabric "Ultrasuede" (U.S.A.), "Alcantare" (Europe), "Ecsaine" (Japan) produced by Toray Industries, Inc. The fabric is brushed to provide a velvetlike feel to the surface. Accordingly, the vascular grafts fabricated from such very fine fibers are soft and pliable, regardless of their dense textures, and have a low porosity. Permeability of the grafts (water permeability: about 93 ml) is much lower than that of conventional high-porosity grafts (water permeability: 1,200 to 2,00 ml) made from ordinary polyester fibers (OPF, about 16 μm in thickness). Therefore, the grafts are expected to solve some of the problems found with implantation of conventional low-porosity grafts. For example: 1) Difficulty in fitting and suturing to an aorta which is stiffened due to arteriosclerosis and calcification, which can lead to severe bleeding at the anastomotic line. 2) Low-porosity grafts which ravel easily at their cut ends because of their plain woven structure and can be brushed to prevent raveling as a result of its fiber entanglement created thickness. 3) Formation and stabilization, or the healing process is significantly delayed compared with that of high-porosity grafts because the tightly woven structure reduces the number of interstices among the fibers, which play an important role as scaffording for cell adhesion and infiltration.

The most important property of a vascular graft is rapid and uniform vascularization, which closely correlates with initial clot formation, and subsequent adhesion and infiltration by endothelial and other cells. In the high-porosity grafts, the quick change of initial clot to a fibrin layer occurs by the elution of erythrocytes. This rapid shift is considered to cause much speedier healing of the high-porosity graft. The color change in the initial clot layer from red to white at three days after implantation of the graft, which indicated the change of the clot to a fibrin layer, took place as promptly as in the high-porosity graft, but in the graft the initial clot formation remained in the brush layer, and no further growth was noticed. Consequently, the clot layer of the graft was very thin (10 to 100 μm). In the control graft, however, the color change from red to white was significantly delayed, and overgrowth of the initial clot always occurred. These findings suggest that the extremely pliable UFPF on the inner surface of the graft effectively prevents overgrowth of the initial clot, leading to rapid endothelialization. However, the mechanism has not as yet been clarified. In the control grafts, endothelialization took place only in the vicinity of the anastomotic line but did not move to the central area, even after long-term implantation, while in the grafts endothelialization was noticed in the middle part of the grafts early on (42 days).

A conspicuous characteristic of the graft is shown in Fig. 8a, in which numerous fibroblasts can be seen infiltrating into the fibrin layer over the brushed UFPF, something not seen without UFPF. This finding suggests that UFPF provides a suitable microenvironment for the infiltration and proliferation of these cells, which accompany capillary formation on the graft surface, leading to induction of endothelial cell colonies. The reddish surface at 7 to 35 days shows this capillary formation. In addition, it seems that formation of a neointima, i.e., infiltration and proliferation of fibroblasts and other cells, significantly accelerate the proliferation of endothelial cells by releasing cell growth factors. On the other hand, in conventional low-porosity grafts, such healing was markedly delayed, even after long-term implantation. It is therefore concluded that the extremely fine fiber used in the graft improves the softness, pliability and the healing seen in conventional low-porosity grafts.

Furthermore, vasa vasorum was clearly seen in the neointima of the graft surface implanted for 375 days. Such formation is thought to prevent degenerative changes such as calcification, hyalinization, and arteriosclerosis and leads, as well, to long-term stabilization of the neointima. It seems clear, therefore, that the healing process strongly depends on the thickness of the fiber. Studies on the correlation between the cross-sectional shape of the fiber and the healing process are in progress.

These findings suggest that fine fibers, which provide a suede-like fabric, are a promising material applicable to not only vascular grafts, but to a variety of artificial organs as well.

SUMMARY

Four kinds of vascular graft prostheses were demonstrated in this communication. One is a graft with an artificial antithrombogenic property. The other three have a common concept, i.e., natural antithrombogenicity of endothelialization on their luminal surfaces. For a small diameter vascular graft, antithroambogenicity is essentially required. During the past two decades, antithrombogenic synthetic polymer prostheses like a heparinized polymer graft displayed above have been studied. Some of them showed marked ability to prevent platelets adhesion and fibrin deposit, however, the properties were not long-term. The polymer materials were covered with some blood components, such as proteins or cells. The materials were attacked by these cells and enzymes. It has been very difficult to maintain the initial property against the blood stream, because most of the materials cannot face the blood stream directly after implantation. After the covering with such substances, the materials have to act as the matrix for cell migration and proliferation, however, many of the antithrombogenic polymer materials were not designed for this purpose and the cells cannot obtain a stable anchoring on the surface. The floating pannus observed in the antithrombogenic graft will be developed in this way.

Except as vascular prostheses for temporary use, most of the prostheses have to be implanted for a very long-term, sometimes, more than 30 years. In the implantation period, the initial stage just after implantation is very short. Therefore, the vascular prostheses have to be designed to have a stable condition for the long-term. To maintain the antithrombogenicity of vascular prostheses, it will be very difficult with the properties of the synthetic polymer materials. Permanent antithrombogenicity of endothelial cells is the best and natural method for this purpose.

In this communication, three grafts were demonstrated to have stability of neointima, which were covered with endothelial cells. Artificial antithroambogenicity have to be effective within only very short period of time after implantation. Heparinization in the biological materials is one example of a temporally used antithrombogenicity. After the release of heparin, the graft has to provide the best condition in structural and mechanical properties for cell migration and proliferation. EPTFE graft is one of the samples with its incomplete structure. It had been believed to have an antithrombogenic property on the luminal surface. PTFE is a very hydrophobic material with poorly adhesive properties. Expanding gives it a flexible property and microporous structure, which was expected to contribute the cell anchoring. Ideally, antithrombogenic properties will prevent the cell adhesion, while microporous structure will help the cell adhesion.

Ultrafine polyester fibers accelerate the cell migration and proliferation, resulting in a rapid neointima formation with endothelial cell lining. The biological graft crosslinked with polyepoxy compounds can give a natural anchoring site for cell habitation. The tissue fragments transplantation technology will contribute to rapid neointima formation on the prosthesis. These technologies will be useful for the development of vascular prostheses in future.

CONCLUSION

There are two concepts of the design of vascular prostheses with respect to antithrombogenicity. One is artificial and another is natural. Our experimental data showed that the former is suitable for the short-term use and the latter is for permanent use. Some temporary antithrombogenicity will be helpful for the very short period of time after implantation to maintain early stage patency before endothelialization, however, basic structure of the vascular prostheses have to be designed as a best matrix for the host cell migration and proliferations.

REFERENCES

1. Noishiki Y, Nagaoka S, Kikuchi T, Mori Y: Application of porous heparinized polymer to vascular graft, Trans. Am. Soc. Artif. Intern. Organs, 27:213, 1981.
2. Noishiki Y, Miyata T: Successful animal study of small caliber heparin-protamine-collagen vascular grafts, Trans. Am. Soc. Artif. Intern. Organs, 31:102, 1985.
3. Noishiki Y, Yamane Y, Satoh S, Niu S, Okoshi T, Tomizawa Y, Wildevuur CHR: Healing process of vascular prostheses seeded with venous tissue fragments, Trans. Am. Soc. Artif. Intern. Organs, 37:478, 1991.
4. Noishiki Y, Watanabe K, Okamoto M, Kikuchi Y, Mori Y: Evaluation of a new vascular graft prosthesis fabricated from ultrafine polyester fiber, Trans. Am. Soc. Artif. Intern. Organs, 32:309, 1986.
5. Tanzawa H, Mori Y, Miyama H, Hori M, Oshima N, Idezuki Y: Prevention and evaluation of a new a thrombogenic heparinized hydrophilic polymer for use in cardiovascular system, Trans. Am. Soc. Artif. Intern. Organs, 19:188, 1973.
6. Idezuki Y, Watanabe H, Hasegawa M, Kanasugi K, Mori Y, Nagaoka S, Hagio M, Yamamoto I, Tanzawa H: Mechanism of antithrombogenicity of a new heparinized hydrophilic polymer: Chronic in vivo studies and clinical application, Trans. Am. Soc. Artif. Intern. Organs, 21:436, 1975.
7. Mori Y, Nagaoka S, Masubichi Y, Itoga M, Tanzawa H, Kikuchi Y, Yamada Y, Yonaha T, Watanabe H, Idezuki Y: The effect of released heparin from the heparinized hydrophilic polymer (H-RSD) on the process of thrombus formation, Trans. Am. Soc. Artif. Intern. Organs, 24:736, 1978.
8. Mori Y, Nagaoka S, Itoga M, Tanzawa H, Yamada Y, Watanabe H, Idezuki Y: The effect of heparin release from a heparinized hydrophilic polymer (H-RSD) on antithrombogenicity, Trans. Am. Soc. Artif. Intern. Organs, 2(Suppl):66, 1978.
9. Noishiki Y, Miyata T, Kodaira K: Development of a small caliber vascular graft by a new crosslinking method incorporating slow heparin release collagen and natural tissue compliance, Trans. Am. Soc. Artif. Intern. Organs, 32:114, 1986.
10. Noishiki Y, Miyata T: A simple method to heparinize materials, J. Biomed. Mat. Res., 20:337, 1986.

11. Wang CL, Miyata T, Weksler B, Stenzel KH: Collagen induced platelet aggregation and release, J. Biomchem. Biophys. Act., 544:555, 1978.
12. Hayashi K, Handa H, Nagasawa S, Okamura A, Moritake K: Stiffness and elastic behavior of human intracranial and extracranial arteries, J. Biomechanics, 13:175, 1980.
13. Miyata T, Noishiki Y, Matsumae M, Yamane Y: A new method to give an antithrombogenicity to biological materials and its successful application to vascular grafts, Trans. Am. Soc. Artif. Intern. Organs, 29:363, 1983.
14. Noishiki Y: Pattern of arrangement of smooth muscle cells in neointima of synthetic vascular prostheses, J. Thorac. Cardiovasc. Surg., 75:894, 1978.
15. Mori Y, Nagaoka S, Takiuchi H, Kikuchi N, Tanzawa H, Noishiki Y: A new antithrombogenic material with long polyethyleneoxide chains, Trans. Am. Soc. Artif. Intern. Organs, 28:459, 1982.
16. Burger K, Sauvage LR, Rao AM, Wood SJ: Healing of arterial prostheses in man; its incompleteness, Ann. Surg., 175:118, 1972.
17. Herring MB, Gardner AL, Glover J: A single technique for seeding vascular graft with autologous endothélium, Surg., 84:498, 1978.
18. Noishiki Y, Yamane Y, Tomizawa Y, Okoshi T, Satoh S, Wildevuur CHR: Endothelialization of vascular prostheses by transplantation of venous tissue fragments, Trans. Am. Soc. Artif. Intern. Organs, 36:346, 1990.
19. Bell E, Ehrlich HP, Buttle DJ, Nakatsuji T: Living tissue formed in vitro and accepted as skin-equivalent tissue of full thickness, Science, 211:1052, 1981.
20. Wesolowski SA: Evaluation of tissue and prosthetic grafts, Charles C. Thomas Publisher, Spring Fields, 1962.
21. Wesolowski SA, Fries CC, Karlson KE, DeBakey ME, Sayer PN: Polosity primary determinant of ultimate fate of synthetic vascular grafts, Surg., 50:91, 1961.
22. Yates SG, Aires MS, Barros AB, Burger K, Fernandes LG, Wood SJ, Rittenhouse EA, Davis CC, Mansfield PB, Sauvage LF: The preclotting of porous arterial prostheses, Ann. Surg., 188:611, 1978.
23. Snooks SJ, Groft RJ, Chir CM, Wagner C: How should we preclot knitted Dacron graft? J. Vasc. Surg., 7:538, 1988.
24. Wuerflein RC, Campbell GS: Analysis of preclotting technique for prosthetic arterial grafts, Am. Surgeons, 29:179, 1963.
25. Sauvage LR, Berger K, Wood SJ, Sameh AA, Wesolowski SA, Golaski WM, Dedomenico M, Hartmenn JR: A very thin, porous knitted arterial prosthesis: Experimental data and early clinical assessment, Surg., 65:78, 1963.
26. Okamoto M: Ultra-fine fiber and its application, Preprints Japan-China Related Symposium on Polymer Science and Technology, 26 Tokyo, October 1981.

CONTRIBUTORS

Klaus Affeld, Biofluidmechanik Laboratory, UKRU, Freie Universitat Berlin, 1 Berlin 19, Germany

Manuel Anguita, Heart Transplantation Unit, Hospital Regional "Reina Sofia," University of Cordoba, 14004 Cordoba, Spain

Jose M. Arizon, Heart Transplantation Unit, Hospital Regional "Reina Sofia," University of Cordoba, 14004 Cordoba, Spain

Juan Jose Badimon, Unidad de Investigacion Cardiovascular, Centro de Investigacion y Desarrool, Consejo Superior de Investigaciones Cientificas, Barcelona, Spain; and Cardiovascular Biology Research, Cardiac Unit, Massachusetts General Hospital, Harvard Medical School, Boston, MA, USA

Lina Badimon, Unidad de Investigacion Cardiovascular, Centro de Investigacion y Desarrool, Consejo Superior de Investigaciones Cientificas, Barcelona, Spain; and Cardiovascular Biology Research, Cardiac Unit, Massachusetts General Hospital, Harvard Medical School, Boston, MA, USA

Bernard Brami, Centre d'Etudes Coeur et Fluides, 22, rue Boulainvilliers, 75016 Paris

Manuel Concha, Heart Transplantation Unit, Hospital Regional "Reina Sofia," University of Cordoba, 14004 Cordoba, Spain

Winnie Cui, Department of Biomedical Engineering, Memphis State University, Memphis, TN 38152, USA

Paul Didisheim, Biomaterials Program, Devices and Technology Branch, Division of Heart and Vascular Diseases, National Heart, Lung and Blood Institute, National Institutes of Health, Bethesda, MD 20892, USA

Y. C. Fung, Department of AMES-Bioengineering R-012, University of California, San Diego, La Jolla, CA 92093, USA

Harry L. Goldsmith, McGill University Medical Clinic, Montreal General Hospital, Montreal, Quebec, Canada

Patrick Havlik, Laboratoire de Recherches Chirurgicales, University of Aix-Marseille II, France

Ned H. C. Hwang, Cardiovascular Engineering Laboratory, Department of Biomedical Engineering, Memphis State University, Memphis, TN 38152, USA

Amelia Jimenez, Heart Transplantation Unit, Hospital Regional "Reina Sofia," University of Cordoba, 14004 Cordoba, Spain

Takeshi Karino, McGill University Medical Clinic, Montreal General Hospital, Montreal, Quebec, Canada

Stephen G. Kovacs, University of South Florida, Division of Cardiovascular Surgery, Department of Surgery, MDC 0016, 12901 Bruce B. Downs Blvd., Tampa, FL 33606, USA

Jose M. Latre, Heart Transplantation Unit, Hospital Regional "Reina Sofia," University of Cordoba, 14004 Cordoba, Spain

Fernando Lopez-Rubio, Heart Transplantation Unit, Hospital Regional "Reina Sofia," University of Cordoba, 14004 Cordoba, Spain

Gordon D. O. Lowe, University Department of Medicine, Royal Infirmary, Glasgow G31 2ER, UK

Peter P. McKeown, University of South Florida, Division of Cardiovascular Surgery, Department of Surgery, MDC 0016, 12901 Bruce B. Downs Blvd., Tampa, FL 33606, USA

Thierry Mesana, Laboratoire de Recherches Chirurgicales, University of Aix-Marseille II, France

Adel A. Mikhail, Cardiovascular and Thoracic Surgery Division, University of Sherbrooke, Sherbrooke, Quebec, J1H 5N4, Canada

Anastasio Montero, Heart Transplantation Unit, Hospital Regional "Reina Sofia," University of Cordoba, 14004 Cordoba, Spain

Jean-Raoul Monties, Laboratoire de Recherches Chirurgicales, University of Aix-Marseille II, France

Yasuharu Noishiki, First Department of Surgery, Yokohama City University, School of Medicine, 3-9, Fukuura, Kanazawa-ku, Yokohama, 236, Japan

Mirjam G.A. Oude Egbrink, Cardiovascular Research Institute Maastricht, University of Limburg, Maastricht, The Netherlands

Travis J. Phifer, Department of Surgery, Louisiana State University School of Medicine, Shreveport, LA 71130, USA

Robert S. Reneman, Cardiovascular Research Institute Maastricht, University of Limburg, Maastricht, The Netherlands

H. Reul, Helmholtz-Institute for Biomedical Engineering at the RWTG Aachen, Pauwelsstr. 30, D-5100 Aachen, Germany

Helge E. Roald, Biotechnology Centre of Oslo, University of Oslo, P. O. Box 1125, 0317, Oslo, Norway

Kjell S. Sakariassen, NYCOMED BIOREG AS, N-0372, Oslo, Norway

Jose Aznar Salatti, Biotechnology Centre of Oslo, University of Oslo, P. O. Box 1125, 0317, Oslo, Norway

Klaus Schichl, Biofluidmechanik Laboratory, UKRU, Freie Universitat Berlin, 1 Berlin 19, Germany

Dick W. Slaaf, Cardiovascular Research Institute Maastricht, University of Limburg, Maastricht, The Netherlands

Steve S. Slack, Department of Biomedical Engineering, Memphis State University, Memphis, TN 38152, USA

Geert Jan Tangelder, Cardiovascular Research Institute Maastricht, University of Limburg, Maastricht, The Netherlands

F. Javier Teijeira, Cardiovascular and Thoracic Surgery Division, University of Sherbrooke, Sherbrooke, Quebec, J1H 5N4, Canada

Vincent T. Turitto, Department of Biomedical Engineering, Memphis State University, Memphis, TN 38152, USA

Federico Valles, Heart Transplantation Unit, Hospital Regional "Reina Sofia," University of Cordoba, 14004 Cordoba, Spain

Shi-Kang Wang, Institute of Thermal Science, Tianjin University, Tianjin, China

Bea Woldhuis, Cardiovascular Research Institute Maastricht, University of Limburg, Maastricht, The Netherlands

Michael R. T. Yen, Department of Biomedical Engineering, Memphis State University, Memphis, TN 38152, USA

Andreas Ziemann, Biofluidmechanik Laboratory, UKRU, Freie Universitat Berlin, 1 Berlin 19, Germany

Benjamin W. Zweifach, AMES-Bioengineering, University of California, San Diego, La Jolla, CA 92093-0412, USA

INDEX

ABIOMed BVS system, 367, 371
Adhesion, 129, 135, 136
Alveolar sheet, 66, 68, 69
Angina, 175, 178, 180, 182
Angiography, 175
Angioplasty, 180, 182, 183
Aggregation, 46–55, 57, 58, 60, 129, 134, 135
Anastomosis, 387, 391–398, 402–404
Anastomotic sites, 417, 425–427, 429
Aneurysm, 127, 143, 144
    saccular, 143, 144
    incipient, 144
Annular vortex, 132, 134, 135, 137
Antigen-antiserum bond, 45
Antigen-antibody bond, 45
Antigenic type B cross linked, 45
Aorta, descending, 142, 143
Aortic arch, 142, 143
Aorto-celiac junction, 139
Arcades, 9, 13, 17
Arterial compliance, 251
Arterial pressure, systolic, 328, 329
Arterioles, 9–13, 15, 17–19, 22
    arcade, 9, 17, 22
    mid-sized, 22
    transverse, 9, 11, 13, 17–19
    terminal, 3, 7, 9–11, 13, 18, 19, 22
Artificial heart valve, 299, 300, 306, 307, 317, 322
Assistance
    short-term, 350, 351
    prolonged or permanent, 350, 351
Atherogenesis, 98
Atherosclerosis, 127, 129, 141, 144, 148, 176, 183, 184
    and wall thickening, 144–148
    lesions, 145–147
    plaque, 175–177, 183, 184
Atrophy, 111
Autologous artery, 387
Autologous vein, 385–387
Autopsy, 419
Autoregulation, 15, 19
Axial dependence phenomenon, 157
Azimuthal angle, 42, 55

Backwash, 220
Baroreceptor control, 241

Basal tone, 13–17, 19, 22
Bifurcation, 129, 132, 133, 137, 139–141, 144–146
Bileaflet system, 229
Biocompatibility, 189, 192, 195
Biomaterial, 189–195
Blood
    apparent viscosity, 26, 36, 38
    coagulation, 192
    contacting surfaces, 250
    and material interactions, 190, 191
    minute, 1, 3
    regional distribution of, 10
    rheology, 92
    tissue exchange, 8, 9, 16, 19, 20
Bradyarrhythmia, 326, 335, 336
Brain damage, 240
Brownian motion, 50, 55

Capillary
    circulation, 240
    perfusion, 10, 11, 19
    pulmonary, 65, 66, 70
    viscometer, 103, 104
Captopril, 327–330
Capture frequency, 46
Cardiac failure, 326, 327, 329, 330, 333
Cardiac index, 327, 328
Cardiogenic shock, postoperative, 364
Cardiopulmonary bypass (CPB), 189, 190, 193, 364
Cardiovascular system, 1
    disease, 103–105
    models, 260
Carotid bifurcation, in humans, 140
Carotid sinus, 140, 141
Cavitation, 260, 273, 274
Cellular metabolism, 240
Cell-vessel interaction, 65
Channels, small pore, 4
Circle of Willis, 143, 144, 148
Circuits, in-parallel, 10, 19
Clinical performance, 197, 206
Closing angle, 204
Coagulation, 91, 92, 96, 98, 99
Coating, 250
Collagen, 151, 153, 154, 157–168
Collision
    frequency, 46, 47, 51, 52, 55

Collision (cont'd)
  two-body, 41, 42, 44, 46, 47, 50–53, 55
Colloid stability theory, 44, 46, 47
Compact design, 251
Compatibility, 353, 358, 359
Compliance chamber, 372, 375, 378, 380
Complication, 326, 331, 333–336, 338, 339
Computer Aided Design (CAD), 246–248, 250, 256
Computer Numerically Controlled (CNC) manufacturing, 247
Continuum mechanics, 110
Corpuscles, 129
Cross-bridges, 45
Cross-flow, secondary, 132
Cross-linking reagent, 421

Dacron, 395
Denudation, 390, 392
Diastolic function, 337
Diffusion, 92, 94, 95
Disc clearance, 204, 221
Disc curvature, 202
Donor, 330, 331, 340
Doppler shift, 263, 264
Doublet, 42, 44–46, 51, 52, 55
Durability, 216, 218, 353, 354, 359
  testing, 250, 256
Dynamic range, 264, 268, 269

Echocardiography, 326, 330, 331, 337
Eddy formation, 128
Edge-on configuration, 86, 87
Elements, various, 251
Endothelial cells, 2–4, 8, 14–17, 19–22
  autologous, 194, 195
Endothelialization, 417, 424, 425, 427, 428, 430–433
Endothelial matrix, 158
Endothelial organellae, 4
Endothelium, 97, 98, 129, 136
Erosion, surface, 273
Extracellular matrix, 110, 122
Extracorporeal membrane oxygenation (ECMO), 364

Fahraeus-Lindquist effect, 65, 70–72, 78
  inversion of, 65, 72
Features
  clinical, 326, 327
  hemodynamic, 326, 327
Feed artery, 9, 17, 19, 22
Finite Element Methods (FEM) analysis, 247
Fibrin, 155, 156, 158–162, 167, 168
Fibrinogen, 47, 52, 53, 57, 91, 93–95, 97, 98, 155–157, 160, 168
Fibrinogen molecule, bivalent, 52, 53
Fibrinolytic mechanism, 192
Fibroblast, 426–428, 430, 431
Flow
  backward, 204, 205
  central, 198–201, 205, 206, 212, 213, 216
  channel diameter, 201
  creeping, 44
  counter-rotating double helicoidal, 140, 141

Flow (cont'd)
  forward, 202–207, 220
  laminar, 130
  lateral, 198, 199, 206
  leakage, 261, 273
  model, 280–284, 287, 292, 299–305, 307, 309, 310, 322, 398–400
  non-central, 198
  non-pulsatile, 237, 239
  oscillatory, 134
  particulate, 66
  pattern, 127, 133, 136–147
  Poiseuille, 42, 43, 46–48, 85, 127, 128, 130, 131
  pulsatile, 237, 239, 240, 250
  regurgitant, 262, 271
  secondary, 128, 132, 133, 138–147
  separation, 128, 130, 132, 133, 136, 137, 139, 142, 146–148, 306, 310, 313–318, 322, 323
  spiral secondary, 138, 140, 141, 142, 144, 145
  squeeze, 260, 273, 274, 279, 281
  turbulent, 127, 130, 139, 140
  unsteady and disturbed, 127
  visualization, 306–308, 310, 317, 322
Fluid dynamics, 91–93, 95, 97, 99
Fluid mechanical stresses, 127
Flutter, 277
Force
  attractive, 44, 46, 51
  normal, 44
  repulsive, 44, 47
  shear, 44, 45
Frequency shift, 257, 263–267, 269, 270, 283–285, 290, 291
Functional adaptation, 110

Geometrical irregularity, 127
Geometric friction factor, 69
Glycocalyx, 47
Glycoprotein
  Ib, 157, 159, 164
  Ib-IX, 95, 98
  IIb-IIIa, 52, 95, 157, 164
Graft
  tissue, 386, 390
  vascular, 189, 190, 193, 194, 385, 388–393, 404, 408

Hamaker constant, 44, 47
Heart-lung machine, 364
Heart transplantation, 325, 326, 329, 332, 333, 335, 338–340
Hematocrit, 65, 66, 68–70, 72, 75, 77–83
Hemodilution, 106
Hemolysis, 365, 369, 375, 376, 378, 379
  Index of (IH), 254
  testing, 250, 254, 255, 347, 348, 358, 360
Hemopump, 365, 367
Hemorheology, 103
Hemostasis, 91, 96
Heparin, 418, 420, 422–426
High altitude, 109–115
Homeostasis, 1, 2, 5, 7, 8, 11, 18, 19, 21
Hybrid feedback signal, 266

Hydraulic load, 251
Hyperbaric, 109
Hypertension, 109, 118, 120, 122
Hypoxia, 109, 115

Immobilization, 111
Immunologic mechanism, 192
Impedance, 300–302
　input, 256
　output, 261
Implantable devices, 370, 375, 370
　Novacor, 367, 370, 372, 374, 375, 377, 379
　Thermedics "Heart Mate", 370, 372
Implantation, 345–347, 350–352, 354–356, 359
Inductive flow probe, 254
Infarction, myocardial, 175, 178, 182
Inotropics, 327–330, 334, 335
In-series, 11
Interstitial tissue, 11
Intervalveolar septa, 66
Intima, 387, 390, 392, 396, 398, 406
Intra-aortic balloon pumping (IABP), 364, 365, 367, 369
Intracellular reticulum, 110
Intrinsic controls, 2

Jarvik-7 TAH, 363
Jiggling, 240

Kidney function, 239

Laser Doppler Anemometry (LDA), 230, 232, 257, 260, 261, 263–273, 279, 282–285, 288, 289, 291
Leakage, 198, 202, 204, 205, 217, 220–222
Leak back, 78
Leukocyte rolling, 29, 34, 38
Liotta heart, 374
Liquid
　Newtonian, 66, 128
　non-Newtonian, 66
Lung parenchyma, tethering of, 113

Mainstream (frictionless layer), 130–134, 137–140, 142, 144
Manifestation, 5
Mechanical failures, 229
Mechanical support devices, 363, 364
Media, 129, 133–135, 137–143, 146, 147
Microcirculation, 1, 2, 4, 5, 8–11, 17, 22, 65, 80, 84, 87
　topology, 4
Microvessels, velocity distribution in, 65, 78
Mock circulation, 299–301, 307
Mock loops, 250–256
Modules, 2
Molecular mechanism, 151
Monoclone, 45
Monodisc, 230
Morbidity, 326, 333, 335, 338
Mortality, 331, 333, 335, 336, 339, 340
Muscle cells, smooth, 8–10, 13–18, 21, 22, 96, 98, 99, 424–428

Myocardial infarction, acute, 364
Myocardial preservation, 333, 335, 336

Nervous system, autonomic, 1
No-load condition, 113, 115

Occlusion, 419, 420
Opening angle, 109, 114, 116–120, 201–203
Optico-electro-hybrid feedback-LDA, (OEHF-LDA), 259, 260, 266–272, 288, 289, 292
Orthokinetic aggregation, 46
Oxygen consumption, 240

Pannus, 419, 420, 421, 423, 424, 432
Parenchymal tissue, 8, 11
Partide migration, 134
Patchiness, 66
Patent, 419, 423, 426, 429
Péclet number, 50, 57
Pellet, 65, 66, 68, 69, 73–75, 78, 80, 81, 83–87
Perfusion chamber, 151–156, 159–161
　annular, 152, 153, 155, 156
　cylindrical, 154
　parallel-plate, 153, 156, 159–161
　semi-cylindrical, 154–156
Pericytes, 2, 3, 10, 16, 21
Permeability, 9, 20, 21
Photocurrent, 270
Physiological load, 250, 251
Piezoelectricity, 111
Platelet, 91–99, 175, 184, 190–193
　adhesion, 97, 98, 181, 182
　aggregation, 95, 96, 98, 177, 182
　concentration distribution of, 26
　exclusion, 29
　focal deposition of, 127
　orientation of, 26
　and platelet bond, 51, 52
　and platelet interaction, 151, 156–158
　storage granules, 157
　surface adhesion, 151, 156, 157, 154
　thrombi, 129
Plugging configuration, 87
Polyclonal, 45
Polystyrene, 44
Polyvinylchloride (PVC), 247
Porosity, 389–391, 393
Precapillary, 3
Precapillary sphincter, 13
Preload-filling pressure, 351, 352, 354, 360
Principal orientation, 112
Prognostic markers, 326
Prostaglandin metabolism, 157, 158
Prosthesis, 387, 388, 389–391
　vascular 418, 425, 426, 428, 430
Pseudo-elastic strain energy function, 120
Pulmonary artery balloon counterpulsation (PABCP), 365
Pumps
　centrifugal, 345, 347–349, 351, 356, 364, 368, 369
　characterization of, 250
　and parts, flexible, 247

Pumps (cont'd)
  peristaltic, 347
  pneumatic displacement, 241, 248
  pneumatic sac, 369
  pusher plate, 244, 245
  roller, 368
  rotational, 246, 253
  spiral, 347, 348, 350, 352
Pyrolitic carbon, 229

Radial components, 128, 138
Radionuclide study, 338
Ramification, 8, 10, 11, 15, 18, 22
Recipient, 325, 326, 330–337, 340
Recirculation, 130, 137–142, 144–148
Recruitment, 2, 3
  spatial, 3
  temporal, 4
Regurgitation, 202, 219, 222
Rejection, 325, 326, 333, 335–340
Reliability, 353, 354, 360
Remodeling
  morphological, 109, 113
  soft tissue, 111
  structural, 109, 113
Resistance, peripheral, 240, 251, 254, 256
Reynolds number, 68, 78, 130, 132–136, 138–140, 142–144, 307, 308
Rouleaux, 58

Self-wash, 220
Shear rate, 92–94, 97–99, 175–178, 180–183
  mean tube, 46, 48, 50–53, 58
Shear stress, 92, 93, 95–97, 99
  of wall, 25, 26, 36–38
Side-arm offshoots, 7, 9–11, 18
Sigmoid aggregation curve, 52
Sinus-shaped housing, 247
Sludging, 60
Smooth cone entry, 73
Sphincters, 3
Stagnation point, 136, 140, 141, 146
Stenosis, 96, 99, 129, 130, 132, 136
  concentric, 180
  eccentric, 178, 180
  vascular, 279
Stochastic, 45
Stokes-Einstein equation, 50
Strain-growth relationship, 122
Strain invariants, 123
Streamlines, 127, 130, 132–136, 139–141, 143
Stress
  and growth relationship, 122
  and invariants, 123
  principal, 113
Subendothelium, 151, 153, 154, 156–159
Symbion, 364, 367, 371, 374
Systolic function, 337, 338

Thromboembolism
  complication, 197, 199, 209, 369, 376, 378
  failures, 229

Thrombogenesis, 158, 168
Thrombogenic surface, 151–159, 161, 163
Thrombosis, 91, 92, 97–99, 127, 129, 141, 175, 178, 180, 181, 183, 184, 192, 193, 195
Thrombotic mechanism, 151, 155, 158
Thrombus, 47, 175–178, 180–184, 417, 420, 421, 423, 424, 426, 427
Throttle clamp, 253
Thyroid function, 326, 331
Tissue
  engineering, 110, 123, 124
  proper, 2, 7, 10
  remodeling, 109–111
T-junctions, 129, 133, 137, 138–140, 146
Total artificial heart (TAH), 355, 363, 364, 374, 378–380
Trajectory equation, 46, 47
Transcapillary fluid exchange, 4
Transit time, mean, 47, 49, 50, 52
Translocation, 386
Transplant
  heterotopic, 325
  orthotopic, 325
Trifurcation, 146, 147
Tube, flat–ended, 74
Tubular expansion, 131, 134, 136

Valve
  aortic, 300, 302, 317
  ball–caged, 230
  bileaflet, 197, 200, 201, 206–208, 211, 212
  Björk-Shiley convexo-concave 60°, 231–233
  Hufnagel, 229
  mechanical, 197, 205, 206, 216, 218, 220–222, 229–231, 233, 234
  monoleaflet, 208, 211, 213–215
  prostheses, 229
  Starr-Edwards, 229, 231–233
  St. Jude, 229
  testing chamber, 272
  tissue, 229
  venous, 136
Van der Waals attraction, 44, 53
Vascular bed, terminal, 2
Vascular resistance, afterload-total, 351, 352, 354
Vascular wall motions, 260, 292
Vasoconstriction, 178, 183, 184
Vasomotor activity, 3
Velocity
  distribution, 130, 132, 143, 144, 147, 148
  profile, 25, 26, 29–38
Venous system, 251
Ventricle, 345–348, 350–356, 358–360
  artificial, 340, 350, 352, 354–356, 359, 360
  assist devices (VADs), 367, 368, 370–374, 376–379, 380
Venturi effect, 273, 274
Vessel
  arterial, 1, 7, 9, 10, 20
  curved, 127, 142
  sinusoidal, 3
Viscometer, 95, 103–106

Viscosity
   apparent, 60, 65, 66, 68–71, 78, 86
   relative, 65, 66, 69
Von Willebrand factor (VWF), 95, 97, 98, 157, 159, 164

Wall adhesion, 57
Washing test, 229, 230, 232

Water hammer, 273, 274
Windkessel element, 251
Womersely number, 68

Y-bifurcation, unique, 140

Zero gravity, 109
Zero-stress state, 109, 113, 114, 116–122

MIX
Papier aus verantwortungsvollen Quellen
Paper from responsible sources
FSC® C105338

If you have any concerns about our products,
you can contact us on
ProductSafety@springernature.com

In case Publisher is established outside the EU,
the EU authorized representative is:
**Springer Nature Customer Service Center GmbH
Europaplatz 3, 69115 Heidelberg, Germany**

Printed by Libri Plureos GmbH
in Hamburg, Germany